P9-BIA-240

UNDERSTANDING SOCIAL STATISTICS

UNDERSTANDING SOCIAL STATISTICS

GENE M. LUTZ, Ph. D.
University of Northern Iowa

MACMILLAN PUBLISHING CO., INC.
New York
COLLIER MACMILLAN PUBLISHERS
London

To Dawn and Heather

Macmillan Publishing Co., Inc.
866 Third Avenue, New York, New York 10022

Collier Macmillan Canada, Inc.

Library of Congress Cataloging in Publication Data

Lutz, Gene M.
Understanding social statistics.

Includes index.
1. Social sciences—Statistical methods.
2. Statistics. I. Title. II. Title: Social statistics.
HA29.L96 1983 300′.1′5195 82-21708
ISBN 0-02-372980-5

Printing: 1 2 3 4 5 6 7 8 **Year:** 3 4 5 6 7 8 9 0

ISBN 0-02-372980-5

PREFACE

Like most Prefaces, this one is written after the project has been finished. So while it is a "Preface" to you, it is a "Postscript" to me. There is a parallel in this to the way we learn about social statistics. We are attempting to understand a world that is already in motion. The researchers have been at work for some time now. We cannot expect them to wait for us to catch up. But we can examine the final products of their work (the reports, articles, and books) and retrace their steps to discover the ideas and decisions they made along the way. It is through this re-construction of the research activity that we can best come to appreciate what social statistics means. This has the further advantage of lessening the alienation many people feel from statistics. That alienation stems both from a lack of understanding of how social statistics are generated and the perception that they are fundamentally impersonal. In truth, social statistics are based on the very human qualities of objectivity *and* subjectivity that can be seen by examining the larger arena of research activity as a whole. I have expressed more of the rationale for this approach to our topic in the first chapter, fearing that you may not find it here in the Preface. If you are reading these introductory notes anyway, so much the better.

What I have attempted is to introduce social statistics from the point of view of a reader and user, a "consumer," if you will, not a producer. That does not mean you cannot or should not learn to be a producer of social statistics. It simply means that you are more likely to see yourself currently in the consumer role. I believe it is more sensible and advantageous to recognize this than to pretend otherwise. Increasingly, I have been teaching this way in past years. My students have been learning with more ease and depth. Their prejudices and fears about statistics have been more successfully challenged and altered. I hope you will have the same experience.

This book contains several features to assist the reader. The most central of these is the inclusion of excerpts from published social research to help you learn and practice your consumer skills. These are called *Research Examples.* They do not always show the preferred application of social statistics. Rather, they show some of its variety, including both admirable and questionable practices.

The chapters will include a mix of intuitive and computational discussions. You need both. The conclusion to each chapter is presented in a section called *Keys to Understanding.* This focuses your attention on the more useful ideas you should know, the best strategies for getting to the heart of the matter, and other gems of helpful advice pertinent to the material in the chapter.

A list of *Key Terms* is provided with each chapter. Take care to know the meanings of these as you go along.

A set of *Exercises* is included for all chapters. Doing the exercises will help you to understand and to test your mastery of the material. I recommend that you do them faithfully. Answers to selected problems from these sets are shown in Appendix E.

Most chapters will conclude with a brief section called *Related Statistics.* This will cover some of the more advanced methods or concepts which have not been presented in detail. They are included to help you recognize these techniques when you encounter them in research reports. If you understand the statistics which are covered in a chapter, you can assume that those in the *Related Statistics* section share some of the same basic ideas. You should find this feature useful as a handbook for future reference. Sources are given so that you can learn more about each of these on your own if you wish.

Because there is some variation in student preparation for studying the material to be covered, two additional features are included in the book. First, to respond to individual differences, there is a *Review of Basic Math* in Appendix A. Some may need only to skim this, others will need to concentrate on recapturing lost skills. Second, there are differences in the level at which entire first courses are taught. In response to this the book has been written so that the instructor can alter the depth along a continuum from the most elementary to the intermediate. This is accomplished by selective use of those more advanced sections marked optional*. These can be omitted without doing violence to the continuity of the chapters.

Occasionally, a feature called *On the Side* appears in the book. Each of these note a point of historical interest, the thoughts of another author, or some other comments that may stimulate you to think about an issue related to the topics being discussed in a chapter. They provide a kind of "time out" for catching up or idle musing as you study the regular material.

Writing a book requires the assistance of many people to bring it to a successful conclusion. Most of these people remain anonymous because they are so numerous. While some of their contributions are relatively small, not all of them are and in the aggregate they are enormous indeed. Among those in this group I wish to give first acknowledgment to the many helpful ideas I have gathered from my students over several years. These undergraduate and graduate students have given me the feedback and direct assistance I needed to sharpen my own approach to teaching and writing about statistical methods in social research. I am most grateful to them.

Nearly every writer has been significantly influenced and guided by one or more teachers during his or her own days as a formal student. I have had the good fortune to have been exposed to two such people. The first was a high school teacher of mathematics, Ronald Moehlis, who I found particularly skilled at mixing wit and encouragement while teaching his subject. The second was a university professor, Dr. Richard D. Warren at Iowa State University, who brought the largest dose of human kindness to statistics that I have ever witnessed. This combined with a deep knowledge of his material has helped nearly a generation of students in the social sciences to understand and be able to teach the subject with success. He now continues this tradition for students in professional education. I am greatly indebted to these two people.

Special thanks are due to Dr. Rodney Brod, University of Montana, Dr. Larry Rosen, Temple University, and Dr. William Fleischman, University of Minnesota, Duluth, for their valuable comments on the early versions of the manuscript and to other reviewers who are unknown to me. Helpful assistance was provided by the staff of Macmillan Publishing Company including Senior Editors Kenneth Scott and James Anker; Production Editor, John Travis; Ronald Nurmi, the sales representative who encouraged me throughout the adventure; copyeditors; the people in production and marketing; and others I do not know personally but who worked on the project.

Indispensible assistance in preparing the manuscript (through its many retypings) was provided by Ethelyn Snyder, Lois Pittman, and Vera Sullivan at my university's Word

Processing Center and by Julie Rathbone and Beverly Kollman through the helpful assistance of my college Dean, Robert E. Morin. The university also provided a timely benefit by granting me a Professional Development Leave in the early stages of the project.

Lastly, I wish to thank the many authors and publishers who granted me permission to reprint materials for which they hold the copyrights. They are noted individually at the appropriate places in the text. The inclusion of these materials was essential to giving the book the unique features which I desired.

Cedar Falls, Iowa G.M.L.

CONTENTS

CHAPTER 1

INTRODUCTION TO SOCIAL STATISTICS

The Need for Statistical Understanding

The statistical analysis of social data has so permeated modern life that it is commonplace. The mass media as well as the academic media present, describe, and analyze social data routinely. Consider five common examples. "The cost of living rose 1.1% last month." "The average American is 2 inches taller and 20 pounds heavier than 100 years ago." "Only one in 25 students at Pleasant Valley College votes in local elections." "PCBs significantly endanger the environment." "Wastes from nuclear power generation last for 10,000,000 years."

Such statistics seem at once precise and mysterious. Many people have only a limited understanding of what is being presented. Take each of the examples above. With the first many people misunderstand that a "small" percentage change like 1.1% is a "large" change to an economist. In the second they fail to recognize the ambiguity of the word "average." For the third they do not have the basis for a fair comparison. With the fourth they have little idea what "significance" means to a statistician. Finally, they are dumbfounded by very large numbers.

It is a feature of everyday life that many consumers are unprepared to comprehend more than the simplest aspects of statistical presentations. The problem results from a situation in which the producers of these statistics grow increasingly sophisticated in analysis techniques at a much faster pace than do members of the consuming public. As a consequence many are in the predicament of wanting and needing to understand social statistics but finding it difficult to do so.

Some people adopt a defeating strategy to cope with the problem. They pretend they can get along well enough without knowing about statistics. Others act as though they understand when they do not. They argue, "Anyone can lie with statistics," or "It's people who matter, not numbers." Both strategies are attempts to dignify ignorance. But ignorance has a way of causing some very negative experiences: embarrassment, poor exam scores, lost job opportunities, and distorted views of reality, for example.

Some people appear to flee from statistics because of their insecurities. They feel that they are unable to learn about statistics because their math backgrounds are weak or they do not have an "analytical mind." A few who say this may be correct. They really will have difficulty with the subject until their math ability and thought processes are improved. But for most people, such an excuse is unwarranted. What they lack is specific experience and instruction with statistics. Their general preparation is entirely adequate. A review of basic math appears in Appendix A. You may check your preparation by reading through this.

In today's world a beginning understanding of social statistics has become a basic skill. Obviously, one who wants to be an expert at research needs the skill. But even at the other extreme, some understanding is needed for one who merely wants to avoid being manipulated by data hucksters. This book is designed to help you to achieve the goal of understanding statistics. It is written in a personal and conversational style that should ease your understanding. It neither glorifies nor vilifies statistical methods. As topics are discussed, the key ideas are highlighted and the critical questions you should be asking are noted. It emphasizes the interpretation of statistics over artful dexterity with formulas. This should make learning about social statistics more meaningful and successful.

Textbooks in social statistics have been around for many years. They represent one of the most significant efforts to create an informed audience for social data analysis. Each book uses its own strategy for achieving the understanding sought by its readers. This book is different yet. It combines two main ideas. First, it asserts that the majority of the readers can learn best from the perspective of professional consumers of social statistics, not producers. Most people studying or working in the social sciences do not need the skills of a professional data analyst. Instead, they need to know how to interpret critically the analyses that are presented to them. This book adopts a professional consumer orientation as it explains the fundamentals of social statistics.

Second, the book contends that one cannot be a competent consumer of statistical information without knowing something about the world of those people who produce that information. One needs to know how social researchers think, why they ask research questions in the ways they do, and what their goals are when they present information. When one understands the perspectives of social research, its basic vocabulary, and the techniques it uses, one is prepared to interpret and apply the information being presented.

There is no need to fear the formalities of social statistics. It does appear in a highly stylized format, including the use of mathematical symbols. But this is no more peculiar than the rituals and symbols of other aspects of social life. They all have both an overriding structure and individual differences. We can study social scientists and their analytical work much as we can any other set of people. Once we de-mystify their language and work we can comprehend social statistics and learn how to ask the right kind of questions of it. So although we approach social statistics from the point of view of its consumers, we also study the perspectives of its producers. Actually, the two ideas are quite compatible.

How Researchers Think

People who analyze social data using statistical methods are part of a long tradition of philosophical thought. To understand those methods you must be aware of the historical context in which they developed and the modern tradition of thought in which they flourish. From the early Greeks through the Middle Ages, the Enlightenment, and on into modern times, great debates have raged about the nature of human thought and its application to producing knowledge about the world. Highlights relevant to social statistics include such ideas as the importance of making a distinction between an object and its observer, the uses of early mathematics and geometry to develop a basis for rational proof, and the establishment of a formalized procedure to conduct inquiries. Because these are such basic ideas, they are both difficult and stimulating. You should examine works in the history of philosophy and science to understand them thoroughly. For now, a brief orientation will have to suffice. (See Reichenbach, 1959; Nagel, 1961; and Hempel, 1966 for more complete discussions.)

During a long period of Western history, theology was displaced as the source of secular truth claims, the natural sciences became distinct from philosophy, and the social sciences broke away from the natural sciences. In the scientific method that emerged with these developments, *empirical* data were established as the grounds for evidence when one wanted to determine the validity of an idea. Empirical evidence is based on direct experience. It is not derived from theorizing, speculating, or consulting an "authority." It therefore relies on information that can be independently verified. The central role of empirical data in science, natural or social, is one of the most fundamental points of which you must be aware in order to understand social statistics.

Traditional Research Model

Although there are many continuing controversies over the specifics, the model commonly followed by social researchers from the traditional scientific perspective can be summarized as consisting of five parts, as illustrated in Figure 1.1. In its most rigid form,

FIGURE 1.1 Traditional Research Model

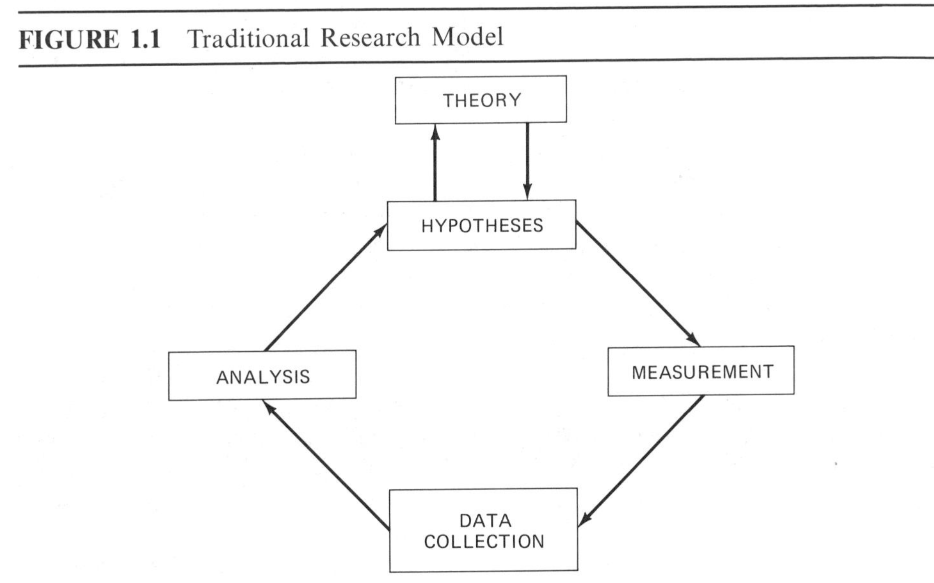

this model is often called the "positivistic model" in recognition of its heritage with the "positive sciences": those approaches to science that depend on tested and systematized experience.

The Traditional Research Model shows the logical sequence of steps in a research project. It is a formalization or idealization of the ways those steps fit together. However, in the actual conduct of research the steps may follow in a different order. Research is often published in a format that is a compromise between the formal model and the real sequence of research activities.

Theory is the beginning point of the Traditional Model. It can be defined as an integrated set of general statements which together offer an explanation for some portion of reality as it is experienced. Suppose that we follow the steps of a traditional researcher who has a theory of stratified political participation in a democracy, as an example. Such a theory would point to the stratified character of society wherein some people have greater resources than others. It would argue that in a stratified society political decisions ordinarily work to the benefit of those in the upper strata and to the detriment of those in lower strata. It would also claim that those in the upper strata have an enhanced position from which to attain and maintain control of political decisions made in a democratic form of government. Therefore, the theory would suggest that members of the upper strata have a vested interest in participating in political decisions in order to protect their standing, whereas those of the lower strata feel powerless to better their positions. The actual description of a theory would be much more detailed, but this should give you the idea of a theory's general character. It is very abstract and potentially wide ranging.

The theory that is chosen guides the entire research process. It influences the selection of a researchable topic and the specific features of social life to be investigated. Because the theory is composed of very general statements it cannot be tested directly. The first step toward more testable ideas is to derive particular hypotheses from the theory. An *hypothesis* is a specific (not general) statement proposing what certain aspects of the entire research problem are like or how they are related to each other. In our example, the researcher might use the general theory of stratified political participation to generate the specific hypothesis that social class is related to political activity. Even though hypotheses use statements that are less general than those in a theory, they still use concepts. A *concept* is an abstraction representing empirical reality. It is not capable of being observed or measured directly. You can think about social class but you cannot see it directly. Our specific hypothesis proposes a connection between two concepts, social class and political activity.

The next step according to the model is to decide how the hypotheses are to be measured (or operationalized). This changes them from statements concerning abstract concepts to statements concerning variables. A *variable* is a measured characteristic that can take on different values for the research units. Having been measured, a variable is therefore concrete, not abstract. A *research unit* is the individual, social group, community, or society that is the object of study. For our researcher, the unit is most likely to be an individual adult in the population being examined. Our example's concept of social class could become the variable that is formed by using a standardized scale that assigns social class scores to people based on such characteristics as income and occupation. Since these kinds of scales usually carry their inventor's name we might whimsically suggest the hypothetical Frank Friendly Social Class Scale as our measuring device. Our concept of political activity could be operationalized more simply as the number of times one has voted in local elections over the past five years based on self-reports.

The entire process of operationalization (turning concepts to variables) is referred to as the *measurement* step in the research model. It includes decisions about such issues as what exact questions will be asked, how information will be collected, and if and how numbers will be assigned to the information collected.

Data collection occurs next in the model. It is the process of applying the measurement decisions to the research units in order to gather *observations.* An observation is any empirical item of information that has been collected. The term applies whether the data collection method was literally visual observation or any other method, such as questionnaire, interview, or the use of secondary sources. Ideally, observations are based on direct experience and can be verified by outsiders.

Ordinarily, numbers are assigned to represent each observation according to a code. These coded observations are known as *data.* More about this aspect of the measurement process will be explained in Chapter 2. In our example the data would be the social class scores and the voting frequencies for each of the persons studied.

The researcher must determine the patterns in the data and decide what they mean according to some set of procedures. This is the analysis step, and it is the main topic of

FIGURE 1.2 Progression of Steps in the Traditional Research Model

Level of Abstraction	Step	Example
General (abstract) ↓	Theory ↓	Several statements which could be summarized to argue that in a democracy a stratified society creates a vested interest in politics for those of the upper strata.
	Hypothesis ↓	Social class is related to political activity.
	Measurement ↓	Scores on the Frank Friendly Social Class Scale will be used to measure social class. The number of times one has voted in local elections will be used to measure political activity.
Specific (concrete) ↓	Data collection ↓	(see data below)
	Analysis ↓	Use Pearson's coefficient of correlation to measure association: r +.65 between FF scores and times voted. Conclude that there is a rather strong, positive relationship between the scores.
	Hypothesis ↓	Conclude that the hypothesis is supported given its measurement.
General (abstract)	Theory	Conclude that the theory is valid insofar as it has been tested.

Respondent	FF Score	Times Voted
A	5	10
B	3	5
C	1	2
.	.	.
.	.	.
.	.	.

this book. *Analysis* is the systematic examination of data. We will be focusing on *statistical methods* of analysis. These are procedures for interpreting data based on principles of mathematics and scientific inquiry. A technique such as the Pearson's coefficient of correlation (r) (introduced in Chapter 6) might be used in our example to determine whether the social class scores and voting frequencies were related. This statistic expresses the degree of relationship as a decimal between 0 (no relationship) and ± 1 (maximum relationship). Let us suppose that it was $+.65$, meaning that the two variables in our example had a positive relationship of rather high strength.

Following the analysis the researcher compares the data to the initial hypotheses in order to reach conclusions about their accuracy. The researcher may decide that the hypotheses are supported or not. In our example, the hypothesis likely would be supported. Social class was related to political activity. Finally, the hypotheses as tested by the data are reexamined in terms of the theory from which they were derived so that the researcher can evaluate the theory's integrity. It may have withstood the test of the data or it may need modifications of some kind. Our theory of stratified political participation appears to have been verified in the limited way that it was tested.

This progression from theory to data and back to theory is illustrated in Figure 1.2 for our example. It shows how the very general ideas of a theory of stratified political participation in a democracy were specified in an hypothesis linking the concepts of social class and political activity. Through the measurement decisions these two concepts became the two variables of (1) scores on the Frank Friendly Social Class Scale and (2) voting frequency in local elections. The data collected were concrete observations on these two variables for the specific individuals selected for study. The analysis used Pearson's coefficient of correlation to find a relationship between the two variables. Thus the hypothesis and theory were concluded to have tentative support given the measurement and analysis decisions that had been made.

A more extended example may help you follow this general description of the world of the traditional researcher. Research Example 1.1 presents some exerpts from a published article showing how two authors conducted their study following the general form of the Traditional Model. The example concerns the ability of schizophrenic patients to control the impressions they can make on hospital staff members. Although you may not be familiar with some of the terms used in the example, study it for its structure—the organization of its ideas. Note how the steps fit together in the tight logic that characterizes the Traditional Research Model.

RESEARCH EXAMPLE 1.1 An Example of Social Research That Follows the Traditional Research Model

From Benjamin M. Braginsky and Dorothea D. Braginsky, "Schizophrenic Patients in the Psychiatric Interview," Journal of Consulting Psychology, *31 (6), December 1967:543–547. Reprinted by permission.*

The present investigation is concerned with the manipulative behavior of hospitalized schizophrenics in evaluative interview situations. More specifically, the study attempts to answer the question: Can schizophrenic patients effectively control the impressions (impression management,

A brief statement of the problem and objective are often presented first.

Goffman, 1959) they make on the professional hospital staff?

Typically, the mental patient has been viewed as an extremely ineffectual and helpless individual (e.g., Arieti, 1959; Becker, 1964; Bellak, 1958; Joint Commission on Mental Illness and Health, 1961; Redlich & Freedman, 1966; Schooler & Parkel, 1966; Searles, 1965). . . . In this context one would expect schizophrenics to perform less than adequately in interpersonal situations, to be unable to initiate manipulative tactics, and certainly, to be incapable of successful manipulation of other people. This statement is explicitly derived from formal theories of schizophrenia and not from clinical observations.

The authors review the main theoretical perspectives from which they are working, both those with which they disagree and their own. The extensive list of other researchers is a convenient method to let you know where to find more detailed information. Limited space in the journal does not allow a more complete discussion of the theory.

In contrast to the above view of the schizophrenic, a less popular orientation has been expressed by Artiss (1959), Braginsky, Grosse, and Ring (1966), Goffman (1961), Levinson and Gallagher (1964), Rakusin and Fierman (1963), Szasz (1961, 1965), and Rowbin (1966). Here schizophrenics are portrayed in terms usually reserved for neurotics and normal persons. Simply, the above authors subscribe to the beliefs that: (a) the typical schizophrenic patient, as compared to normals, is not deficient, defective, or dissimilar in intrapsychic functioning; (b) the typical schizophrenic patient is not a victim of his illness; that is, it is assumed that he is not helpless and unable to control his behavior or significantly determine life outcomes; (c) the differences that some schizophrenic patients manifest (as compared to normals) are assumed to be more accurately understood in terms of differences in belief systems, goals, hierarchy of needs, and interpersonal strategies, rather than in terms of illness, helplessness, and deficient intrapsychic functioning. This orientation leads to the expectations that schizophrenic patients do try to achieve particular goals and, in the process, effectively manipulate other people.

Expectations for empirical research results are developed from the theoretical perspective.

If long-term patients are both motivated to live on open wards and to remain in the hospital and if, in addition, they effectively engage in impression management in order to realize these desires, then the following would be expected:

1. Psychiatrists will rate patients in the discharge and the mental status conditions as being similar with respect to psychopathology and need for hospital control. . . . Patients in both condi-

The exact hypotheses to be tested are presented. Concepts are identified. Here they are *induction conditions* speci-

tions will give the impression of being "sick" and in need of hospital control in order to decrease the probability of discharge.

2. Psychiatrists will rate the patients in the open ward condition significantly less mentally ill and less in need of hospital control than patients in the discharge and mental status conditions. That is, patients in the open ward condition will give the impression of being "healthy" in order to maximize their chances of remaining on an open ward.

A sample of 30 long-term (more than 2 continuous years of hospitalization) male schizophrenics living on open wards was randomly selected from ward rosters. Two days prior to the experiment, the patients were told that they were scheduled for an interview with a staff psychologist. . . . Each patient was escorted to the interview room by an assistant, who casually informed the patient in a tone of confidentiality about the purpose of the interview (preinterview induction). Patients were randomly assigned by the assistant to one of three induction conditions (10 to each condition). The interviewer was unaware of the induction to which the patients were assigned, thereby eliminating interviewer bias.

Discharge induction. Patients were told: "I think the person you are going to see is interested in examining patients to see whether they might be ready for discharge."

Open ward induction. Patients were told: "I think that the person you are going to see is interested in examining patients to see whether they should be on open or closed wards."

Mental status induction. Patients were told: "I think the person you are going to see is interested in how you are feeling and getting along in the hospital.". . .

[During the interview] the patients' responses were tape-recorded. The interview was terminated after 2 minutes, whereupon the purpose of the experiment was disclosed.

Three staff psychiatrists from the same hospital separately rated each of the 30 tape-recorded interviews during two 40-minute sessions. The psychiatrists had no knowledge of the experiment. . . . The psychiatrists rated the patient on the following dimensions: (a) the patient's degree of psychopathology, using a five-point scale ranging from "not at all ill" to "extremely ill"; (b) the

fied as three types (discharge, open ward, and mental status), *psychopathology,* and *need for hospital control.*

The measurement section describes the sample members and how each concept will be measured and thereby transformed into a variable consisting of observations. For example, the discharge condition is measured as the situation in which patients were told the interviewer is interested to see if the patient might be ready for discharge. Psychopathology is measured as the scores assigned by the three psychiatrists on the five-point scale ranging from "not at all ill" to "extremely ill."

amount of hospital control a patient needed, ranging on an eight-point scale from complete freedom ("discharge") to maximum control ("closed ward, continual observation"); and (c) the structural or qualitative aspects of the patient's speech characteristic was based on the sum of the psychiatrist's rating of 14 Lorr scale items (Lorr, 1953). Each item was rated on an eight-point scale ranging from not at all atypical to extremely atypical verbal behavior.

The mean age of the patients was 47.4 years (SD = 8.36). The mean educational level of the group was 8.05 years of schooling (SD = 3.44). The median length of hospitalization was 10 years. In terms of diagnostic categories, 43% of the sample. was diagnosed as chronic undifferentiated schizophrenic, 37% as paranoid schizophrenic, 10% as catatonic, and the remaining 10% as simple schizophrenic. There were no differences between the three experimental groups on any of the above variables.

Beyond what was described in the measurement section, the sample is reviewed in terms of key characteristics that allow you to judge its representativeness. Any further details about the data collection process would be given here, if needed. Because the actual data are often very complex, they are usually summarized in tables and presented with the analysis. In this article one table did nicely.

Mean Psychopathology and Need-for-Hospital Control Ratings by Experimental Condition[a]

	Open Ward		Mental Status		Discharge	
Rating	**M**	**SD**	**M**	**SD**	**M**	**SD**
Psychopathology	2.63	.58	3.66	.65	3.70	.67
Need for hospital control	2.83	1.15	4.10	1.13	4.20	1.42

[a]Table 1 in the original.

The means of the psychopathology ratings by experimental condition are presented in [the table]. The ratings ranged 1–5. The analysis of variance of the data yielded a significant condition effect ($F = 9.38$, $p < .01$). The difference between the open ward and discharge conditions was statistically significant ($p < .01$; Tukey multiple-range test). In addition, the difference between the open ward and the mental status condition was significant ($p < .01$). As predicted, there was no significant difference between the discharge and mental status conditions.

The analysis section indicates the statistical techniques used (analysis of variance, Tukey's test, etc.) It also describes the results of those statistical analyses.

On the basis of these analyses it is clear that patients in the open ward condition appear significantly less mentally ill and in less need of hospital control than patients in either the discharge or mental status conditions. Obviously the patients in these conditions convey different impressions in the interview situation. . . .

In comparing the structural or qualitative aspects of patient speech no significant differences were obtained between experimental conditions. . . . Patients "sounded" about the same in all three conditions.

In summary then, the hypotheses were confirmed. It is clear that patients responded to the inductions in a manner which maximized the chances of fulfilling their needs and goals. When their self-interests were at stake patients could present themselves in a face-to-face interaction as either "sick" or "healthy," whichever was more appropriate to the situation.

The analysis is linked back to the hypotheses to reveal how accurate they were. Here they came out as expected.

The traditional set of assumptions concerning schizophrenics, which stresses their irrationality and interpersonal ineffectuality, would not only preclude the predictions made in this study, but would also fail to explain parsimoniously the present observations. It is quite plausible and simple to view these findings in terms of the assumptions held about people in general; that is, schizophrenics, like normal persons, are goal-oriented and are able to control the outcomes of their social encounters in a manner which satisfies their goals.

Finally, the theory is reexamined to see what effect the study had on it. Here the authors' theory is unchanged, but the alternative ideas are concluded to be inadequate. Had the authors' hypotheses not been supported, their own theory would have been questioned. This would have required considerable discussion to explain.

The Rise of Statistical Methods

Because the traditional research approach based its strategy for producing knowledge about the world on empirical data, it created a need for systematic procedures by which those data could be examined. This spurred development of the field of statistics, both basic and applied.

One of the first applications in modern time was the update of a very old concern. John Graunt (1620–1674) published some "political arithmetic" about the English population which was used for taxation purposes and other goals of the politicians of the time. The same interest in statistics is shown in ancient records from early civilizations and the days of the Roman Empire. Indeed, the root of the term *statistics* is *state,* which reveals its originally political meaning. Other late Renaissance applications were for insurance purposes. Life insurance companies needed reliable mortality tables to set premium rates. Shipping companies and their insurers needed data on the experience of shipping ventures to establish the expected incidence and value of goods lost at sea. The work of Englishman Edmund Halley (1656–1742) was useful to both.

The uses of probability theory were based on the early work of such notables as the Italian Galileo Galilei (1564–1642), the Frenchmen Pierre de Fermat (1608–1665) and Blaise Pascal (1623–1662), and the Swiss James Bernoulli (1654–1705). Their ideas about probability still keep gambling both a popular and profitable activity. In the 1800s statistics based on the standard normal distribution were enhanced through their application to astronomy. Major contributors included the Frenchman Marquis de Laplace (1749–1827) and the German Carl Gauss (1777–1855). The Belgian Adolphe Quételet (1796–1874) advanced this work to show its more widespread uses. Especially interesting to us are his applications to such social problems as crime. His work was then called "moral statistics," which sounds rather quaint to our ears.

Interest in the association between variables thereafter blossomed in the genetics work of Francis Galton (1822–1911) and Karl Pearson (1857–1936). Several contemporary techniques still retain Pearson's name in theirs, such as the correlation coefficient mentioned earlier. Concern with cause and effect prompted the design of experimental studies to develop next. A primary interest was with the analysis of different methods for agricultural production to feed a rapidly expanding population. People were asking such questions as: "Is the application of 'natural' or 'man-made' fertilizer a superior technique?" Ronald Fisher (1890–1962) was a dominant figure of this time. His tables of expected values for several types of theoretical distributions are still used. The entire topic known as analysis of variance, used with experimental data, is one of his greatest contributions.

This abbreviated history of statistics clearly reflects its applied character in the social sciences, as well as all sciences. But while statistical techniques evolved from within the guiding framework of the Traditional Research Model, not all of its contemporary uses are part of that same approach. It may surprise you to learn that statistical methods are often used by social researchers who reject the traditional perspective. Those methods are thus seen as being more universal than the research philosophy which spawned them. Yet the objections to that philosophy are important to consider.

Criticism of the Traditional Research Model

The Traditional Research Model has dominated research activity in the social sciences for the past 75 years or so. But it has not been without its critics. The extreme "rationalism" of the model has been especially troubling to some. The formal process of moving from a general theory to hypotheses, measurement, data, analysis, and back again to hypotheses and theory makes the later stages of the research process seem entirely determined by the decisions made in the first stages. Critics argue that this forces the researcher to be insensitive to the views of the people who are being studied. Rather than listening to how the people themselves characterize what they are doing and feeling, the researcher examines them from the point of view of the general theory that began the process. The critics suggest that one should displace theory from its potentially prejudicial position in research by beginning the process with direct observations instead. Explanation in the form of the "natural language" of those being studied may then lead to hypotheses and even theory, perhaps.

It is true that most general theories do not conceptualize things in the same way, or in the same terms, as do the people being studied. As a criticism, this is somewhat overstated when you recognize that the general theories used by researchers have ordinarily been suggested by observations of everyday life made in everyday terms. Remember that the Traditional Model is a formalization of the actual process that research follows. Research is often conducted on multiple fronts, with decisions being made and remade about all parts of the process until the very end. Indeed, much traditional research begins with unsystematic observations which subsequently cause the researcher to seek a general explanation (theory).

Traditionalists further argue that it is not possible to empty one's mind of preconceived theories and hypotheses before collecting data. It is very difficult to leave one's own experience and culture behind when entering a new situation. It is easy to see "new" behaviors as examples of "old" familiar behaviors. The traditionalists argue that it is better to recognize this by deliberately comparing the new research situation to the old. That is the intended role of theory in guiding such research.

Despite these rejoinders by the traditionalists, criticisms of their approaches to social research continue to be heard and are probably gaining in adherents. To review some of these criticisms and alternative research models see Motwani (1967) and Schwartz and Jacobs (1979).

A final word of caution about all models of research is required so that you are not misled. Whatever process a social researcher uses, individual judgment is exercised at all steps. As a consumer of social statistics, you must anticipate that each example of research will be somewhat different from all the others. Individual investigators make unique decisions about how to conduct the specifics of their studies. Not even fellow traditionalists will make the same decisions about the same research problem. This is an indication of legitimate differences in opinion, not that some researchers are right and others are wrong. These "inconsistencies" are evidence of the vitality of social research, not its weaknesses.

Measurement Issues

Validity and Reliability

Regardless of the method used to collect data, the researcher and consumer need to be concerned with the issues of validity and reliability. *Validity* is the extent to which a measurement truly represents an intended characteristic. For example, when someone says "This information indicates that the public approves of private enterprise," we want to know if it is true. Does this information really represent that public attitude? This is the validity question. Although there are some special techniques used to determine validity, they are left to more specialized or advanced texts. (See Bailey, 1982:68–72 and Kerlinger, 1964:444–462.) Nonetheless, you should be sensitive to the issue and not accept claims for validity without some justifications being offered.

The issue of reliability is of concern as well. *Reliability* is the extent to which a measurement consistently represents an intended characteristic. This time when someone says "This information indicates that the public approves of private enterprise," we want to know if they would get the same information each time they or anyone else collected it in the same way from the same or similar people. Perhaps they have used a technique that would cause people to give different answers each time they were asked. The questions used may have been ambiguous and interpreted in various ways. Obviously, if data are not reliable, they cannot be valid; if they are not consistent, they cannot measure any single characteristic.

Researchers often speak of their procedures and data as being *objective*. One meaning of this term is identical to reliability. If a measuring instrument yields the same information from the same people no matter who the researcher is, that instrument is said to be objective. If it does not do that, the procedures and data that result are said to have *bias*.

As with validity, the more technical means of assessing reliability are beyond the scope of this book. (See Bailey, 1982:73–75 and Kerlinger, 1964:429–443.) But be a cautious believer in data's reliability and demand some explanation before accepting such an assertion.

Level of Measurement

One of the most important decisions a researcher makes when deciding how to handle all the issues that arise at the measurement step of the research process is the level of measurement to be used for each variable. It is important because it is a decision that affects much of what it is possible and appropriate to do later in the analysis step. A consumer is in a better position to understand why a researcher chooses to do some kinds of analyses and not others once the level of measurement that has been used is known. In addition, one is better able to spot abuses of statistics in which certain analysis techniques have been used in inappropriate measurement situations.

Level of measurement concerns the extent of mathematical meaning that is implied in the categories used to measure any variable. It is a rather large and complicated topic that relies heavily on what is called measurement theory. We will examine an overview of this including some of the more relevant details.

There are four basic levels (or scales) of measurement: nominal, ordinal, interval and ratio.* They give increasing mathematical meaning to the categories used in the measurement moving from the nominal to the ratio level.

Nominal Level

The lowest level of measurement is *nominal*. Nominal measurement simply classifies the observations by kind. Each observation is classified and labeled by the appropriate category name. As a result of nominal measurement, we know which observations are similar and dissimilar to each other. For example, you might classify your classmates' identification with political parties as being "Democratic," "Republican," "other party," or "none." If so, you are using nominal measurement. The categories identify people according to their party identification by kind but not by any degree of identification. As a consequence, the only mathematical meaning implied by the categories is that they are different from each other. You cannot perform ordinary arithmetic operations such as addition, subtraction, multiplication, or division between the categories. Adding two Democratic identifications together does not create a Republican identification. You can only distinguish between such nominal categories. Other common examples of social measurement at the nominal level are marital status ("married," "never married," "divorced," "separated," "other"), ethnic identification ("white," "black," "Hispanic," "Native American," "other"), and religious affiliation ("Protestant," "Catholic," "Jewish," "other").

As the political party example illustrates, nominal measurement typically uses verbal phrases as the categories. But a researcher may also assign numbers to those categories.

*There are other approaches to measurement levels including a well-known one which includes a fifth level, the absolute. This level refers to measurements in which one records the number of times an event occurs. (See Stevens, 1951.) In this book the absolute type is combined with the ratio level because the same kind of statistical procedures are appropriate to both, generally.

For example, the following labels could be used: 1 = Democratic, 2 = Republican, 3 = Other Party, and 4 = None. Numbers may be used to label nominal categories for at least two reasons. First, the numbers are often simpler to use than some verbal categories, which can be quite wordy. Second, most data analysis uses computer programs and statistical methods that require the data to be in the form of numbers, not words. The assignment of numbers to nominal categories does not transform the measurement to a higher level. No new arithmetic manipulations are suddenly possible because of the new labels. Subtracting "Democratic" (1) from "none" (4) does not equal an identification with some "other party" (3). It should be clear to you that the numbers chosen for assignment to nominal categories are arbitrary. Any set of numbers would serve the labeling purpose. It is merely conventional to begin numbering at "1" and continue with consecutive whole numbers.

When using nominal measurement, the major concerns are to ensure that the categories are exhaustive and mutually exclusive. Exhaustive means that enough categories have been used so that every observation can be assigned to one of them. Mutually exclusive means that there is only one correct category into which each observation can be placed. If an observation cannot be placed within any one of the categories, there is a violation of the exhaustive rule. If any observation can be placed into more than one category, the exclusive rule is broken.

For example, if we tried to classify off-campus residences as "expensive," "run-down," or "fun," we might have a problem assigning some place to only one category because it fits several of them (it's both expensive and run-down). This is the exclusive issue. We might also have difficulty assigning another place to at least one of these categories because it might not fit into any of them. This is the exhaustive issue. It is not always easy to design a set of categories that avoid these problems. Nominal measurement should not be taken for granted merely because it is the simplest level.

Ordinal Level

The second level of measurement is *ordinal.* Ordinal measurement classifies the observations by kind and it also indicates a degree or order that the categories have relative to each other. The categories are viewed as belonging along a dimension ranging from a set of extremes such as "low" to "high" or "more" to "less." For example, traditional parents may informally classify dating partners of their teenage children as "highly acceptable," "acceptable," or "unacceptable." If so, they are using an ordinal scheme of measurement. The categories classify various dating partners as nominal measurement does. But they also indicate an order of acceptability that the parents have toward those partners. Those in the "highly acceptable" category are not only different from those in the "acceptable" category, they are also more acceptable to the parents.

If numbers are assigned to the ordinal categories, they should reflect the ordering. Here, for example, a logical method would be: 1 = unacceptable, 2 = acceptable, and 3 = highly acceptable. In this scheme, the higher the category of acceptability, the higher the number that is assigned. The reverse could be done as well. The classifying and ordering meanings to the numbers are the only intended mathematical interpretations that are to be made when seeing this. One can determine whether the observations are greater than, less than, or equal to each other. But no further mathematical meaning is implied. We still cannot add, subtract, multiply, or divide ordinal categories. Thus we cannot specify the extent of differences between observations.

Common examples of ordinal measurement are social class ("upper," "middle," "lower"), place of residence by size ("rural," "small town," "suburb," "central city"), and some measures of attitude that place individuals into such categories as "high," "moderate," and "low."

Interval Level

The *interval* level of measurement goes one step beyond ordinal to indicate how much difference there is between categories. It classifies by kind like nominal measurement and indicates order like ordinal measurement. But it also specifies a numerical interval between the categories, called a *unit of measurement*. This identifies the smallest difference that is being quantified by the measurement. Obvious examples of interval measurement are not so easy to cite. There is considerable disagreement in the social sciences about what kinds of measurement qualify as being interval.

To illustrate, suppose an interview contains the following statement: "The government should set aside more places for preservation in their natural state." The respondent is asked to choose a value from the following set that best represents his or her opinion about the statement:

Strongly Disagree	Disagree	Uncertain	Agree	Strongly Agree
1	2	3	4	5

Suppose one respondent answers "4, Agree" and a second respondent answers "2, Disagree." What do we know about how they compare? We know that the five categories allow the respondent opinions to be classified and thereby reveal that they are different. So it is at least nominal measurement. We also know that the first respondent agrees more with the statement than does the second one, so it qualifies as ordinal measurement.

But do we know how much more one agrees than the other? Can it be quantified numerically? Some researchers will say that the difference is two points of agreement ($4 - 2 = 2$) and therefore this is an interval measurement. Others will be unwilling to attach such mathematical meaning to these values (and phrases) used as categories. They will argue that the differences between the positions are not known in a numerical, quantitative way, only that the order of the categories is certain. For example, they will ask whether the difference between "Disagree" and "Strongly Disagree" ($2 - 1 = 1$) is really the same as the difference between "Agree" and "Uncertain" ($4 - 3 = 1$)? After all, both do equal 1. To conclude that this kind of measurement is interval requires us to accept the numerical, quantitative differences in the numbers as they ordinarily would be understood when doing addition or subtraction. A difference of "1" is the same regardless of the two consecutive numbers being compared. We are not required to agree that these particular numbers (1, 2, 3, 4, 5) are the ones that must be used. Any set of equally spaced numbers would be acceptable. But we must agree that the numerical *differences* ("intervals") between those numbers accurately represent the extent of differences between the categories. Thus in interval measurement, the usual mathematical processes can be applied to the differences between categories (although not to the category score values themselves).

Since this debate over interval measurement has not been resolved, researchers must make their own decisions. Consequently, the reader must be prepared to find contrary opinions on virtually identical measurement questions depending on the researcher. What should you believe? Believe that it is an unresolved issue!

Most statisticians and measurement theorists consider the kind of example we are discussing to be an ordinal measurement. However, in practice a majority of researchers tend to treat it as an interval measurement! They argue that the numerals in the response set imply the quantitative interval differences that can be calculated. If not, what was the purpose of the numerals? Further, since this type of five-point (strongly disagree to strongly agree) scale, called the Likert method after its founder, and its variations (seven-point scales, 11-point scales, and so on) are so common, they may consider any of them to be interval even if the numerals were not presented to the respondents but only added afterward for analysis purposes.

Why is there this disparity between theory and practice? Researchers tend to push the measurement assumption to the higher level because statistics at the interval plateau yield much more information about those data than do statistics at the ordinal level. Studies of the effects of this "exaggeration" tend to show that it does not lead to erroneous conclusions in general (Labovitz, 1970). Of course, it may be incorrect in any particular situation.

The example above illustrated the measurement of attitude using a single item. There is more likely to be agreement among both theorists and researchers that the measurement of an attitude using a series of such items which have undergone a "standardization" process may constitute an interval scale. This process is itself a detailed statistical analysis which sorts through a battery of items to determine which ones measure the intended attitude and how the responses to them can be combined into a total score for each respondent. Examples with which you are familiar are the kinds of tests you take to assess your intellectual ability ("IQ") and preparation for college (SAT, ACT, GRE, and others). These have been examined to determine how they perform as measurements and their total scores (e.g., IQ = 115 or SAT = 750) are often regarded as interval statistics. The techniques of such scale construction are an advanced topic we cannot cover in this book. However, you may have heard some of their names, especially factor analysis and scale scores. The hypothetical Frank Friendly Social Class Scale earlier illustrates measurement using several indicators combined to form a scale score that would be considered by many to have interval quality.

As a guide in understanding this book, we will ordinarily consider any situation such as the one above from the point of view of the researcher. This means that we will be seeing more use of interval statistics than the measurement theorists wish.

As mentioned, one of the main features of an interval measurement is that it incorporates a unit of measurement. Above, we referred to a difference of "two points of agreement" between the two respondents on the attitude scale. In that example, the unit of measurement was a single point of agreement on the scale. If we take a standardized test of general intelligence which has a range from 50 to 150 where only whole numbers are possible, the unit of measurement there is one point of intelligence (however intelligence is defined by that test). Measurements at or beyond the interval level must have a specified unit of measurement.

Ratio Level

The last of the four main levels of measurement is the *ratio* level. It has the features of interval measurement with an addition especially important to the measurement theorists and researchers. This is the added characteristic of a true (or absolute) zero point which represents the complete absence of the characteristic being measured. In discussing interval measurement we noted that we need not accept the assignment of any specific numbers to categories, only that the differences between them are correct. In ratio measurement we must accept both. The values *and* their interval differences are the ones that should be used. That is because the categories correspond to specific numbers in their full mathematical meaning. They are cardinal, not arbitrary. The numbers themselves can be added, subtracted, multiplied, and divided. It is the absolute zero point in a measurement scheme that makes this possible. Like any other number used as a category, no respondent is required to have the score of a true zero. It is necessary only that such a numbered category exists in the measurement scheme.

Suppose that you and some of your friends list the populations of your hometowns as exactly as you can determine them. This would be a ratio measurement. We would know the categories to which each of you belong (all positive whole numbers are the categories), the order of those categories (larger numbers mean larger hometown populations), and the numerical differences between the categories (simply subtract any two population sizes in which you are interested). But more than that would be known. If you said your hometown has a population of 9203 and someone else's is 36,812, those numbers do differ by 27,609, as they would in an interval measurement. But additionally, your hometown has one-fourth of the population size of the other hometown. This ratio comparison of the scores is only possible with ratio measurement. There is a true zero point in this method of measuring hometown populations. Zero means no population at all.

To reiterate, the interval scales above are not ratio measurements because the numbers assigned to the categories are arbitrary. It would not matter whether the five categories were numbered 1, 2, 3, 4, 5 or 10, 11, 12, 13, 14 or any other set as long as the differences in the numbers represented the differences the researcher intended. They could even be numbered 0, 1, 2, 3, 4 but that would not make it a ratio measurement. Here the zero is just as arbitrary as all the other numbers. No category represented absolutely no agreement with the statement.

In most measurements whereby researchers count people (population sizes), their behaviors (how many times they break a law), or other characteristics (how many years old they are), they are using a ratio measurement. Do not confuse this with counting how many people belong to a particular category of a nominal, ordinal, or interval variable. These frequency numbers have ratio quality regardless of the measurement level for the categories. Rather, we are referring to situations in which the categories themselves represent the counts that are made.

Common examples of ratio measurements in social science are age, years of education, height, exact incomes, number of employees, and size of state budgets in dollars. A common natural science example you may misunderstand to be ratio measurement is temperature. Degrees in either Farenheit or Celsius are interval scales, not ratio. The numbers (15°C or −10°F) are arbitrary. There is no true zero point on either scale. 0°C represents the freezing temperature of water and 0°F is even more unconnected to any physical absolute. In contrast, the Kelvin scale of temperature is regarded as being ratio. 0°K corresponds to the point of no molecular motion and scale values above this represent relative degrees of such molecular activity. Hence it is true that 50°K is 25° warmer than 25°K and that the first is twice as warm as the second. However, 50°F is not twice as warm as 25°F, although the difference is legitimately 25°.

Figure 1.3 summarizes the distinctions that we have been making about level of measurement. As the figure shows, the levels are cumulative; ordinal has the characteristics of nominal plus its own, interval has the characteristics of nominal and ordinal plus its own, and so on. Be sure to note this because many students new to the topic will make hasty judgments on the level of a measurement for a particular variable without going

FIGURE 1.3 Properties of the Four Levels of Measurement

	Property of the Measurement			
Level of Measurement	Classified by Kind	Categories Have Order	Differences Between Categories Are Known	Has True Zero Point
Nominal	Yes			
Ordinal	Yes	Yes		
Interval	Yes	Yes	Yes	
Ratio	Yes	Yes	Yes	Yes

through the sequential classification procedure. For example, a ratio measurement must have all the previous characteristics plus a true zero point, not just an apparent zero point.

One must be able to determine the level of measurement for a variable in order to proceed smoothly with most of what follows in social statistics. Concepts can be measured as variables with any level of measurement. It is up to each researcher to decide. The reader cannot assume that researchers will make the same decision about any one concept because they often have different theoretical perspectives and analysis plans. One should not try to memorize that any particular concept will always be measured at one level of measurement. It will not be.

Just as exaggerating the level of measurement may cause errors, there is a potential cost for underrepresenting the level. The use of nominal-level statistics for interval data, for example, will reveal only nominal characteristics of those data. The ordinal and interval differences between the categories will be ignored. This disregards information that is contained in the data. Clearly, the best practice under ideal conditions is to use statistics that extract the most information possible without distorting its true character.

Continuity and Precision

Continuity

Two further measurement issues that are important to understand are continuity and precision. *Continuity* refers to a variable's use of a necessarily limited (discrete) or unlimited (continuous) set of numerical values in its representation.

A continuous variable has no minimum interval difference between the potential scores that can be given to the observations being made. A continuous variable such as our heights can be expressed as 70 inches tall, 70.5 inches tall, 70.57334 inches tall, and so on. There is no theoretical limit in the detail of the height characteristic as it naturally exists. There is, of course, a limit to how exactly we are able to measure it. Even the most sophisticated instrument used will have its limits. But height itself is not limited in this way, so it is referred to as a continuous variable.

On the other hand, we have many variables that are limited in the set of numbers that can represent them regardless of the sensitivity of the measuring instrument. You can have 0, 1, 2, 3, or even 20 sisters, but you cannot have 1.5 or 3.3337 sisters. It is not a problem of the instrument used to measure this that keeps us from having tenths or ten-thousandths of a sister. It is a limitation within the variable itself. The same is true of the amount of change we may carry (minimum detail is 1 cent) or number of times we have been legally married (minimum detail is 1 time). These types of variables are necessarily discrete.

Because all measuring devices have limited accuracy, all measurements have the appearance of being discrete. There is a limit to the number of places that can be determined in the scores used. This is true even of continuous variables. You need to consider the underlying variable, not its measurement, to distinguish discrete from continuous.

Although a variable may be discrete when measured one observation at a time, the social researcher will often make calculations from data that can cause the reader to think the variable is continuous. The average (mean) number of sisters may be 1.78 or the average amount of change we have may be 65.3 cents. This is perfectly acceptable within the world of social statistics because we understand that these types of statements do not mean that the original data were continuous. Only the statistic being presented is continuous. The statistic creates a new variable in which the research units are aggregates of the original units. The calculated average for one data set becomes a single score value in the new variable of average numbers of siblings or average amounts of change for several potential data sets.

A consumer of social statistics can be easily confused by this. As we progress into discussions of actual statistical techniques, you will be able to see how the discrete–continuous distinction influences the applications of those techniques.

Precision

The computations of all statistical measures require the researcher to decide how much precision is to be used. *Precision* refers to the extent of accuracy that is maintained within a measurement or its subsequent analysis. Generally, researchers express their final calculations with one or two more places of accuracy than the original data had.

With a discrete variable such as years of formal education completed, we might summarize the scores of 10, 12, and 13 as having a mean (average) of 11.7, using one extra place of precision. Likewise, we might represent the mean income for the scores of $5555.55, $7788.80, $14,010.05, and $19,000.34 as $11,588.685. There are some other kinds of statistics that typically use two extra places of accuracy beyond the original. Standardized scores such as "*z* scores" usually look like 1.96 or −2.33. Statistics such as correlation values often appear as .44 or −.78. Generally, accuracy is not taken any further than the one- or two-place rule. To know when to do which, one must observe the cultural traditions within statistical analysis.

The Organization of Statistics

One of the best ways to understand anything is to know its organization. The field of social statistics is organized according to two basic types of concerns. The first is problems of description and the second is problems of inference. *Descriptive statistics* summarize data. The summary attempts to reveal the overall patterns of the data or to highlight some of their special aspects. Many common statistics such as baseball averages, record-book statistics, and graphs of economic trends are descriptive statistics. Each represents a summarization of many individual observations so that we have a convenient and simple way to think about the data.

A big concern when doing descriptive analysis is to select the kinds of summaries that achieve a useful reduction in the original complexity of the data but which remain faithful representations of that complexity. One can summarize to such an extent that important features of the original data are lost. It requires considerable skill to make the right decisions, and some individual creativity is often helpful. Part I of this book looks at the major types of descriptive statistics used in the social sciences.

Inferential statistics make generalizations (inferences) from sample data to population data. A descriptive summary of that data is ordinarily a preliminary step to the inferences. More often than not, social researchers gather data from a smaller number of people, groups, cities, or whatever (a sample) than they want to know about (a population). They do this for reasons of economy, time, and also because they know the statistical advantages of this procedure. After summarizing the data in these samples they move on to make projective statements which they believe are representative of the larger population that they are studying. Knowing how to do inference problems depends on knowing how to evaluate the likelihood that erroneous generalizations will be made. This is the use to which social researchers put the whole topic of probability. After an interim section on probability (Part II), this book will proceed through the more common types of inferential statistics (Part III).

Keys To Understanding

There are three general keys to understanding social statistics that apply in nearly all the situations to be covered. It is therefore appropriate that they are presented in this first chapter.

Key 1.1. Try to approach social statistics as a study of an "other culture." This will broaden your mind; it is more difficult to hold prejudices and stereotypes and to be ethnocentric (believing your own views are superior) once you experience the world from a different perspective. Further, it will give you more realistic expectations about how quickly you can learn new information and what tactics are likely to be successful in mastering it.

This first chapter has included references to the classical participant-observation method used by some social scientists when they try to understand cultures different from their own. New consumers should not attempt to reconstruct the world of social statistics to conform to their present worlds. The two usually are not alike, and one or both are often distorted in the attempted reconstruction. Rather than demanding that the social researcher speak your language, learn some of the language of social research. This book will try to mediate the problems you have with this by providing some translations. Actually, the number of specialized words and symbols that one needs to learn is very small compared to the number in a foreign language course. A list of these "Key Terms" appears at the end of each chapter and in a glossary in Appendix B at the back of the book. You will not be able to "speak" statistics like a native after this book any more than you could speak Russian after one exposure. But you should have a good beginning.

Key 1.2. Adopt a critical attitude. Read research reports carefully. Get a general overview and then examine all details. Keep asking yourself what is being said and whether you are convinced that it is correct. Has enough information been given so that you can make your own judgment? What must you assume in order to agree with the author? Are you willing to agree? Why or why not? If you fail to ask these kinds of questions, you leave yourself dependent on the persuasive arguments that others can make. This is a prime source of the alienation people have from research. They cannot interpret it themselves. But a self-sufficiency attitude and capability is a key to becoming an "educated consumer" of social statistics.

Key 1.3. Be tentative in reaching a conclusive judgment about research. Often research will be presented in a highly condensed fashion to conserve space or time. If your critical questions leave you saying that you do not understand or you do not agree, consider the possibility that it is missing information that is to blame, not erroneous information. Perhaps you will have to dig beyond the specific article you have at the moment to find the information you need. Maybe you need to talk to more people, even write the author. That is what other social researchers have to do. Do not misunderstand—you should pull down any halos that you may have given to social researchers. Some do make mistakes. Some erroneous ideas and results are published. But usually there are different types of truth to a research problem depending on the perspectives taken. Do not label strange-sounding ideas as nonsense until you have reason to be rather confident of your judgment. So, on to descriptive statistics!

ON THE SIDE 1.1 Statistics as a Language

A student in my social statistics course told me of an encounter with his econometrics professor:

"I told Professor Smith that his approach to statistics differed greatly from yours. He puts a formula on a board, tells you to get it into your notes, memorize it, and then expects you to know it for the test. I told him that my other professor puts a formula on the board and asks the class whether they understand it or not. If the class says 'Yes,' he erases it and says, 'Ok, it's in the book.' I asked Smith, 'Why the difference?' He didn't answer . . . He didn't really like the question."

This student was being exposed to two different approaches to statistics. The econometrician viewed statistics as a discipline, valuable in its own right, with facts to be learned and rules to be followed.

I like to look upon statistics as a language. A language is a collection of symbols with a widely accepted set of rules for their manipulation. Knowing a language allows one to communicate with others who know it. Recognizing those aspects of statistics which make it simply a means of communicating with others suggests a pedagogical approach to statistics that differs from that of the econometrician mentioned above. . . .

Pedagogically, the implications of viewing statistics as a language number three. First, just as one needs a reason to learn a language, and learn it well, one needs a reason for learning statistics. Except for a few people who find joy and fulfillment in the development of the discipline, most students must have an interest in a substantive question, a large amount of relevant data amenable to statistical manipulation, and others who are interested in the student's interpretation of the data. It has been my experience that most students perform poorly not for innate lack of ability, but because they have no foreseeable reason for learning what the professor is trying to teach. Changing this situation is often difficult, and responsibility for doing so must be seen to rest with both professor and student. But, just as with language, efficient learning will occur when the criterion of "having a reason for learning" is met.

Second, viewing statistics as a language suggests stressing *understanding* procedures rather than "learning" techniques. The emphasis should be on concepts rather than formulae, on underlying models and assumptions rather than specific tests and mechanics of implementation. Just as it is easier to develop one's own writing style if one "knows" the rules of grammar in some

nonmechanistic sense, so that one does not have to check each sentence for split infinitives and dangling participles, it is much easier to become felicitous with statistics if one "knows" the field without having to recall or search out the right formula in each set of circumstances.

Finally, a nonpedantic approach to statistics would emphasize the use of various resources in the development of statistical facility. There is always someone around to take a complicated sentence, break it down, cut out the fat, and show you how to say what you want to say with much less effort than you are exerting. One always has a dictionary accessible to find out how to spell a word. And one usually can find a thesaurus to perk up a phrase. Why, then, we must ask, do we insist that students devote an inordinate amount of time to memorization, application, and mindless recall when there are libraries full of statistical texts and faculties full of people schooled in the subtleties of the field to whom one can appeal for technical assistance? I do not think there is a reasonable answer to that question. Students should expend their energy in educationally sound activity—in conceptual learning, in practicing and refining their skills, and in the development of resources which can be used to solve problems beyond one's ability.

It seems reasonable to think of statistics as a language. It puts the problem of motivation in the student's lap; it focuses attention on the intellectual infrastructure of the field instead of on its superficial manifestations—the formulary; it redirects pedagogical energy from pedantic inculcation of facts to the establishment of a situation in which students have the reason and the opportunity to learn the trade. And for me, all of this makes a potentially frustrating, alienating subject approachable and fun.

Adapted from William Ray Arney, "Statistics as a Language," ***Teaching Sociology,*** **5 (2), January 1979:173–178. Beverly Hills, Calif.: Sage Publications, Inc. Reprinted by permission.**

KEY TERMS

Empirical
Traditional Research Model
Theory
Hypothesis
Concept
Variable
Measurement
Research Unit
Data Collection

Observation
Data
Analysis
Statistical methods
Validity
Reliability
Objective
Bias

Level of Measurement
- Nominal
- Ordinal
- Interval
- Ratio

Unit of Measurement
Continuity
Precision
Descriptive Statistics
Inferential Statistics

EXERCISES

1. Consider the following ways that educational attainment could be measured. Indicate the level of measurement for each method.

(a) None
Primary completed

Secondary completed
Postsecondary completed
(b) Graduated from a public high school
Graduated from a private, sectarian high school
Graduated from a private, nonsectarian high school
(c) no formal education
1 grade completed
2 grades completed
⋮
12 grades completed
⋮
16 grades completed
⋮
(d) Educational level

Very Low	Modest	Substantial	Very High
1	2	3	4

2. Consider the following ways that age could be measured. Indicate the level of measurement for each method.
(a) Young
Middle-aged
Old
(b) Whole numbers of years rounded to last birthday
(c) Whole numbers of years rounded to nearest birthday
(d) Young Old

1 2 3 4 5

3. Consider the following ways that sexual orientation could be measured. Indicate the level of measurement for each method.
(a) Female
Male
(b) Masculine
Somewhat masculine
Androgynous
Somewhat feminine
Feminine
4. For each of the following, determine the level of measurement and whether it is a discrete or continuous variable:
(a) elapsed time for runners in a cross-country race (measured with a stopwatch to the nearest second);
(b) order of finish for runners in a cross-country race (1 = first place, 2 = second place, and so on);
(c) number of times people have moved to a new residence within the past two years (measured as actual number of new addresses);
(d) condition of housing (measured on a 1–6 scale where 1 = very poor and 6 = excellent).
5. Identify the name of the variable and the level of measurement in each of the following examples taken from published research (tables a–d).

(a) Type of Sexual Offense ($n = 30$)

Primary Offense	Number of Clients
Obscene phone call	4
Indecent exposure	9
Solicitation of minor females	6
Child molestation	4
Rape	2
Homosexual activity	5

Source: Adapted from Steven N. Silvers, "Outpatient Treatment for Sexual Offenders," *Social Work,* 21 (2), March 1976: 135, Table 2.

(b) Frequency Distribution of Social Index Items with Insufficient Information

Number of Items Missing	*N*	%
None	806	82.4
One	164	16.8
Two	7	.7
Three	1	.1
Total	978	100.0

Source: Albert F. Osborn and Tony C. Morris, "The Rationale for a Composite Index of Social Class and Its Evaluation," *British Journal of Sociology,* 30 (1), March 1979:50, Table 3. Reprinted by permission.

(c) Reported Causes of Death in Sample Wanitabe Genealogies, c. 1900–1962

Cause	Males	Females	Total	Percent
Kuru	16	131	147	31.1
Other sorceries	66	36	102	21.6
Unknown	41	34	75	15.6
War injuries	42	13	55	11.6
Undiagnosed sickness	16	18	34	7.2
Dysentery	13	14	27	5.7
Old age	9	7	16	3.4
Burns	6	1	7	1.5
Influenza	2	2	4	.8
Infant malnutrition	3	1	4	.8
Suicide	—	2	2	.4
Totals	214	259	473	100.0

Source: Shirley Lindenbaum, *Kuru Society: Disease and Danger in the New Guinea Highlands.* Palo Alto, Calif.: Mayfield Publishing Co., 1979:65, Table 2. Reprinted by permission.

(d) Juvenile Court Dispositions for Those Arrested for a Violent Offense, Franklin County, Ohio, 1956–1960 (percents)

Disposition	Percent
Unsupervised probation	43.9
Supervised probation	20.8
Detention facility	18.6
Institutionalization	16.7
Total	100.0
	(N = 3316)

Source: Charles David Phillips and Simon Dinitz, "Labelling and Juvenile Court Dispositions: Official Responses to a Cohort of Violent Juveniles," *Sociological Quarterly,* 23 (2), Spring 1982:270. Adapted from Table 1.

6. Are the following examples of descriptive or inferential statistics?
(a) A newspaper conducts a poll of 200 randomly selected citizens to estimate how many people will support a school bond issue in the upcoming election.
(b) A researcher lives in a nursing home for a month to determine what living conditions are like in that home.
(c) A presidential commission examines data from 10 rural counties to prepare national policy recommendations on rural poverty.
(d) A student examines all publications by an instructor in order to compare them with the class lectures he has heard.

PART

DESCRIPTIVE STATISTICS

CHAPTER 2

DESCRIPTION THROUGH TABLES AND GRAPHS

Two of the most common ways in which the data for social research are presented to consumers are tables and graphs. These are the tools used by researchers to begin the process of data analysis. In some studies the researcher will look for patterns in the data and then describe them using the techniques of descriptive statistics discussed in this first section of the book. In other studies the researcher will go beyond these descriptions to make generalizations about the larger social reality outside the data collected using inferential statistics. In either case, the first step for the researcher is to describe the data at hand.

Tables and graphs, like all descriptive techniques, have three major goals: (1) *organize* the data into meaningful formats, (2) *summarize* the data into simpler accounts, and (3) *emphasize* those features of the data that are especially relevant to the study. The overall concern of the researcher is to reduce the complexity of the original data and thereby reveal those features of greatest interest. Any data reduction technique reduces some of the precision or detail found in the original data for the individual research units. But with care an appropriate reduction can be found which achieves the three goals of description without distorting the original data's character. The researcher seeks a delicate compromise between the two.

Types of Tables

Some examples of the major types of tables that are found in social research are presented in Research Examples 2.1 to 2.12. Although this variety may seem confusing at first, it can be organized quite simply by use of a two-part taxonomy. We can categorize tables by (1) the number of variables that are included and (2) the statistical form of the data that is being displayed. Using the first criterion, we may find that a table is *univariate* (one variable), *bivariate* (two variables), or *multivariate* (three or more variables). By the second criterion, we may encounter a table in which the data are shown as frequencies, proportions, percentages, cumulative frequencies, cumulative percentages, ratios, rates, or some other form. Any of these data forms may appear in a table alone or in a combination with the other forms.

Using this two-part taxonomy we find that some tables show a "univariate frequency distribution," as do Research Examples 2.1 to 2.4. Others may be a "multivariate percentage distribution," as are Research Examples 2.10 and 2.11. The research examples show how the taxonomy works for several types of tables.

There is more to the taxonomy of table types than merely naming tables. Scrutinizing a table for the two criteria used in classifying helps one to begin reading, rather than ignoring, the table. Many consumers of social research do not really read tables. They only give them a quick glance. But learning to read a table is the best way to understand the researcher's interpretation of it and the remainder of the analysis that has been performed.

Beyond learning to use the taxonomy, the other steps in table interpretation are better examined separately for each of the major types of tables as illustrated in the research examples. In the process we can discuss the procedures that a researcher uses when constructing each type. We can also get an idea of the broad range of topics that are commonly studied by social researchers.

Univariate Frequency Distributions

Visualize the task before a researcher who has finished the data collection phase of a study and now wants to describe the observations using tables. If the data were collected with a questionnaire or interview schedule, the researcher's office floor is probably covered with stacks of completed (and some half-completed) forms. If the method was observation, the desks and tables are cluttered with copious notes, rough tallies of noted behaviors, or stacks of recording tapes each threatening an avalanche. No matter what method was used, there is likely to be a mass of information that seems overwhelming in this initial form.

One of the researcher's first tasks is to *code* the observations. Most codes are numerical in order to be compatible with computers and statistical analysis. Thus the code tells how numbers are to be assigned to each observation. For each variable a code is devised so that every unique answer has its own numerical equivalent. In some traditional research these variable codes have been predetermined and may even appear on the data collection instrument in small type to the side of the questions. In other types of research the codes are devised after the range of responses is known.

Once the observations are coded they can be prepared for processing by a computer or another means. This is done by transfering the data to computer cards, magnetic tape, magnetic disk, or some other medium. The processing can then proceed to tabulate, cross-classify, or inspect the coded data according to the wishes of the researcher.

RESEARCH EXAMPLE 2.1 Univariate Frequency Distribution—Nominal Variable with Categories Ordered by Frequencies

Race of Respondent

Race	Frequency
White	13
Black	3
Oriental	1
Hispanic	1
Total	18

Source: Dan A. Lewis and Rob Hugi, "Therapeutic Stations and the Chronically Treated Mentally Ill," *Social Service Review*, 55 (2), June 1981:206–220. Adapted from Table 1.

Imagine what the researcher may have been dealing with in Research Example 2.1. The races of 18 people who had made extensive use of mental health facilities have been determined. In this study race was considered to be a nominal variable divided into four categories. Each category could be assigned a code from 1 to 4. (See Figure 2.1.) Since the variable is at the nominal level this assignment is arbitrary. Each case also would be given a *case number* (a code from 1 to 18) to maintain case identity and separation during the data processing. Case numbers are commonly assigned according to the sequence in which the cases were selected for study or the sequence in which the information was gathered. But there are other possibilities as well.

This information could be displayed in a *raw data distribution.* This is a list of the data in the order of the case numbers. In this example, we would show the coded racial identities for each of the 18 persons listed from case number 1 to number 18. A sophisticated description of the data is generally not possible from this type of distribution. It retains too much of the original complexity and does not have a useful organization.

For this reason the data are commonly relisted in an *ordered array.* The ordered array displays the data in the order of its coded values on the variable, not the order of the case numbers. The result is that those cases having the same variable codes are clustered together and we can begin to see any patterns that might exist in the distribution of scores. In our example, it is immediately apparent that the scores primarily represent only one of the four categories of race: white.

Our description task is even simpler if we tally the number of responses for each category and then present them in a *univariate frequency distribution,* such as the table in R.E. 2.1. This table also shows a *simple frequency distribution,* meaning that the different categories of the variable have not been combined into grouped categories. Grouped distributions will be discussed shortly.

Note the basic parts to the R.E. 2.1 table. There is a title identifying the variable and type of cases and a footnote giving the source of the information. The first column lists the four categories of the variable and it is headed by the variable's name. Because the variable in this example uses nominal measurement the categories can be listed in any order.

FIGURE 2.1 Coding Raw Data and Determining a Frequency Distribution

1. Coding Raw Data

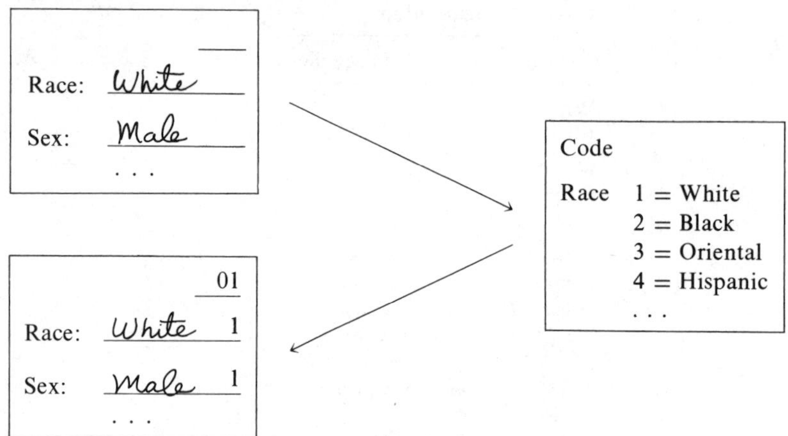

2. Raw Data Distribution

Case Number	Race
01	1
02	1
03	4
04	1
05	1
06	2
07	1
08	1
09	1
10	1
11	2
12	1
13	3
14	1
15	2
16	1
17	1
18	1

3. Ordered Array

Race
1
1
1
1
1
1
1
1
1
1
1
1
1
2
2
2
3
4

4. Frequency Distribution

Race of Respondent

Race	f
White	13
Black	3
Oriental	1
Hispanic	1
Total	18

This table follows a common practice of ordering the categories from the one having the greatest frequency to those having the least frequency. Residual categories (such as "other" or "no data") would have been listed last, if they existed. The second column lists the number of times each occupation occurred among all the cases. This column is commonly headed by the words "Frequency" or "Number" or with their initials "f" or "N," respectively. The last row of the table shows the total number of cases represented in the distribution. It too is sometimes labeled "N" rather than "Total." The preferred meaning of N is the number of research units which generally corresponds to "Total." It is also known as the *sample size*, assuming the data represent a sample rather than a population.

Looking at this table we can quickly see the pattern in the distribution of races among these people. The table shows us exactly the extent to which white dominated the racial backgrounds among this set of individuals.

Research Example 2.2 also shows a univariate frequency distribution. This one presents the occupations of some mental health board members. If you compare it to the R.E. 2.1 table, you should notice that the basic format is the same: title and source footnote, left column lists the variable and its categories, right column lists the frequencies for each category. There are some subtle differences, though. The categories of occupation are not listed in order according to their frequencies. This may have been done for one of two reasons. First, this table may have been prepared for comparison with other tables using this ordering. Second, in this study occupation may have been considered to represent ordinal measurement, not nominal. It is common for social researchers to view occupation

RESEARCH EXAMPLE 2.2 Univariate Frequency Distribution—Ordinal Variable with Categories in High-to-Low Order

Occupations of Mental Health Board Members

Occupation	Frequency
Professional	39
Managerial/ administration	13
Sales	3
Clerical	5
Craftsman	4
Operative	2
Homemaker	9
Student	2
Unemployed	12
Total[a]	89

[a] Excludes 13 no data cases.

Source: Donald P. Bartlett and Robert J. Grantham, "A Regional Evaluation of Citizen Mental Health Boards," *Journal of Community Psychology*, 8 (4), October 1980:291–301. Adapted from Table 2.

as an ordinal variable in which the categories are ordered according to the relative prestige given the specific occupations by the general public. If a variable is measured at the ordinal level or higher, the table should present the categories in their order. Research Example 2.2 uses a high-to-low ordering but it may be reversed if the researcher wishes.

Another difference you may find in similar tables is that the total is given at the top, not the bottom. This is not a mistake, only a difference in style. It is common for tables that are especially long or that appear as part of a long series in such places as government publications to present the total at the top. This makes it easier to locate.

A quick review of the R.E. 2.2 table indicates that this distribution has a somewhat different pattern from the one in R.E. 2.1. In R.E. 2.2 the frequencies are more evenly spread across the categories than in R.E. 2.1. The primary clustering for R.E. 2.2 is in the categories of professional, managerial/administration, and unemployed. This is a typical occupational distribution for members of mental health boards.

Grouped Data

Research Examples 2.3 and 2.4 illustrate a slight variation in table construction that is used for variables which are continuous or which have several categories in their measurement. Many interval and ratio variables are like this. In both R.E. 2.3 and 2.4 the scores are reported in a *grouped data* format. Grouped data means that the individual scores have been combined into a smaller number of wider categories. This is done to provide a more useful summary of the distribution. Data presented in a grouped format are treated as being continuous, whether the raw scores were discrete or continuous. If summarized in a frequency table, the presentation is called a *grouped frequency distribution.*

Research Example 2.3 shows the elapsed time ("recall interval") between pregnancies and subsequent interviews about the use of fertility planning methods in a U.S. national study. These data have been grouped into four categories. The ages of our set of

RESEARCH EXAMPLE 2.3 Univariate Grouped Frequency Distribution—Closed Categories

Recall Interval Between Pregnancy and Interview of Fertility Planning Methods

Recall Interval (years)	Frequency
0–4	1850
5–9	1575
10–14	1111
15–19	346
Total	4882

Source: Norman B. Ryder, "Consistency of Reporting Fertility Planning Status," *Studies in Family Planning,* 10 (4), April 1979:115–128. Adapted from Table 3.

RESEARCH EXAMPLE 2.4 Univariate Grouped Frequency Distribution—Open Categories

Ages of Mental Health Board Members

Age	Frequency
Under 30	8
30–39	14
40–49	26
50–59	34
60+	11
Total[a]	93

[a] Excludes 9 no data cases.
Source: Donald P. Bartlett and Robert J. Grantham, "A Regional Evaluation of Citizen Mental Health Boards," *Journal of Community Psychology,* 8 (4), October 1980:291–301. Adapted from Table 2.

mental health board members (from the same study as R.E. 2.2) have been grouped into five categories in R.E. 2.4. If the grouping had not occurred, both tables would have been extremely long and cumbersome to present and interpret. There simply are too many different scores to be handled individually.

There are three main considerations that determine the categories that are used in grouping data. First is deciding the number of categories to use. The overall goal is to choose a number that reduces the original complexity of the data while retaining the nature of the actual distribution. In data sets of ordinary size (less than 1000) it is typical to have between 3 and 15 categories when grouping. Extremely large data sets such as those collected by the U.S. Bureau of the Census may use many more categories that have become standardized for hundreds of tables. Obviously, the researcher has considerable discretion when deciding on the number to use.

A second factor is the size or width of the grouped categories. The wider the categories are made, the less exactly you know the scores of individual cases. For example, you have a more precise idea of a case's true age when it is in a grouped category of 15–19 than one of 0–49. Hence researchers attempt to use categories that contain no less precision than they are willing to tolerate for interpretation purposes. It may be acceptable to know a person's age only within 5 years, but to know it only within 50 years usually is not very useful.

A last consideration is the ease with which the table can be read and remembered. This is best accomplished when the categories have equal widths of 10 units (or some fraction or multiple of 10). Since our number system is 10-based it is routine for people to approximate and cluster in tens. We can recognize distributions using widths of 10 at a glance, but it takes extra effort if the widths are 4 or 13, for example. However, it is not always possible or wise to use widths such as 5 or 10, or to use equal widths at all. A distribution may naturally cluster at intervals other than 5 or 10 or at irregular intervals.

This natural clustering usually should be followed when grouping scores. Otherwise, serious distortion of the distribution's true character may occur.

Both R.E. 2.3 and 2.4 conform to these three considerations in grouping data as far as we can determine. Research Example 2.4 also illustrates the common practice of using *open categories* when grouping. In this example the lowest and highest categories are open, whereas those in between are closed. *Closed categories* have defined endpoints, open categories do not. Open categories are used when there are large gaps between scores. Without them there could be several categories with very low or zero frequencies. This would not create a concise summary of a distribution.

Some Technical Considerations About Categories of Measurement

This brings us to some other distinctions about quantitative categories of measurement, grouped and ungrouped. The endpoints of the categories as they appear in the table are called *tabular* or *apparent limits.* But the exact beginning and ending points are known as the *true limits.* To determine the tabular limits one simply reads the table. To determine the true limits, one must know how the variable in the table was measured and how the scores were rounded, if at all.

The distinction between tabular and true limits is important for understanding the computations based on these kinds of measurement categories. Such computations treat the data as though they were continuous. In this view, the measurement categories cover all score values that are imaginable (not merely those that occurred), including those that apparently fall between the categories listed in a table. Thus we can think of placing the tabular categories on a continuous number line and then making a decision about the lowest and highest possible scores that could be represented by each tabular category so that there are no possible scores omitted from the range of values being considered. Figure 2.2a shows some examples.

In the case of the recall intervals reported in years in R.E. 2.3, we can view its most exact measurement to be to the nearest whole number, the nearest year. Thus the tabular limits of 5–9 have true limits of 4.5–9.5, for example. Most whole-number measurements are like this one. The true lower limit of such a category is .5 below its tabular lower limit and the true upper limit is .5 above its tabular upper limit.

FIGURE 2.2a Comparison of Tabular and True Limits of Measurement Categories

Recall Interval (R.E. 2.3)

	"0–4"		"5–9"		"10–14"		"15–19"		Tabular Limits
−0.5		4.5		9.5		14.5		19.5	True Limits

Age (R.E. 2.4)

"Under 30"		"30–39"		"40–49"		"50–59"		"60+"	Tabular Limits
Undef.	$29.\overline{9}$		$39.\overline{9}$		$49.\overline{9}$		$59.\overline{9}$	Undef.	True Limits

There are a few social measurements that differ from this. Research Example 2.4 may be one of these. Age can be rounded to the *nearest* whole year, but usually it is rounded to the *last* whole year instead. This is the customary way in which people record age and some other variables, such as marriage anniversaries or calendar years. In these instances the true lower limits of categories coincide with the tabular lower limits, whereas the true upper limits are a fraction less than the next higher tabular lower limit. In R.E. 2.4 the tabular age limits of 30–39 have true limits of 30.0 and $39.\overline{9}$,* respectively. In practice this means that anyone who is at least 30 but not yet 40 would be included in this category. Those who are nearly 30, even less than one-half year younger, would be in the next lower category. Those who have recently had their fortieth birthday would be in the next higher category. Open categories, of course, have no tabular or true limits at their open ends. They are said to be "undefined" in the vocabulary of mathematicians.

Knowing the true limits of a category allow one to formally define its width. Category *width* (w) is the difference between a category's true upper and true lower limits. In a formula it is

$$w = l_u - l_l \qquad \text{Category Width Formula}$$

where w = category width
l_u = true upper limit of a category
l_l = true lower limit of a category

Using this formula, the width of the 5–9 recall interval in R.E. 2.3 is 5 years $(9.5 - 4.5 = 5)$. The width of the 30–39 age category in R.E. 2.4 is 10 years $(39.\overline{9} - 30.0 = 9.\overline{9} = 10)$. In both instances a careless examination of the categories as they appear in the tables would cause one to say that the widths are 4 and 9 years, respectively. This is not correct.

The last technical feature of data of interest here is the *midpoint* (m) or *class mark* of a category. A midpoint is the value exactly halfway between the true upper and lower limits of a category. Knowing the true limits allows one to find the midpoint by adding those limits and then dividing by 2. In a formula, it is

$$m = \frac{l_u + l_l}{2} \qquad \text{Category Midpoint Formula}$$

where m = midpoint of a category
l_u = true upper limit of a category
l_l = true lower limit of a category

Following the formula, the midpoint of the 5–9 recall interval in R.E. 2.3 is 7.0 years $[(9.5 + 4.5)/2 = 14/2 = 7.0]$. The midpoint of the 30–39 age category in R.E. 2.4 is 35.0 years $[(39.\overline{9} + 30.0)/2 = 69.\overline{9}/2 = 34.\overline{9} = 35.0]$.

Knowing these technical features of grouping helps in interpreting tables that present grouped data. Some of these tables mistakenly show tabular limits that overlap. If so, some ambiguity has been created about the classification of scores when the same value is used as the shared tabular limits for adjacent categories. For example, if two adjacent tabular categories are listed as 0–5 and 5–10, where were cases classified with scores of 5?

*The line over the number indicates that it is repeated in all succeeding decimal places.

FIGURE 2.2b Technical Features of Categories of Measurement

Tabular Limits	True Limits	Width (w)	Midpoint (m)
	Recall Interval (R.E. 2.3)		
0–4	−0.5–4.5	5	2
5–9	4.5–9.5	5	7
10–14	9.5–14.5	5	12
15–19	14.5–19.5	5	17
	Age (R.E. 2.4)		
Under 30	Undefined–29.$\overline{9}$	Undef.	Undef.
30–39	30.0–39.$\overline{9}$	10	35
40–49	40.0–49.$\overline{9}$	10	45
50–59	50.0–59.$\overline{9}$	10	55
60+	60.0–Undefined	Undef.	Undef.

This same knowledge also explains why some apparent gaps between categories in a table have not really omitted any scores. For example, there is no real gap between the 0–4 and 5–9 recall intervals in R.E. 2.3 even though it may have appeared that way at first. Being able to find the true limits and midpoints is also necessary for making or verifying calculations based on grouped data tables. This is discussed more fully in later chapters. Figure 2.2b shows all the technical features of grouped data that we have been discussing for R.E. 2.3 and 2.4.

Univariate Percentage Distributions

The researcher's task in producing a table that shows a univariate percentage distribution requires only one additional step beyond a univariate frequency distribution. The same earlier steps of observation coding, raw data distribution, ordered array, and even a frequency distribution are followed. The added step is to convert the frequencies to percentages. But "For what purpose?", you may ask.

Knowing the frequency for each category of a variable is useful and even necessary for data analysis. But knowing the percentage may be even better. Return to the frequency distribution in R.E. 2.1. We can see that 13 respondents were white and 3 were black. But to reach the conclusion that these were the predominant racial backgrounds requires us to consider these frequencies in comparison to the 18 total number of cases. Had there been 500 cases instead, these same frequencies would have had a much different meaning. So although the frequencies are important in themselves as findings from the study, comparisons between the categories and to the entire distribution require a statistic that is less absolute and more relative in character. The percentage is just such a statistic.

Further, frequencies do not usually allow us to make valid comparisons between different distributions. Suppose that we had a second study reporting the racial backgrounds of 1000 users of mental health facilities in a different location. Both the R.E. 2.1 actual distribution and the second hypothetical distribution are presented in Figure 2.3. Compare the frequency distributions of race for the two sets of individuals. In which one are we more likely to find a person whose racial background is white? Or black? This kind of question will frustrate an attempt to answer it because the distributions do not have the same number of cases. We need a statistic in which the number of cases has already been considered; one that is "standardized" for case size, as statisticians say. Again, the percentage is an answer.

The percentage statistic is based on a simpler statistic, the proportion, which also makes relative comparisons possible. A *proportion* is a frequency divided by the total number of cases. It compares a part of a distribution to the whole. In a formula it is

$$P = \frac{f}{N} \qquad \text{Proportion Formula}$$

where P = proportion
f = frequency
N = number of cases

Proportions are expressed as decimals. Therefore, the proportion of individuals in the first study (R.E. 2.1) who were white was .72 ($P = 13/18 = .72\overline{2} = .72$). The proportion who were black was .17 ($P = 3/18 = .1\overline{66} = .17$). These proportions more easily describe the predominance of whites and blacks within the entire distribution than do the frequencies alone.

In the second study (the hypothetical one) the proportion of individuals who were white was also .72 ($P = 722/1000 = .722 = .72$). The proportion who were black was .13 ($P = 126/1000 = .126 = .13$). To return to our earlier question, it is now clear that one is equally likely to find a person who is white in either study but more likely to find a person who is black in the first even though the frequency was higher in the second. In looking over Figure 2.3 note how the proportions add to 1.00 (except for rounding errors) when all categories are included. This should always be true when the entire distribution is being shown.

A percentage can be used for the same purpose as a proportion. It is a statistic that indicates the proportion per every 100 cases. A *percentage* is a frequency divided by the

FIGURE 2.3 Comparing Distributions by Frequencies, Proportions, and Percentages

	By Frequencies		By Proportions $\left(P = \frac{f}{N}\right)$		By Percentages $\left[\% = \frac{f}{N}(100)\right]$	
Race	Study 1[a] f	Study 2[b] f	Study 1 P	Study 2 P	Study 1 (%)	Study 2 (%)
White	13	722	.72	.72	72.2	72.2
Black	3	126	.17	.13	16.7	12.6
Oriental	1	52	.06	.05	5.6	5.2
Hispanic	1	100	.06	.10	5.6	10.0
Total	18	1000	1.01[c] (18)	1.00 (1000)	100.1[c] (18)	100.0 (1000)

[a]Data from R.E. 2.1.
[b]Hypothetical data.
[c]Totals do not add to 1.0 or 100.0 due to rounding error.

total number of cases (a proportion) and multiplied by 100. In a formula it is

$$\% = \frac{f}{N}(100) \qquad \text{Percentage Formula}$$

where % = percentage
f = frequency
N = number of cases

As Figure 2.3 indicates, the percentage of individuals who were white in the first study was 72.2 [% = (13/18)(100) = 72.2%]. The percentage who were black was 16.7 [% = (3/18)(100) = 16.7%]. In the second study the same respective percentages were 72.2 [% = (722/1000)(100) = 72.2%] and 12.6 [% = (126/1000)(100) = 12.6%]. The slight differences between the proportions and percentages (other than the effect of multiplying by 100) are due only to the varying number of significant digits used in the calculations.

The proportion and percentage distributions shown in Figure 2.3 are similar to the format used for univariate frequency distributions. The relative concentration of scores in each category is shown. Proportions generally are reported to two or three places to the right of the decimal point and percentages to one or two. The column heading for the proportions may be "Proportions" or simply "*P*." For percentages the heading is commonly "%," "Percentage," "Pct," or "Percent." A total is shown for both types of distributions. Some researchers prefer to add the actual proportions or percentages in the tables and to report that total even though rounding errors may prevent it from being 1.00 or 100.0, respectively. Others prefer to always show the total to be 1.00 for proportions or 100.0 for percentages in order to indicate that the table includes all the scores. It is standard practice to show the number of cases (N) in parentheses below each total. This makes it possible to reconstruct the frequencies for the categories by solving either the proportion or percentage formula for the unknown f value. For example, we can determine that the frequency of blacks in the first study is 3 by the following manipulation:

$$\% = \frac{f}{N}(100) \qquad \text{so } f = \frac{\%}{100}(N)$$
$$= \frac{16.7}{100}(18)$$
$$= \frac{300.6}{100}$$
$$= 3$$

Caution must be exercised in interpreting proportions and percentages that are based on a small number of cases. There were only 18 cases in our data from R.E. 2.1. Thus a change of only a few cases could dramatically alter the proportions or percentages. For example, a switch of only two persons in that study would change a proportion by .11 (2/18) and a percentage by 11.1 [(2/18) × 100]. However, in our comparative, hypothetical study such a switch would change a proportion by only .002 (2/1000) and a percentage by .2 [(2/1000) × 100]. This demonstrates the greater stability of larger data sets and allows us to draw conclusions from them with more confidence. Be sure to note the case size when interpreting relative statistics such as proportions and percentages.

Research Example 2.5 presents a univariate frequency and percentage distribution

RESEARCH EXAMPLE 2.5 Univariate Frequency and Percentage Distribution

General Acceptance and Nonacceptance of the Alcoholic Relative to the Sick Role by Number and Percentage

Acceptance-Nonacceptance Pattern	*N*	%
Strongly accepting	5	4.2
Accepting	46	39.0
Ambivalent	17	14.4
Nonaccepting	43	36.4
Strongly nonaccepting	6	5.1
No answer	1	0.8
Total	118	100.0

Source: H. Paul Chalfant and Dorinda N. Noble, "The Transition to Medicalized Views: Alcoholism and Social Workers," *Journal of Sociology and Social Work,* 6 (6), November 1979:792–804. Table 1. Reprinted by permission.

concerning the reactions of some social workers to alcoholics in a sick role. It is very common for the frequencies and percentages to be presented together in one table like this. In fact, R.E. 2.2 and 2.4 appeared this way in their original sources. The relevant part merely was extracted for illustration purposes earlier. The combined format allows you to take advantage of the descriptive information contained in both forms of the data at once. From R.E. 2.5 we can see not only how many social workers accepted alcoholics in the sick role to each degree, but we can also note the standardized comparisons of those frequencies in their percentage equivalents. Notice the very divided pattern among these social workers. Most were either accepting or nonaccepting and only a few were either ambivalent or extreme.

Research Example 2.6 presents three statistical forms of a distribution in one table. The example concerns the age distribution of a set of Filipino women who had not moved from their rural places of residence in a study of migration and fertility. In addition to the frequencies and percentages, the table shows the cumulative percentages. Cumulative data forms can be based on either frequencies or percentages; cumulative frequencies (*cf* or *cumf*) or cumulative percentages (*c%* or *cum%*). *Cumulative data forms* indicate how many (or what percentage) of the scores fall into and below, or into and beyond, each of the categories of a variable. Because the scores are accumulated across the categories, cumulative forms are appropriate only for variables at the ordinal or higher levels of measurement.

In R.E. 2.6 the cumulative percentages indicate the total percentage of women who are in each age category plus all categories representing younger ages. For example, it indicates that 56.3% of the women were under 30 years of age. This type of accumulation is known as a *less-than cumulative form.* As suggested by the definition above, you may also encounter a *more-than cumulative form,* which indicates how many (or what percentage) of the scores are in and beyond each category. If applied to the same example, a more-than cumulative percentage distribution would indicate the percentage of women who were 30 years old and older. When you read a table that shows a cumulative distribution you obviously need to know which of the two types (less-than or more-than) is being displayed. If not, serious confusion will surround you.

The process of making cumulative computations begins with the category frequencies. These are added in the desired direction (less-than or more-than) with the cumulative frequency being found for each successive category. The last entry will be the sum of all the frequencies, which is also the number of cases (N). In a formula this is

$$cf = f_1 + f_2 + \ldots + f_k$$

Cumulative Frequency Formula

where cf = cumulative frequency
f_1 = frequency for first category
f_2 = frequency for second category
f_k = frequency for last category

If the cumulative percentages are desired, they are found by converting the cumulative frequencies to percentage form. The appropriate formula is

$$c\% = \frac{cf}{N}(100)$$

Cumulative Percentage Formula

where $c\%$ = cumulative percentage
cf = cumulative frequency
N = number of cases

RESEARCH EXAMPLE 2.6 Univariate Grouped Frequency, Percentage, and Cumulative Percentage Distribution

Age of Rural Sedentary (Nonmigrating) Women, Philippines

Age	f	%	$c\%$
15–19	547	23.5	23.5
20–24	394	16.9	40.4
25–29	369	15.9	56.3
30–34	354	15.2	71.5
35–39	274	11.8	83.3
40–44	217	9.3	92.6
45–49	173	7.4	100.0
Total	2328	100.0	—

Source: Virginia A. Hiday, "Migration, Urbanization, and Fertility in the Philippines," *International Migration Review,* 12 (3), Fall 1978:370–385. Adapted from Table 1.

FIGURE 2.4 Finding Cumulative Distributions (Data from R.E. 2.6)

A. Less-Than Cumulative Distributions

Age of Rural Sedentary Women

Age	f —(1)→	cf —(2)→	$c\%$
15–19	547	547	23.5
20–24	394	941	40.4
25–29	369	1310	56.3
30–34	354	1664	71.5
35–39	274	1938	83.3
40–44	217	2155	92.6
45–49	173	2328	100.0
Total	2328	—	—

1. Find the less-than cumulative frequencies:

$$cf_l = f \text{ for a category} + f \text{ for all categories with lower scores}$$

2. Find the less-than cumulative percentages:[a]

$$c\%_l = \frac{cf_l}{N}(100)$$

B. More-Than Cumulative Distributions

Age of Rural Sedentary Women

Age	f —(1)→	cf —(2)→	$c\%$
15–19	547	2328	100.0
20–24	394	1781	76.5
25–29	369	1387	59.6
30–34	354	1018	43.7
35–39	274	664	28.5
40–44	217	390	16.8
45–49	173	173	7.4
Total	2328	—	—

1. Find the more-than cumulative frequencies:

$$cf_m = f \text{ for a category} + f \text{ for all categories with higher scores}$$

2. Find the more-than cumulative percentages:[a]

$$c\%_m = \frac{cf_m}{N}(100)$$

[a]It is not advisable to find cumulative percentages by adding separate percentages because rounding errors may be compounded.

Figure 2.4 demonstrates the process of computing cumulative frequencies and cumulative percentages using the data from R.E. 2.6. Both the less-than and the more-than types of cumulative data forms are illustrated.

Ratios and Rates

In addition to proportions and percentages, ratios and rates are common standardized data forms that are often calculated from frequency distributions. A *ratio* is a comparison between two quantities. Those quantities may be in the form of frequencies or something else. Proportions and percentages are also ratios because they compare two quantities, one of which is a total. But the common meaning of ratios in social science is of a comparison between quantities that are not totals. When working with frequencies a ratio compares two parts of a distribution to each other; the total is not involved in the calculation. The following formula can be used for finding ratios from frequency distributions:

$$\text{ratio} = \frac{f_a}{f_b} \qquad \text{Ratio Formula}$$

where f_a = frequency for one part of a distribution

f_b = frequency for another part of a distribution

We can calculate ratios from the data presented in R.E. 2.6. For instance, the ratio of women aged 25–29 to those aged 30–34 is 369/354 = 123/118, 123:118, 123 to 118, 1.04:1.00, 1.04 to 1.00, or simply, 1.04. Ratios are typically expressed in any of the last six ways shown in this example. In the first three of these the original fraction has been reduced to its simplest terms. In the last three the fraction has been converted to a quotient. Like proportions and percentages, ratios have been standardized for the number of cases included. Thus they also allow one to make valid comparisons within and between distributions.

Ratios have been used extensively by demographers to describe aspects of population distributions, such as the ratio of males to females. The demographers have added *multipliers* to these computations to ease public understanding of them. Multipliers are powers of 10 which move the decimal point in a ratio so that the result is ordinarily a whole number. These multipliers may be 100, 1000, or even larger, depending on the comparison being made. The standard *sex ratio* used by demographers illustrates this. It is defined as the ratio of males to every 100 females in a population:

$$\text{sex ratio} = \frac{f_m}{f_f}(100) \qquad \text{Sex Ratio Formula}$$

where f_m = frequency of males

f_f = frequency of females

The sex ratio in a population having 1500 men and 1600 women is 94 [(1500/1600)(100) = (.9375)(100) = 94]. This means that the population had 94 males per every 100 females. The multiplier 100 moves the decimal point to avoid a presentation using only fractional parts of people. It seems more humanistic that way but, of course, it makes no difference statistically.

Rates are a special form of the ratio. They compare the frequency with which a particular event actually occurred to the frequency with which that event potentially could have occurred over a given period of time. Rates generally use multipliers selected to clear the decimals in the answer like ratios. Birth and death rates are typical examples from demography. The *crude birthrate* compares the number of live births to the size of the population and is multiplied by 1000 for a known period. In a formula it is

$$\begin{array}{c}\text{crude birthrate}\\ \text{(during a year)}\end{array} = \frac{\text{number of live births}}{\text{total population at midyear}}(1000)$$

Because the total population is not at risk, more refined birthrates are also used. One example is the *general fertility rate,* which compares the number of live births to the number of women of childbearing age. In a formula it is

$$\begin{matrix}\text{general fertility rate}\\ \text{(during a year)}\end{matrix} = \frac{\text{number of live births}}{\begin{matrix}\text{population of women between}\\ \text{15 and 44 at midyear}\end{matrix}}(1000)$$

Imagine a society that had a total population of 200,000 of which 60,000 were women between 15 and 44. If there were 3600 live births during a year, the crude birthrate would be 18 [(3600/200,000)(1000) = 18]. The general fertility rate would be 60 [(3600/60,000)(1000) = 60]. These two rates mean that this society had 18 live births per every 1000 people in its total population and this represented 60 live births per every 1000 women between 15 and 44 years old.

Other common rates, such as the crime rate and suicide rate, use 100,000 as the multiplier in their computations because the incidence is so much lower than it is for births. The 100,000 multiplier transforms these rates to a whole number or mixed number just as the 1000 multiplier does with births.

To interpret ratios and rates one needs to know the components of their comparison and the multiplier (if any). Then think about the limitations to the comparisons that are being made. For example, with rates you should ask yourself how well the denominator represents the true potential for an event to occur. Maybe a more refined version is needed to answer the questions that are being asked by the researcher.

With both ratios and rates you must guard against the assumption that large answers are necessarily impressive. Like proportions and percentages, ratios and rates have been standardized for case size so you do not know whether they represent a pattern across 20 people or 20,000,000. Remember that even small variations in very small data sets result in very dramatic changes in these kinds of statistics. Large data sets are much more stable and are not influenced as easily by a few erratic cases. This is a subject we will examine more closely in the third section of the text on inferential statistics. But you can start now to give attention to the size of the data sets as you go about interpreting standardized statistics. You especially must refrain from imputing statistical significance to findings until you understand the role of probability in analysis.

Multiresponse Variable Distributions

Now that we have looked at a few examples of univariate distributions and the computations that they foster, examine the table in R.E. 2.7. It is another univariate frequency and percentage distribution, but it has a new feature. If we add the percentages, we see that they do not add to 100%. This is not a mistake; these percentages should not add to any number in particular. Unlike the previous examples, the cases in R.E. 2.7 could each have more than one score on the variable. Here people were asked to indicate their sources of information in learning about a potential recreational development in their area. Some learned about it only from television reports or their friends. But others had obtained information from several sources.

The N shown in this table indicates the number of respondents, not the number of information sources used by the respondents. Thus 80.2% of the 81 respondents used the newspaper as a source of information about the development. The table does *not* tell us that 80.2% of the information sources cited were newspaper reports. Be sure that you understand the difference in these two interpretations. This kind of table will use a foot-

RESEARCH EXAMPLE 2.7 Univariate Frequency and Percentage Distribution—Multiresponse Variable

Sources of Information Concerning Recreational Development Plans (Percentages)

Source	f[a]	%
Newspapers	65	80.2
Radio or TV	33	40.7
Family members or relatives	24	29.6
Friends or neighbors	28	34.6
Public officials	7	8.6
Organized groups	7	8.6
Other	10	12.3
(*N*)		(81)

[a] Multiresponse item.
Source: Robert E. Kramer and Gene M. Lutz, *Public Perception of Recreational Needs in the Brushy Creek, Iowa, Service Area.* Cedar Falls, Iowa: Center for Business and Behavioral Research, University of Northern Iowa, 1980. Adapted from Table G-5.

note or some other means to tell a reader that it shows the distribution for a *multiresponse variable* and that it should be interpreted accordingly.

Bivariate Frequency Distributions

A *bivariate frequency distribution* shows how the cases are classified according to two variables at once. This is not to be confused with a table that might display the frequency distributions for two variables separately. A bivariate distribution shows how many cases fall into all combinations formed by pairing the categories of one variable with those of another. For this reason it is also called a *cross-classification table.*

This may be clearer in R.E. 2.8. The two variables shown there are union membership and sex. Like other tables there is a title identifying the variables and there is a source footnote. One variable is placed vertically in the leftmost column with a heading and its categories listed below it. But the remainder of the table differs from univariate forms. A second variable is placed horizontally at the top of the table with a heading and its categories listed in a row below it. The frequencies are placed in the body of the table, each one filling a *cell. Marginal totals* are given for one (or both) of the variables separately at the edges of the table. The *overall* or *grand total* showing the number of cases is also commonly included in the lower right corner.

The frequencies in a bivariate table can be interpreted only by referring both to the row and column categories to which they belong. For instance, in R.E. 2.8 there were 213 union members who were males and 278 nonunion members who were females. But there were more than 213 union members overall and more than 278 females in the entire study.

The cross-classification nature of these frequencies would be even more obvious if you had constructed the table yourself, as the researcher must do. Figure 2.5 shows this process for the table in R.E. 2.8. After assembling a raw data distribution for the two variables the researcher tallies the cases one by one according to their sex and union membership simultaneously. The tallies are recorded in a bivariate tally table. For example, case 001 is shown as a male and a member of a union, so a tally mark is made in the upper left corner cell of the tally table. Case 002 is a female who did not belong to a union, and a tally mark is made in the lower right corner cell. After all the cases have been tallied, the tallies are converted to frequencies and entered in the formal table showing the bivariate distribution. This handwork is often performed by a computer in social research. But the computer process is analogous to the hand procedure just described.

A reasonable question to ask is whether it matters which variable is placed on the side and which at the top of a bivariate table. Clearly, it is possible to read the table no matter which way it is constructed. But there are some general guidelines about variable placement in one particular kind of research. In these studies the researcher has conceptualized one variable to be an *independent variable* and another to be a *dependent variable.* An independent variable is a variable that is thought to influence or predict another variable, but no outside or previous influence on itself is being investigated. A dependent variable is a variable that is thought to be influenced or predicted by another variable (the independent variable).

In R.E. 2.8 the researcher was examining the idea that sex may influence one's likelihood of belonging to a union. Hence sex was thought of as an independent variable and union membership as a dependent variable. In such instances it is typical for the dependent variable to be placed at the side of the table with its categories identifying the rows. The independent variable is put at the top of the table with its categories identifying the columns. It is also common for the dependent variable to be named first in the title of the table. These conventions were followed by the researcher in presenting his table in R.E. 2.8.

However, these customs are not always followed. In some bivariate tables the placement of the variables is purposely reversed. This may be done when the independent variable has a large number of categories but the dependent variable has only a few. Reversing the placement makes it easier to compare the frequencies between the categor-

RESEARCH EXAMPLE 2.8 Bivariate Frequency Distribution

Union Membership by Sex

Union Membership	**Sex** Male	Female	Total
Yes	213	68	281
No	289	278	567
Total	502	346	848

Cells

Marginal Totals

Overall or Grand Total

Source: John R. Sutton, "Some Determinants of Women's Trade Union Memberships," *Pacific Sociological Review,* 23 (4), October 1980:377–391. Adapted from Table 1.

FIGURE 2.5 Constructing a Bivariate Frequency Distribution

1. Raw Data Distribution

Case No.	Sex	Union Membership
001	Male	Yes
002	Female	No
003	Male	Yes
004	Male	Yes
.	.	.
.	.	.
.	.	.
848	Male	No

2. Bivariate Tally Table

Union Membership	Sex: Male	Sex: Female
Yes	𝍸 𝍸 |||...	|||. . .
No	𝍸 |. . .	𝍸 𝍸 |. . .

3. Formal Table Showing the Bivariate Frequency Distribution

Union Membership	Sex: Male	Sex: Female	Total
Yes	213	68	281
No	289	278	567
Total	502	346	848

ies of the dependent variable. This arrangement also fits most page sizes better. Either might be the reason that the variables are reversed in a bivariate table. It is better to rely on the text, not the table, to know which variable is dependent and which is independent.

Bivariate Percentage Distributions

Moving from a bivariate frequency distribution to a bivariate percentage (or proportion) distribution raises one of the more complex problems in table construction and interpretation. The problem is to decide how the percentages are calculated. A bivariate frequency distribution gives three choices for percentaging a table. There are totals for the rows, totals for the columns, and the overall total representing all the cases. A reader must know which one of these has been used by the researcher in order to understand a bivariate percentage distribution table.

Consider R.E. 2.9. In this study the educational levels of people (divided into four categories) were being examined according to their attitude about the right of a patient or family to discontinue medical treatment for terminally ill persons (divided into three categories). Because the researcher wanted to compare the categories of educational level to each other according to the attitude variable, the percentages were based on the total number of persons in each of these four educational groupings. We can verify that this was the method of percentaging by noticing that the numbers add to 100% (or nearly so with rounding errors considered) in the columns, but not in the rows or overall. Further, these are the only percentage totals shown in the table.

Given this method of percentaging the table we can see how persons from the four education categories compare within any one of the three attitude categories. For example, among which education category is there the most agreement that treatment can be discontinued? Answer: It is highest among those in the "some college" category (82.9% is the highest percentage in the first row). Or, what is the educational level of those who most disagree with a right to discontinue treatment? Answer: It is those with less than a high school degree who are most negative about the right (22.6% is the greatest percentage in the third row).

Suppose that we asked which attitude category is most dominated by those who have completed college. Answer: The table is not percentaged in a way that allows us to say. Since this question asks us to compare the three attitude categories to each other we need a table showing percentages calculated on the basis of the totals for each of these categories (i.e., the row totals). Specifically, we need to know what percentage of the cases in each row fall into the fourth column (the "college graduate" column). This means that we must reconstruct the bivariate *frequency* distribution, calculate the row totals, and then calculate the row percentages. This process is shown in Figure 2.6. With the row percentages we can see that the correct answer to our last question is that college graduates most dominate the ambivalent attitude category (23.7% versus 18.8% and 6.8% for agreement and disagreement, respectively). This probably is not the answer you gave if you tried to guess it from the column-percentaged table in R.E. 2.9. If you are confused at this point, please read this section again very carefully.

The general rule for percentaging tables is to base the percentages on the totals for the groups that are to be compared. To compare groups represented in the column variable, use the column totals to convert the frequencies to percentages. To compare groups represented in the row variable, use the row totals to find the percentages. Another way to say this is that the table is percentaged in the opposite direction of the desired comparison between frequencies. To compare frequencies in a row to each other, they are percentaged on the basis of their column totals and the reverse.

Because it is so easy to be confused about the kinds of questions that can be answered in either row- or column-percentaged tables, one must study the examples very

RESEARCH EXAMPLE 2.9 Bivariate Percentage Distribution

Patient/Family Right to Discontinue Treatment in Terminal Illness by Educational Level (Percentages)

	Educational Level				
Have Right to Discontinue Treatment	Less Than High School	Complete High School	Some College	College Graduate	
Strongly agree and agree	70.7	75.6	82.9	81.9	
Ambivalent	6.8	11.9	5.7	12.5	
Disagree and strongly disagree	22.6	12.6	11.4	5.6	
Total	100.1	100.1	100.0	100.0	
	(133)	(135)	(70)	(72)	(410)

Source: Marie Haug, "Aging and the Right to Terminate Medical Treatment," *Journal of Gerontology*, 33 (4), July 1978:586–591. Adapted from Table 3.

closely. Be convinced that the questions are not "double-speak" and that there is a real difference between them.

The third way in which a bivariate frequency table can be percentaged is to use the overall total (N) as we do in a univariate percentage table. In the bivariate situation we may want to know how the row–column combinations compare with each other. For example, in R.E. 2.9 we might want to know how the "disagreement–less than high school" combination compares in relative size to the "agreement–college graduate" combination. This is identical to the type of question we ask with bivariate frequency distributions except that we want the data in the form of percentages. Once we have these percentages we can compare bivariate combinations either internally to others in the same table or with those in similar tables regardless of the number of cases involved.

FIGURE 2.6 Re-Percentaging a Bivariate Distribution (Data from R.E. 2.9)

1. Bivariate Distribution Percentaged by Column Totals

	Education				
Attitude	<H.S.	H.S.	<Col.	Col.	
S. Agree/Agree	70.7	75.6	82.9	81.9	
Ambivalent	6.8	11.9	5.7	12.5	
Disagree/S. Disagree	22.6	12.6	11.4	5.6	
	100.1	100.1	100.0	100.0	
	(133)	(135)	(70)	(72)	(410)

2. Convert Percentages to Frequencies

$$f = \frac{\%}{100}(N)$$

	Education				
Attitude	<H.S.	H.S.	<Col.	Col.	
S. Agree/Agree	94	102	58	59	313
Ambivalent	9	16	4	9	38
Disagree/S. Disagree	30	17	8	4	59
	133	135	70	72	410

3. Recalculate Percentages by Row Totals

	Education				
Attitude	<H.S.	H.S.	<Col.	Col.	
S. Agree/Agree	30.0	32.6	18.5	18.8	99.9 (313)
Ambivalent	23.7	42.1	10.5	23.7	100.0 (38)
Disagree/S. Disagree	50.8	28.8	13.6	6.8	100.0 (59)
					(410)

FIGURE 2.7 Percentaging a Bivariate Distribution on the Overall Total (Data from R.E. 2.9)

1. Frequency Distribution

Attitude	Education <H.S.	H.S.	<Col.	Col.	
S. Agree/Agree	94	102	58	59	313
Ambivalent	9	16	4	9	38
Disagree/S. Disagree	30	17	8	4	59
	133	135	70	72	410

Convert to percentages using the overall total (N = 410):

$$\% = \frac{f}{N}(100)$$

2. Percentage Distribution

Attitude	Education <H.S.	H.S.	<Col.	Col.	
S. Agree/Agree	22.9	24.9	14.1	14.4	
Ambivalent	2.2	3.9	1.0	2.2	
Disagree/S. Disagree	7.3	4.1	2.0	1.0	
					100.0 (410)

Using the same data from R.E. 2.9, we would divide each of the bivariate frequencies by 410 (the overall total) and multiply by 100. The results are shown in Figure 2.7. Now we can see that there are approximately one-half as many persons in the "disagreement–less than high school" combination than in the "agreement–college graduate" combination (7.3% and 14.4%, respectively).

Multivariate Distributions

Research Example 2.10 presents a multivariate distribution in the percentage format. A *multivariate distribution table* is one showing how the cases are classified on at least three variables concurrently. The R.E. 2.10 table has three variables: marital status, highest degree attained, and sex for a national study of college professors. The researcher was investigating the differential effects of educational attainment on marital status for male and female professors.

Notice that the percentages are based on the column totals. Some multivariate percentage tables do not show the totals for the percentages, so we must do our own addition to determine the method of percentaging that was used. Once we know the basis of the percentages we can proceed to interpret the table. By comparing the degree attainment

RESEARCH EXAMPLE 2.10 Multivariate Percentage Distribution

Marital Status of Professors by Highest Degree Attained and Sex, 1969 (Percentages)

Marital	Male			Female		
Status	Ph.D.	Master's	Bachelor's	Ph.D.	Master's	Bachelor's
Married (once)	81.2	79.0	74.4	35.0	40.7	59.3
Remarried	7.3	5.7	7.0	6.2	4.7	1.0
Separated	.8	.5	.9	.7	1.3	.6
Single, never married	8.5	12.6	15.9	46.4	41.6	32.6
Single, divorced	1.7	1.6	1.4	7.4	6.9	4.5
Single, widowed	.5	.5	.3	4.3	4.7	2.1
Total	100.0	99.9	99.9	100.0	99.9	100.1[a]
N	152,155	94,145	15,744	17,099	40,968	6,195

[a]Corrected from cited figure.

Source: Carnegie Commission faculty survey, 1969, as cited in Michael A. Faia, "Discrimination and Exchange: The Double Burden of the Female Academic," *Pacific Sociological Review,* 20 (1) January, 1977:3–20. Table 2. Reprinted by permission.

categories in each row we see that there are several large differences for females, but few for males. Notice how higher degree attainment tends to work against marriage for females. This is the same conclusion that the author reached.

Even with only three variables, a reader needs patience to interpret a multivariate table. After identifying the variables, one should connect each number in the table with the appropriate categories of those variables. For example, the number 8.5 in the fourth row and first column of the R.E. 2.10 table means that 8.5% of males having a Ph.D. never married. When a table has five or six variables, as some do, this connecting process may be time consuming, but it is the only sure method to understand the information that is being displayed.

A final hint: Like bivariate tables, if one variable is considered to be a dependent variable, it often will be placed vertically along the left side of the table and it will be mentioned first in the table title. But because of the variations in style used, this must be verified by carefully reading the text surrounding the table.

Partial Distributions

Not all tables present an entire distribution. Some contain only those parts that are of interest to the researcher's analysis. These are called *partial distributions.* Research Example 2.11 presents another multivariate percentage table, but it has this new complication. Try to determine the method of percentaging that was used in this table. No matter which way you add the percentages you will not get totals of 100%. The explanation is that only parts of the complete multivariate distribution have been included. While the distributions for the marriage duration and sex variables in this table are complete, that for the third variable, thinking about divorce, is not. Only the percentages who were thinking of

RESEARCH EXAMPLE 2.11 Partial Multivariate Percentage Distribution

Percentage Thinking About Divorce by Duration of Marriage and Sex

Years Married	Sex			
	Total	Male	Female	*N*
One or less	16	10	21	50
2–5	18	15	20	151
6–10	19	16	22	153
11–15	9	4	14	117
16–20	7	6	9	120
21–30	8	7	10	267
Over 30	4	2	5	350

Source: Alan Booth and Lynn White, "Thinking About Divorce," *Journal of Marriage and the Family*, 42 (3), August 1980:605–616. Adapted from Table 1.

divorcing their spouses are included; those not thinking of it are omitted. If we had both percentages they would add to 100 for each marriage duration–sex combination.

There are many tables of this same kind in the social science literature. The data may be in some form other than percentages, too. Partial tables showing rates and ratios seem to be especially common. These tables need not be multivariate, either. Attention to the table will disclose the relevant information you need to make the correct interpretation, but it is very easy to miss the message at first glance when one is only beginning a statistical education.

Time-Series Distributions

One last type of table merits special mention: the *time-series table.* A time-series distribution is one that uses time as one of its variables. The data may be in the form of frequencies, percentages, or any of the others already discussed. There also may be any number of variables included. The time-series distribution is important because it introduces the dimension of time to the research, unlike the other types we have examined. A special set of statistics have been developed for the analysis of these distributions, including one to be discussed here.

Data that have been collected over time are part of the more general category of *longitudinal studies.* These allow for an analysis of change. This analysis is most easily accomplished when the same cases are measured at each of the points in time being considered. Such study designs are referred to as panel studies. When this is not possible or practical, researchers may measure different sets of cases at each point. This is called a separate samples study. If you want to study individual shifts, such as the changes in attitude toward one's spouse, only a panel study of individuals will be completely satisfying. But in studies where only the overall total changes in a population are of interest, such as voter approval of a political candidate in successive months of a campaign, the separate samples design is appropriate. A research methods textbook will provide more about the details of longitudinal research; e.g., see Simon, 1969:284–291.

Research Example 2.12 presents a time-series distribution. It shows the number of workers from the Alsace region of France who daily commuted to West Germany and

Switzerland from 1962 to 1976. The table shows the yearly fluctuations within the overall trend toward greater numbers of commuters over the period. Perhaps the most commonly used descriptive statistic designed for time-series data is the *percentage change.* It is a special ratio which compares the amount of change that has occurred between two points in time to the absolute amount at the first point and is then multiplied by 100. If the data are in frequencies, the formula is

$$\text{percentage change} = \left[\frac{f_{t2} - f_{t1}}{f_{t1}}\right](100)$$

Percentage Change Formula

where f_{t1} = frequency at the first time point being compared
f_{t2} = frequency at the second time point being compared

The formula yields a statistic that describes percentage change in reference to the starting point. Decreases will be identified with a minus sign; increases with a plus sign. The largest possible negative change is -100%. But the maximum positive change may exceed $+100\%$; it is not limited by the formula.

RESEARCH EXAMPLE 2.12 Time-Series Distribution

Alsatian Workers Commuting to West Germany and Switzerland

Year	West Germany	Switzerland	Total
1962	4,079	3,840	7,919
1963	4,167	4,440	8,607
1964	5,081	5,038	10,119[a]
1965	5,783	5,107	10,890
1966	7,151	4,879	12,030
1967	5,262	6,053	11,315
1968	5,294	7,369	12,663
1969	6,302	8,553	14,855
1970	7,101	10,056	17,157
1971	8,741	11,454	20,195
1972	10,330	13,647	23,977
1973	11,199	16,393	27,592
1974	13,456	17,600	31,056
1975	13,251	16,805	30,056
1976	12,742	14,639	27,381

[a]Corrected from cited figure.

Source: Direction Régionale du Travail and Direction Régionale de la Sécurite Sociale, Strasbourg, Switzerland, as cited in J.N. Tuppen, "A Geographical Appraisal of Trans-frontier Commuting in Western Europe: The Example of Alsace," *International Migration Review,* 12 (3), Fall 1978:386–405. Adapted from Table 2.

Using data from R.E. 2.12, we can calculate the percentage change in transnational commuters between any two of the reported years. For instance, from 1963 to 1964 there was a 21.9% increase in the numbers commuting to West Germany: [(5081 − 4167)/4167](100) = (914/4167)(100) = (.2193)(100) = +21.9%. But from 1966 to 1967 there was a 26.4% decrease in the numbers commuting to West Germany: [(5262 − 7151)/7151](100) = (−1889/7151)(100) = (−.2642)(100) = −26.4%. Of course, changes between any two years can be calculated; they need not be consecutive years.

Like other descriptive statistics discussed in this chapter, the percentage change standardizes for the number of cases. Thus it shares the resulting benefit of being comparable across distributions with unequal N's. But it also shares the caution that the true practical importance of a specific result may be disguised. A large percentage change may or may not represent a big shift in absolute numbers. A change from 500 to 1000 is a 100% increase, but so is a change from 4 to 8. Only the context of a research problem can tell one whether they have the same or any practical importance.

Keys to Understanding Tables

Several specific points have been made so far in this chapter about the construction and interpretation of the different kinds of tables commonly found in social research. But there are some overriding guidelines that can be summarized here to help you understand tables of all kinds. These are based on two major objectives: identifying the type of table and interpreting the table.

There are four keys for identifying the type of table. *Key 2.1.* Read the table's title and explanatory footnotes, including the source note. These will help you identify the context and general content of the table. *Key 2.2.* Determine how many variables are being displayed and then identify their names. *Key 2.3.* Identify each variable's categories of measurement and its level of measurement. *Key 2.4.* Identify the statistical form of the data that is being presented. You should now be ready to categorize the table according to the taxonomy introduced in this chapter.

There are five additional keys for interpreting a table once you know its type. *Key 2.5.* Look for the totals and determine how they were calculated. Check their computation if possible. *Key 2.6.* Determine the number of cases (and the number of scores, if there is more than one score per case). Watch for residual categories ("no data," "refused," etc.) which may be listed separately but indicate cases that you might otherwise miss. *Key 2.7.* Choose one number from each row and each column of the table and give it a verbal interpretation. Say what it means in words according to the variable(s) in the table. *Key 2.8.* Construct some tentative interpretations of the table's contents on your own. Characterize the patterns you see using your own words. *Key 2.9.* Compare the researcher's interpretations of the table with your own interpretations and with the actual contents of the table. Try to resolve any conflicts by more carefully reading the table and the researcher's prose. But do not make any hasty judgments about the inadequacy of your own interpretations. Some research reports can be wrong. Be humble but retain your self-confidence!

These keys to understanding encourage you to be an active consumer of tables. After going through these steps many readers are surprised at how little they first noticed about a table. But with practice, table-consuming skills improve markedly.

Types of Graphs

Many kinds of graphic techniques are used to provide visual displays of social data. Graphs are especially useful to give an overview of an entire distribution in a way that is concise and has quick impact. Many people find that it is easier to remember information presented in a graph because it uses symbols that are less abstract than the numbers found in a table. But the capability of a graph to convey an impression quickly also makes it vulnerable to misinterpretation. A graph may seem to say much either because it is packed with information or because it is ambiguous. Thus, like tables, graphs must be constructed and interpreted carefully.

There are two principles to consider when viewing graphs. First, the graph is trying to maximize communication with its intended audience. Second, the graph is trying to represent information with precision. Usually, these two principles do not conflict when data are graphed. But there are situations in which there is tension between the two and a researcher may need to give one precedence over the other. This may cause some confusion for a consumer of the graph. Some researchers can be very creative in the flourishes they give to their graphs. They do this to highlight some ideas for the reader. But sometimes a technically exact interpretation of their work will reveal imprecisions that confound the reader. It all depends on the data, the researcher, and the audience.

We will look at some of the major kinds of graphs used by social researchers. As a consumer of graphs you should try to understand the purposes and limitations of each graphic technique.

Pictographs

Pictographs use figures or likenesses to illustrate distributions. They are most commonly used to show comparisons between a few groups or a few points in time. Of all the types of graphs they tend to be the least precise but do have a very dramatic visual effect on the reader. For these reasons they must be consumed with great caution.

There are two major types of pictographs. In the first type a single figure is used for each distribution or category being displayed. The size of the figure is adjusted to show the comparison; the larger the differences are, the more the sizes of the figures differ. The (a) part of R.E. 2.13 is this type of pictograph.

The correct interpretation of the (a) graph is achieved by noting the heights of the two figures according to the scale shown on the side. The number of Fair Housing complaints received by the Department of Housing and Urban Development was approximately 1000 in 1970 and 3300 in 1977. Thus the number of complaints more than tripled. A limitation of this type of graph is that the visual impact exaggerates the real difference. You may have been surprised to discover that the 1977 figure represents only three times as many complaints as the 1970 one. You may have noted that the 1977 figure covers nine times the area of the 1970 one. This discrepancy happens because the figures are two-dimensional, whereas the comparison is only in one dimension (the height). This problem is difficult for the maker of the graph to avoid. If the figures are not kept to scale in both their heights and widths, they are not usually acceptable to our aesthetic tastes.

The second type of pictograph does not have the scale problem of the first type. Instead, the same-sized figure is used repeatedly to show the comparison. It is the relative number of figures rather than their sizes that show the differences. Part (b) of R.E. 2.13 illustrates this method. Although the visual impact of this graph may not be as immediate as the first one, it is more accurate. We interpret the graph by counting the number of

RESEARCH EXAMPLE 2.13 Pictographs

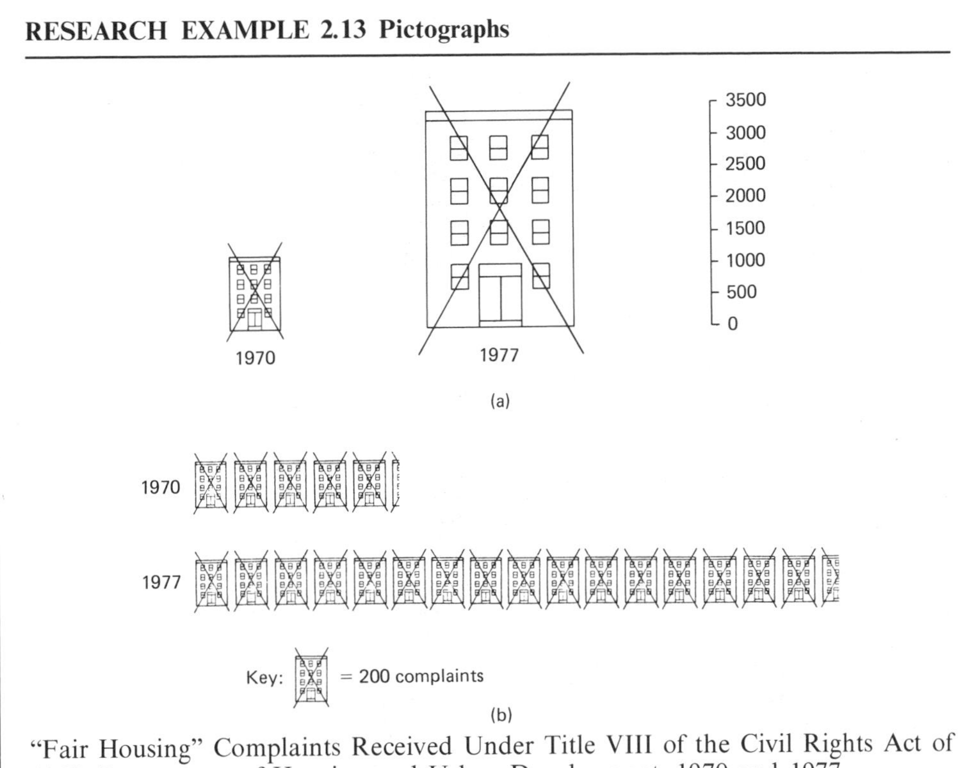

"Fair Housing" Complaints Received Under Title VIII of the Civil Rights Act of 1968, Department of Housing and Urban Development, 1970 and 1977.

Source: Department of Housing and Urban Development, *1970 HUD Statistical Yearbook.* Washington, D.C.: U.S. Government Printing Office, 1970, Table 83, p. 85; Department of Housing and Urban Development, *1977 HUD Statistical Yearbook.* Washington, D.C.: U.S. Government Printing Office, 1977, Table 1, p. 24.

repeated figures and converting them to frequencies according to the key given. Again we see that there were over three times as many complaints in 1977 as in 1970.

Both types of pictographs have only limited precision. In the first type we must approximate according to the scale shown at the side. In the second type we must approximate according to the fractional parts of the last repeated figures. Only the general pattern of differences can be shown in any pictograph.

Area Graphs

Area graphs are less subject to misleading interpretations than are pictographs. *Area graphs* are used to show the relative dispersion of cases for a variable having a few categories, especially those measured at the nominal or ordinal level. Area graphs may use any shape, but the most common are the circle and rectangle.

Research Example 2.14 shows two circle graphs being used to compare the residential distributions of blacks to whites in the United States. The graphs show the rather dramatic differences, especially for the residential categories of metropolitan rings and central cities.

RESEARCH EXAMPLE 2.14 Circle Graphs

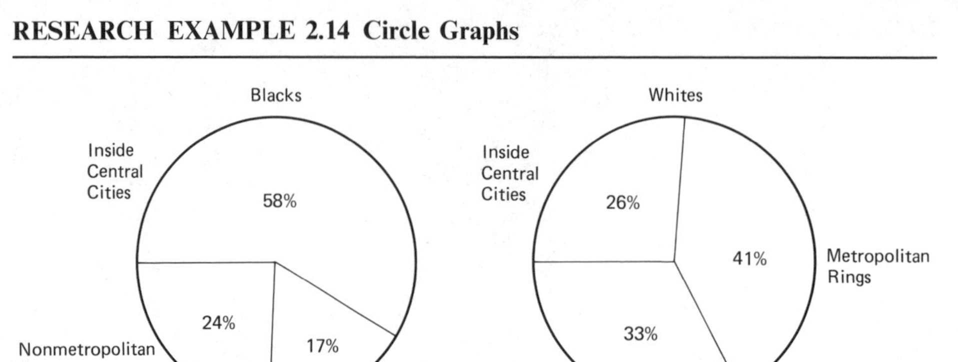

Residential Distributions of Blacks and Whites in the United States, 1974

Source: Bureau of the Census, *The Social and Economic Status of the Black Population in the United States,* Current Population Reports, Special Studies, Series P-23, No. 54, Washington, D.C.: U.S. Government Printing Office, 1974, Figure 2.

The size of each part of the area in a circle graph is determined by dividing the total 360 degrees for any circle proportionately to the number of cases in each of the categories. This is accomplished by multiplying each category's proportion by 360 to determine the central angle for that part. For example, a category with 58% of the scores in it would have a central angle of 208.8 degrees [(.58)(360) = 208.8]. A protractor can then be used to mark the boundaries for that part of the circle.

Research Example 2.14 demonstrates a successful application of the circle area graph. It quickly communicates a visual impression of how the black and white residential distributions compare. Each part of the circle is labeled by its category name and percentage. Many readers would find it simpler to remember the general nature of these two distributions from these graphs than if the same information had been presented in a table.

Bar Graphs

Bar graphs use a series of separated rectangles or "bars" to illustrate the distribution for a variable that has discrete measurement. Like area graphs, variables at the nominal or ordinal measurement levels are usually the most appropriate. But the bar graph will handle variables with many more categories. The length of the bars indicates the relative concentration of the scores in the categories. The bars are arranged according to an approximation of the classical two-dimensional Cartesian coordinate graphing system. Figure 2.8 shows this system, which was devised by René Descartes, a seventeenth-century mathematician and philosopher. One dimension is described by a horizontal axis and is known as the X-axis. The second dimension is described by a vertical axis and is known as the Y-axis. The two axes cross at the center, which is referred to as the point of origin and has the value of zero for both axes.

Bar graphs use one of the two axes to rest the bars against and the second as a scale to show the relative number of scores in each of the categories. *Vertical bar graphs* rest the bars on the X-axis; *horizontal bar graphs* set the bars against the Y-axis. Vertical bar graphs are the more common type in social research.

Research Example 2.15 presents a vertical bar graph for a nominal variable whose distribution is represented in frequencies. The graph indicates the types of food resources that were used by a sample of Yakutat Tlingit Indians in Alaska. The prevalence of seafood types is clear from the figure. Although the same information could be displayed in a circle graph, it is easier to interpret it from the bar graph. A circle graph with this number of categories (13) is beyond the limits of what is usually practical. The parts of the circle would be too small and similar in size to be easily distinguished.

Research Example 2.16 shows a vertical *multibar graph* for part of the data that appeared in R.E. 2.10. In R.E. 2.16 the marital status distributions for male and female professors with the Ph.D. degree are illustrated. The comparison is striking in the figure. The male professors are overwhelmingly concentrated in the married category, whereas the female professors are more dispersed across several categories of marital status.

Notice that both vertical bar graphs have descriptive titles, each axis has a verbal label, each Y-axis has a set of axis numerals beginning with zero and continuing with a consistent set of numbers referenced to the axis by tick marks, and the bars are separated between the categories on the X-axis. Had either bar graph concerned an ordinal variable, the categories would have been presented in their low to high ordering beginning at the left end of the X-axis.

Research Example 2.17 presents yet another type of bar graph. This one is a horizon-

FIGURE 2.8 Cartesian Two-Dimensional Graphing System

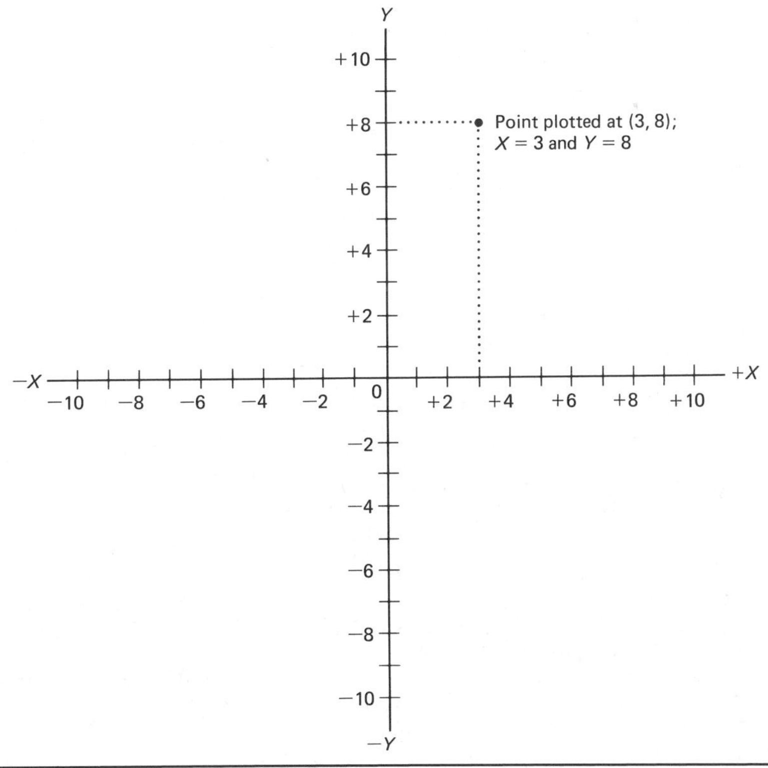

RESEARCH EXAMPLE 2.15 Vertical Bar Graph

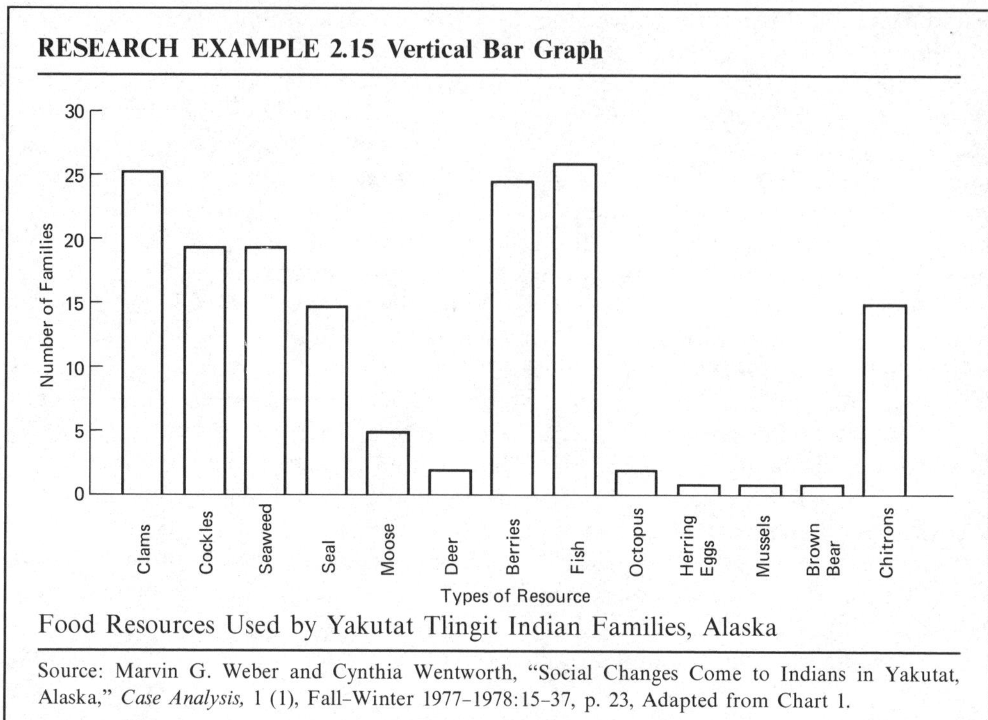

Food Resources Used by Yakutat Tlingit Indian Families, Alaska

Source: Marvin G. Weber and Cynthia Wentworth, "Social Changes Come to Indians in Yakutat, Alaska," *Case Analysis*, 1 (1), Fall–Winter 1977–1978:15–37, p. 23, Adapted from Chart 1.

RESEARCH EXAMPLE 2.16 Vertical Multibar Graph

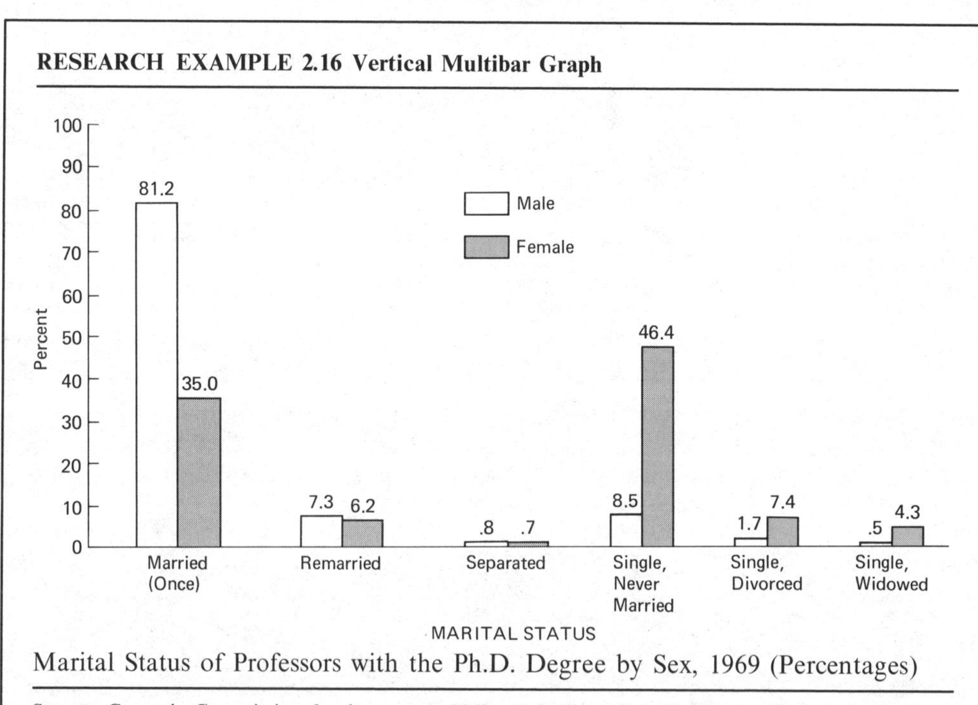

Marital Status of Professors with the Ph.D. Degree by Sex, 1969 (Percentages)

Source: Carnegie Commission faculty survey, 1969, as cited in Michael A. Faia, "Discrimination and Exchange: The Double Burden of the Female Academic," *Pacific Sociological Review*, 20 (1), January 1977:3–20.

RESEARCH EXAMPLE 2.17 Horizontal Compound Bar Graph

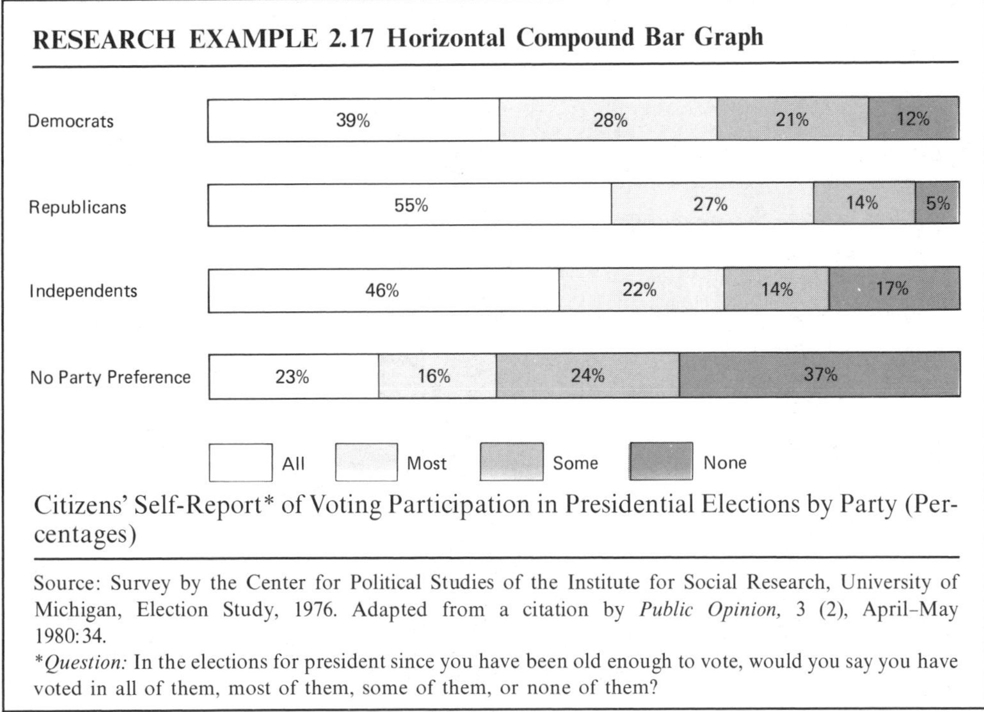

Citizens' Self-Report* of Voting Participation in Presidential Elections by Party (Percentages)

Source: Survey by the Center for Political Studies of the Institute for Social Research, University of Michigan, Election Study, 1976. Adapted from a citation by *Public Opinion,* 3 (2), April–May 1980:34.

**Question:* In the elections for president since you have been old enough to vote, would you say you have voted in all of them, most of them, some of them, or none of them?

tal *compound bar graph.* It shows the extent of voter participation in U.S. presidential elections according to the political party affiliation of the citizen. From it we can see that Republicans indicated that they voted in the elections more often than the other groupings. Those without any party affiliation said they voted less frequently. Notice that the bars in a compound graph are all the same length when the data are in percentages. Internal divisions for the bars are marked to show the distributions. This graphing technique is often found in reports of opinion polling.

Histograms

Histograms use a series of connected bars to show the distribution of a variable measured at the interval or ratio level. The bars are connected to emphasize visually the continuity of the categories in the variable. The area defined by both the height and width of the bars is proportional to the distribution of the scores across those categories.

Vertical histograms (the more usual type) place the variable on the *X*-axis and show the distribution on the *Y*-axis. The categories can be located on the *X*-axis with true graphing precision because of the higher measurement qualities of interval and ratio variables. But this precision is sometimes compromised if the researcher fears that the reader is not highly sophisticated. Some examples will show this.

Research Example 2.18 depicts the distribution in number of years served by a sample of mental health board members. The data are for another variable from the same study as represented in R.E. 2.2. The histogram indicates that the majority of the board members had served on their boards for relatively short periods. Notice that the five categories of years are equally spaced along the *X*-axis in the top figure (a). This is always done when the categories are the same width, but here we see that it was done even with

RESEARCH EXAMPLE 2.18 Variations in Histograms of Ungrouped Data

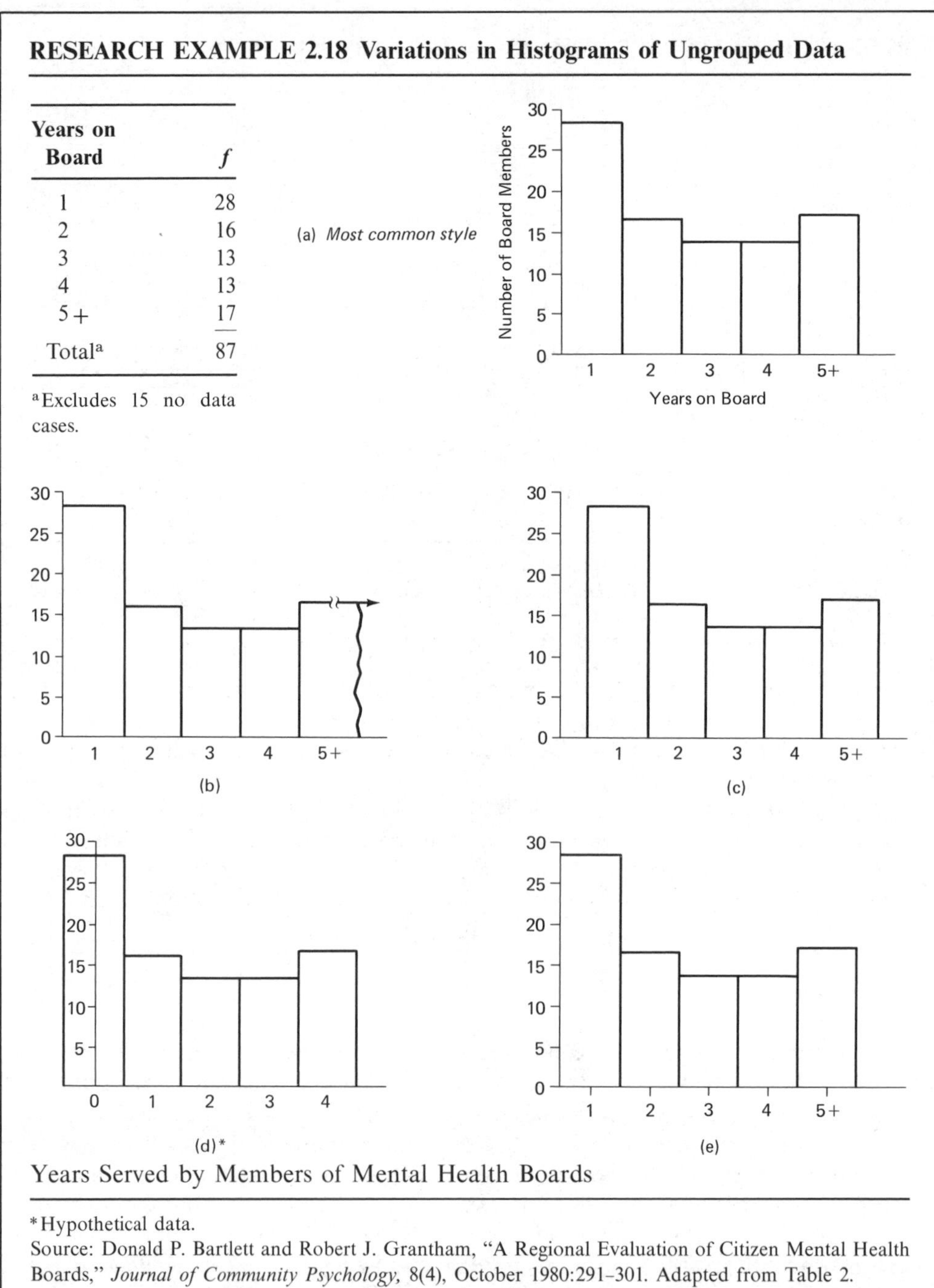

Years on Board	f
1	28
2	16
3	13
4	13
5+	17
Total[a]	87

[a] Excludes 15 no data cases.

Years Served by Members of Mental Health Boards

*Hypothetical data.

Source: Donald P. Bartlett and Robert J. Grantham, "A Regional Evaluation of Citizen Mental Health Boards," *Journal of Community Psychology,* 8(4), October 1980:291–301. Adapted from Table 2.

the last category, which was open. This is a common practice that is followed in histograms, although it is not strictly accurate. A more technically precise method of graphing an open category is shown in the (b) figure, but it is seldom found in social research. Most researchers consider it to be an unnecessary complication to communicating with the reader.

The first bar of the histogram in the (a) figure uses the Y-axis as its left boundary. If we consider the true limits of the first category, this too is a technical error. The first category has true limits of .5 to 1.5. Except for the first lower limit, the categories in a histogram are usually graphed at their true limits. If this first lower limit is also graphed exactly, the X-axis begins with 0. The result for this same data set is the (c) figure. It leaves a small (proportional) open space to the right of the Y-axis. This is a precise and correct method, although it does not occur as often as the method used in the (a) figure.

Now imagine that the first category's true lower limit had been a negative number, such as $-.5$. Graphing this histogram exactly would cause the first bar to overlap the Y-axis and its numerals. As the (d) figure shows, this would likely confuse many readers. ("What do you mean, someone served $-.5$ year on a board?") Most readers would object to the graph on aesthetic grounds as well.

Some researchers will also place tick marks on the X-axis at the midpoints of the categories above their labels, as the (e) figure shows. This is an optional and acceptable practice if it is done exactly. However, it adds little to a correct interpretation of the histogram when compromises of the kinds just illustrated have already been made.

Research Example 2.19 presents a histogram that represents grouped data. This is probably the most common use of histograms. The data are from the same study used for R.E. 2.6. Here the age frequency distribution of the rural sedentary women in the Philippine study of migration is being graphed. With grouped data, there are two widely acceptable methods for placing the categories on the X-axis. The most precise method uses tick marks to indicate the boundaries of the categories at their true limits. With this age distribution, the true limits can best be shown at the lower limits because they are exact (15.0, 20.0, etc.) The (a) figure shows this method. It is appropriate when the scores have been rounded to the last whole number, as we are assuming has been done here with age.

When scores have been rounded to the nearest whole number, the true limits of the categories will be at points halfway between whole numbers. Labeling the boundaries at the halves when the data are whole numbers is usually not done in histograms because it might confuse the reader. Instead, the tick marks are omitted, although the boundaries are placed with precision at the exact X-axis values. The tabular (not true) limits are simply centered below the appropriate bars. The (b) figure shows this practice. The second method is sometimes used with data rounded to the last whole number as well. It is a matter of preference by the researcher. Whichever method is used there are two general guidelines. When tick marks are shown on a graph they should be placed and labeled with exact precision. If the researcher fears that such precision will confuse most readers, tick marks should not be shown.

You can also notice that the (a) figure in R.E. 2.19 shows that a segment of the X-axis has been omitted because there were no scores below 15. If this were not done, a large empty space would be left at the beginning of the axis, as shown in the (c) figure. It is common to find histograms like the (a) figure where the axis has been shortened. Not all histograms will do this so explicitly. They may begin by simply marking the X-axis with the first category's true lower limit, as shown in the (d) figure. Or they may leave a small undefined space before the first bar as the (e) figure does. Although both of these last two practices are technically incorrect, they have become conventional in social research. You can expect to find many graphs like them.

Polygons

A polygon is another method used to display the distributions of variables that have continuity across their categories of measurement. A *polygon* is a multisided closed figure. It is formed by joining points plotted above the midpoints of categories which represent

RESEARCH EXAMPLE 2.19 Variations in Histograms of Grouped Data

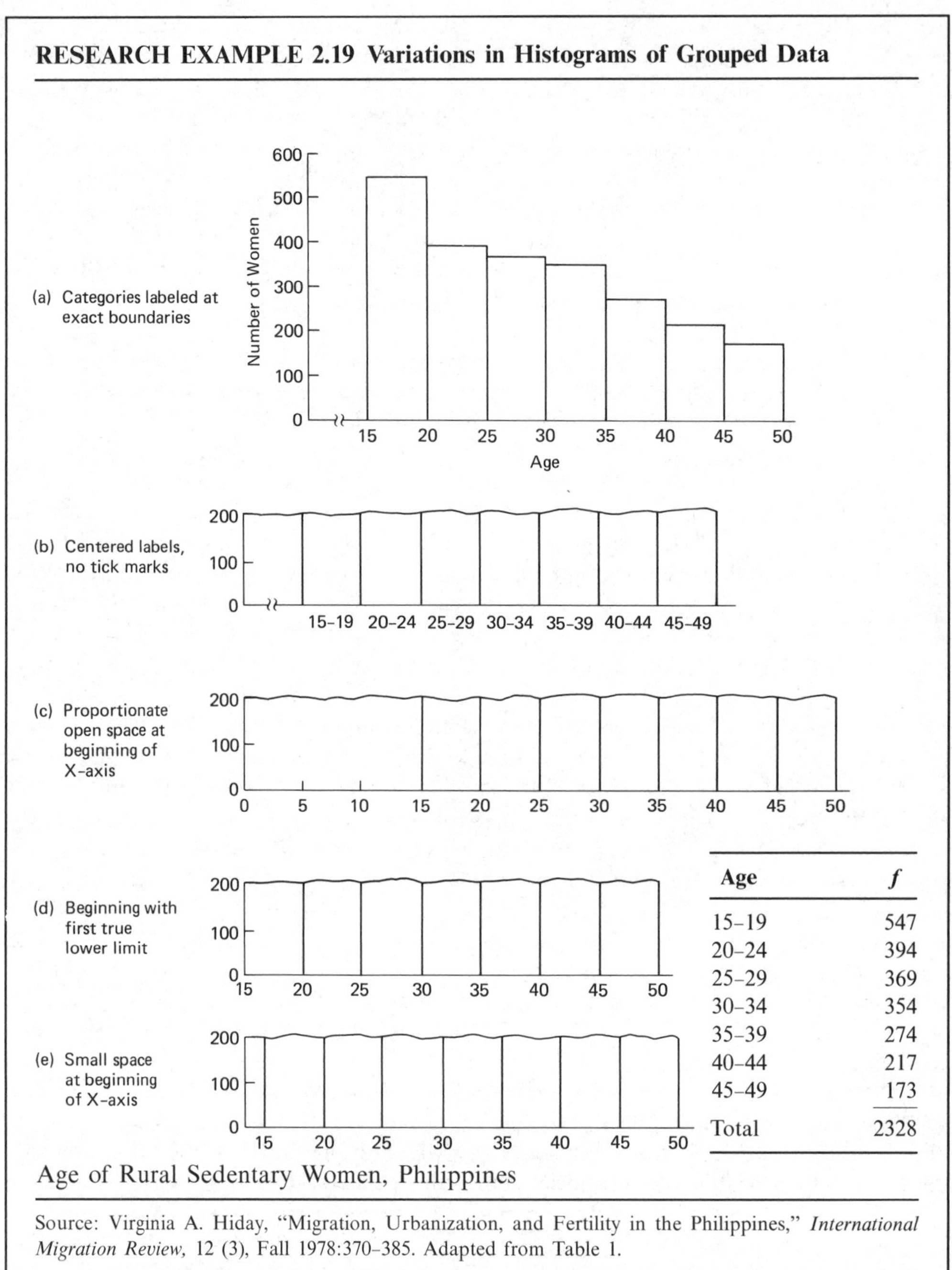

Age	*f*
15–19	547
20–24	394
25–29	369
30–34	354
35–39	274
40–44	217
45–49	173
Total	2328

Age of Rural Sedentary Women, Philippines

Source: Virginia A. Hiday, "Migration, Urbanization, and Fertility in the Philippines," *International Migration Review,* 12 (3), Fall 1978:370–385. Adapted from Table 1.

their frequencies or percentages. The *X*-axis is marked at the midpoints plus one additional point set an equal distance below the first midpoint and another set above the last midpoint. The frequencies (or percentages) are marked on the *Y*-axis, exactly as is done in the histogram. The two additional *X*-points are plotted at zero frequency. Once plotted, the points are connected by a series of lines and joined to the baseline (*X*-axis) to form the

RESEARCH EXAMPLE 2.20 Polygon of Grouped Data

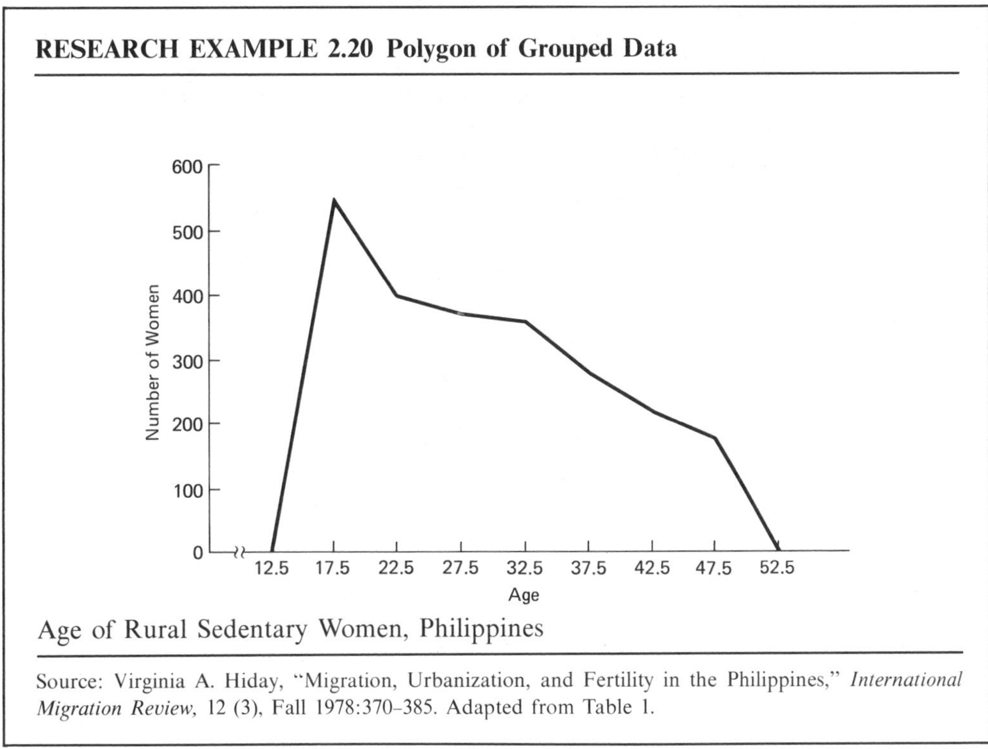

Age of Rural Sedentary Women, Philippines

Source: Virginia A. Hiday, "Migration, Urbanization, and Fertility in the Philippines," *International Migration Review,* 12 (3), Fall 1978:370–385. Adapted from Table 1.

figure. The result is a graph that emphasizes the continuity and overall shape of the distribution.

Research Example 2.20 shows a frequency polygon for the same age distribution of the women in the Philippine migration study used in R.E. 2.19 to illustrate the histogram. Note how the added plotting points at the beginning and end of the figure serve to close it so that it is not suspended above the *X*-axis. This practice conveys the idea that the figure covers the entire distribution; no parts have been omitted. The figure clearly shows the greater concentrations in the younger age categories for this sample.

It is easy to distort the impression created by a histogram or polygon by manipulating the length of the two axes. If the *X*-axis is shortened or the *Y*-axis is lengthened, the categories are squeezed together. This creates the impression that slight differences between the frequencies or percentages are really large differences. If the *X*-axis is lengthened or the *Y*-axis is shortened, large differences are minimized.

The examples shown in Figure 2.9 illustrate this distortion for hypothetical data. The accurate representations in Figure 2.9 (a) and (d) are shown to be distorted in parts (b), (c), (e), and (f), although all represent the same data. Because this manipulation is a serious issue in graphic interpretation, researchers have adopted a standard for the relative lengths of the two axes. The *Y*-axis should be approximately three-fourths the length of the *X*-axis. This standard is widely accepted in social research, but it is often violated in the popular press and in such image management areas as advertising and politics. A consumer of graphs must carefully read the axes labels and numerals to interpret figures. Do not be overly influenced by the relative sizes of bars in a histogram or by the comparative slopes of line segments in a polygon. Instead, think in terms of the numbers on which the graph is based.

FIGURE 2.9 Distortions in Graphing Histograms and Polygons (Hypothetical Data)

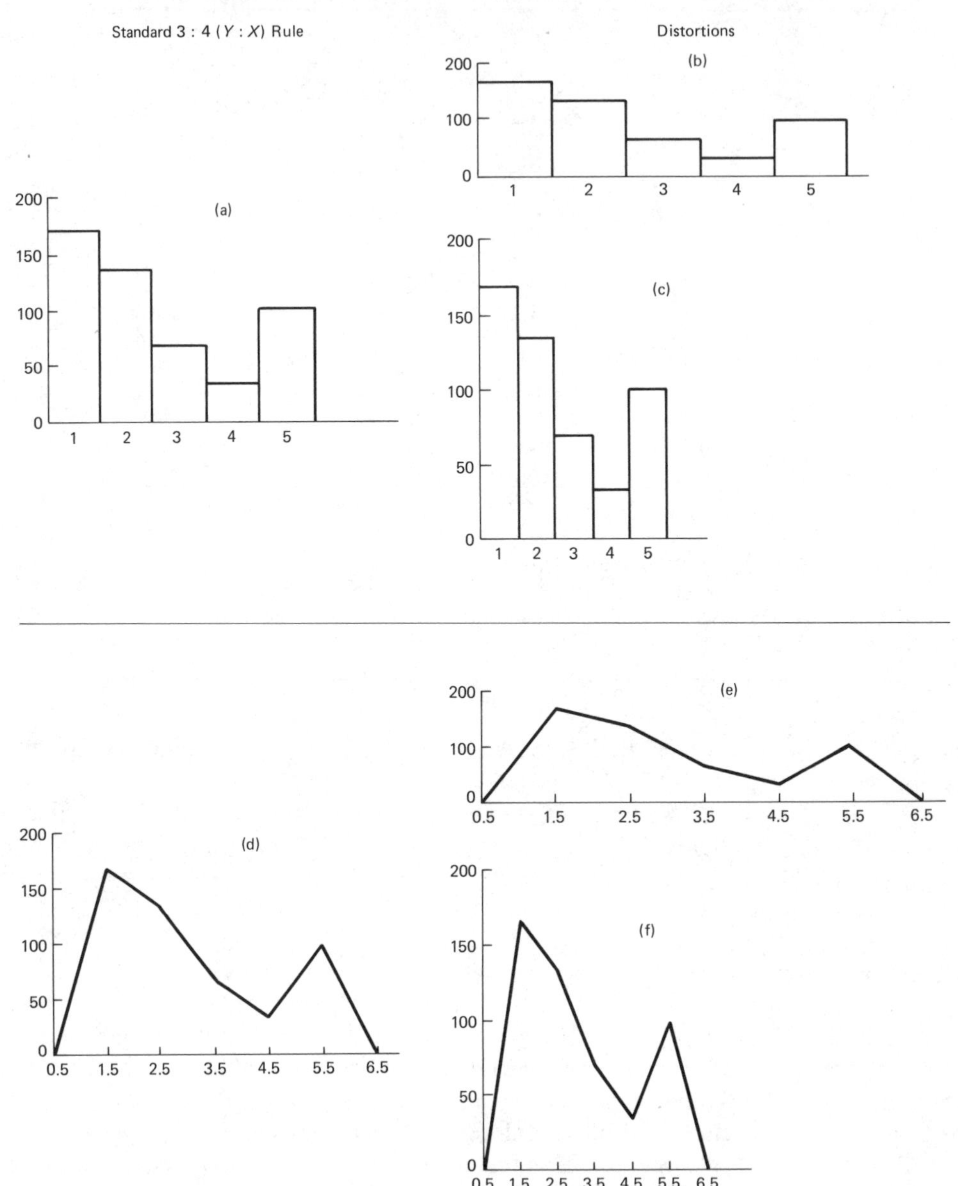

Forms of Continuous Distributions

Correctly prepared histograms and polygons for variables that are continuous (or nearly so) reveal the *form or shape* of distributions. A description of this form is itself a type of statistical summary. The form can be described by focusing on three features: symmetry, kurtosis, and modality. These will be discussed in a qualitative way here. At the end of the chapter, in the *Related Statistics* section, some quantitative definitions are presented. To understand those you will need to read Chapters 3 and 4 first. Some of the classical shapes are sketched in Figure 2.10.

Symmetry is the extent to which a distribution can be divided into two mirror images. In a completely symmetrical distribution a line can be drawn which divides the distribution into two halves that are exact reverse images of each other. The same number of scores are distributed in reverse order in each half. If a distribution is not symmetrical, it

FIGURE 2.10 Shapes of Continuous Distributions

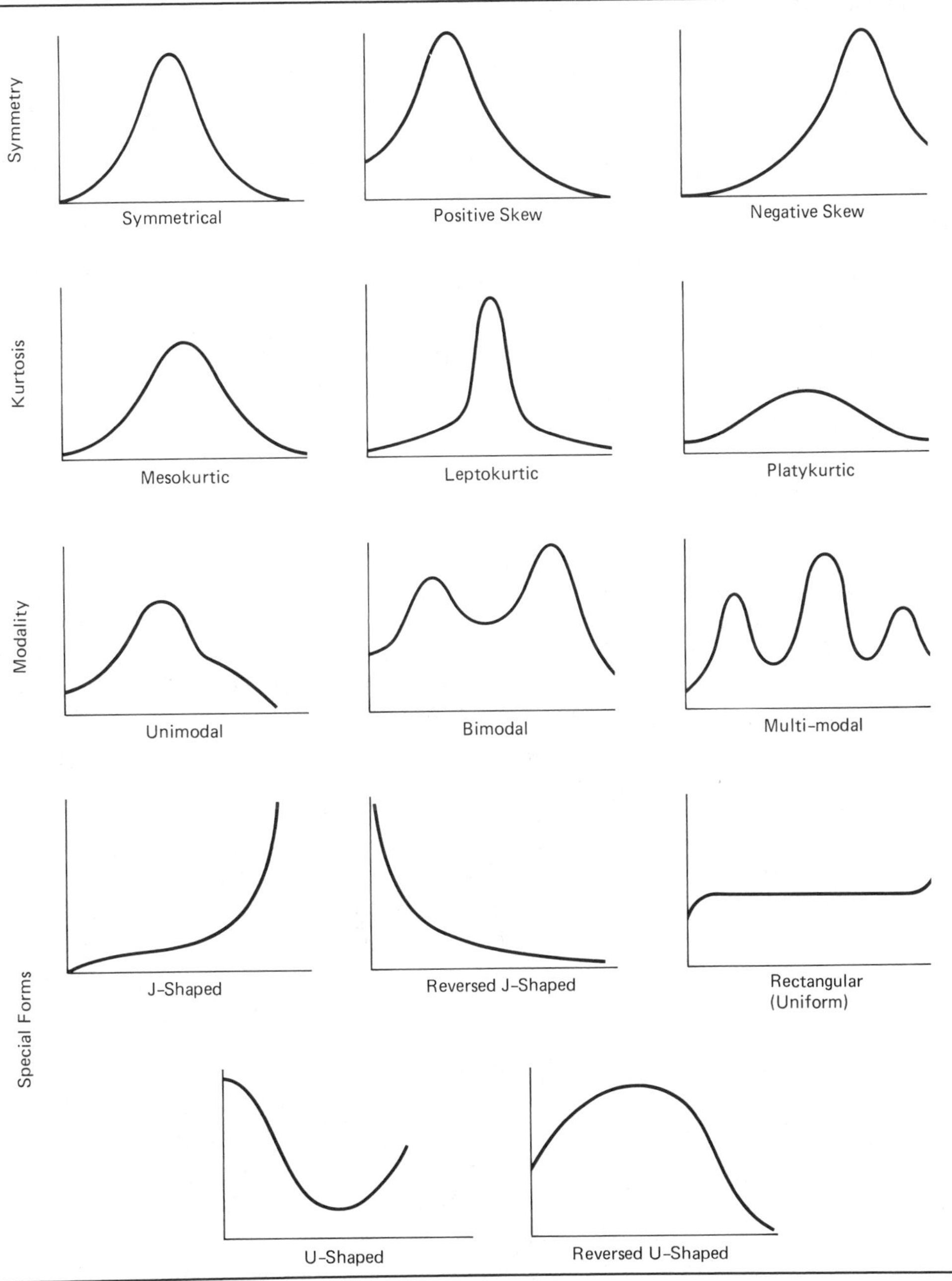

is said to have *skew*. A skewed distribution has a few scores that are especially higher or lower than the majority. When a few scores are much lower in value than the others, the distribution has *negative* or *left skew*. When a few are much higher than the others, it has *positive* or *right skew*.

Kurtosis is the extent of clustering or peakedness in a distribution. Distributions that have a very large proportion of their scores clustered together are referred to as being *leptokurtic*. Those with very little clustering are called *platykurtic*. Those with the usual degree of clustering are called *mesokurtic*. Despite the tendency for these terms to sound as though they were describing horrible diseases, they can be quickly learned.

Modality is the number of points in a distribution where scores are highly concentrated. Each of these points is called a *mode*. Distributions with only one point of concentration are called *unimodal*. Those with two points of concentration are called *bimodal*. Those with three or more points of concentration are called *multimodal*.

Beyond these three separate descriptions of form, some distributions have characteristic shapes with special names. These include *J distributions, U distributions,* and *rectangular distributions.*

Being familiar with the vocabulary of distribution shapes allows one to quickly understand these short descriptions when they are used by a researcher. They do provide a concise summary of a distribution's overall form, although they are not always highly precise when left in qualitative terms. For instance, how flat must a distribution be before it is called platykurtic? Because of this limitation one may need more information when interpreting a distribution's character.

Scatter Diagrams

Bivariate distributions are often graphed as scatter diagrams. *Scatter diagrams* use the Cartesian graphing system to plot each individual case according to its score on two variables. Each variable is represented on one of the two axes. The plotted points are not connected by lines, unlike a polygon.

Research Example 2.21 presents a scatter diagram. It shows the case-by-case plots for age and the number of occasions on which persons participated in some form of outdoor recreation during a year. The prevailing tendency across the plots is for those who are older to recreate less often than those who are younger. The data come from the same study as that represented in R.E. 2.7.

Scatter diagrams are useful to visualize the overall pattern in bivariate distributions. One can see whether or not there is a consistent pattern across the entire distribution and whether there are any unusual cases. These "deviant" cases are of interest because they are unlike the majority of cases and they require special attention to be explained. The examination of scatter diagrams is important in the selection and computation of various statistics used in the analysis of bivariate distributions. More will be said about this later in the book.

Line Graphs

Line graphs are similar in appearance to polygons, except that the line connecting the plotted points does not drop to the X-axis at the ends. One major use of line graphs is to show time-series data. The time points are generally placed on the X-axis and the frequencies (or other data forms) for a variable are shown on the Y-axis. The line joining

RESEARCH EXAMPLE 2.21 Scatter Diagram

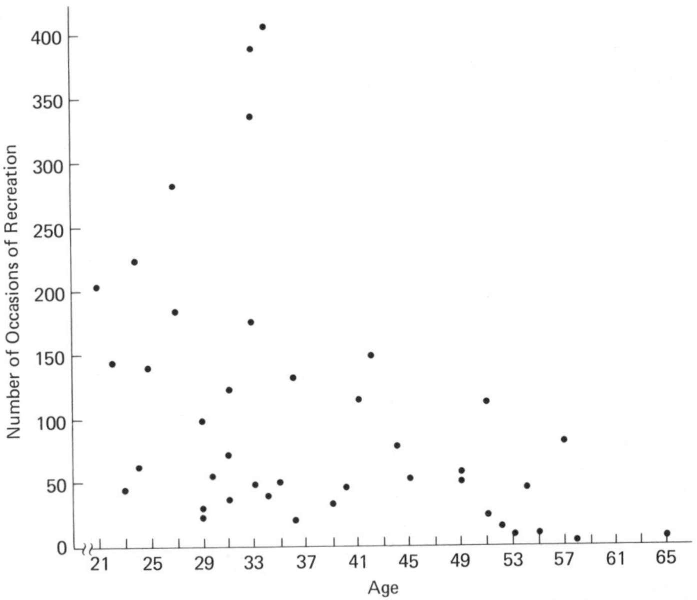

Total Annual Number of Occasions of Participation in Outdoor Recreation by Age, Webster County, Iowa, 1979–1980 ($N = 41$)

Source: Robert E. Kramer and Gene M. Lutz, *Public Perception of Recreational Needs in the Brushy Creek, Iowa, Service Area.* Cedar Falls, Iowa: Center for Business and Behavioral Research, University of Northern Iowa, 1980. Base data.

RESEARCH EXAMPLE 2.22 Line Graph

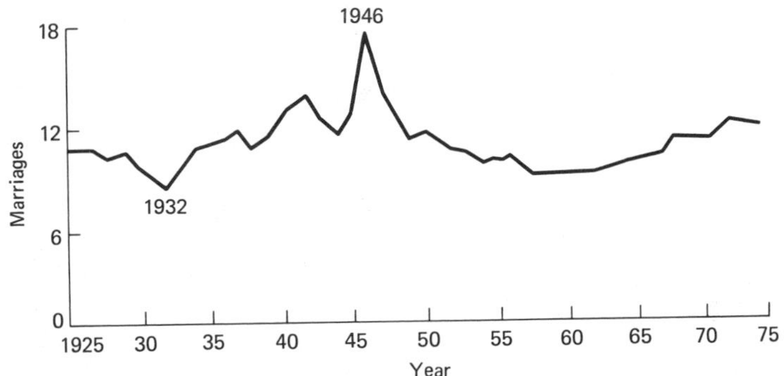

Marriages per 1000 U.S. Population, 1925–1975

Source: Monthly Vital Statistics Report, 26 (2) (Supplement) as cited in Elwood Carlson, "Divorce Rate Fluctuations as a Cohort Phenomenon", *Population Studies,* 33 (3), November 1979:523–536, Figure 2.

RESEARCH EXAMPLE 2.23 Multiline Graph

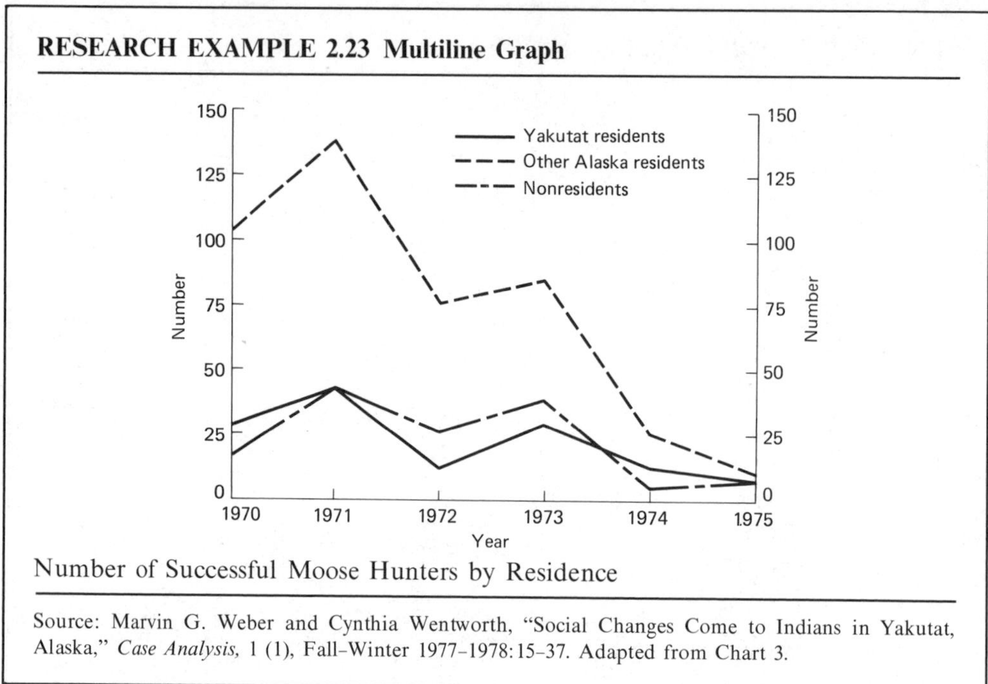

Number of Successful Moose Hunters by Residence

Source: Marvin G. Weber and Cynthia Wentworth, "Social Changes Come to Indians in Yakutat, Alaska," *Case Analysis,* 1 (1), Fall–Winter 1977–1978: 15–37. Adapted from Chart 3.

the points is called a *trend line* because it indicates the change through time for the characteristic on the *Y*-axis.

Research Example 2.22 is an illustration of a line graph showing the changes in the marriage rate from 1925 to 1975. The earlier trend in marriage rates is shown to be associated with historical cycles in the economy and in war and peace. The more recent trend does not have these kinds of apparent association.

Research Example 2.23 shows a *multiline graph.* It depicts the number of successful moose hunters in Alaska by their residence. It is from the same source as shown in R.E. 2.15 concerning the native Indians of Yakutat. The graph depicts the convergence in the number of successful hunters of all types to a relatively low level. Notice that this graph duplicated the *Y*-axis numerals on both sides in order to ease the identification of exact plotting points. This is especially useful when the graph is very wide.

Keys to Understanding Graphs

The keys to understanding graphs are much like those for tables. Again, we can organize them according to two major goals: determining the type of graph and interpreting the graphs.

Key 2.10. To identify the type of graph begin by reading the title and footnotes to the figure. This gets you oriented. *Key 2.11.* Note the axes labels and then identify the number and names of the variables being displayed. *Key 2.12.* Note the axes numerals and determine the categories and levels of measurement for the variables. *Key 2.13.* Identify the statistical form of the data that is being presented.

Key 2.14. To begin interpreting the graph, ask what the purpose of the graph is. One of the most basic distinctions is whether the graph is attempting to show continuity or differences. The answer will usually set you on the right track to a correct interpretation. *Key 2.15.* Check for possible sources of distortion in the graph's construction. Note the proportional lengths of the axes. Look at the numerals used where the axes are started. Are there any "artistic" flairs that distract from a straightforward interpretation of the graph? *Key 2.16.* Choose some parts of the graph's contents and give your interpretations of them. Connect each of these selected parts to the variables in the graph. For example, "28 board members had served only one year on their mental health boards" in R.E. 2.18 or "46.4% of the female professors with Ph.D.s had never married" in R.E. 2.16. *Key 2.17.* Give your own tentative interpretation to the entire figure. Use your own words to characterize what you see. *Key 2.18.* Compare your interpretation to the researcher's. If there are conflicts in the interpretations, try to reconcile them. Remember that behind every graph there is a table. If the graph is confusing, return to the table (if it is available) for clarification.

In this chapter we have looked at entire distributions as they are reported in tables and graphs. In the next we look only at those features of distributions that seem to be most characteristic of their clustering or center. Being able to interpret tables and graphs is necessary to this next activity.

KEY TERMS

Univariate
Bivariate
Multivariate
Code
Case Number
Raw Data Distribution
Ordered Array
Univariate Frequency Distribution
Simple Frequency Distribution
Sample Size (*N*)
Grouped Data Distribution
Open Categories
Closed Categories
Tabular Limits
True Limits
Category Width (*w*)
Category Midpoint (*m*)
Univariate Percentage Distribution
Proportion (*P*)
Percentage (%)
Cumulative Data Form
- Less-Than Cumulative
- More-Than Cumulative
- Cumulative Frequency
- Cumulative Proportion
- Cumulative Percentage

Ratio
Multiplier
Sex Ratio
Rate
Crude Birthrate
General Fertility Rate
Multiresponse Variable
Bivariate Frequency Distribution
Cross-Classification Table
Cell
Marginal Total
Grand or Overall Total
Independent Variable
Dependent Variable
Bivariate Percentage Distribution
Percentaging a Bivariate Distribution
Multivariate Distribution
Partial Distribution
Time-Series Distribution
Longitudinal Study
Percentage Change
Graph
Pictograph
Area Graph

Bar Graph
- Vertical Bar Graph
- Horizontal Bar Graph
- Multibar Graph
- Compound Bar Graph

Histogram
Polygon
Form of a Distribution
Symmetry
- Negative Skew
- Positive Skew

Kurtosis
- Leptokurtic
- Platykurtic
- Mesokurtic

Modality
- Mode
- Unimodal
- Bimodal
- Multimodal

Scatter Diagram
Line Graph
- Trend Line
- Multiline Graph

EXERCISES

1. Construct a table showing the percentage distribution for the data presented in R.E. 2.2. Include all appropriate labels for your table.
2. (a) Would your percentage distribution (based on R.E. 2.2) be better graphed as a bar graph or as a histogram? Why? (b) Construct the type of graph you choose and label it completely.
3. Calculate the following ratios based on the data in R.E. 2.2: (a) the ratio of professionals to clerical workers; (b) the ratio of craftsmen to operatives; (c) the ratio of students to unemployed.
4. The following scores represent the number of instances in the past semester that a group of college professors forgot to bring their notes to class. (a) Arrange the scores in an ordered array. (b) Construct a table showing the frequency and percentage distribution of these scores. Completely label your table. (c) Write a one-paragraph narrative summarizing the distribution as revealed in your table.
   5 2 0 1 1 1 0 2 0 8 1 4 0 1 3 1 6 0 0 1 2 4 1 0 3 1 0 2 3 1 2
5. (a) What is the official demographer's sex ratio in a population of 6250 women and 6340 men? Suppose that this same population was composed of 3000 women between 15 and 44 years old and that there were only 50 live births during the year. (b) What is this population's crude birthrate? (c) What is its general fertility rate?
6. Use the following test scores from a social work licensing exam to construct a grouped frequency distribution table having five grouped categories each with a width of 10. Label your table completely.
   25 36 22 41 57 32 65 29 44 36 51 48 23 37 39 45 46 55 52 24 38 35 61
   68 28 64 54 47 33 67 59 25 27 42 56 45 41 63 42 35 27 52 44 48 25 33
   66 21 48 53
7. Construct a chart showing the following for your grouped frequency table from Exercise 6: the tabular limits, true limits, widths, and midpoints of the grouped categories.
8. (a) Construct a histogram for the data in R.E. 2.3 and completely label your graph. (b) Construct a polygon for the same data and completely label this graph. (c) Which graph gives the greater emphasis to the continuity of the scores in the distribution?
9. Construct four circle graphs (one for each political party identification) showing the data from R.E. 2.17. Completely label your graph.
10. (a) Would the data in R.E. 2.5 be better graphed as a bar graph or a histogram? Why? (b) Construct the type of graph you chose and label it completely.
11. Construct three separate bivariate percentage tables for the data in R.E. 2.8 (a) by percentaging on the row totals; (b) by percentaging on the column totals; (c) by percentaging on the overall total. (d) Which percentaged table tells you how males and females compare in belonging to a union? (e) Which percentaged table tells you how those who are male and do not belong to a union compare to those who are female and do belong to a union? (f) Which percentaged table tells you how union and nonunion members compare in being females?
12. Construct a multiline graph for the data in R.E. 2.12 showing the number of commuters to West Germany and Switzerland separately and combined. Label your table completely.
13. Use the data in R.E. 2.12 to answer the following: (a) What was the percentage change from 1962 to 1976 for those commuting to West Germany? (b) What was the percentage change from 1962 to 1976 for those commuting to Switzerland? (c) What was the largest negative percentage change in the total number of commuters between any two consecutive years? (d) What was the largest positive percentage change in the total number of commuters between any two consecutive years?
14. Use the "years married" variable and the frequency shown as "N" in the R.E. 2.11 table to construct (a) a less-than cumulative frequency distribution and (b) a more-than cumulative frequency distribution. (c) Interpret your *cf* entry for the "11–15" category in each cumulative distribution.
15. Use a current issue of a social science journal to locate an example of one of the types of tables or graphs discussed in this chapter. Use the *Keys to Understanding* to write a short narrative identifying and interpreting this table.

RELATED STATISTICS

Ogive

An ogive is a graph of a cumulative distribution. It may show either the less-than or more-than cumulative distribution for data in any form. It is similar in appearance to a polygon except that the plotting points on the *X*-axis are not category midpoints. For less-than cumulative distributions, the true upper limits are the *X*-plotting points. For more-than cumulative distributions, the true lower limits are the *X*-plotting points. The figure is not closed, unlike a polygon.

See: Mueller et al. (1977:77–81).

Population Pyramid

A population pyramid is a special kind of horizontal histogram. It shows the distribution of a population according to its sex and age. The distribution for each sex is shown separately for each age category. These two distributions are set base to base so that the result is an irregularly shaped pyramid.

See: Loether and McTavish (1980:91–93).

Statistical Map

A statistical map is a graph that shows a variable's distribution over a geographical area. The geographical area is subdivided into smaller areas such as states, neighborhoods, or blocks. The variable's distribution is overlaid by various methods of shading or marking of the areas to indicate the differences. It is commonly found in social ecology studies. For example, delinquency rates in a city's neighborhoods could be shown with a statistical map.

See: Loether and McTavish (1980:94–99).

Triangular Plot

A triangular plot is a graph that allows three dimensions to be shown. A third axis (*Z*-axis) is added to the other two (*X* and Y) to form an equilateral triangle. The scales for the three axes are in percentages. The result is a scatter diagram for three variables.

See: Loether and McTavish (1980:104–112).

Log and Semilogarithmic Graph

All previously mentioned types of graphs have used rectangular scales. This means that a particular distance on a scale has the same meaning all along that scale. For instance, .10 inch might equal 100 persons at any place on a *Y*-axis. When the data do not follow a uniform pattern, rectangular scales can be inadequate. For example, rate-of-change scores would appear as an upward-sloping straight line in a rectangular line graph even though the rates of increase were decreasing from year to year. This visual illusion can be corrected if the same data are presented with one axis using a logarithmic scale instead. This would be a semilogarithmic graph. In more complex situations both axes may use logarithmic scales. This would be a log graph.

See: Loether and McTavish (1980:112–116).

Skew Formulas

Rather than evaluating skew in general qualitative terms such as "extremely positive" or "somewhat negative," some researchers will quantify the extent and direction of a distribution's skew. A commonly used measure is Pearson's coefficient of skewness (*S*). It compares the mean and the median (discussed in Chapter 3) to the standard deviation (discussed in Chapter 4) as follows:

$$S = \frac{3(\text{mean} - \text{median})}{\text{sample standard deviation}}$$

Positive skew causes the mean to exceed the median so that the *S* coefficient is positive. The more extreme the skew is, the larger the coefficient becomes. Similarly, the coefficient becomes increasingly negative with negative skew. It is zero when there is no skew in the distribution.

See: Neter et al. (1978:55).

An even more quantitative measure of skew is known as *Sk*. It is based on the deviations around the mean:

$$Sk = \frac{m_3}{S^3}$$

where $m_3 = \Sigma x^3/N$

S = sample standard deviation

These symbols are explained in Chapters 3 and 4.

See: Loether and McTavish (1980:163–164).

Kurtosis Formula

Kurtosis, like skew, can be quantified. One method is known as *Ku.* A normal (mesokurtic) distribution yields a *Ku* value of 3.0. Distributions that are flatter than normal have values less than 3.0. Those that are more peaked than normal have values greater than 3.0. The further the value is from 3.0, the more extreme is the kurtosis. The formula is

$$Ku = \frac{m_4}{m_2^2}$$

where $m_4 = \Sigma x^4/N$

$m_2 = \Sigma x^2/N$

See: Loether and McTavish (1980:164-165).

CHAPTER 3

DESCRIBING CENTRAL TENDENCY

Summaries of social data using tables and graphs seldom achieve all the goals of description and data reduction. Even tables and graphs may contain more information than we want at times. As a consequence we may ask: "What was the average of these scores?" or "What was the typical situation here?" The search for the "average" or the "typical" is the topic of this chapter. Statisticians call this *central tendency* because the average or typical values are generally near the center of a variable's distribution. Thus they are trying to determine where the distribution tends to cluster or center. You can better understand this focus if you contrast it to an attempt to determine how the scores in a distribution tend to diverge or differ from each other. The latter topic is referred to as *variation* and is discussed in the next chapter.

The Meanings of Central Tendency

"Average," "typical," and even "central tendency" are all very ambiguous terms. However, social statistics attempts to be very precise. As a result, data analysis proceeds by clarifying such terms. Exactly what is meant when someone speaks of an "average?"

Examine Research Examples 3.1 and 3.2, which have been reprinted in very limited

RESEARCH EXAMPLE 3.1 Using the Median to Measure Central Tendency

From Breston Berry, Race and Ethnic Relations, *4th ed. Boston: Houghton Mifflin Company, 1978, 61–62. Reprinted by permission.*

. . . Half a century ago psychologists set about testing the intellectual powers of people; and the intelligence test became immediately popular with those interested in the problem of race differences. Here at last, they thought, was an instrument that would enable them to settle this problem of superiority once and for all in an objective, scientific manner. Certain groups like the English, Scotch, Jews, and Germans come very close to the norm. Negroes in general do poorly, and American Indians score lowest of all. Mexicans do slightly better than the Indians; and Italians, Poles, and Portugueses fall somewhere between the Mexicans and the Negroes. Chinese and Japanese, on the other hand, have made high scores, showing relatively little inferiority in this respect to the whites. Ethnic groups from northwestern Europe, such as the British, Dutch, Germans, and Scandinavians, have proved themselves superior on the tests.

Klineberg presents in tabular form the results of a great number of these intelligence tests:

Summary Table of Ethnic Differences in I.Q.[a]

Ethnic Group	Number of Studies	I.Q. Range	Median I.Q.
American control groups	18	85–108	102
Jews	7	95–106	103
Germans	6	93–105	100.5
English and Scotch	5	93–105	99
Japanese	9	81–114	99
Chinese	11	87–107	98
American Negroes	27	58–105	86
Italians	16	79–96	85
Portuguese	6	83–96	84
Mexicans	9	78–101	83.5
American Indians	11	65–100	80.5

[a]Table 3 in the original. From Otto Klineberg (ed.), *Characteristics of the American Negro.* New York: Harper & Row, Publishers, 1944, p. 35.

The results of the intelligence tests, so flattering to some ethnic groups and so disparaging to others, have not gone unchallenged. For example, it is pointed out that not all the groups tested have the same facility with the language of tests. Some people are bilingual; and while the ability to speak two languages equally well is certainly a valuable accomplishment, it is definitely a handicap when one is taking an intelligence test. . . .

form here. Read these over and ask yourself what the authors mean when they write about the "average." In R.E. 3.1 the author discusses the use of intelligence tests to compare ethnic groups, a highly debated issue. He indicates that some groups "come very close to the norm," some "score lowest of all," some "do slightly better," and others "fall somewhere between." He also reports that some "have made high scores" and others "have proved themselves superior on the tests." These are all very ordinary phrases in common speech, but they are based on a specific definition of central tendency. The intelligence test data represent thousands of individuals with widely differing scores. How were they all summarized statistically to arrive at these very simple sounding descriptions? The table that is reprinted contains the answer—the "middlemost" value, the median, was used here.

In R.E. 3.2 the author examines unemployment in two counties of Utah and the state as a whole as part of a study of energy project developments. The unemployment statistics discussed are "average annual percentage rates." This author is not referring to some "middlemost" idea of central tendency as the earlier example was. Rather he is using a calculation of central tendency that adds the unemployment percentages for several specific time periods during each year and then divides this total by the number of periods. It is what the researchers call an "arithmetic mean" kind of average. This same statistic was also calculated for the six years combined and is presented at the bottom of the table opposite the word "Mean."

As these two examples show, researchers do not always use the same measure of central tendency. There are actually three such measures that you will find in general use. In addition to the median and mean, there is the mode, the most frequently occurring score in a distribution. Why does not every researcher use the same measure? Why do some use the mode, others use the median, and still others use the mean? To understand this diversity you must first know exactly how these measures are calculated.

We are going to examine the calculation procedures for these three measures from the three major types of data distributions: raw data, simple (ungrouped) frequency, and grouped frequency. Remember that our purpose is to get back to the question of why there are different types of measures of central tendency. Once you know the answer to that question you can readily consume such statistical information with relative ease. It is one of the simpler topics in social statistics.

RESEARCH EXAMPLE 3.2 Using the Mean to Measure Central Tendency

From Kenneth R. Weber, "Out-Migration and Economic Marginality: A Case Study of Local Demographic and Economic Conditions Prior to Proposed Energy Development in Southern Utah," The Social Science Journal, *16 (2), April 1979:9ff. Reprinted by permission.*

. . . With the desire for increased national independence in fuel sources, energy development and mineral exploitation in the arid West have become increasingly important national goals, yet the social impact and social costs of such development are only beginning to be understood. While mineral exploitation and economic development have become almost foregone conclusions for much of this region, often too little attention has been paid to the potentially impacted area's recent socio-economic history. A knowledge of local socio-economic conditions is critical, for it is into this milieu that the new projects are thrust. The new projects are, in one sense, appendages to the existing local structure, yet, due to the vast scale of the projects and the relatively undeveloped nature of the impacted localities, these appendages often threaten to envelop and even reconstitute the local economy and society. Thus, any attempt to provide for the harmonious integration of industrialization into these rural areas must be cognizant of both the extent, nature, and demands of the industry *and* the social and economic history of the projected impact area. This paper uses the primary impact area of the proposed Kaiparowits power project—Kane and Garfield counties, Utah—as a case study to describe the social and economic characteristics of one Inter-Mountain area prior to proposed energy development. By analyzing previously and presently existing conditions, the extent of change and its social impact on the local population may be assessed more accurately. . . .

Unemployment and underemployment have long been critical and continuing problems in Garfield and Kane counties, and unemployment figures from 1950 to 1974 well exemplify the economically marginal position of the area [see the table]. Statistics show both counties have recorded unemployment rates of up to 20 percent—rates three times larger than the highest recorded by either the state or nation during this 25-year period. The unemployment rates of the two counties have been consistently higher than those of the state or nation. Only during the Glen Canyon Dam construction period (1957–1965) and in 1973 and 1974 did Kane County's unemployment rate drop into single digit percentage figures. Even with the massive employment associated with this construction, however, Kane County's rates were still up to twice as large as the corresponding state figures. Unemployment in Garfield County was even greater than in Kane County. In 19 of 25 years of this period, Garfield County's unemployment rates exceeded that of Kane County. Only in a single year, 1965, was Garfield County's rate below 11 percent.

Average Annual Percentage Unemployment Rate, Garfield and Kane Counties, Utah, and United States, Selected Years, 1950–1974[a]

Year	Garfield	Kane	Utah	U.S.
1950	20.8	19.1	5.3	5.3
1955	14.4	12.4	3.9	4.4
1960	12.9	7.7	4.5	5.5
1965	9.2	8.2	5.8	4.5
1970	17.1	16.2	5.8	4.9
1974	15.4	7.5	6.1	5.6
Mean 1950–1974	14.1	11.8	4.8	4.8

[a]Table 5 in the original. From Department of Employment Security, Salt Lake City, Utah; Department of Labor, Bureau of Labor Statistics, *Economic Report of the President, 1971.* Washington, D.C. U.S. Government Printing Office, 1971; Utah Department of Employment Security, Annual Reports, Vol. III, *Labor Market Information,* 1973–1975; *Monthly Labor Review,* 98 (7), 1975:75.

Seasonal unemployment and underemployment also characterize the local occupational situation. According to the 1970 Census, Utah occupied the sixth lowest position among states in the percentage of employed persons working 50 or more weeks a year. Within Utah, Kane and Garfield counties ranked third and seventh lowest, respectively, in this same category. The two counties thus rank in the bottom quarter of counties in a state that is in the lowest eighth nationally in the percentage of employees working 50 or more weeks a year. . . .

Finding Central Tendency from Raw Data Distributions

Mode (*Mo*)

The *mode* (*Mo*) is the most frequently occurring score in a distribution—to repeat from above. It is the simplest measure of central tendency to calculate and interpret, but it usually provides the least amount of information about a distribution of the three measures. It can be found in data at any one of the four levels of measurement (nominal, ordinal, interval, or ratio). With raw data presented in either the tabular or graphic form it is found by straightforward inspection. Suppose that seven family members had the following political party affiliations: Democrat, Republican, Democrat, Socialist, Democrat, Independent, and Democrat. The mode would be Democrat because there were more Democrats than any other one affiliation.

Maybe 11 college friends have the following student classifications: freshman, sophomore, freshman, junior, sophomore, sophomore, junior, senior, sophomore, junior, and sophomore. If so, the mode is sophomore. Or suppose that the 11 take a quiz in Seminar on the Occult and receive the scores 3, 5, 4, 6, 10, 6, 5, 5, 6, 8, and 6. What is the mode? It is 6. Notice that in giving the mode each time we said nothing about the exact frequency with which the modal value occurred. We did not say that the mode was "four Democrats," only that it was "Democrat." Nor did we say that the mode was "five sophomores," only that it was "sophomore." Some researchers will provide the extra frequency information with the mode, some will not. All we know for sure when we see a mode reported is that the score that is given occurred more often than any one other score. We do not even know whether a majority of the scores in the distribution were at the mode. It is a common mistake to assume that the mode identifies the value of a majority of the scores. Look back at each of the examples just given. Only in the first one (political affiliation) did a majority of the scores happen to fall on the mode.

Consider the three possible distributions of ages for residents of apartment houses being shown in Figure 3.1. The first polygon (a) illustrates a distribution that has two points of equal concentration. If the scores had merely been listed in an array, we would have noted that there were an equally high number of the ages 10 and 25. On the polygon this is immediately apparent. What would you say the mode is here? There are actually two modes, ages 10 and 25. Unlike the other two measures of central tendency, there can be one, two, or more modes in a distribution. These distributions would be referred to as *unimodal, bimodal,* and *multimodal,* respectively, as noted before in Chapter 2. Thus the first polygon shows a bimodal distribution.

The second polygon (b) is similar to the first except that the two points of concentration do not have equal frequencies. Because there are still two clearly predominant points of clustering, the distribution is yet described as being bimodal. This may be specified as a *major mode* at age 10 and a *minor mode* at age 25.

Instead of having two or more modes, a distribution may not have any mode. That would happen if all, or nearly all, of the scores occurred the same number of times. Such a situation is shown in the third polygon (c) of Figure 3.1. Since there is no real concentration of scores anywhere in the distribution, naming a mode would not be a meaningful way to describe the data. Instead, indicating that there was "no mode" would more accurately communicate the absence of age clustering in this distribution.

Median (*Mdn*)

The second measure of central tendency is the *median* (*Mdn*). It identifies the middlemost value in a distribution.* This value is either at the middle or as close as one can get to the middle if the scores are put in order. The median may be identical to

*A more technically precise definition of the median is that point on a measurement scale below which exactly 50% of the scores fall. However, this definition requires an interval scale and treats the data as being continuous. The definition presented here, while less precise, additionally allows the median to be found for ordinal data using categories of measurement that are either verbal or discrete and numerical, which better matches common practices.

FIGURE 3.1 Distributions with Two Modes or No Mode

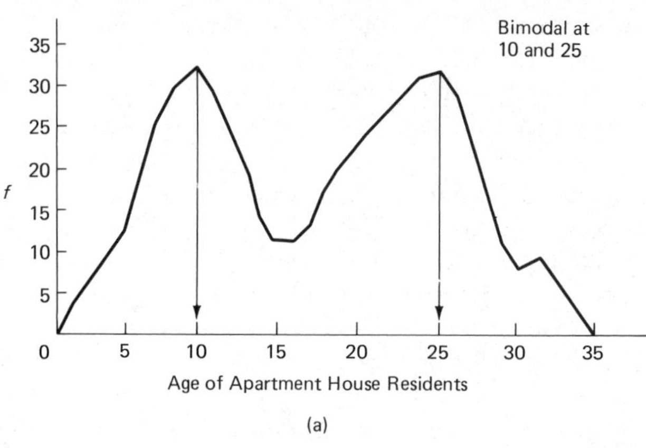

(a)

(b)

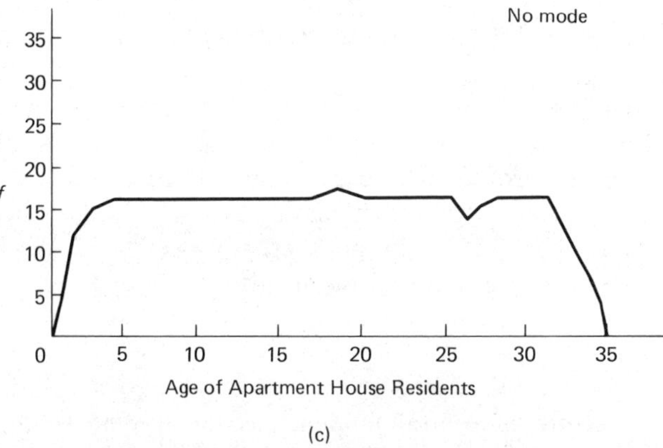

(c)

one of the actual scores that occurred or it may be equal to some value between two actual scores. In comparison, the mode will always be identical to a score that actually occurred in raw data distributions. To speak of the middle of a distribution requires that the distribution has an order to it. Therefore, the median requires at least ordinal level of measurement.

Finding the median from raw scores takes three steps. First, put the scores in order. Second, find the position of the median in the ordered distribution. Third, determine what value (verbal or numerical) it has. For example, if we had the raw scores 3, 6, 5, 2, 7, 1, and 6, we would first put them in order; that is, put them into an ordered array. Once they are in order (7, 6, 6, 5, 3, 2, 1) we are ready to find the position of the median. This is done by dividing the number of scores plus one ($N + 1$) by the number 2. In this example we would find that the median is at the $(7 + 1)/2 = 8/2 = 4$th position. Counting from either end of the ordered distribution we would find that the score in the fourth position is 5. Thus the median is 5. The score of 5 is the middlemost score in this distribution.

The number of scores was an odd number in the last example. When it is even, one small adjustment is needed in the procedure. After ordering the scores and finding the position of the median we will find that it falls between two scores rather than at one of them. We add those two scores and divide by 2 to determine the value of the median.* For example, with the ordered scores 19, 21, 23, 24, 25, 26, 29, and 30, find the position of the median as usual with the $(N + 1)/2$ formula: $(8 + 1)/2 = 4.5$th position. The fourth score has the value 24 and the fifth is 25, so the value of the median is $(24 + 25)/2 = 49/2 = 24.5$. Although no actual score is exactly in the middle of this distribution, 24.5 is the middlemost value.

Look at the examples in Table 3.1 to make sure that you understand how to find the median regardless of the type of variable with which you are working. Some of the examples are verbal and some are numerical.

The three-step process we are discussing for finding the median is the same one that would have been followed in R.E. 3.1. The IQ "averages" given for the various ethnic groups are medians. Each is the middlemost score of the averages achieved by each group.

The median is also referred to as the 50th percentile. As nearly as possible, the same number of scores are above it as are below it. Whether or not exactly 50% are below and above depends on the distribution itself. If several scores have the same middle value, less than half will be lower in value and higher in value. Only when the median is at a value between two actual scores that differed from each other would the median have exactly 50% of the scores below and above it. The example above with eight scores was such a situation. In all other examples mentioned less than 50% were above and below since the median score accounted for some percentage of the distribution itself.

Mean ($\overline{X}$)

The third measure of central tendency is the *arithmetic mean* ($\overline{X}$), *arithmetic average*, or *mean*, for short. It is probably the most commonly used measure of central tendency in social statistics. This is because its mathematical properties make it attractive for both descriptive and inferential statistical purposes. Its definition is solely arithmetic, so it is often unclear to its newest consumers. The mean is defined as the sum of the scores in a distribution divided by the number of scores. If a variable is labeled as the "X" variable, its arithmetic mean is found by the formula on page 82.

*The position of the median may place it between two adjacent categories of an ordinal variable measured with verbal, as well as numerical, categories. In this event, one would express the median as a hyphenated combination of those two categories; e.g., lower-middle class.

TABLE 3.1 Examples of Finding the Median from Raw Data

Respondent	Age	Social Class	Years of Education
A	22	High	12
B	25	Low	13
C	19	Low	11
D	23	Middle	14
E	33	High	16
F	28	Low	12
G	41	Middle	12
H	39	Middle	13
I	27	Low	10
J	21	High	No response

Age:
1. Order the distribution:
   19 21 22 23 25 27 28 33 39 41
2. Find the position of the median:
   $(10 + 1)/2 = 5.5$th position
3. 5th score = 25 and 6th score = 27
   $(25 + 27)/2 = 26$
   median = 26 years

Social Class:
1. Order the distribution:
   low low low low middle middle middle high high high
2. Find the position of the median:
   $(10 + 1)/2 = 5.5$th position
3. 5th score = middle and 6th score = middle
   median = middle class

Year of Education:
1. Order the distribution:
   10 11 12 12 12 13 13 14 16
2. Find the position of the median:
   $(9 + 1)/2 = 5$th position
3. 5th score = 12
   median = 12 years

$$\bar{X} = \frac{\Sigma X}{N} \qquad \text{Mean Formula for Raw Data Distribution}$$

where, $\bar{X}$ = "X bar," the mean
Σ = summation process, symbolized by the Greek capital letter "sigma"
X = individual raw score
N = number of raw scores

To find the mean for the raw scores 21, 33, 15, 28, and 27, we first find their sum (124) and then divide it by the number of scores (5). The mean is 24.8. If six delinquents have been arrested one time, two times, four times, three times, two times, and three times each, the mean number of arrests is $(1 + 2 + 4 + 3 + 2 + 3)/6 = 15/6 = 2.5$ times. If you have three siblings, I have two, and someone else has five, then the mean for our little group is $10/3 = 3.3$ siblings.

This is the same process that was being used in R.E. 3.2, discussed earlier. We do not have the raw data to work from to verify the calculations. Nor can we merely add the annual averages and divide by 6 to find the mean for the 1950–1974 period that is reported at the bottom of the table. Averaging averages often gives erroneous results because it compounds rounding errors. It is accurate only when we know the number of scores on which each separate average was based. That information is not shown in R.E. 3.2. Let us examine this issue further.

Suppose we knew that an audience was composed of 200 females and 100 males. If the females had a mean age of 25 and the males had a mean age of 30, the mean age of the entire audience would *not* be $(25 + 30)/2 = 55/2 = 27.5$ years. Rather we would have to find the *weighted mean* that gives proportional influence to the two groups based on their relative case sizes. We must multiply each separate mean by its respective number of cases, add the two products, and then divide by the total number of people:

$$\bar{X}_w = \frac{N_a(\bar{X}_a) + N_b(\bar{X}_b)}{N_a + N_b} = \frac{200(25) + 100(30)}{300} \qquad \text{Weighted Mean for Two Groups}$$

$$= \frac{5000 + 3000}{300} = \frac{8000}{300} = 26.7$$

where $\bar{X}_w$ = weighted mean
N_a = number of cases in the first group
N_b = number of cases in the second group
$\bar{X}_a$ = mean of the first group
$\bar{X}_b$ = mean of the second group

This may or may not have been the procedure that was followed in R.E. 3.2 to find the 1950–1974 mean. However, it could have been used if the researcher had wished. The procedure saves time and effort over going back to all the individual raw scores to calculate the mean in the usual way.

Unlike the median, it is not necessary to put the raw scores into an ordered array before doing the calculations to find the mean. Like the median, though, the final answer may be a value that is not the same as any one of the original raw scores. In fact, the mean usually is not the same. Only the mode will always equal the same value as any of the raw scores.

Calculating the mean requires the arithmetic processes of addition and division. Recalling the discussion of levels of measurement in Chapter 1, we know that these processes require at least interval data. To add and divide scores assumes that their differences are known numerically. We cannot find the mean for a variable that has been measured at the nominal or ordinal level. Knowing how to calculate the mean may yet leave one asking "But what does the mean really mean?" The statistician's answer to this question is that the mean represents a "center of gravity" or the "center of the deviations." This interpretation does not alter the fundamentally mathematical meaning of the mean; it merely expands it. There is no common phrase that substitutes as an explanation such as "most frequently occurring" or "middlemost" do for the mode and median, respectively.

The "center of gravity" analogy for the mean refers to an elementary physics demonstration wherein small weights are put on a board at various points and we determine where to place a fulcrum so that the board balances at perfect level. (This only works

FIGURE 3.2 Mean as the Center of Gravity or Center of Deviations

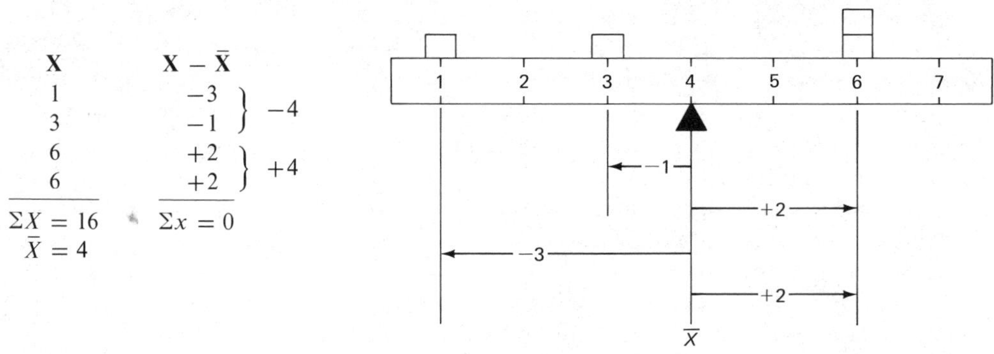

exactly if we assume that the board weighs nothing.) Faced with this problem most of us first use the "trial-and-error" method until we discover the principle involved. If we merely add the distances that the weights are placed from one end of the board and then divide by the number of weights, we can find the correct location for the fulcrum in one try. The arithmetic mean is a type of average that tells us this balancing point in a set of numbers. Examine Figure 3.2.

The same idea is behind the "center of deviations" explanation. Suppose that we have the raw scores listed in Table 3.2. We first find their mean (17). Then we transform the raw scores into deviation scores. A *deviation score* indicates how far a raw score differs from the mean and whether it is above or below that mean. The commonly used symbol for a deviation score is the lowercase x (assuming that the variable is labeled as X). The formula is

$$x = X - \bar{X}$$ Deviation Score Formula

where x = deviation score
X = raw score
$\bar{X}$ = mean score

If we sum those deviations in Table 3.2 which are positive (+14) and those which are negative (−14), we notice that the sums are the same except for the sign. And if we add these two sums together, we get zero. We could not find the deviations around any other number except the mean for a distribution and still have the positive and negative deviation sums be equal (disregarding their signs). Nor would the addition of those two sums result in a total sum of zero. Only the sum of the deviations around a mean will always be zero, $\Sigma x = 0$. Hence the mean is said to be the center of the deviations.

The other two measures of central tendency do not have such an interpretation. Neither the mode nor the median require information about the exact differences between scores in their calculations. Thus deviation scores are not relevant to their meaning. While the "center of the deviations" idea is still strictly a mathematical interpretation of the mean, it is a comparatively simple one.

TABLE 3.2 Deviations from the Mean, x

Commuting Distance from Residence to Workplace, Round Trip (Miles)

X	$x = X - \bar{X}$		Deviations from Other Values $Y = 15$		$Y = 20$
25	+8		+10		+5
21	+4	+14	+6		+1
19	+2		+4		−1
17	0		+2		−3
15	−2		0		−5
12	−5	−14	−3		−8
10	−7		−5		−10
$\Sigma X = 119$	$\Sigma x = 0$		+14	$\neq 0 \neq$	−21
$N = 7$					
$\bar{X} = 17$					

Finding Central Tendency from Simple Frequency Distributions

Researchers do not need to have their data in the form of raw scores in order to calculate the measures of central tendency. Although it is the best data form from which to understand the meaning of the measures, it is easier to calculate them from simple (ungrouped) frequency distributions. More algebra skills are required for these computations (and for those in the grouped frequency form that follow).

Mode

In Table 3.3 there is an example of finding the mode from a simple frequency distribution. The process depends on being able to read the table. Just for practice, what does the first row mean? Answer: "The predominant ethnic background of 25 clients was

TABLE 3.3 Finding the Mode from a Simple Frequency Distributions

	Predominant Ethnic Background of Agency Clients		
	Ethnicity	***f***	
	Hispanic	25	
	Italian	17	
	German	5	
(2) mode ⟶	Black	37	⟵ highest frequency (1)
	Polish	11	
	Lithuanian	2	
	Total	97	

Hispanic." The simple frequency distribution has merely condensed the raw data into a convenient form. Whereas a raw data distribution would have listed 25 separate scores of "Hispanic," the simple frequency distribution uses a single row, with one column indicating the ethnic background category and a second column indicating the number of respondents belonging to that category.

To find the mode we first locate the highest number in the frequency column. The category of the variable associated with that number is the mode. Thus in this example the mode is black since that category had the highest frequency. There were more blacks in the distribution than any one other ethnic background. The procedure is the same regardless of the level of measurement.

Median

Finding the median from a simple frequency distribution follows a pattern like the one for raw data, except that five steps are used rather than three.

1. Be sure that the data categories are in order. Most simple frequency distributions with variables at the ordinal or higher levels of measurement will already be in order.
2. Find the position of the median using the previous $(N + 1)/2$ formula.
3. Calculate a less-than cumulative frequency distribution. Begin with the category that is lowest in quantitative value, regardless of whether it is at the top or bottom of the table.
4. Find the median category. Look through the cumulative frequency numbers to locate the one that is either equal to or just greater than the position value. That *cf* value identifies the category that contains the median.
5. Determine the value of the median. Since all the scores within each category of a simple frequency distribution are identical, the median equals the score of the median category.

Examine Table 3.4 for an illustration. It shows a distribution of social class for parents of heroin addicts. The categories are in the high-to-low order. The $(N + 1)/2$ position is the 41.5th score. The 26th through the 56th scores are included in the category having "56" in the *cf* column. Since this includes the 41.5 position, the upper-middle class is the median category. And since all 31 of the scores in that category are the same, the median equals upper-middle class.

TABLE 3.4 Finding the Median from a Simple Frequency Distribuion

Social Class of Parents for a Sample of Heroin Addicts

	Social Class	f	cf	
(1) check	Upper–upper	10	82	
order	Lower–upper	16	72	$(N + 1)/2 = 41.5$ (2)
(5) median ⟶	Upper–middle	31	56 ⟵	median category (4)
	Lower–middle	12	25	
	Upper–lower	9	13	
	Lower–lower	4	4	
	Total	82	↑	

(3) calculate a less-than *cf* distribution

TABLE 3.5 Finding the Mean from a Simple Frequency Distribution

Number of Traffic Tickets Received by Two Groups

	Public Officials		General Public	
Number of Tickets	f	fX	f	fX
10	0	0	1	10
9	0	0	2	18
8	0	0	4	32
7	0	0	3	21
6	1	6	3	18
5	0	0	2	10
4	2	8	2	8
3	2	6	1	3
2	3	6	1	2
1	4	4	0	0
0	8	0	1	0
	$N = \Sigma f = 20$	$\Sigma fX = 30$	$N = \Sigma f = 20$	$\Sigma fX = 122$
$\bar{X} = \frac{\Sigma fX}{N}$	$\bar{X} = \frac{30}{20}$		$\bar{X} = \frac{122}{20}$	
	$= 1.5$		$= 6.1$	

Mean

The mean, too, uses a procedure that is only slightly modified from the one for raw data. One still finds the sum of the scores and divides this by N. But because every individual score is not listed in a simple frequency distribution, one cannot simply add the values appearing in the first column to find that sum. Each of the listed values has occurred as many times as the frequency shown beside it. The statistician realizes that if we first multiply the frequency entry by the score value and then add these products, the result is the sought-after sum. This sum is now properly labeled as ΣfX. This method is a simpler one than relisting the individual scores and adding them directly. The new method works because multiplication is the same as a series of additions. Three times four yields the same answer as adding four to itself three times. Once the ΣfX sum is known, we divide it by the number of cases to determine the mean. The number of cases is found by summing the frequency column. That is, $\Sigma f = N$. The formula, then, is

$$\bar{X} = \frac{\Sigma fX}{N}$$ Mean Formula for Simple Frequency Distributions

where $\bar{X}$ = mean
X = raw score value
fX = product of a raw score value and its frequency of occurrence for each category
Σ = summation process
N = number of raw scores found by summing the frequencies (Σf)

ON THE SIDE 3.1 Statistical Symbols

Researchers make a distinction between characteristics of a sample and a population. A sample characteristic is referred to as a *statistic,* whereas a population characteristic is called a *parameter.* To indicate which is being discussed researchers will often use different symbols for each. Typically, the symbols of the modern Roman alphabet are used for sample data and symbols from the Greek alphabet are used for population data. There are some special uses of certain symbols in the Greek alphabet that do not pertain to this sample-population distinction. The Greek capital sigma (Σ) is used to indicate the summation process with any data. The Greek lowercase pi (π) refers to the circumference of a circle divided by its diameter. But ordinarily the Roman versus Greek alphabet distinctions are used for sample and population characteristics, respectively.

The arithmetic mean is one example of this distinction. The mean of a sample uses the Roman capital $\overline{X}$ (assuming that the variable was labeled X). The mean of a population uses the symbol μ, the Greek lowercase mu. Both of these have the same formula, though. Each sums the scores and divides by the number of scores.

Some authors do not follow this convention. They may use the symbol M for a population mean. But others may use M for a sample mean! If you find M used, you will need to seek further information to clarify its intended meaning. Generally, we will be using conventional symbols (sometimes simplified) to maximize the transference of what you learn here to research reports. However, we may mention both the sample and population symbols for various statistical measures only when we need to make the distinction for some immediate purpose.

Table 3.5 illustrates the process. The total number of tickets (ΣfX) received by the public officials was 30. Thus the mean number per official was 1.5 ($30/20 = 1.5$). For the general public group a total of 122 tickets were received for a mean of 6.1 per person ($122/20 = 6.1$). Whether or not this difference can be attributed more generally to public officials and the general public is a matter of statistical inference. It is not appropriate to suggest a generalization on the sole basis of descriptive statistics, such as the mean.

*Estimating Central Tendency from Grouped Frequency Distributions

Social data presented in the form of a grouped frequency distribution have less precision than as either a raw data or simple frequency distribution. Grouping means that different scores have been combined and they are no longer distinguishable from each other. This imprecision is the sacrifice researchers knowingly make when they decide to summarize data in the grouped type of distribution.

The exact mode, median, and mean cannot be found from grouped distributions because of the imprecision. Only estimates can be made. The estimating procedures used are designed to minimize any systematic error whereby the results are consistently higher or lower than the exact results. Each of the procedures is based on one or more assumptions about the data that have been summarized. The estimates can only be as accurate as the assumptions allow. This point is of great importance in interpreting such calculations. Beyond the usual questions that should be asked about the reliability and validity of the data and the appropriateness of the measure selected, one should pay special attention to the assumptions being made when these estimating procedures are being used.

To distinguish the estimates of central tendency from the exact values, a circumflex (ˆ) is added to their representative symbols. Both the circumflex and prime mark (′) are used in statistics to denote an estimate. We will use the circumflex.

Mode ($\widehat{Mo}$)

As always, the mode is the easiest of the three measures of central tendency to calculate. First, locate the highest frequency. Second, determine the category having that highest frequency. Third, determine the midpoint of that category. The midpoint value is the estimated mode for grouped data. The procedure assumes that the ungrouped data follow a smooth, continuous distribution and that the midpoint of the modal category represents the highest concentration of scores. If there are a large number of cases and the

TABLE 3.6 Estimating the Mode from a Grouped Frequency Distribution

Monthly Incomes of College Students

	Monthly Income	f	
	$751–1000	45	
	$601–750	284	
	$451–600	401	
(2) modal category ⟶	$301–450	1123	⟵ highest frequency (1)
	$151–300	322	
	$1–150	61	
	Total	2236	

(3) $\widehat{\text{Mo}} = \dfrac{300.50 + 450.50}{2} = \$375.50$

Size of Communities Receiving Urban Renewal Funds

	Population	f	
	350,001–400,000	3	
	300,001–350,000	0	
	250,001–300,000	5	
	200,001–250,000	9	
(2) modal category ⟶	150,001–200,000	14	⟵ highest frequency (1)
	100,001–150,000	6	
	50,001–100,000	0	
	1–50,000	2	
	Total	39	

(3) $\widehat{\text{Mo}} = \dfrac{150{,}000.5 + 200{,}000.5}{2} = 175{,}00.5$

variable's level of measurement is ratio or interval, the assumptions may be justified. But with a small number of cases or data at the lower measurement levels, the estimated mode may be very inaccurate. Examine the distribution carefully before making an assessment of the estimated mode.

Table 3.6 illustrates the process of estimating the mode in one situation where the assumptions are reasonable and one where they are not. In the first example of monthly incomes for college students, the highest frequency (1123) occurs for the category $301–450. That category's midpoint is found by adding its true lower and upper limits and then dividing by 2. The estimated mode is therefore $375.50. This estimate deserves some credibility since there is a large number of scores in this example, the distribution appears smooth and continuous, and the level of measurement of the variable before grouping would have been ratio. The true mode may be very close to the $375.50 estimate.

In the second example in Table 3.6 the computational procedure is the same. The mode for the size of communities receiving urban renewal funds is estimated to be a community of 175,000.5 persons. However, the number of cases is small compared to the width and number of categories in the variable. The distribution is not continuous or smooth; the frequency even drops to zero at two places in the middle of the distribution. If we had the raw scores to examine, it is not at all unlikely that there may not have been any true mode at all. Or if there was one, it could easily have been at some value greatly different from the 175,000.5 estimate. These speculations illustrate the difficulty of estimating the mode for many small data sets. Remember that the estimating procedure used is the best one available. But one should be very cautious about such calculations unless the assumptions behind those calculations are reasonably well met in each instance.

Median ($\widehat{Mdn}$)

It is possible to estimate the median and mean from grouped data with more confidence than it is for the mode. This is because the assumptions for the calculations are more often justified. The estimated median assumes that the scores are spread evenly throughout the median category. Think back to some comments made about constructing grouped distributions in Chapter 2. It is best if the grouped categories are devised so that scores do not cluster at either the upper or lower ends but instead conform to the natural concentrations. If this is true, then assuming that the scores are spread evenly throughout the median category will not involve a great distortion. Even when this assumption is not true, the estimated median will always be in the correct category. If the categories are narrow in width, even the worst estimate will not be excessively inaccurate. As with all grouped data, the distribution is assumed to be continuous. The formula used to estimate the median from grouped data is a combination of a known part and an estimated part:

$$\widehat{Mdn} = l_m + \left[\frac{N/2 - cf_{bm}}{f_m}\right]w_m$$

Estimated Median Formula for Grouped Frequency Distributions

where $\widehat{Mdn}$ = estimated median
l_m = true lower limit of the median category
N = number of cases
cf_{bm} = less-than cumulative frequency immediately below the median category
f_m = frequency in the median category
w_m = width of the median category

TABLE 3.7 Estimating the Median from a Grouped Frequency Distribution

Stockholdings Among Members of the Fat Cats Club

Stockholdings	f	cf
\$400,001–500,000	21	122
\$300,001–400,000	35	101
\$200,001–300,000	30	66
\$100,001–200,000	26	36
\$1–100,000	10	10
Total	122	

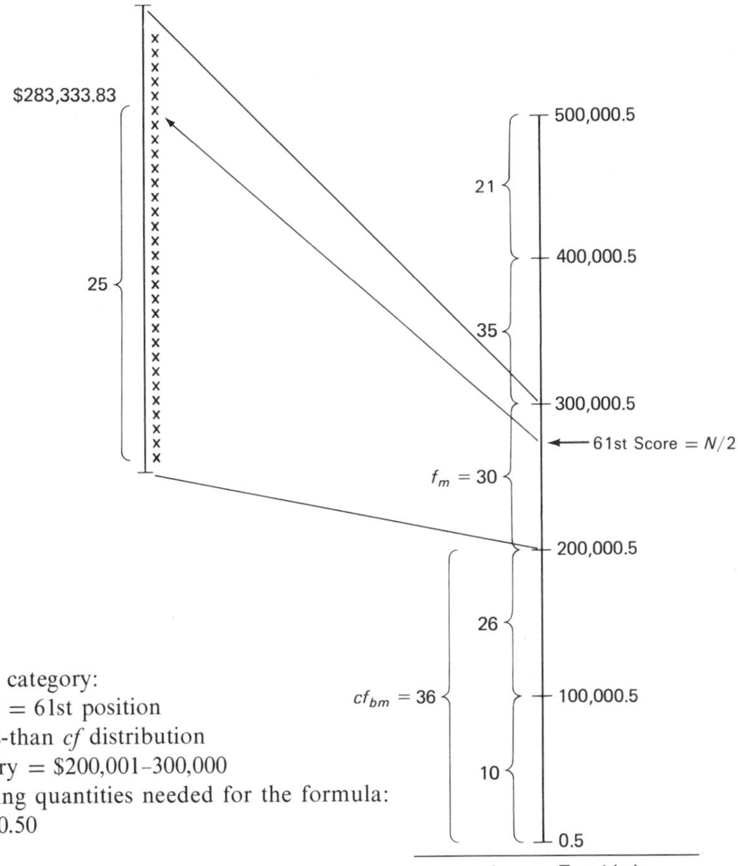

1. Find the median category:
   $(N/2) = 122/2 = 61\text{st}$ position
   calculate a less-than cf distribution
   median category = \$200,001–300,000
2. Find the remaining quantities needed for the formula:
   $l_m = \$200{,}000.50$
   $cf_{bm} = 36$
   $f_m = 30$
   $w_m = \$100{,}000$
3. Do the computations in the formula:

$$\widehat{\text{Mdn}} = \$200{,}000.50 + \left[\frac{61 - 36}{30}\right](\$100{,}000)$$

$$= \$200{,}000.50 + \frac{25}{30}(\$100{,}000)$$

$$= \$200{,}000.50 + \$83{,}333.33$$

$$= \$283{,}333.83$$

Let us go immediately to an example to see the formula used. Table 3.7 illustrates the process for a grouped distribution of stockholdings by members of the Fat Cats Club. The first concern is to identify the median category. This is done by finding the position of the median using the formula of $N/2$ rather than the $(N + 1)/2$ formula we used with raw data and simple frequency distributions. The new position formula is used because the overall formula for estimating the median assumes continuity of the distribution which is comparable to the $(N + 1)/2$ calculation for ungrouped data. In this example we see that the 61st fat cat (122/2) would be the middlemost person in an ordered distribution of the stockholders. Next, the less-than cumulative frequency distribution is calculated. From it we see that 66 is the *cf* entry closest to the number 61 without being under it. Hence the 37th through the 66th scores (including the 61st) are in the $200,001–300,000 category. We then proceed to determine the remaining quantities needed for the median formula as the work below the table demonstrates. From there we make the appropriate substitutions into the formula and perform the computations. The estimated median is a score of $283,333.83 in this example.

One can better understand how the formula works from the diagram below Table 3.7. It shows that 36 scores are known to be below the median category. In consequence, the median must be at least $200,000.50 but less than $300,000.50 (the true lower and upper limits, respectively). The 30 scores in the median category are assumed to be evenly spread across its $100,000 width. The formula determines what 25/30th of that category would represent in dollars since we need 25 of those 30 scores to reach the 61st score. That amount is $83,333.33. And when added to the $200,000.50 we have the estimated value of the median. To the extent that those 30 scores in the $200,001–300,000 category are actually spread evenly across it, the estimated median is accurate. As an approximation the estimate can be used with cautious confidence.

Mean ($\hat{\bar{X}}$)

The mean, too, can be estimated from a grouped frequency distribution. The estimating procedure again makes some assumptions about that distribution. The data are assumed to be at the interval level of measurement. It is acceptable to approximate this with ordinal measurement using numerical categories. The data are assumed to be continuous. Generally, this is approached whenever N is large. The midpoint of each category is assumed to represent the actual scores in that category. This will be true when the actual scores are distributed evenly throughout the category.

The formula used to estimate the mean from grouped data is very similar to the one used to find the mean from simple frequency distributions. It is

$$\hat{\bar{X}} = \frac{\Sigma fm}{N}$$ Estimated Mean Formula for Grouped Frequency Distributions

where $\hat{\bar{X}}$ = estimated mean
Σ = summation process
fm = product of each category's midpoint (m) and its frequency (f)
N = number of cases, found by Σf

TABLE 3.8 Estimating the Mean from a Grouped Frequency Distribution

Annual "Sick Day" Absences at Explosives Unlimited

Number of Sick Days	f	(1) m	(2) fm
40–49	2	44.5	89.0
30–39	3	34.5	103.5
20–29	10	24.5	245.0
10–19	7	14.5	101.5
0–9	4	4.5	18.0
Total	26		557.0

1. Find the category midpoints.
2. Find the fm products.
3. Estimate the mean with the formula

$$\hat{X} = \frac{\Sigma fm}{N} = \frac{557}{26} = 21.4 \text{ days}$$

The formula uses the fm products as an approximation of the fX products that were calculated in the simple frequency formula. The exact score values (X) are not known in a grouped frequency distribution. So the midpoint of each category is used as the best approximation of what the exact scores might have averaged.

An example of using the formula is shown in Table 3.8 for a grouped distribution of "sick day" absences at Explosives Unlimited. First the category midpoints are determined (m). Then each midpoint is multiplied by its category frequency (fm). These products are added together to estimate the sum of the scores (Σfm). This product is divided by the number of scores, as usual. Here the estimated mean is 21.4 days. We can imagine various raw score distributions that could have resulted in Table 3.8. It is not difficult to think of many distributions which would have exact means close to our estimated mean. Only very discontinuous raw score distributions would have had exact means which were obviously at an extreme difference from the estimated mean.

Even in the best situations, the estimated mean will probably differ some from the exact mean. The problem for the reader as well as the researcher is that one cannot determine how accurate the estimate is unless the raw scores themselves are available. It is not a good practice to impute precision to the estimated mean unless the distribution is very large and obviously continuous.

Keys to Understanding

Let us emphasize the major points about understanding central tendency statistics. It is important to know how statistics are calculated if you are going to evaluate them. This knowledge helps to de-mystify the statistics you see used. The mechanics of calculation reveal some of the assumptions that the researcher is making. That gets us back to the question of why there are different measures of central tendency. Different measures force the researcher to make different assumptions about the data and they also reflect different purposes for the analysis.

Earlier in this chapter we looked at two research examples in which different measures were used. In R.E. 3.1 the author used the median. In R.E. 3.2 the author used the mean. Why were those decisions made? How do we know whether they were appropriate decisions? There are three keys to understanding why certain measures of central tendency are used over others. Ask yourself:

1. What was the level of measurement for the variable being described?
2. What was the form of the variable's distribution?
3. What were the purposes of the research?

Level of Measurement

Key 3.1. Determine the level of measurement. Each of the measures of central tendency requires a certain minimal level of measurement in a variable before it can be calculated. This was mentioned earlier as each measure was explained. To summarize, the mode needs only nominal measurement, so any of the four levels will be sufficient. The median needs ordered data, so anything at the ordinal level or above will do. The mean requires one to add and divide scores, so the measurement must be either interval or ratio.

If no other criteria were to be considered, the level of measurement would prevent a researcher from finding means on anything except interval or ratio variables. Medians could only be found for ordinal, interval, or ratio variables. But the mode could always be found, as long as the variable met the minimal standards of nominal measurement.

The authors in both R.E. 3.1 and 3.2 seem to be justified in their choices of central tendency. IQ scores as determined by standardized tests are generally agreed to represent at least ordinal measurement. Some would attribute even interval quality to such data. In R.E. 3.2, percentage unemployment rates (like all rates) are ratio measurements. Both researcher's appear to be using measures appropriate to the level of measurement.

Form of the Distribution

Key 3.2. Determine the form of the distribution. There are two features of a distribution's form that affect the selection and interpretation of a measure of central tendency. The first is modality and the second is skew. Recall that modality refers to the number of modes in a distribution. Unless told otherwise, most people assume distributions are unimodal. If a variable is distributed in any way other than unimodal, the researcher should communicate this. Unfortunately, this is not always practiced. As a consumer of statistics you must watch for this possibility. Look at whatever information is given about the data's distribution to see whether it can be assumed to be essentially unimodal. Remain cautious in your assessment of the data's central tendency unless you know its modality. Research Examples 3.1 and 3.2 are like most because they give us very little information on which to judge modality. We are forced to assume it. But we can do so consciously rather than unthinkingly.

Skew is the second feature of a distribution's shape that interests us. Recall that skew is the lack of symmetry in a distribution. Examine the examples in Figure 3.3. The first (a) is symmetrical; it is not skewed. In unimodal, symmetrical distributions like this one, the mode, median, and mean all equal the same value. The other examples illustrated show

FIGURE 3.3 Skew and the Relative Positions of the Mode, Median, and Mean

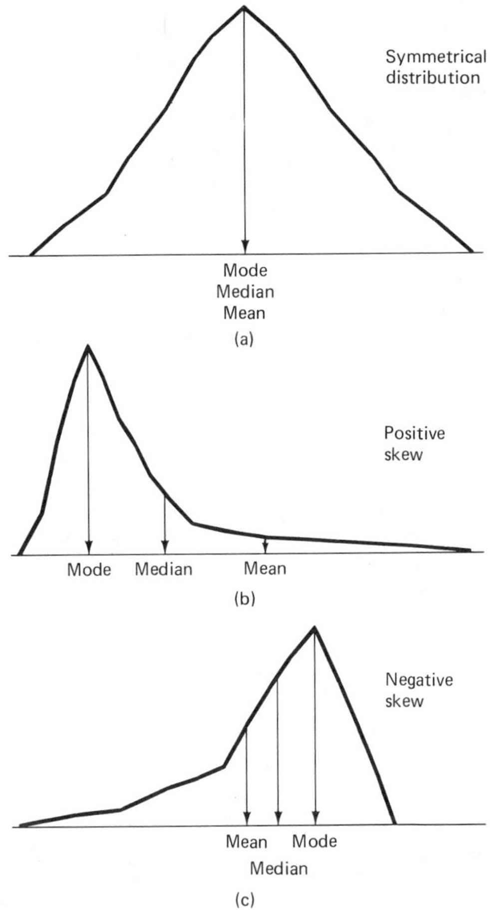

positive (b) and negative skew (c). As discussed in Chapter 2, a distribution with positive skew has a few scores that are much higher in value than are the majority. It is skewed toward the positive end of the horizontal axis, along which the score values are arranged. Notice that the median is now higher in value than the mode and that the mean is now higher than both the median and the mode. In negative skew the distribution has a few scores that are much lower in value than are the majority. It is skewed toward the negative end of the horizontal axis. Now the median is lower in value than the mode and the mean is lower than both of the other measures. This pattern in skewed distributions is characteristic as long as they are unimodal and approximately continuous.

We can demonstrate these patterns by taking a simple example. Suppose that an elementary school teacher promises her pupils that they can go on a field trip to the ice cream factory if they do well on an upcoming geography quiz. The quiz is given with the following results (arranged in an ordered array):

$$8 \quad 7 \quad 6 \quad 5 \quad 5 \quad 4 \quad 3 \quad 2$$

The mode, median, and mean are equivalent at the value 5. The teacher tells her pupils that she is in doubt about whether this performance warrants the field trip. The students pressure her to postpone judgment until two classmates return from illness. She agrees, and they return to score 20 and 19 on the quiz. In the resulting distribution (20, 19, 8, 7, 6, 5, 5, 4, 3, 2) the mode is still 5, but the median has moved up slightly to 5.5, and the mean has increased to 7.9. Despite the teacher's suspicion of possible collusion between the students, the field trip is taken.

The two distributions demonstrate the relative effects of skew on the three measures. The mode is not influenced, the median is somewhat influenced, and the mean is most influenced. Statisticians often describe this pattern as one in which the mean is most "sensitive" to extreme scores. Indeed, it does react to skew more than do the other two measures.

Because of the mean's sensitivity, the median is the most preferred measure of central tendency for extremely skewed distributions. It does not ignore the skew as the mode does, nor does it "exaggerate" the skew like the mean. It is inappropriate for a researcher to describe a highly skewed distribution without the median. Look for skew when evaluating measures of central tendency.

Purposes of the Research

Key 3.3. Determine the purposes of the research. The final key to understanding why researchers choose different measures of central tendency is to ask what their purposes were in doing the research. Let us consider four major types of purposes that might be operating.

First, the researcher may be comparing one set of data to another. Like most features of such a research project, the selection of a central tendency measure would be determined by the data with which one wanted to compare. If the previous research used medians, the present research must use the median. We cannot compare a mode to a median, or compare any two different measures to each other. They do not measure the same feature of central tendency. Hence we may find a researcher using a measure that seems less than optimum until we realize that we are looking at an example of comparative research.

Second, the researcher may be influenced by some rather ignoble motives. Because few actual distributions are perfectly symmetrical and unimodal, the three measures will yield different answers. If one has a vested interest in the outcome of the analysis (and it is difficult ever to be completely disinterested), a strategic selection of one measure over another might create a more desired impression of central tendency.

For example, if a distribution of annual salaries in a company is extremely positively skewed, a management representative may choose to summarize its central tendency with the mean since it gives the highest value. To counter, a labor representative may use the mode since it would provide the lowest value. Advertising is another area where such infamous selection of the most "useful" measure occurs. Even in most legitimate social research, there are subtle pressures to produce information that pleases someone.

When you evaluate central tendency statistics, search for such possible sources of influence on the researcher. Some researchers will state their personal feelings so that you can evaluate their work with that background information. Most will not be so open.

Third, a researcher may select a particular measure simply because of the type of descriptive summary that is preferred. A researcher may be interested in finding the one kind of situation that happens more often than any other and is therefore most likely to recur in the future. This researcher would choose the mode. Another may want to know when a situation is at the halfway point. This may have a policy implication about when an intervening strategy should be implemented, for instance. The median would be the appropriate choice for this purpose. The mean would be used when one wanted to give complete numerical weight to all the scores in a distribution regardless of how extreme any of them might be.

Fourth, the researcher may be looking ahead to some inferential statistics and generalizations that are desired. Any detailed discussion of this is getting ahead of ourselves at this point. But to state it simply, we need to be concerned with the relative stability of the measures when doing inferential statistics. If used over and over again with different samples drawn from the same population, which measure would tend to give the most reliable information? It is the mean. Thus the mean might be used for some descriptive purposes at one point in a research project because it is also going to be used later for inference purposes.

This discussion about evaluating the measures of central tendency may have caused you to think in terms of selecting the *one* best measure for particular situations. That would be a mistake, generally. There is no rule in social statistics that requires you to choose only one measure. In fact, most situations can be more accurately described by using two or three measures to summarize a distribution's central tendency.

Research Examples 3.3 and 3.4 illustrate just such multiple selections of measures. In R.E. 3.3 the author uses both the median and the mean to describe the duration of breast-feeding in Asian countries. As can be determined from the reprinted statistics, the distributions for many places were skewed, generally in a negative direction. Several of the medians and means are so different that we would have gotten a much different impression of the distributions had only one measure been reported. The median is an especially good choice in this example because the data were collected by many people and agencies without the uniform control of one researcher. Since the median is less sensitive to the exact score values than is the mean, the median will not be as affected by data that have less than the highest reliability.

In R.E. 3.4 the author has examined information on young city planners. Notice that he uses all three measures of central tendency in both the text and table. "The mean age

RESEARCH EXAMPLE 3.3 Using Two Measures of Central Tendency

From Louise Williams, "WFS Reports Confirm Changes in Asian Breastfeeding Patterns," Intercom, *Population Reference Bureau, 7 (4), April 1979:7ff. Reprinted by permission.*

Results of World Fertility Surveys in Asia now published support mounting evidence that women in developing countries who have more education or who live in urban environments are less likely to breastfeed. The change in breastfeeding practices is manifested in two ways: not only are fewer women choosing to breastfeed, but those who do breastfeed do so for shorter periods. This disturbing evidence comes at a time when more and more of the beneficial effects of breastfeeding are being spelled out. In 1979, the International Year of the Child, the importance of breastfeeding cannot be exaggerated since it is so highly correlated with the enhancement of the overall welfare of young children. . . .

The tables which appear in the WFS First Country Reports present breastfeeding duration only for the penultimate child. This has been done because women will have finished breastfeeding the next-to-last child at the date of interview so all breastfeeding durations are completed durations. [see the table]. . . .

Breastfeeding Duration by Place[a]

Country	Duration of Breastfeeding (Months) Median	Mean
Bangladesh	24	19.0
Fiji	—	8.4
Indonesia	24	16.7
Korea, South	23	15.6
Malaysia	4	5.3
Nepal	25+	18.6
Pakistan	17	16.5
Sri Lanka	24	14.2
Thailand	24	11.1

[a] Adapted from Table 1 in the original.

Mortality is much higher among bottle-fed infants. It has been estimated that the present risk of mortality to bottle-fed infants in developing countries is about the same as it was 50 to 100 years ago in the developed countries. Formerly, in the now developed countries, the risk of mortality for bottle-fed compared with breast-fed infants was between 2 to 1 and 55 to 1. Along with higher mortality rates, higher morbidity rates are associated with infants who are not breastfed. Health care facilities in developing countries generally are inadequate and under-staffed; thus, bottle-fed infants who require additional health care represent a high and unnecessary cost both financially and in terms of added human suffering. . . .

RESEARCH EXAMPLE 3.4 Using Three Measures of Central Tendency

From Robert J. S. Ross, "The Impact of Social Movements on a Profession in Process," Sociology of Work and Occupations, *3 (4), November 1976:437ff. Reprinted by permission.*

. . . Who are the men and women who have chosen to work with clients in ways new to the city planning process? The first set of facts about them is relatively straightforward: they are younger people who are largely products of mid- to late-sixties graduate schools.

The mean age ($N = 106$) of the people interviewed was 33 years, 50% were below 31 years old, and the modal group (15 persons) was 28 years of age.

The planners graduated college, largely, in the mid-sixties. 46% of them left college between 1963 and 1967, the median year is 1964, the mode 1966, and the average 1962 ($N = 109$).

Of the 112 respondents for whom we have this personal data, nine had four years (7) or less (2) of college, ten had one year of graduate work, 55 had two or three years of graduate work, and 38 had four to six years of graduate work.

This produces a pattern of leaving graduate school in the late sixties. For the 101 planners from whom we obtained this data, the median year of leaving graduate school was 1967, the mode (15 persons) was 1971, and the average was 1966.

This pattern sustains an important part of the interpretation offered of the new role in regard to the impact of the New Left on advocate planners: the great bulk of those interviewed were "on campus" at the height of the period when the "Movement" was campus-based and highly involved in issues of race and poverty. The youthfulness of the group and its recent completion of graduate work also supports the general proposition which expects new professional segments, when they are based in social movements, to recruit from among new practitioners. . . .

The respondents were asked: "Do you feel part of any movement(s) for social change?" Eighty-six percent ($N = 94$) responded positively. Then, in the same self-administered section of the schedule, the planners were asked about a series of potentially influential social movement and professional organizations. [The table] presents the question and the results for the sample as a whole.

Question: Thinking back on the period of approximately 1963–1969 here are some organizations and movements which might have influenced you. Would you rate the influence of each on your activity and/or thinking from 1, no influence at all, to 7, an extremely important influence.

Organizations Influential with Advocate Planners[a]

Organization (Ranked in Descending Order of Average Influence)	Average Rating (Range 1-7)	Mode (%)	Median
Welfare Rights Organization (WRO)	4.1	5 (25)	4
Student Non-Violent Coordinating Committee (SNCC)	3.9	4 (21)	4
Southern Christian Leadership Conference (SCLC)	3.8	4 (20)	4
Students for a Democratic Society (SDS)	3.6	1 (22)	4
Congress on Racial Equality (CORE)	3.6	5 (19)	4
Planners for Equal Opportunity (PEO)	3.2	1 (32)	3
National Association for the Advancement of Colored People (NAACP)	2.9	2 (24)	3
Urban League	2.7	1 (36)	2
Urban Coalition	2.6	1 (42)	2
Americans for Democratic Action (ADA)	2.5	1 (41)	2
American Institute of Planners (AIP)	2.2	1 (45)	2
Young Americans for Freedom (YAF)	1.4	1 (86)	1

[a]Table 3 in the original.

A number of interesting patterns appear. First, the professional organization of planners, the AIP, is the lowest of all of the genuine options—only Young Americans for Freedom, a rightist group formed in 1960 under the aegis of publicist William F. Buckley, is lower than the AIP in influencing the planners. Second, the advocate-oriented PEO, partly professional and partly an expression of social movement concerns, is in the middle of the influence cluster. The most influential organizations, on the average, are the more militant action-oriented ones.

Finally, SDS, the most characteristic New Left organization, is in the upper cluster of influencers; but its sources indicate the polarities among the respondents: the difference between its modal response of "no influence" and its relatively high median suggests the contrasting experience and orientation among the advocates.

These five relatively militant and radical groups [Welfare Rights Organization, Student Non-Violent Coordinating Committee, Southern Christian Leadership Conference, Students for a Democratic Society, and Congress on Racial Equality] were rated in the highest four positions in influencing the advocates. Despite the fact that the absolute averages of the ratings are not strikingly high, their relative position, especially when constrasted to the AIP and PEO, suggests the validity of seeing advocate planners as influenced by, and part of, social movement activity originating outside the planning profession itself. Nevertheless, this overview is only part of the picture, for the planners were divided in the ratings which produced these averages according to their choices among professional or radical outlooks. While it is logically possible that there were radicals who took up nonadvocate roles and who would also have attributed influence to militant groups, it is concretely demonstrable that radical identity had a coherent impact on the advocate's reports of organizational influence. . . .

. . . was 33 years, 50% were below 31 years old [the median], and the modal group . . . was 28 years of age." Assuming this age distribution was continuous, would you say it was skewed positively or negatively? Answer: Positively. Read the entire example closely for other illustrations of the way the multiple selection of measures allows you to get a more complete impression of how the variables were distributed. Like R.E. 3.2, this author uses the term "average" at times to refer to the mean. Although it is an unnecessarily ambiguous term for either author to use, with careful reading one can determine that they are referring to the arithmetic mean.

You should practice using all three keys to understanding central tendency statistics on the four research examples included in this chapter. Ask yourself about the level of measurement, the form of the distribution, and the purposes of the research and researcher. See if you understand the selection of measures and whether you agree with those selections. If not, be careful to distinguish between what you believe are mistakes and what needs more information for a confident evaluation. The latter problem is much more likely given the limited material that has been reprinted here for each example.

KEY TERMS

Central Tendency	Multimodal	Mean ($\bar{X}$)
Mode (*Mo*)	Major Mode	Weighted Mean ($\bar{X}_w$)
Unimodal	Minor Mode	Deviation Score (x)
Bimodal	Median (*Mdn*)	*Estimated Value ($\widehat{Mo}$, $\widehat{Mdn}$, $\hat{\bar{X}}$)

EXERCISES

1. The following scores represent the percent of time some social workers spent in direct contact with clients during a one-week period: 19, 10, 55, 31, 22, 9, 18, 12, 27, 14, 20, 15, 19, 17, 22, 40, 19, 33, 5. Determine the (a) mode, (b) median, and (c) mean percentage of time spent in direct client contact.
2. The following scores represent the number of errors made by 22 subjects in an experiment when asked to spell some often misspelled words: 5, 9, 0, 7, 2, 3, 5, 6, 3, 1, 4, 5, 4, 5, 2, 1, 8, 6, 5, 4, 7, 5. Determine the (a) mode, (b) median, and (c) mean number of errors.
3. The following table shows the frequency distribution of the number of rooms in all year-round occupied housing units in the United States rounded to the nearest 1000. Calculate the (a) mode and (b) median number of rooms.

Number of Rooms in Year-Round Occupied Housing Units (thousands), United States, 1970

Number of Rooms	f
9 or more	2,191
8	3,269
7	6,376
6	13,482
5	16,874
4	14,131
3	7,570
2	2,459
1	1,306
All units	67,658

Source: Bureau of the Census, *Housing Characteristics for States, Cities, and Counties, 1972, United States Summary*. Washington, D.C.: U.S. Government Printing Office, 1972, 1:22, Table 4.

*4. Consider the following distributions of business establishments within two communities.

Distributions of Businesses by Employment Size

Employment Size	Community A	Community B
20 or more paid employees	25	8
8–19 paid employees	51	20
4–7 paid employees	117	85
1–3 paid employees	203	52
No paid employees	44	35
All businesses	440	200

(a) Find the estimated modal employment size for both communities and interpret each estimate.
(b) Find the estimated median employment size for both distributions and interpret each estimate.

(c) Explain why the estimated mean employment size cannot be calculated for either distribution in its present form.
(d) Assume that the highest employment size category ended at 50 paid employees and then calculate the estimated mean for community A.
(e) Assume that the highest employment size category ended at 500 paid employees and then calculate the estimated mean for community A.
(f) Compare your answers in parts (d) and (e) and comment on their difference in relation to the mean's sensitivity to extreme scores.

*5. In the following table the value of all owner-occupied housing units in the United States is displayed. Find the (a) estimated mode and (b) estimated median. (c) What affect do you think the unequal width categories might have on the estimated mode? On the estimated median?

Value of Owner-Occupied Housing Units (thousands), United States, 1970

Value	*f*
Less than $5000	1,934
$5000–7499	2,293
$7500–9999	2,673
$10,000–12,499	3,309
$12,500–14,999	3,093
$15,000–17,499	3,319
$17,500–19,999	3,116
$20,000–24,999	4,674
$25,000–34,999	4,436
$35,000–49,999	2,045
$50,000 or more	996
Total	31,888

Source: Bureau of the Census, *Housing Characteristics for States, Cities, and Counties, 1972, United States Summary.* Washington, D.C.: U.S. Government Printing Office, 1972, 1:28, Table 5.

6. Consider a distribution in which religious affiliation was measured using the four categories Protestant, Catholic, Jewish, and other. Suppose that this distribution's mode was reported to be Protestant. Which of the following would you know to be a true conclusion about that distribution? Explain why each of the other conclusions would not be known to be true.
(a) Most people in this distribution were Protestants.
(b) More people were Protestants than all other categories combined.
(c) More people were Protestants than any other category.
(d) Protestants are in the middle of the distribution.

7. Consider a distribution in which social class has been measured using the categories of lower class, middle class, and upper class. Suppose that this distribution's median was reported to be middle class. Which of the following would you know to be a true conclusion about that distribution? Explain why each of the other conclusions would not be known to be true.
(a) Most people were in the middle class.
(b) The middlemost score was a score of middle class.
(c) Half of the people were in social classes below the middle class.
(d) More people were in the middle class than any of the other social classes.

8. Consider a distribution in which family size has been measured to the exact whole number. Suppose that this distribution's mean was reported to be 4.5. Which of the following would you know to be a true conclusion about that distribution? Explain why each of the other conclusions would not be known to be true.

(a) The value 4.5 divides the distribution into two equal halves.
(b) The sum of all family sizes divided by the number of families was 4.5.
(c) Most families had either 4 or 5 members.
(d) The sum of the deviation scores below 4.5 would equal the sum of the deviation scores above 4.5.

9. The distribution of family income in most countries, including the United States, is unimodal with extreme positive skew. A relatively few have very high incomes. The U.S. Bureau of the Census summarizes that distribution using the median. Evaluate this practice considering the shape of the distribution, the level of measurement, and the research purpose.

10. Grades are typically summarized with the mean (the grade-point average). Evaluate this practice both as it applies to individuals and to student bodies as a whole.

11. What would be the advantages and disadvantages to designing an antipoverty program around the modal poverty family income level compared with the median or mean poverty level?

12. Find two research articles in social science journals that use any of the measures of central tendency discussed in this chapter. Explain, interpret, and evaluate the measures as they are used.

RELATED STATISTICS

Midrange

The midrange is an average of the highest and lowest scores in a distribution. It identifies the point that is exactly halfway between the two most extreme scores:

$$\text{midrange} = \frac{\text{highest score} + \text{lowest score}}{2}$$

See: Johnson (1980:25).

Percentile

The median can be described as the 50th percentile; the point that is closest to having exactly 50% of the scores below it. Other percentiles can be found by the general formula

$$p\text{th percentile} = l_p + \left[\frac{pN/100 - cf_{bp}}{f_p}\right] w_p$$

where p = desired percentile
l_p = true lower limit of the category containing the desired percentile
N = number of cases
cf_{bp} = less-than cumulative frequency immediately below the pth percentile category
f_p = frequency within the pth percentile category
w_p = width of the pth percentile category

For example, the 95th percentile would be found by the formula

$$95\text{th percentile} = l_{95} + \left[\frac{95N/100 - cf}{f}\right] w$$

The calculated answer would indicate the value below which 95% of the scores occurred. See: Neter et al. (1978:55).

Moving Average

The usual measures of central tendency (mode, median, and mean) either ignore or destroy the time sequence of the scores in a time-series distribution. One measure that preserves time sequence while still indexing central tendency is the moving average. It also has the benefit of smoothing fluctuations in time-series data. A suitable time interval is selected, such as 5- or 10-year periods, and scores for each successive and overlapping period are used to calculate the mean for that period. For example, if 5-year periods are selected, a series of means is calculated for the first 5 years, the second through the sixth years, the third for the seventh years, and so on. The technique is useful to reveal overall trends which might be disguised when the data have large fluctuations across single time points in the series.

See: Neter et al. (1978:606–608).

Geometric Mean

Some distributions that researchers analyze consist of ratios between scores rather the scores themselves. For instance, you might be working with the ratio of clients at the beginning of one period compared to the second period, the ratio of the second to the third, and so on. A series of ratios calculated at various points across time may lend themselves to the geometric mean as a measure of central tendency. It uses the formula

$$G = (R_1 \times R_2 \times \cdots \times R_n)^{1/N}$$

where R_1 = first ratio

See: Neter et al. (1978:56–57).

Harmonic Mean

Other special distributions may consist of rate of change scores. The time it takes an interviewer to complete a series of interviews could be such an example. The first interview may take much longer than the later ones because the interviewer learns to be more efficient. Rates of change from the first interview to the second, from the second to the third, and so on, could be the kind of data suitable for the harmonic mean. It uses the formula

$$H = \frac{N}{1/X_1 + 1/X_2 + \cdots + 1/X_n}$$

where X_1 = first rate of change

See: Neter et al. (1978:57–58).

CHAPTER 4

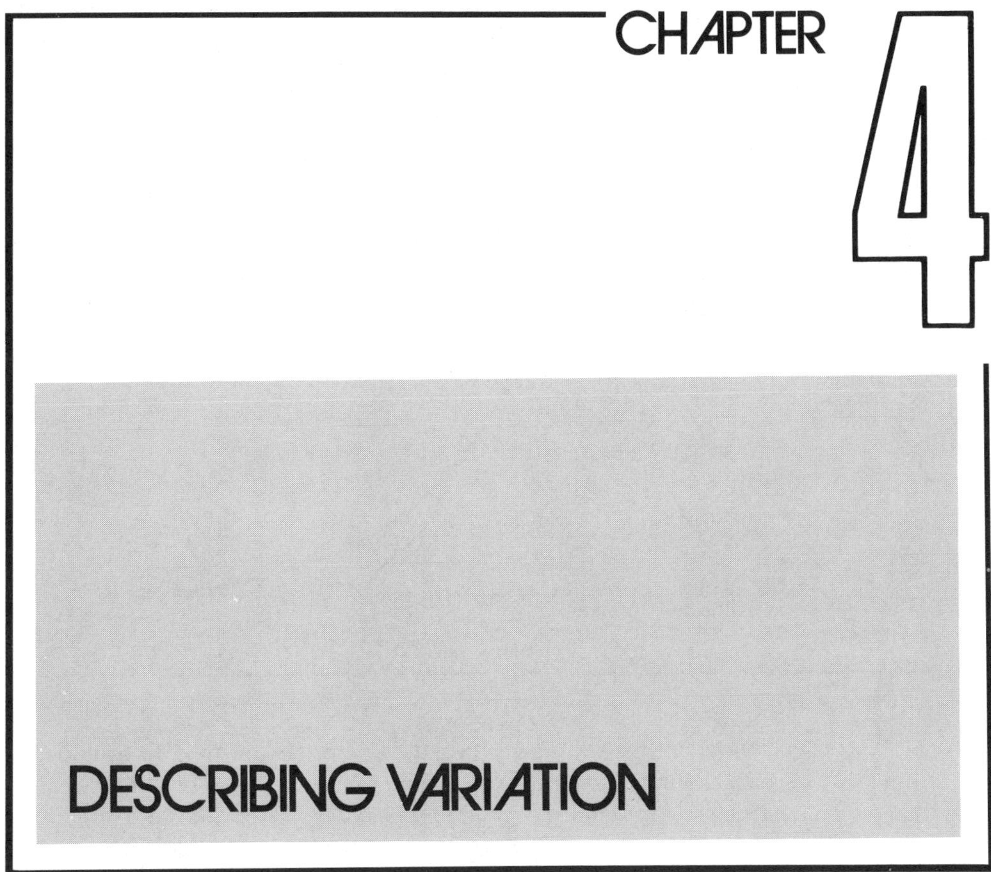

DESCRIBING VARIATION

Variation is crucial to all data analysis. If there is one idea that is more important than all the others, this is it. *Variation* is the extent and manner in which the scores in a distribution differ from each other. Reflect a moment on the questions that people commonly ask about the social world. "Why is there such a large difference between the richest and poorest persons in our society?" "Why do some students succeed in school and others fail?" "How many different occupational patterns do people follow?" "Why doesn't everyone vote?" "Why does it seem that there are so many kinds of deviance around today?"

All of these are questions about the differences in the ways that people behave and think. If there were no variations, there would be little to interest us, little to wonder about or to study. The main concern of most social research is to describe and explain the differences in social behavior. As we saw in Chapter 3, statistics of central tendency tell us about the ways in which scores in a distribution are clustered. Statistics of variation tell us about the dispersion or scattering of those scores.

The Meanings of Variation

Distributions may be similar in central tendency but different in variation. Comparing them only by their central tendencies is inadequate. Two construction workers may have the same mean arrival time at the job site. But Joe may vary only 2 to 3 minutes from the starting time each day while Charlie comes as much as an hour late some days and an

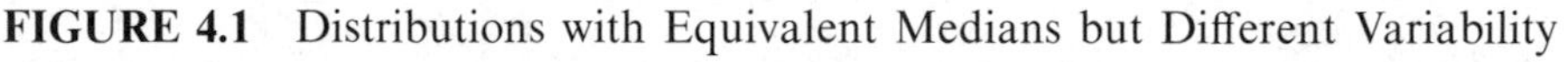

FIGURE 4.1 Distributions with Equivalent Medians but Different Variability

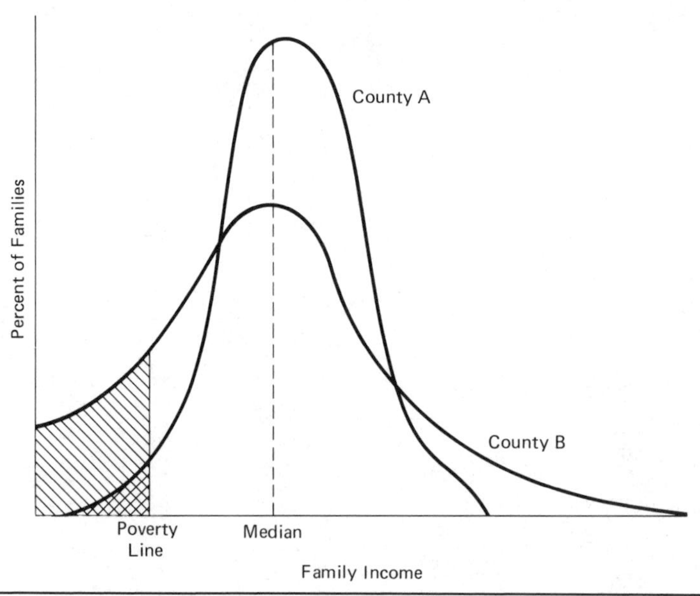

hour early on others. Only Charlie will be needing the union steward to keep from being fired. Similarly, two counties may have the same median family income but that does not mean that they have the same percentage of families below the poverty line and in need of assistance. As Figure 4.1 illustrates, county B varies more than county A in family income and it will have more families in poverty even though both have the same median.

These two quick examples demonstrate that variation can be useful to describe and compare distributions. Another use of measures of variation is to draw inferences about populations from samples. Although this chapter will concentrate on the descriptive applications of variation measures, it will also briefly point to some of their inferential uses.

Like central tendency, there are different ways in which variation is commonly measured. Look ahead now to the research examples in this chapter. In Research Example 4.1 the range is reported as a measure of variation to accompany the mean as a measure of central tendency. In this example two samples of elderly persons in New York State are being compared in age and education. The samples appear to be very similar both in central tendency and variation on these two characteristics.

In R.E. 4.2 we see a report on efforts to estimate fertility rates from worldwide pregnancy data. In the figure shown, a measure called the interquartile range is used as a measure of variation for the percentage of all pregnancies that go underreported. In this instance that measure is paired with the median as an indicator of central tendency. These two statistics are presented for three cohorts of pregnancy duration. *Cohorts* are statistical groups of persons with similar characteristics but are not necessarily social groups in which the members interact.

Research Examples 4.3 and 4.4 show the use of a measure called the standard deviation. Both examples also use the mean to describe central tendency. In R.E. 4.3 the researchers are examining the willingness of the public to use psychiatric social work services. As a part of this research the authors report the mean and standard deviations for prestige ratings given to various occupations by a sample of adults in New Orleans. The standard deviations appear to be very similar and thus to suggest that the variation in the ratings is rather consistent. Later you will discover that this is not necessarily so.

Research Example 4.4 concerns a study of race differences in sentences given to criminals in Florida. In the description of the cases (criminals), the authors report the means and standard deviations for several different characteristics. Many of the standard deviations for the characteristics are different from each other, but that may not mean that the extent of variation really differs. Research Examples 4.3 and 4.4 illustrate one of the limitations in comparability for some measures of variation that we will be discussing.

Research Example 4.5 presents two measures of variation: the standard deviation and the coefficient of variation. Like R.E. 4.4, this example represents an issue in criminology. The authors are looking at reliability problems associated with official crime reports using data from Chicago. Again, the mean is the indicator of central tendency. Together, the three statistics allow a meaningful comparison in the reporting records between the several types of crime being examined.

These research examples demonstrate some of the variety and multiple uses of measures of variation in social research. There are many measures in use but we will examine only the major ones here. As a reader of research reports you should be asking the same type of questions that we asked in Chapter 3. Why do the researchers use the particular kinds of variation measures that we see reported? What is the meaning of each measure? What are the advantages and limitations of each measure? This time we will organize our discussion according to the level of measurement of the variable that is being described. We start with nominal variables and move through ordinal, interval, and ratio variables in turn. Most of the examples use raw data distributions. The formulas to be used with either simple or grouped frequency distributions that are not illustrated for each of the variation measures are shown in the section *Related Statistics* at the end of the chapter. Therefore, we will be concentrating more on the meaning of the measures than on their computations in all possible data situations.

A Nominal-Level Measure—The Index of Dispersion (*D*)

There are only a very few measures of variation in wide use for data at the nominal level of measurement. One of these is the *index of dispersion* (D). For any nominal variable the D statistic has a minimum value of 0, meaning that the distribution has no variation, and a maximum of 1.0, meaning that it has the greatest possible variation. Usually, the statistic is some value between 0 and 1 which indicates the extent to which the scores are dispersed between the minimum and maximum extremes. Its formula is

$$D = \frac{k(N^2 - \Sigma f^2)}{N^2(k - 1)} \qquad \text{Index of Dispersion Formula}$$

where D = index of dispersion
k = number of categories in the variable
N = number of cases
f^2 = squared frequency within a category
Σf^2 = sum of the squared frequencies for all categories

TABLE 4.1 Finding the Index of Dispersion for Nominal Data

Religious Orientation	Group 1		Group 2		Minimum		Maximum	
	f	f^2	f	f^2	f	f^2	f	f^2
Western	8	64	3	9	12	144	3	9
Eastern	3	9	4	16	0	0	3	9
Other	1	1	2	4	0	0	3	9
None	0	0	3	9	0	0	3	9
Total	12	74	12	38	12	144	12	36

$$D = \frac{k(N^2 - \Sigma f^2)}{N^2(k - 1)}$$

Group 1:

$$D = \frac{4(144 - 74)}{144(3)} = \frac{4(70)}{432} = \frac{280}{432} = .65$$

Group 2:

$$D = \frac{4(144 - 38)}{144(3)} = \frac{4(106)}{432} = \frac{424}{432} = .98$$

Minimum Dispersion:

$$D = \frac{4(144 - 144)}{144(3)} = \frac{4(0)}{144(3)} = \frac{0}{432} = 0$$

Maximum Dispersion:

$$D = \frac{4(144 - 36)}{144(3)} = \frac{4(108)}{144(3)} = \frac{432}{432} = 1.0$$

Consider the example shown in Table 4.1. At the left it displays the distributions in religious orientation for two groups. We might think of the first group as representing middle-class urban adults and the second as representing middle-class university students. Glancing at the frequencies in the two distributions you will notice that the first group is somewhat more homogeneous (alike) in religious orientation than the second. But exactly how much less variation does the one have compared to the other?

Working through the D statistic formula we find that $k = 4$ (there are four categories of religious orientation) and $N = 12$ (there are 12 individuals) for each group. To find the Σf^2 quantity notice that each separate category's frequency is first squared and then these values are added for each distribution. This produces the ratio of 280/432 and a D value of .65 for the first group. It produces the ratio of 424/432 and a D value of .98 for the second group. We can now conclude that the university students vary considerably more in religious orientation than do the urban adults. In fact, the students are near the maximum possible variation given this measurement of religious orientation. On the other hand, the urban adults are only dispersed 65% of the maximum possible across the four categories.

Although this explanation is sufficient for a workable interpretation of the D statistic, a little more information will make it even more useful. Suppose that we focus our attention on the two numbers that formed the ratios preceding the final value of D for each group as calculated in Table 4.1. What do you suppose they represent? If they are reduced to simplest terms, the numerator indicates the number of unique (unlike) pairs of scores that could be created from the 12 cases as they are actually dispersed across the four categories of measurement. The simplified denominator indicates the number of unique pairs of scores that could be created for the 12 scores if they had been spread to the maximum extent across the four categories. Let us see what this means.

The right side of Table 4.1 shows how the distributions could have been dispersed to the minimum and maximum possible degrees. In the minimally dispersed distribution the scores have no spread; they are all in one category. In the maximally dispersed distribution the scores are spread as much as possible; there is the same number in each category.

Pairs of unique scores are created by linking each case in a category with each case in a different category. Thus, for the maximum distribution each of the three "Western" scores can be paired with each of the nine scores in the other three categories to create $3 \times 9 = 27$ pairs. Each of the three "Eastern" scores can be paired with the six in the "other" and "none" categories to create $3 \times 6 = 18$ pairs. Each of the three scores of "other" can be paired with the three in the "none" category to create $3 \times 3 = 9$ pairs. No other unique pairs can be created from this distribution, so the total is $27 + 18 + 9 = 54$ pairs if the 12 scores are maximally dispersed. In the minimum distribution no unique pairs can be formed because all the scores are in the same category.

The number of pairs for the group 1 and group 2 actual distributions are found as follows. For group 1 there are 35 unique pairs: $(8 \times 3) + (8 \times 1) + (3 \times 1) = 35$. For group 2 there are 53 unique pairs: $(3 \times 9) + (4 \times 5) + (2 \times 3) = 53$. The ratio shown for the computation of D for group 1 was 280/432. Reduced to its simplest terms this would be 35/54, which is identical to the numbers of pairs that we have just calculated. The 424/432 ratio for group 2 reduces to 53/54, which also matches the calculations just presented.

The technique of examining distributions on the basis of the number of unique pairs that are produced is not used only with the D statistic. It is also used with measures of association for nominal and ordinal variables as described in Chapter 5.

Dual-Level Measures for Ordinal and Interval Data—The Range-Based Statistics

There are no widely accepted measures of variation for ordinal-level variables. As a result many researchers prefer to use a nominal-level measure for those ordinal variables that have only verbal categories and to use an interval-level measure for those that have numerical categories. Several measures based on the range are designed for use with interval data, but they can also be applied to numerical ordinal data with a special restriction. If the restriction is made, only a minimum of distortion to the actual measurement quality of the ordinal variable occurs. We will examine these.

Range (*R*)

The *range* (*R*) identifies the highest and lowest scores in a distribution. Suppose that a student asks her friends how much studying they did last week. After surveying everyone she can calculate the range simply by noting the most study time and the least study time. Suppose that these were 25 hours and 1 hour. The range is 25 to 1, or 24 hours. If she continued this informal research project for a semester, she could compare the weekly ranges as a rough indication of the trend in variation that study time had for her group of friends. The same procedure can be used for any numerical ordinal or higher level variable to provide a quick measure of variation.

The range uses the simple formula

$$R = X_h - X_l \qquad \text{Range Formula}$$

where R = range
X_h = highest score in a distribution
X_l = lowest score in a distribution

Computation of the range is very easy once you have inspected a distribution. One need only exercise care in locating the two most extreme scores. Especially long raw data distributions may need to be put into an ordered array to ensure that neither the highest nor lowest score is overlooked. If the range merely notes the highest and lowest scores (e.g., 25 to 1 hours), it requires only ordinal data. If it subtracts the lowest score from the highest score (e.g., 24 hours), it requires interval data.

Let us look again at the example from Table 3.2 in Chapter 3 concerning the round-trip commuter distances in miles between one's residence and workplace. These scores were: 25, 21, 19, 17, 15, 12, 10. According to our two levels of interpretation the range is 25 miles to 10 miles, or 15 miles. The ordinal interpretation notes that the longest commuting distance was 25 miles and the shortest was 10 miles. The interval interpretation means that the persons in this distribution had commuting distances that differed by no more than 15 miles from each other.

Some researchers would report the range for this same distribution to be either 25.5 to 9.5, or 16 miles. They are considering the possibility that the highest score of 25 might be rounded down from a value of nearly 25.5. Similarly, the lowest score of 10 could have been rounded up from a value of 9.5. In other words, they are calculating the range on the basis of the true limits of the highest and lowest scores in the distribution. We call this the *true limits range* (*TLR*). It is also known as the *inclusive range.* Its formula is

$$TLR = U_h - L_l \qquad \text{True Limits Range Formula}$$

where TLR = true limits range
U_h = true upper limit of the highest score
L_l = true lower limit of the lowest score

The true limits range is often used where scores are continuous and have been rounded or combined into grouped categories. Unfortunately for the reader, it is not standard practice for writers to distinguish between these two versions of a range when one is reported. You must remind yourself to look for this possible ambiguity in the meaning of a presented range.

There are two major limitations to the range. First, unlike the index of dispersion, it cannot readily be compared between distributions for different variables. Whereas the index of dispersion always has a minimum value of 0 and a maximum value of 1.0, the range will have highest and lowest values which depend on each particular distribution. As a result we are limited to comparing distributions for the same variable. For example, we cannot compare meaningfully a range of 10 miles in commuting distances with a range of 50,000 in hometown populations. They differ too much in what statisticians call *scale.* This means that the number and values of the score categories for one variable are greatly unlike those for the other. The one with the larger scale will tend to have a larger range, regardless of its relative variation. The range can be used only to compare distributions for the same variable with the same scale, which is usually indicated by the distributions having similar means.

The second main limitation of the range as a measure of variation is a consequence of its simplicity. By focusing only on the two most extreme scores the range has the potential to seriously distort our impression of a distribution's true variation. Suppose that the distribution in our commuting distances example above had contained one additional person who commuted 100 miles. The new distribution would be: 100, 25, 21, 19, 17, 15, 12, 10. The new range would be 100 to 10, or 90 miles, not 15. If someone told us that a distribution of commuting distances had a range of 100 to 10, or 90 miles, would we be likely to imagine that it was like our new one which contained one very extreme score? Not likely. We would have expected the individual commuting distances to be more evenly spread across the 90-mile difference between the longest and shortest distances.

This illustrates how the range can be a misleading variation summary for a highly skewed distribution. For this reason it is a useful measure of variation only for those variables which have a distribution that is approximately continuous and symmetrical. For other variables the range should be supplemented with additional information about the shape and variation of the distribution. Its status has similarities to the mode and mean. Like the mode it is easy to compute but gives only an approximate summary. Like the mean it is sensitive to extreme scores. The range is a rough description of variation which must be used and interpreted with caution.

Research Example 4.1 uses the range to measure the variation in age and education for two samples of elderly persons. Because the end points of the ranges are reported as whole numbers in this example we might speculate that each of these represents the range and not the true limits range. But we could easily be wrong.

RESEARCH EXAMPLE 4.1 Using the Range as a Measure of Variation

From Emory L. Cowen, Stephen H. Davol, Gunars Reimanis, and Alfred Stiller, "The Social Desirability of Trait Terms: Two Geriatric Samples," Journal of Social Psychology, *56, (2), April 1962:217ff. Reprinted by permission.*

The importance of formal and stylistic components of behavior as determinants of verbal responses on personality assessment procedures has received increasing consideration in recent years. . . . Rapidly cumulating evidence now points to the likelihood that responses by many individuals on such measures may be governed less by the specific content of the item, and more by response characteristics such as the response set tendency, operant acquiescence, and the use of social desirability (S-D) stereotypes, each largely intrinsic to the respondent himself. . . . During the past several years, a research group at the University of Rochester has been engaged in the study of one such stylistic variable, social desirability (S-D), i.e., the S's [subject's] tendency to respond so as to put himself in the best possible light. [Others] have demonstrated a very high degree of correlation between this variable and the probability that propositions relevant to the description of one's personality will be endorsed. . . .

[O]ur initial research efforts have been guided by such questions as: What are the S-D perceptions that characterize various populations? To what extent are such perceptions constant, and in what ways do they vary across groups? . . . For the most part our work thus far has been restricted either to college students or young adults. . . . One of the present gaps in our program, is the lack of available data with respect to the S-D stereotypes on other age groups—particularly the very young, and oldsters. . . . The present study seeks, in part, to remedy this deficiency by providing basic information about the S-D stereotypes of geriatric Ss . . . .

Each of the two geriatric groups included 40 white males Ss. The institutional Ss were all drawn from the domiciliary at the V.A. Center, Bath, New York. The community sample was taken from three centers for oldsters in the Rochester area, each of which provides, either formally or informally, opportunities for varying types of social, recreational, or discussion activities. [The table] summarizes some basic descriptive characteristics of the two samples. Educationally they are quite close. . . . [A]s might be expected, these two groups are considerably less well educated than any other thus far studied. The community group averages about six years older than the domicilary group. That there are only two Ss in the former who are under 67 years of age reflects the fact that 65 is pretty much of a functional minimum age for membership in this locale. On the other hand the domiciliary sample parallels the population mean on this dimension; hence the two groups while somewhat different from each other are probably quite representative ones. . . .

Descriptive Characteristics for the Two Geriatric Samples[a]

		Domiciliary ($N = 40$)	**Community ($N = 40$)**
Age	$\bar{X}$	64.8	71.0
	Range	57–78	59–79
Education	$\bar{X}$	9.2	9.8
	Range	7–16	6–18

[a]Table 1 in the original.

It is clear from the information presented in R.E. 4.1 that the two samples were similar in both the mean and range for age and education. However, this does not mean that the variation of the scores between the highest and lowest ages or levels of education were similar. Remember that the range does not examine the pattern of variation inside the highest and lowest scores. Some other variation measure is needed for this information.

The range's complete reliance on the two most extreme scores led to the development of other range-based measures which examine only parts of the total range. These measures belong to a family of percentile and interdecile ranges that can be used to neutralize the effect of extreme scores on summaries of variation. Percentile ranges identify that part of a distribution which falls between any two less-than cumulative percentages. For example, the 5–95 percentile range excludes the extreme 5% of the scores from each end of the distribution. Interdecile ranges identify that part of a distribution which falls between any two deciles (percentage multiples of 10). For example, the 2nd–8th interdecile range excludes the extreme 20% of the scores from each end of the distribution. The basic formulas for these measures are presented in the section *Related Statistics* at the end of the chapter. If these range measures perform the subtraction between two points in a distribution, they require interval data. If not, only ordinal data are needed. We will look at one of these measures here, the interquartile range (Q).

Interquartile Range (Q)

The *interquartile range* (Q) identifies that part of an ordered distribution which contains the middle 50% of the scores. It locates the first quartile (Q_1) and the third quartile (Q_3) in the distribution. The first quartile is the point at and below which 25% of the scores occur. The third quartile is the point at and below which 75% of the scores occur. You are already familiar with the second quartile (Q_2); it is also called the median. The fourth quartile (Q_4) is the highest score in a distribution, the point at and below which 100% of the scores occur. The formula for the interquartile range is

$$Q = Q_3 - Q_1 \qquad \text{Interquartile Range Formula}$$

where Q = interquartile range
Q_3 = third quartile
Q_1 = first quartile

Imagine that we were examining eight job applicants drawn from a larger pool who had competed in doing a timed task. The order of finish for the eight selected people is shown in Table 4.2. We can describe the variation of these scores with the interquartile range. There are six small steps in the process that are similar to those we used with the median in Chapter 3.

1. The first step is to put the scores into an ordered array.
2. We find the position of the first quartile using the formula $N/4$. Q_1 will have two scores below it in the ordered distribution ($8/4 = 2.0$).

TABLE 4.2 Finding the Interquartile Range for Ordinal Data

Order of Finish on a Timed Task

Applicant	Order of Finish
A	2
B	4
C	1
D	5
E	14
F	9
G	7
H	8

1. Order the distribution:

 1 2 4 5 7 8 9 14

2. Find the position of Q_1:[a]

 $N/4 = 8/4 = 2$ scores below it

3. Determine the value of Q_1:

 2nd score = 2 and 3rd score = 4

 $Q_1 = (2 + 4)/2 = 3$

4. Find the position of Q_3:[a]

 $3N/4 = (3 \times 8)/4 = 6$ scores below it

5. Determine the value of Q_3:

 6th score = 8 and 7th score = 9

 $Q_3 = (8 + 9)/2 = 8.5$

6. Find the interquartile range:

 $Q = 8.5$ to 3 (or 5.5)

 1 2 | 4 5 7 8 | 9 14 (with Q_1 = 3 marked between 2 and 4, and Q_3 = 8.5 marked between 8 and 9)

[a] Computational note: If the positions of Q_1 or Q_3 are not whole numbers, round them to whole numbers before continuing.

3. We determine the value of Q_1. Since the second score is a 2 and the third is a 4, Q_1 is given their mean, a value of 3 [(2 + 4)/2 = 3].
4. The fourth and fifth steps repeat the process for Q_3. Its position is found by the formula $3N/4$. Q_3 will have six scores below it [(3 × 8)/4 = 6].
5. The value of the sixth score is 8 and the value of the seventh score is 9. The value given to Q_3 is 8.5 [(8 + 9)/2 = 8.5].
6. The final step is to substitute the values of Q_1 and Q_3 into the formula for the interquartile range: $Q = 8.5$ to 3, or 5.5. We can interpret this to mean that the middle 50% of the selected group finished between the 8.5th and 3rd places on the task. A second interpretation might be to say that the middle 50% of the group were spread across 5.5 places of finish.

Let us again look more closely at these two interpretations of Q. The first interpretation merely notes the values of Q_3 and Q_1. The second interpretation subtracts Q_1 from Q_3. Subtracting between the two values requires the assumption that an interval-level difference exists between them. This would be appropriate if the variable had been measured at the interval level. But the order-of-finish variable in Table 4.2 has only ordinal-level quality. We do not know how much time elapsed between the order of finishes, only their order. Thus our ordinal level interpretation of Q is the appropriate one for this example.

Like the range, Q cannot be compared across distributions for variables having different scale. For example, a Q for a distribution of ages is not comparable to one for a distribution of family sizes. Nor could two distributions of ages be compared in variation by the interquartile range unless they had the same medians since Q effectively identifies the points that on average are 25% above and 25% below Q_2 (the median).

Research Example 4.2 uses the interquartile range to describe the variation in the percentage of underreporting in pregnancies that occurs for women in each trimester of pregnancy. Its use was restricted to a graphic presentation and it showed only the values of Q_1 and Q_3, not their subtracted difference. In this way it illustrates the ordinal use of the interquartile range that we have been discussing.

RESEARCH EXAMPLE 4.2 Using the Quartile Deviation as a Measure of Variation

From Noreen Goldman and Charles F. Westoff, "Can Fertility Be Estimated from Current Pregnancy Data?" Population Studies, *34 (3), November 1980:535ff. Reprinted by permission.*

The recent history of developments in demographic methods has been characterized by numerous, frequently ingenious attempts to estimate current fertility in populations with limited or defective data. . . . [One solution proposed] in the last few years is the possibility that current fertility . . . can be estimated from information on current pregnancy collected in fertility surveys as a matter of routine. . . . With the increasing availability of information for developing countries participating in the World Fertility Survey, it is now possible to evaluate the usefulness of current pregnancy data across a broad range of countries in which standard questionnaires were employed. In this analysis, we examine such data for 15 countries in Asia and in Latin America. These surveys were large national probability samples of women of reproductive age (generally 15-49). . . . The sample size in these 15 interviews ranged from about 3,500 to 9,000. . . .

The status of current pregnancy was determined in all countries by the simple question "Are you pregnant now?" For women who replied 'Yes,' a follow-up ques-

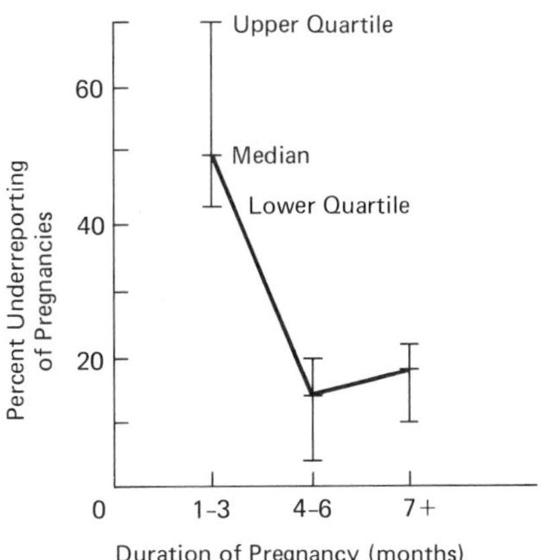

Distribution of Percentage of Underreported Pregnancies by Trimester, on Assumption That the Mean Annual Number of Births During the Past 24 Months Is Correct Level of Recent Fertility, for 15 Countries, World Fertility Survey. Figure 1 in original.

tion: 'When is the baby due?' was asked and the response recorded in month and year. In some countries, an alternate version of this question was asked: 'In what month of pregnancy are you now?'. . . Women who were not certain whether they were pregnant were regarded as replying 'No' to the question.

The detailed pregnancy data assembled for this analysis allow us to assess the reports on duration of pregnancy. Since neither a re-interview survey nor a multiround survey was available to us . . . we can evaluate the data only by comparing them with reports of births. For example, on the assumption that the reported number of births during the past 24 months . . . is the correct level of fertility, we can estimate the expected number of pregnancies in each trimester. . . . The percentage difference between the reported number of pregnancies in each trimester and the estimated number . . . is shown in [the figure], in which the median, and upper, and lower quartiles based on the 15 countries are also shown. On the assumption that the number of births is correctly reported and unchanging, the figures imply a median percentage underreporting of pregnancies during the three trimesters of 50 per cent, 14 per cent and 17 per cent respectively. The very high rate in the first trimester is in large part due to women who were not yet aware of their pregnancies. Probably because of premature births, the underreporting of pregnancies is less widespread in the second than in the third trimester. . . .

Two Interval- and Ratio-Level Measures—The Mean Deviation and Standard Deviation

Mean Deviation (*MD*)

The *mean deviation* (*MD*) calculates the arithmetic average of the absolute value of the deviation scores in a distribution. It is also called the *absolute average deviation* or simply the *average deviation.* The deviation score x was introduced in Chapter 3 when we discussed the mean. The central idea to the mean deviation is to include every score in the measure of variation. This avoids the problem of having only two scores completely dominate the statistic as we saw with the range and measures based on it.

There are several mathematical ways to allow every score to contribute to a measure of variation. One is to make pairs of all the scores and then summarize their differences. That is basically what we did with the index of dispersion, but with nominal data. Although this approach would work with interval data, it would be very laborious because of the high number of score categories. The mean deviation uses the more efficient method of comparing all scores to a single central value, the mean. These differences are then summarized. In a formula it is

$$MD = \frac{\Sigma\,|x|}{N} \qquad \text{Mean Deviation Formula}$$

where MD = mean deviation
x = deviation score $(X - \overline{X})$
$|x|$ = absolute value of each deviation score
Σ = summation (of the absolute values of the deviation scores)
N = number of cases

The first column of Table 4.3 presents the data for the seven commuter distances that we were examining earlier. In the second column their deviation score equivalents have been calculated. Remember from our discussion in Chapter 3 that the deviation scores always sum to zero for any distribution. This will happen for a distribution whose scores are very much alike as well as for a distribution whose scores are very different from each other. Therefore, the sum of the deviations cannot be used as an indication of variation. But if we focus only on how far each raw score deviates from the mean and not on the direction of its deviation (above or below the mean), we will get a nonzero sum that does reflect the dispersion of the scores. This is the result produced by taking the absolute value of the deviations. The deviations below the mean that have negative signs are changed to have positive signs. In Table 4.3 the absolute deviations sum to 28. This means that the seven commuting distances differed collectively by a total of 28 miles from the mean commuting distance of 17 miles.

This total must be standardized for the number of cases to make it comparable to other distributions. Otherwise, distributions with more scores would tend to have a larger total, even though their scores might be less varied. This standardization is accomplished by dividing the total by N. The result for the Table 4.3 distribution is a mean deviation of $28/7 = 4$ miles. This is interpreted to mean that the arithmetic average of the seven deviations is 4 miles. Some persons differed from the mean by more and some by less, but the arithmetic average of the differences is 4 miles.

Like the range and other measures based on it, the mean deviation is sensitive to differences in scale and means and to extreme scores. A mean deviation of 4 miles around a mean of 17 miles does not represent the same variation that a mean deviation of 4 miles around a mean of 500 miles represents. The latter is clearly a less varied distribution considering the greater distances involved. As a result we must be very careful when we compare mean deviations between distributions for different variables or even for different distributions on the same variable. A few extreme scores will cause the mean deviation to change markedly, like the mean itself. Consequently, we should look for extreme scores when the mean deviation is used even to describe only a single distribution.

The mean deviation is a sound measure of variation for describing distributions if

TABLE 4.3 Finding the Mean Deviation for Interval/Ratio Data

Commuting Distance from Residence to Workplace, Round Trip

Miles X	$x = X - \bar{X}$	$\lvert x \rvert$
25	8	8
21	4	4
19	2	2
17	0	0
15	−2	2
12	−5	5
10	−7	7
119	0	28

1. Find the mean:

$$\bar{X} = \frac{\Sigma X}{N} = \frac{119}{7} = 17$$

2. Find the deviations:

$$x = X - \bar{X}$$

3. Determine the absolute value of each deviation:

$$\lvert x \rvert$$

4. Sum the absolute values of the deviations:

$$\Sigma \lvert x \rvert = 28$$

5. Find the mean deviation:

$$MD = \frac{\Sigma \lvert x \rvert}{N} = \frac{28}{7} = 4.0$$

the cautions noted above have been considered. But it was never developed for use in inferential statistics. This, combined with its limited capacity for comparisons, caused it to be a little-used statistic. It is seldom reported. Despite this rarity, its logic provides a good introduction to the widely-used standard deviation, at which we will look now.

Standard Deviation (σ, S, and s)

The *standard deviation* is the square root of the arithmetic average of the sum of the squared deviations around the mean in a distribution! That sounds more formidable than it really is. There are two points you should note from this definition of the standard deviation. First, it is similar but not identical to the mean deviation. Second, its meaning is entirely mathematical (like the mean) and one should not attempt to redefine it with some common verbal interpretation.

The standard deviation is an extremely versatile statistic. It can be used as a descriptive summary of variation for a single variable, as can the other measures in this chapter. It can be used to compare the variation of different variables under certain conditions. It can be used to draw inferences from a sample's variation to a population's variation. It is also the basis for many other inferential statistics with interval and ratio variables, such as those that test for significant differences and association between variables.

There are three standard deviation measures that we learn about in this chapter. (There will be even more later.) One is a descriptive measure for population data, called the *population standard deviation,* σ (Greek lowercase sigma). The second is a descriptive measure for sample data, called the *sample standard deviation,* S (Roman capital letter). The third is an inferential measure for a population but calculated from sample data, called the *estimated or inferential population standard deviation,* either s (Roman lowercase letter) or $\hat{\sigma}$ (Greek lowercase sigma with a circumflex). These three measures differ not only in symbol but sometimes also in concept and formula. Further, each has two equivalent formulas that are commonly used. One is for a definition and the other is used to simplify its computation. We look first at the two descriptive measures.

Descriptive Standard Deviation (σ and S)

Strictly speaking, the verbal definition given above is for a descriptive measure of standard deviation, σ or S. These measures are descriptive (not inferential) summaries of the variation in a distribution. The formulas used as a mathematical definition are the same for both σ and S:

$$\sigma = \sqrt{\frac{\Sigma x^2}{N}} \qquad \text{Definitional Formula for } \sigma$$

$$S = \sqrt{\frac{\Sigma x^2}{N}} \qquad \text{Definitional Formula for } S$$

where σ = descriptive standard deviation for a population
S = descriptive standard deviation for a sample
x^2 = squared value of each deviation score $(X - \bar{X})$
Σx^2 = sum of the squared deviations
N = number of cases (in the population for σ or in the sample for S)
$\sqrt{\ }$ = positive square root value

Let us look again at our commuter distances example to see how the two descriptive standard deviations are calculated and interpreted. The distances and appropriate computations are presented in Table 4.4. Follow the six steps shown to find either σ or S.

Like the mean deviation, the standard deviation measures variation according to the deviations around the mean score. Thus the first step is to find the mean and the second step is to find the deviations. In the third step the deviations are squared. This serves the same purpose that determining the absolute values did for *MD;* it gets rid of any negative signs on the deviations so that when they are summed we get a nonzero total. Indeed, the fourth step is to sum the squared deviations. In Table 4.4 it is 162. This sum (Σx^2) is called the *total variation.* We will use it again in future statistics.

The fifth step standardizes for case size by dividing the total variation by N: $162/7 = 23.14286$. This quantity is called the *variance.* It is labeled σ^2 for a population and S^2 for a sample. The variance is often reported as a measure of variation itself. The final step in finding the standard deviation is to take the positive square root of the variance. This returns the calculations to their original units of measurement. The units (miles, in this example) were squared as well as the scores themselves back in the third step. Without taking the square root of the variance we would be left with 23.14286 miles squared in our example. Since miles squared is not a useful notion for describing differences in commuter distances, we prefer its square root. This allows us to speak of the variation as having a standard deviation of 4.81 miles. If we consider these scores to be an entire population that we are describing, we label this answer σ. If we consider these scores to be a sample that we are describing, we label it S. Again, the formula is the same; only the context and its symbol change.

In this example the mean was a whole number and therefore the deviation scores and their squared values were whole numbers. Of course, this is not always the situation. When the mean contains a decimal part, the deviations will, too. And if the mean is rounded, so will be the deviations. The result is that the calculation can be considerably more difficult and rounding errors can accumulate to produce an inexact answer. To avoid this researchers have derived alternative but equivalent computational formulas that do not use either the mean or the deviations. One computational formula commonly used for

TABLE 4.4 Finding the Descriptive Standard Deviations (σ or S) for Interval/Ratio Data Using the Definitional Formula

Commuting Distance from Residence to Workplace, Round Trip

Miles X	x	x^2
25	+8	64
21	+4	16
19	+2	4
17	0	0
15	−2	4
12	−5	25
10	−7	49
119	0	162

1. Find the mean:

$$\bar{X} = \frac{\Sigma X}{N} = \frac{119}{7} = 17$$

2. Find the deviations:

$$x = X - \bar{X}$$

3. Square the deviations:

$$x^2$$

4. Sum the squared deviations to find the total variation:

$$\Sigma x^2 = 162$$

5. Divide this sum by N to find the variance:

$$\sigma^2 \text{ or } S^2 = \frac{\Sigma x^2}{N} = \frac{162}{7} = 23.14286$$

6. Find the square root of the variance to determine the standard deviation:

$$\sigma \text{ or } S = \sqrt{\frac{\Sigma x^2}{N}} = \sqrt{23.14286} = 4.81 \text{ miles}$$

In summary:

$$\sigma \text{ or } S = \sqrt{\frac{\Sigma x^2}{N}} = \sqrt{\frac{162}{7}} = \sqrt{23.14286} = 4.81 \text{ miles}$$

σ and S with raw data is as follows:

$$\sigma = \sqrt{\frac{\Sigma X^2 - \frac{(\Sigma X)^2}{N}}{N}} \qquad \text{Computational Formula for } \sigma$$

$$S = \sqrt{\frac{\Sigma X^2 - \frac{(\Sigma X)^2}{N}}{N}} \qquad \text{Computational Formula for } S$$

where σ = descriptive standard deviation for a population
S = descriptive standard deviation for a sample

X^2 = squared value of each raw score
ΣX^2 = sum of the squared raw scores
ΣX = sum of the raw scores
$(\Sigma X)^2$ = squared value of the (ΣX) sum
N = number of cases (in a population for σ or in a sample for S)
$\sqrt{}$ = positive square root

Table 4.5 shows how the computational formulas can be applied to our commuter distances. Notice that the formulas do *not* use deviation scores (x). They use raw scores (X) instead. Give very close attention to the algebra of the formulas. Be sure that you see the difference between ΣX^2 and $(\Sigma X)^2$. In the first of these you square the raw scores before summing them. In the second, you sum the raw scores before squaring that total. The subtracted difference between the two quantities in the numerators $[\Sigma X^2] - \left[\frac{(\Sigma X^2)}{N}\right]$ of either formula (σ or S) is the total variation. Dividing it by N gives us the variance (σ^2 or S^2). When we take its square root we have the standard deviation (σ or S). Note that the values for the variance and standard deviation using the computational formulas in Table 4.5 are identical to those from the definitional formulas in Table 4.4. This happens because there were no rounding errors in the calculations. Otherwise, there might have been some small differences. If so, the computational formulas' results would have been the more accurate ones.

A descriptive standard deviation gives us a summary measure of a distribution's variation. Like the mean deviation it has the desired quality of including every score in its computation and it is standardized for the number of cases. Also like the mean deviation it has the limitation of being sensitive to the scale and mean of the scores in a distribution. As a consequence, you can compare one standard deviation to another only in certain circumstances. One of these is when the same scale is used to measure variables having very similar means. Thus we could compare our example's standard deviation of 4.81 miles to another distribution of commuter distances if it had a mean of 17 miles. If such a second distribution had a standard deviation of 10 miles, we would be correct to conclude that the second one was more varied than ours. This simply means that in the second distribution the individual commuting distances were more heterogeneous than were our seven distances, given equivalent means.

Research Examples 4.3 and 4.4 illustrate common situations in which a standard deviation is used to describe variation. In R.E. 4.3 it is used to show the variation for 10 occupations which were rated on the same 1 to 9 scale. Because this measurement scheme was identical for all 10 of the occupations, you may be tempted to compare the standard deviations directly to determine the relative consistency of the ratings. That would be a mistake in interpreting this research. It is not sufficient that the potential measurement scale be constant. The actual measurement scale as indicated by the scores and their means in the distributions must be nearly identical before the standard deviations can be compared directly.

In R.E. 4.3 this means that we can compare the standard deviation of the ratings given to lawyers to that for the psychiatrists with little distortion because their means are so close (7.42 and 7.39, respectively). We would be correct to conclude that there was only slightly more consistency in the ratings given to psychiatrists than to lawyers (standard deviations of 2.08 and 2.18, respectively). We could also compare social workers and teachers (means of 6.40 and 6.38, respectively). This time we would be correct to believe that the ratings for the social workers were more homogeneous than those for teachers

TABLE 4.5 Finding the Descriptive Standard Deviation (σ or S) for Interval/Ratio Data Using the Computational Formula

Commuting Distance from Residence to Workplace, Round Trip

Miles X	X^2
25	625
21	441
19	361
17	289
15	225
12	144
10	100
119	2185

1. Square the raw scores:

   X^2

2. Sum the squared raw scores:

   $\Sigma X^2 = 2185$

3. Sum the raw scores:

   $\Sigma X = 119$

4. Square this sum:

   $(\Sigma X)^2 = (119)^2 = 14{,}161$

5. Divide (4) by N:

   $$\frac{(\Sigma X)^2}{N} = \frac{14{,}161}{7} = 2023$$

6. Subtract (5) from (2) to find the total variation:

   $$\Sigma X^2 - \frac{(\Sigma X)^2}{N} = 2185 - 2023 = 162$$

7. Divide the total variation by N to find the variance:

   $$\sigma^2 \text{ or } S^2 = \frac{162}{7} = 23.14286$$

8. Find the square root of the variance to determine the standard deviation:

   $$\sigma \text{ or } S = \sqrt{\sigma^2} \text{ or } \sqrt{S^2} = \sqrt{23.14286} = 4.81 \text{ miles}$$

In summary:

$$\sigma \text{ or } S = \sqrt{\frac{\Sigma X^2 - \frac{(\Sigma X)^2}{N}}{N}} = \sqrt{\frac{2185 - \frac{(119)^2}{7}}{7}} = \sqrt{\frac{2185 - \frac{14{,}161}{7}}{7}}$$

$$= \sqrt{\frac{2185 - 2023}{7}} = \sqrt{\frac{162}{7}} = \sqrt{23.14286} = 4.81 \text{ miles}$$

RESEARCH EXAMPLE 4.3 Using the Standard Deviation as a Measure of Variation

From Joseph F. Sheley, Janise S. Legeai, Stephen J. Schenthal, and Nancy T. Jamerson, "Public Willingness to Utilize Psychiatric Social Work Services," Journal of Social Service Research, *3 (4), Summer 1980:395ff. Reprinted by permission.*

The study described in this paper was inspired by the frequent complaint of many psychiatric social workers that their field lacks public legitimacy. Some argue that the public has little knowledge about and even less inclination to utilize the services of the social worker in private psychotherapeutic practice. . . . Employing survey data, this paper attempts to shed some empirical light on the accuracy of this self-image and the related possibility of increasing public utilization of psychiatric social work services. . . .

Survey data obtained from residents of the greater New Orleans area are analyzed to determine the extent of their knowledge about and contact with psychiatric social workers. . . . Data employed in this study were obtained through telephone interviews of residents . . . [who] have reached the age of 18 and lived in New Orleans for at least one year.

While psychiatric social workers may rightfully question the degree to which the public understands their work, they are incorrect if they assume the public grants them minimal occupational status. Respondents were given a list of 10 varied occupations (re-randomized for each respondent) and asked to rate each on a prestige scale of 1 (low) to 9 (high). Psychiatric social workers were granted moderately high status. As [the table indicates], they placed fourth behind physicians, lawyers, and psychiatrists, and ahead of social workers, teachers, farmers, electricians, plumbers, and garbage collectors. This prestige score ranking closely approximates those found in larger, national occupational prestige studies. . . .

Prestige Ratings of Selected Occupations[a]

Occupations	Mean Prestige Rating	Standard Deviation
1. Physician	8.14	1.72
2. Lawyer	7.42	2.18
3. Psychiatrist	7.39	2.08
4. Psychiatric social worker	6.76	1.89
5. Social worker	6.40	1.99
6. Teacher	6.38	2.24
7. Farmer	6.05	2.24
8. Electrician	5.82	1.91
9. Plumber	5.23	2.01
10. Garbage collector	4.06	2.46

[a]Table 1 in the original.

(standard deviations of 1.99 and 2.24, respectively). Any other direct comparisons in standard deviations would not be advisable because the means differ too greatly.

In R.E. 4.4 the standard deviation is used to summarize the variation for 13 variables in a study of sentences given to criminals. Again some of these variables used an identical measurement scale. A binary code (0 and 1) was used for the dichotomies: race; sex;

RESEARCH EXAMPLE 4.4 Using the Standard Deviation as a Measure of Variation

From James D. Unnever, Charles E. Frazier, and John C. Henretta, "Race Differences in Criminal Sentencing," Sociological Quarterly, *21 (2), Spring 1980:197ff. Reprinted by permission.*

For nearly fifty years, sociologists and criminologists have been attempting to determine the relationship between race and criminal court decisions. . . . The question of the effect of race on official court decisions has occupied many scholars for so long a time because (1) it is an important social issue, (2) it is widely considered a testing ground for propositions derived from conflict theory, and (3) the research literature is laced with contradictory findings and conclusions. . . .

Data for the present analysis were collected by the second author from presentence investigation reports filed in one six-county judicial district in Florida. The district includes one county with rural outskirts and a central urban area of more than 100,000 population. The other five counties are exclusively rural, the largest community being less than 6,000 residents. The sample was drawn from cases adjudicated, and followed by a presentence investigation, in the urban county over the 12 month period between June 1, 1972 and May 31, 1973. The 229 cases represent nearly 90 percent of the total number of cases receiving a pre-sentence report during that year. . . .

[The table] presents the means and standard deviations for the variables. The codes are as follows: age is coded in years; race (1 = white; 0 = black); sex (1 = male; 0 = female); employed (1 = currently employed; 0 = not employed); and marital status (1 = married, spouse present; 0 = other). Education is coded as number of years of schooling completed. The number of adult arrests is coded as the number of previous arrests for misdemeanors and felonies recorded on the FBI identification

Means and Standard Deviations[a]

Variables	Mean	Standard Deviation
Age	25.60	10.32
Race	.57	.49
Sex	.86	.34
Employment	.57	.48
Marital status	.33	.47
Education	10.67	2.48
N arrests	3.57	6.13
N convictions	1.26	1.61
Severity	123.92	117.06
District attorney [recommendations]	.23	.42
Police [recommendations]	.27	.44
Probation office [recommendations]	.25	.43
Disposition	.73	.44

[a]Table 1 in the original.

record; and number of adult convictions includes sentences of a fine, probation, incarceration in jail or prison (or some combination of these) listed in the disposition column of the same record. Severity is the sum of the maximum sentences (measured in months) for all original charges against the defendant.

The dependent variable, disposition, is a dummy (1 = probation; 0 = incarceration). The recommendations of the police, district attorney and probation officer are coded as follows: 1 = incarceration; 0 = other (probation recommendation or no recommendation). Missing data were replaced by mean values. . . .

Race differences in sentencing outcomes are substantial in these data. We also present some evidence suggesting that when race differences occur, they may originate in the early stages of the sentencing process.

employment; marital status; recommendations by the police, district attorney, and probation officer; and disposition. A *dichotomy* is a measurement that has only two opposite categories; e.g., "white" or "nonwhite," "yes" or "no," and so on. The binary code divides the cases into those who have the characteristic in question (code = 1) and those who do not (code = 0). Among these dichotomies we see that all except sex have nearly identical standard deviations. But the only legitimate comparisons in standard deviations are race with employment and the three types of recommendations with each other because of their similar means. In both of these sets of comparisons the standard deviations are very similar, indicating that the variations are very much alike. But, the means and measurement scales of the other variables are too diverse to make their standard deviations directly comparable.

There are two ways in which standard deviations can otherwise be comparable. One occurs in the special situation where we have distributions shaped like the classic "normal" or "bell-shaped" curve. The second occurs when a standard deviation is used in a measure called the coefficient of variation. Both of these are discussed after we have covered the inferential population standard deviation.

Inferential Standard Deviation (s)

We need to think ahead to inferential statistics. Suppose that a researcher wanted to describe the distribution of commuting distances for an entire metropolitan area. Because of the enormous complexity of gathering the data for the entire commuting population, it is very likely that the researcher would draw a representative sample of commuters instead. Let us imagine that our same example of seven commuting distances was that sample. Acutally, this would be an extremely small sample for a researcher to use. But it will illustrate our discussion nonetheless.

If the researcher wanted to generalize from the sample to the population, we might expect that it would be a simple matter of finding the relevant statistics for the sample and merely projecting them to that population. For example, since our sample's mean is 17 and its standard deviation is calculated to be 4.81, the best guess would be to say that the population's mean and standard deviation have the same two values. As it happens we would be correct about the mean but wrong about the standard deviation. The reasons for this are too complicated to detail here. In short, a sample will provide an unbiased estimate of a population's mean but an underestimate of its standard deviation. That is, a sample's mean might be higher or lower than the population's mean, but its standard deviation will tend only to be lower.

Because of this consistent and known bias, researchers use a special formula for calculating a standard deviation from a sample when they want to use it as an estimate of a population's standard deviation. We call this measure the *inferential or estimated population standard deviation.* It typically uses either s or $\hat{\sigma}$ as a symbol. Some researchers prefer to use $\hat{\sigma}$ because it is a symbol referring to the population value that we are seeking (σ), with an added circumflex (ˆ) to indicate it is an estimate. Others prefer s because it indicates that sample data are being used. We will use s.

The formulas used for s divide the total variation by one less than the total number of cases, rather than by the number itself. This is not an arbitrary change; it has been derived mathematically. The consequence of the change is that s will have a slightly higher value than it would have otherwise. The consistent underestimate is avoided so that s gives us a value that is not systematically biased in comparison to the actual value of a population's standard deviation (σ). This is exactly the result that we want. The definitional formula for s from raw data is

$$s = \sqrt{\frac{\Sigma x^2}{N - 1}}$$ Definitional Formula for s

where s = inferential or estimated population standard deviation
x^2 = squared value of each deviation score
Σx^2 = sum of the squared deviations
N = number of cases in a sample
$\sqrt{}$ = positive square root

Table 4.6 illustrates the application of this formula to our commuter data. Notice that all the steps except step 5 are identical to those used above for the descriptive standard deviations in Table 4.5. That is the point at which we divide the total variation by $N - 1$, not N. The variance (s^2) is now 27 and the standard deviation (s) is now 5.20. As noted, this provides us with an unbiased estimate of the variation in a population of commuter distances from which our seven cases were considered to be a sample. We would offer the estimate that the entire population's true standard deviation (σ) in commuter distances is nearer to 5.20 miles than it is to 4.81 miles.

If our sample had contained more cases, the value of the standard deviation would have been changed less by using the inferential formula than the descriptive formula. For example, dividing the same total variation by 99 versus 100 creates less difference than when we divide it by 6 versus 7. One conclusion you should notice from this is that it becomes increasingly important to make a distinction between the two formulas as the sample size diminishes.

There are equivalent formulas for use in computing s as there are for σ or S. One that is commonly used for raw data distributions is

$$s = \sqrt{\frac{\Sigma X^2 - \frac{(\Sigma X)^2}{N}}{N - 1}}$$ Computational Formula for s

where s = inferential or estimated standard deviation
X = raw score
X^2 = squared value of each raw score
ΣX^2 = sum of the X^2 values

TABLE 4.6 Finding the Inferential Standard Deviation (s) for Interval/Ratio Data Using the Definitional Formula

Commuting Distance from Residence to Workplace, Round Trip

Miles X	x	x^2
25	+8	64
21	+4	16
19	+2	4
17	0	0
15	−2	4
12	−5	25
10	−7	49
119	0	162

1. Find the mean:

$$\bar{X} = \frac{\Sigma X}{N} = \frac{119}{7} = 17$$

2. Find the deviations:

$$x = X - \bar{X}$$

3. Square the deviations:

$$x^2$$

4. Sum the squared deviations to find the total variation:

$$\Sigma x^2 = 162$$

5. Divide this sum by $(N - 1)$ to find the estimated population variance:

$$s^2 = \frac{\Sigma x^2}{N - 1} = \frac{162}{6} = 27$$

6. Find the square root of the variance to determine the inferential standard deviation:

$$s = \sqrt{s^2} = \sqrt{27} = 5.20 \text{ miles}$$

In summary:

$$s = \sqrt{\frac{\Sigma x^2}{N - 1}} = \sqrt{\frac{162}{6}} = \sqrt{27} = 5.20 \text{ miles}$$

ΣX = sum of the raw scores
$(\Sigma X)^2$ = squared value of the (ΣX) sum
N = number of scores in a sample
$\sqrt{\ }$ = positive square root

Table 4.7 shows how this formula is applied to our same commuter distances. Again, be careful to follow the algebra in this formula. Notice, too, that the variance and standard deviation values match exactly those that we had for the definitional formula for s because there was no rounding. Figure 4.2 summarizes the definitional and computational formulas for the three standard deviations when they are being calculated from raw data distributions.

TABLE 4.7 Finding the Inferential Standard Deviation (s) for Interval/Ratio Data Using the Computational Formula

Commuting Distance from Residence to Workplace, Round Trip

Miles X	X^2
25	625
21	441
19	361
17	289
15	225
12	144
10	100
119	2185

1. Square the raw scores:

$$X^2$$

2. Sum the squared raw scores:

$$\Sigma X^2 = 2185$$

3. Sum the raw scores:

$$\Sigma X = 119$$

4. Square this sum:

$$(\Sigma X)^2 = (119)^2 = 14{,}161$$

5. Divide (4) by N:

$$\frac{(\Sigma X)^2}{N} = \frac{14{,}161}{7} = 2023$$

6. Subtract (5) from (2) to find the total variation:

$$\Sigma X^2 - \frac{(\Sigma X)^2}{N} = 2185 - 2023 = 162$$

7. Divide the total variation by $(N - 1)$ to find the estimated population variance:

$$s^2 = \frac{162}{6} = 27$$

8. Find the square root of the variance to determine the inferential standard deviation:

$$s = \sqrt{s^2} = \sqrt{27} = 5.20 \text{ miles}$$

In summary:

$$s = \sqrt{\frac{\Sigma X^2 - \frac{(\Sigma X)^2}{N}}{N - 1}} = \sqrt{\frac{2185 - \frac{(119)^2}{7}}{6}} = \sqrt{\frac{2185 - \frac{14{,}161}{7}}{6}}$$

$$= \sqrt{\frac{2185 - 2023}{6}} = \sqrt{\frac{162}{6}} = \sqrt{27} = 5.20 \text{ miles}$$

FIGURE 4.2 Standard Deviation Formulas for Raw Data Distributions

	Definitional Formula	Computational Formula
Descriptive Standard Deviation σ **or** S	$\sqrt{\frac{\Sigma x^2}{N}}$	$\sqrt{\frac{\Sigma X^2 - \frac{(\Sigma X)^2}{N}}{N}}$
Inferential Standard Deviation s	$\sqrt{\frac{\Sigma x^2}{N-1}}$	$\sqrt{\frac{\Sigma X^2 - \frac{(\Sigma X)^2}{N}}{N-1}}$

The Normal Curve Meaning of the Standard Deviation

For some distributions the standard deviation is merely an alternative measure of variation to the mean deviation. It, too, summarizes the spread in a set of interval or ratio data in such a way that every score is considered and the calculated answer is standardized for case size. In these situations its use is limited to either describing the variation of single distributions or comparing those distributions which have similar scales and means.

For other distributions the standard deviation takes on an expanded interpretation. Here when you know the standard deviation, it is possible to determine the percent of scores in a distribution that are within any distance from that distribution's mean. For example, suppose we know that one of these distributions has a mean of 50 and a standard deviation of 10. If so, we would know that approximately one-third (34%) of the scores are between 50 and 60. How is it possible to know this?

The type of distribution we are considering has a form like the classic "normal" or "bell-shaped" curve. It is a unimodal, mesokurtic, and symmetrical distribution which is often approximated with actual research data. Every part of the area in a normal distribution can be found by formula or a table of recorded values for a theoretical distribution called the *standard normal distribution.* We will wait to learn how to do this until we talk about probability in Chapter 7. For now, let us consider some of these areas, such as the one that contains 34% of the scores.

Figure 4.3 illustrates a normal distribution with a mean of 50 and a standard deviation of 10. Notice that the area between the mean (50) and the mean plus the value of one standard deviation (50 + 10) contains the 34% figure. If we look at the area between one standard deviation above the mean (60) and one standard deviation below the mean (40), we see that it contains 68% of the area. This happens because the curve is symmetrical. Looking at the other areas in the figure, you see that 95% of the scores occur between 30 and 70. This is equivalent to the area between two standard deviations on each side of the mean. Finally, note that more than 99% of the scores (nearly all of them) are between three standard deviations above and below the mean. That corresponds to the scores of 20 and 80 in our distribution. Another way to say this is that six standard deviations in a normal distribution approximate its range ($6S \cong R$). In fact, that is a rule of thumb that can be used to check a distribution's shape compared to a normal curve. If the standard deviation multiplied by 6 is not approximately equal to the range, the distribution is not normally shaped.

Suppose that we have another distribution shaped like a normal curve except that in this one the mean is 100 and the standard deviation is 25. Now 34% of its cases are between 100 and 125. Sixty-eight percent are between 75 and 125, and so on in a parallel fashion to our first standard normal distribution. In any normal distribution the percentage of cases between its mean and its mean plus one standard deviation is always 34%. In the same way, there is always 68% between one standard deviation above and below the mean, and so on. It does not matter what the scores are about, what scale they use, or what their means are. If the distribution has the normal shape, the percentages are constant. Indeed, that is one meaning of the "standard" in the name "standard deviation."

This information about the percentage distribution of cases in a normal distribution is useful for descriptive purposes. Suppose that the scores in a distribution are to be divided into several segments, each containing a specific percentage of the cases. We can use our knowledge of the areas in the normal curve to determine the dividing points. This is what test givers do when they want to "curve" an exam so that a definite percentage of test takers receive grades of A, B, C, D, and F. One of the main features of many "standardized tests" is that they consistently deliver distributions that have the desired proportions of high, middle, and low scores. College entrance tests like the SAT (Scholastic Aptitude Test) and the ACT (American College Testing Program) were designed to have this feature.

The percentage meaning of the standard deviation is also useful for inferential statistics. We will be applying it in this way very often in Parts II and III later in the book.

FIGURE 4.3 Percentage of Cases Within Certain Areas of a Normal Distribution

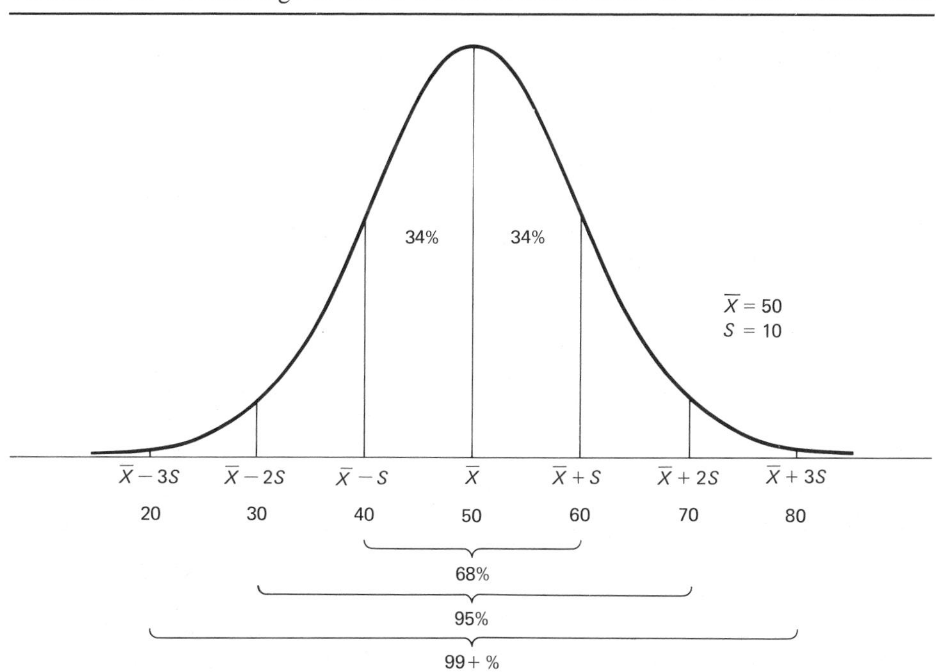

*Finding Standard Deviations from Simple and Grouped Frequency Distributions

To this point we have seen the computations of standard deviation measures from raw data distributions. Of course, it may be necessary to calculate them from the other two common types of distributions: simple and grouped frequencies. There is a full complement of new formulas for these situations parallel to those used with raw data. This includes formulas that define the measures and those that facilitate the computations. A complete set of these formulas is presented in Figure 4.4. They can be illustrated by two examples.

Suppose that we want to find the descriptive standard deviation for a sample (S) from a simple frequency distribution. Such a distribution is shown in Table 4.8. It presents the test scores for an exam given in a course on social change with 45 students. We begin by finding the mean of the distribution. This requires us to find the ΣfX sum (445) and divide it by N (45). The mean is 9.9. Now we determine the deviation score values for each category in the distribution by subtracting this mean from each X value listed. These deviations are now squared and then multiplied by the frequency for their own category. The summation of this column is the total variation (Σfx^2), which equals 208.45 here. This sum is divided by 45 to get the variance, $4.63\overline{2}$. Its square root is our standard deviation, $S = 2.15$.

Now suppose that our same set of scores were put in the grouped frequency distribution shown in Table 4.9 and we wanted to estimate the standard deviation of a larger population of exam scores of which we considered these 45 to be a random sample. That is, we wanted to find s. Again, we begin by finding the mean, although it must be estimated, as are all statistics from grouped frequencies. A column of midpoints is determined. These are multiplied by their category frequencies and the products are summed.

FIGURE 4.4 Standard Deviation Formulas for Simple and Grouped Frequency Distributions

	Definitional Formula	Computational Formula
Simple frequencies σ **or** S	$\sqrt{\dfrac{\Sigma fx^2}{N}}$	$\sqrt{\dfrac{\Sigma fX^2 - \dfrac{(\Sigma fX)^2}{N}}{N}}$
s	$\sqrt{\dfrac{\Sigma fx^2}{N-1}}$	$\sqrt{\dfrac{\Sigma fX^2 - \dfrac{(\Sigma fX)^2}{N}}{N-1}}$
Grouped frequencies σ **or** S	$\sqrt{\dfrac{\Sigma f(m - \hat{X})^2}{N}}$	$\sqrt{\dfrac{\Sigma fm^2 - \dfrac{(\Sigma fm)^2}{N}}{N}}$
s	$\sqrt{\dfrac{\Sigma f(m - \hat{X})^2}{N-1}}$	$\sqrt{\dfrac{\Sigma fm^2 - \dfrac{(\Sigma fm)^2}{N}}{N-1}}$

TABLE 4.8 Finding the Descriptive Standard Deviation (S) from a Simple Frequency Distribution Using the Definitional Formula

Exam Scores in a Social Change Course

Scores X	f	fX	x	x^2	fx^2
15	1	15	5.1	26.01	26.01
14	2	28	4.1	16.81	33.62
13	0	0	3.1	9.61	0
12	4	48	2.1	4.41	17.64
11	10	110	1.1	1.21	12.10
10	13	130	.1	.01	.13
9	7	63	−.9	.81	5.67
8	2	16	−1.9	3.61	7.22
7	2	14	−2.9	8.41	16.82
6	2	12	−3.9	15.21	30.42
5	1	5	−4.9	24.01	24.01
4	1	4	−5.9	34.81	34.81
Total	45	445			208.45

1. Find the mean:

$$\bar{X} = \frac{\Sigma fX}{N} = \frac{445}{45} = 9.8\bar{8} = 9.9$$

2. Find the deviations, their squares, the fx^2 products, and their sum to determine the total variation:

$$\Sigma fx^2 = 208.45$$

3. Divide the total variation by N to find the variance:

$$S^2 = \frac{\Sigma fx^2}{N} = \frac{208.45}{45} = 4.63\bar{2}$$

4. Find the square root of the variance to determine the standard deviation:

$$S = \sqrt{S^2} = \sqrt{4.632} = 2.15$$

This sum (447) is divided by N (45) to obtain our estimate of the mean, 9.9. (Note how closely our estimated mean matched the actual mean here.) Now we find the deviations of each category midpoint from the estimated mean $(m - \hat{\bar{X}})$ and square these differences $[(m - \hat{\bar{X}})^2]$. The squared differences are then multiplied by each category's frequency $[f(m - \hat{\bar{X}})^2]$ and summed to find the total variation, $\Sigma f(m - \hat{\bar{X}})^2 = 218.85$. We now divide this by $N - 1$ (44) to get the estimated variance, 4.97, and take its square root to find our inferential standard deviation. Our best estimate of the standard deviation in exam scores for the hypothetical population of scores we imagine our sample to represent is 2.23. There are two levels of estimating involved in this calculation. First, we were forced to estimate because we were working from grouped frequencies and second, we were calculating a measure that is always an estimate, s. Both sources of estimating could adversely influence the accuracy of our resulting statistic. However, our procedure is the best one available under the assumed circumstances.

TABLE 4.9 Finding the Inferential Standard Deviation (s) from a Grouped Frequency Distribution Using the Definitional Formula

Exam Scores in a Social Change Course

Scores	f	m	fm	$m - \hat{X}$	$(m - \hat{X})^2$	$f(m - \hat{X})^2$
13–15	3	14	42	4.1	16.81	50.43
10–12	27	11	297	1.1	1.21	32.67
7–9	11	8	88	−1.9	3.61	39.71
4–6	4	5	20	−4.9	24.01	96.04
Total	45		447			218.85

1. Find the estimated mean:

$$\hat{X} = \frac{\Sigma fm}{N} = \frac{447}{45} = 9.9\overline{3} = 9.9$$

2. Find the $(m - \hat{X})$ deviations, square these $[(m - \hat{X})^2]$, multiply these squares by each category's frequency $[f(m - \hat{X})^2]$, and sum these products to find the total variation:

$$\Sigma f(m - \hat{X})^2 = 218.85$$

3. Divide the total variation by $N - 1$ to find the estimated variance:

$$s^2 = \frac{\Sigma f(m - \hat{X})^2}{N - 1} = \frac{218.85}{44} = 4.97386$$

4. Find the square root of the variance to determine the inferential standard deviation:

$$s = \sqrt{s^2} = \sqrt{4.97386} = 2.23$$

A Special Scale-Free Measure for Ratio Data—The Coefficient of Variation (CV)

We have noted the scale sensitivity for all the measures in this chapter so far except the index of dispersion. Of the others, only the standard deviation for normal distributions can be compared between variables having different scales and means. Otherwise, distributions of high-value scores ordinarily yield large variation statistics, whereas distributions of low-value scores ordinarily yield small variation statistics. In both instances this is a result of the magnitude of the scores but not necessarily a result of their true degree of dispersion.

One type of measure often used in social data analysis to avoid this problem is the *coefficient of variation* (CV). It standardizes a measure of variation for ratio data by dividing it by its own central tendency value. Since the scale of the scores is represented in both the variation and central tendency measures, the scale effect is negated by the division of the one by the other.

Coefficients of variation are often used for the standard deviation and sometimes for other measures, too. A coefficient of variation for standard deviations uses the following formula:

$$CV = \frac{S}{\overline{X}} \qquad \text{Coefficient of Variation Formula}$$

On the Side 4.1 First-Hand Advice for Math Avoiders

An historian who learned from her own past told fellow faculty members they should do all they can to help students overcome fears about math.

Linda Kerber, professor of history, presented herself as the prototype of a "math avoider."

Besides special sections of math courses at all levels for hesitant students, and a special math clinic, Kerber recommended a publicity campaign to inform students that by avoiding math they are placing arbitrary limits on their future careers.

Kerber, speaking at a meeting of the committee on general education requirements, said:

"The college at which I studied bore some resemblance to this one. It required art or music, English, history and the social sciences, and laboratory sciences. It did not, to my great relief, require mathematics in addition to what I had studied in high school. I took chemistry courses—until I reached the point that math skills were needed—and dropped chemistry. I took experimental psychology and did well in it—until I reached the level that math was needed—and I dropped psychology. I took economics—and did very well in it—until math was needed—and then I dropped economics.

"This process of elimination made my choice of major somewhat easier. It is no accident that I ended in history—that is, a field which seemed to promise I would never again have to contemplate a number. It is an irony I muse upon each week as I join our beginning graduate students in a course on statistics and computers for historians, struggling to learn techniques without which I now risk becoming hopelessly out of date.

"I could not know, when I so studiously avoided math in college, that I was not original. I was only one of thousands of people who were—and continue to be—making similar decisions. Nor did I realize that by avoiding math I was studiously limiting my career options, both within history and outside it.

"For many fields, math is what Berkeley sociologist Lucy Sells has called a 'critical filter.' Certain levels of mathematical literacy are requisite for entry into fields like economics, engineering, geography. Even at relative elementary levels of a subject, mathematical literacy may function as a critical filter.

"It is clear that both men and women are often hesitant to study math. This is a national phenomenon, as a New York survey recently demonstrated. It is a phenomenon we experience here at The University of Iowa, as fewer and fewer of our freshmen enter with four years or more of high school math, and more and more enter with two years or less. In 1966, 14 percent of entering freshmen came with only two years or less in high school math. In 1976, 25 percent had two years or less.

"Both men and women may be 'math avoiders.' But it has come to seem clear that women are more 'successful' math avoiders than men. This trend is apparent on the national scene and at The University of Iowa.

"I think we will serve all our students better when we encourage them to increase their mathematical literacy and if we help even the most fearful math avoiders to overcome their anxieties."

From the University of Iowa *Spectator,* October 1978. Reprinted by permission.

where CV = coefficient of variation
S = standard deviation
$\bar{X}$ = mean

CV can be used to compare standard deviations for ratio variables. In our commuter distances example, CV equals .283 (4.81/17.0). This can now be compared to the CV value of any other distribution of commuter distance to determine which has more variation as measured by the standard deviation. It can also be compared to the CV value for any other ratio variable, including those with entirely different scales and means.

Recall R.E. 4.3, in which prestige ratings were given to 10 occupations. The coefficient of variation *cannot* be used to resolve our earlier difficulties in comparing the standard deviations for these ratings. At best, the ratings represented an interval scale of measurements; a 1 to 9 scale of arbitrary values. But the coefficient of variation requires ratio data.

However, the coefficient of variation can be used to compare the variables in R.E. 4.4. There we were looking at various characteristics associated with criminal sentencing. Those variables with ratio measurement obviously can be examined by the CV statistic. In addition, those variables that are dichotomies can also be examined. The binary coding can be used to create what is called a "dummy variable" which is used in many kinds of more advanced statistics. (See Kerlinger, 1973:116–118.) Dichotomous variables can be treated as ratio variables for purposes of statistical analysis. If we calculate the coefficients of variation for R.E. 4.4, we find that the variation was greatest for the recommendations given by the district attorney ($CV = .42/.23 = 1.826$) and least for the education of the criminals ($CV = 2.48/10.67 = .232$). This is not the conclusion we would have reached had we compared standard deviations directly.

RESEARCH EXAMPLE 4.5 Using the Standard Deviation and the Coefficient of Variation as Measures of Variation

From Michael G. Maxfield, Dan A. Lewis, and Ron Szoc, "Producing Official Crimes: Verified Crime Reports as Measures of Police Output," Social Science Quarterly, *61 (2), September 1980:221ff. Reprinted by permission.*

Most scholars now recognize the limited utility of official crime statistics for measuring the amount of crime in society. This skepticism was enhanced by the development by the Census Bureau of victimization surveys that demonstrated what had long been suspected, that crime is more common than official police records would indicate. . . .

In the analysis that follows, we first discuss some of the ways in which official data on crimes known to police can be used to evaluate police performance. By comparing demand for different kinds of police services, as measured by citizen calls for service, to the number of verified offenses recorded by the police, we are able to analyze the level of "unfounding" of incident reports, and to determine whether or not police decisions to unfound offenses affect particular segments of the population differently than others. . . .

Calls for service and verified crime data were obtained from the Chicago Police Department as beat-level totals for the period from January 8, 1976 through July 21, 1976. . . . [The table] presents descriptive statistics for the recording ratios for Part I personal and property crimes, and for Part II [all other] crimes. Data are presented for individual crime types as well as for aggregations of personal and property crimes. . . . [The data] show the aggregate value of the recording ratio and variation in this indicator across community areas. Overall about 76 percent of calls for service for Part I personal crimes are recorded as verified offenses. Part I property crimes were more likely to be recorded; about 83 percent of expressed demand for these offenses resulted in a verified crime report. Calls for service for Part II crimes were least likely to generate a verified crime; verified Part II offenses are about 64 percent of calls for service for this category over all 74 community areas.

The coefficient of variation in column 3 of [the table] expresses the variation in the recording ratio relative to the mean level of each indicator. There is not much variation in the 74 community areas for personal crimes with a coefficient of variation of .18. Overall, however, there is not much difference in the distribution of values for the recording ratio across the 74 community areas. This finding is important in itself. It suggests that police decisions to record incidents as verified crimes are similar across community areas. To the extent that police are exercising discretion in deciding which offenses to officially record, that discretion is being relatively uniformly applied throughout the city. There are some differences between community areas regarding Part I crimes against the person, but even these differences are not particularly large. . . .

Mean, Standard Deviation, and Coefficients of Variation for the Recording Ratio by Type of Incident[a] ($N = 74$ Community Areas)

Type of Call	Mean[b]	Standard Deviation	Coefficient of Variability
Rape	49.01	36.25	.74
Robbery	67.98	17.54	.26
Assault	99.81	36.24	.36
Burglary	76.50	15.03	.20
Personal crimes	76.05	13.94	.18
Property crimes	83.10	9.52	.11
Part II crimes	64.23	8.57	.13

[a]Table 2 in the original. From Official Chicago Police Department records.

[b]Recording ratio for each incident type $= \frac{\text{verified crimes}}{\text{calls for service}} \times 100$.

Research Example 4.5 used the coefficient of variation to make these kinds of comparisons between variables. The example looks at the reliability in police reports of different types of crime in Chicago. Note again that if we compare the standard deviations directly, we get a different impression of the relative variation in the variables than with *CV*. According to the standard deviations, the recording ratios for rape and assault are approximately equal. But because the incidence of assault is so much greater (look at the means) the standard deviations are not comparable. The coefficients of variation reveal that the variation in reporting rapes ($CV = .74$) actually was far greater than was the variation in reporting assaults ($CV = .36$).

We can examine the other variables in R.E. 4.5 for more examples of how a direct comparison of standard deviations is deceiving. Note that there was the least variation in the reports of property crimes ($CV = .11$), not in Part II crimes (the miscellaneous category of all crimes not listed separately in the table), unlike what the standard deviations may have suggested. Because this table reports the means and standard deviations, we can verify the computations of the *CVs* if we wish.

Keys to Understanding

The keys to understanding the uses of measures of variation can be organized in the same way as the keys for measures of central tendency in Chapter 3. There are three main keys: (1) the level of measurement, (2) the form of the distribution, and (3) the purpose of the research.

Level of Measurement

Key 4.1. Determine the level of measurement. The level of measurement sets a minimum limit for the use of any variation measure. The only measure discussed here that can be applied to nominal level data is the index of dispersion. As long as the categories in a variable are exhaustive and mutually exclusive, the *D* statistic can be calculated. Of course, it can be applied to data at any level above nominal, too. But this is practical only for those variables with a few categories. The common use of *D* is restricted to nominal and ordinal variables. When it is applied to an ordinal variable it does not give any attention to the ordering of the categories; the variable is treated as though it were merely nominal.

There are no widely accepted variation measures designed specifically for ordinal-level variables. But the range and range-based measures, such as the interquartile range, can be applied usefully to ordinal data with a restriction. The restriction is that the end points in these ranges cannot be subtracted; they can be only listed. These same measures can be used without restriction for interval and ratio-level data.

There are several variation measures appropriate for interval-level variables. In addition to the range-based measures, there are the mean deviation and standard deviation. Both provide meaningful summaries of variation for interval data.

There is only one measure discussed that is appropriate for ratio-level variables. It is the coefficient of variation. While it may be used with "strong interval" data, the practice is not generally acceptable. Clearly, it cannot be used with data at the nominal or ordinal level.

All the research examples in this chapter appear to conform to the minimal level of measurement requirement in their selection of variation measures. The only problematic example is R.E. 4.3. There the researchers used the 1 to 9 rating scale to measure occupational prestige. Means and standard deviations were reported for these data, implying that it was of interval quality. Recall our earlier discussions about this type of rating scale. Some researchers consider these measures to have only ordinal quality. They would not use the mean and standard deviation for these data. Others do consider this measurement to have interval quality. Apparently, the researchers in R.E. 4.3 are among the second category.

Form of the Distribution

Key 4.2. Determine the form of the distribution. The form of a distribution refers to its shape. We are concerned with the effect of extreme scores in the distribution of an interval and ratio variable. Some of the interval measures are more sensitive to extreme scores than are others. That sensitivity can cause a measure to distort our impression of the variation of the scores overall. By a wide margin, the range is the one measure most susceptible to this possibility. Because the range relies only on the two most different scores in a distribution, it is completely at their mercy. If one or both of these is greatly different from the other scores, the range will exaggerate the variation.

The other measures based on the range reduce this sensitivity by examining some middle part of the entire distribution. But they can yet distort the variation summary. Since they only examine a part of the complete distribution, they give only a partial description of the overall variation. You must be careful not to project the impression of variation created by a range-based measure to the entire distribution.

The mean deviation and standard deviation have nearly identical indirect sensitivity to extreme scores. Their computations each use deviation scores and the mean, which have direct sensitivity. Ordinarily, this is a significant problem only when a distribution is discontinuous and the number of cases is small. You should look for extreme scores in distributions reporting these measures, if possible. If these scores are present, they will influence the calculated answers and their interpretations. Although important, the sensitivity to extreme scores for these measures is usually far less than for the range and the range-based measures. The coefficient of variation is seldom affected by extreme scores because both of its components (the standard deviation and the mean) have similar sensitivity, which is negated by the division of the one into the other.

Our research examples do not provide us with information about the form of their distributions. This is a common problem with research published in professional journals. We cannot evaluate their use of variation measures by this key.

Purposes of the Research

Key 4.3. Determine the purposes of the research. Let's consider three common purposes for using variation statistics: (1) to describe one distribution, (2) to compare one distribution to another, and (3) to make inferences from a sample to a population.

All the measures in this chapter can be used to make descriptive summaries of how a single variable is distributed. The exact definition of variation changes with each measure, but in its own way each gives you useful information about the homogeneous–heterogeneous character of a set of scores. For example, if a researcher wants to know whether the scores in a distribution could have been more widely spread than they are now, the index of dispersion would provide an answer. Or if a researcher wants to know the arithmetic average of the deviations around a variable's mean, the mean deviation can be used. One common feature of each measure of variation is that larger answers always mean more variation if all other things are equal.

Descriptions of a distribution usually combine a summary of its variation with a summary of its central tendency. Common pairings are to use the index of dispersion with the mode for nominal data, the interquartile range with the median for ordinal data, and the range and/or standard deviation with the mean for interval and ratio data. Often the coefficient of variation is added for ratio data, as we saw in R.E. 4.5.

Much research has the goal of comparing variation between distributions. The index of dispersion and the coefficient of variation each have this type of capability, universally. The D value for a distribution of religious orientations in Ming City is directly comparable to one for Bristol Heights. Similarly, a CV value for the ages of inmates at Hard Rock Penitentiary is directly comparable to one for the ages of inmates at the Campus Compound for White Collar Deviants. Further, each statistic can be compared for distributions on different variables. For example, the D for religious orientations is comparable to the D

for a distribution of political affiliations. In each of these comparisons the larger D value indicates greater variation. The same would be true for the CV statistics.

The standard deviation is comparable across distributions for all variables that are shaped like a normal curve. In these situations it is possible to determine the percentage of scores that fall into any part of the distributions and to compare this information between variables.

Ranges and interquartile ranges can be compared across distributions for the same variable if the means are nearly equal. But comparisons between different variables would seldom be informative. Let us look at two examples. First suppose that we found the ranges in years worked at a foundry separately for its male and female employees. The range for the males was 30 years whereas the range for the females was 10 years, but the mean for both sets was approximately 8 years. It would be reasonable to conclude that females had a smaller difference between their newest and most experienced workers compared to males. Within the range's limited definition of variation, we can say that the females were less varied than the males.

Now suppose that the range for a distribution of family sizes was five members. Knowing this does not tell us anything of importance about its variation relative to variations for the distributions of years worked above. The first two range values cannot be compared to the third because the units of measurement differed (years worked versus number of family members). The same is true of all range-based measures. It is also true for standard deviations when the distributions are not formed approximately like a normal curve or when they do not have the same scales and means. The mean deviation is even less versatile. It is comparable only when calculated for variables having identical scales and means.

Making inferences about a population's variation from sample data is the third of our three purposes. The ability to make accurate inferences depends on the stability of a measure and on knowledge of its behavior over many trials. We have already noted the instability of the range-based measures, especially of the range itself. They would not usually lead to acceptable inferences.

The index of dispersion and the mean deviation have insufficient theoretical backing to allow them to be used for inferences. The probabilities associated with their ability to represent populations when calculated from sample data are not generally known.

Only the standard deviation has both the stability and the theoretical basis required to let the researcher make reliable inferences. There is a known relationship between the standard deviation calculated from a random sample and the standard deviation for the population from which that sample was drawn. We examined that relationship in looking at the formulas for both σ and s above. Researchers who desire to make inferences about variation frequently use the standard deviation (or its relative, the variance).

Each of the research examples in this chapter was attempting to meet all three of the purposes that we have been discussing. Most research does. You may, therefore, have expected each to use the standard deviation because it is the only measure widely accepted for inference purposes. This was the situation for R.E. 4.3 to 4.5. Given the modest reliability noted for the data in R.E. 4.2, it is entirely reasonable that Q was the measure of choice instead. Only R.E. 4.1 appears to have used less than the optimum measure. Recall that it used the range. Actually, the researchers were reporting the ranges (and means) for age and education variables as descriptive (not inferential) summaries of the samples that they used. The main thrust of the article concerned an entirely different issue. It was the tendency of people to present themselves in the best possible light, "social desirability" in the researchers' words. They did use an appropriate inferential measure for this variable.

Overall, there is little negative criticism to raise here about these research examples. Their use of variation measures is generally acceptable.

We can summarize this evaluative information about the measures of variation in the following way:

Index of dispersion (D): a standardized measure at the nominal level that is fully comparable.

Range (R): a rough measure at the ordinal level (if restricted) or the interval level that provides an approximate and potentially unstable indication of variation and is not comparable across different variables.

Interquartile range (Q): a refined range-based measure at the ordinal level (if restricted) or the interval level that is not comparable across different variables.

Mean deviation (MD): a limited use measure at the interval level that is rather stable but not comparable across variables with different scales or means.

Standard deviation (σ, S, s): a highly versatile and rather stable measure at the interval level that is fully comparable for normal distributions and yields unbiased inferences.

Coefficient of variation (CV): a standardized measure at the ratio level that is fully comparable.

One final comment on variation measures should be added. Like measures of central tendency, there is no requirement that only one measure of variation be used to describe a variable. If different purposes are being met and different types of summaries or comparisons are being made, more than one variation measure may be needed and should be reported.

KEY TERMS

Variation
Index of Dispersion (D)
Range (R)
True Limits Range (TLR)
Interquartile Range (Q)
Mean Deviation (MD)
Standard Deviation
  Descriptive Standard Deviation (σ)
  Descriptive Standard Deviation (S)
  Inferential Standard Deviation (s)
Total Variation
Variance
Definitional Formula
Computational Formula
Scale
Coefficient of Variation (CV)
Dichotomy

EXERCISES

1. The following shows the distribution in political party affiliations for a group of students and a group of business executives.

Political Party Affiliation	Students (f)	Business Executives (f)
Democrat	45	31
Republican	26	70
Socialist worker	17	0
Libertarian	7	14
Total	95	115

Find the index of dispersion for (a) the students and (b) the business executives. (c) Which group has greater variation in political party affiliation?

2. The following shows the distribution in occupations for a group of village residents gathered at a local store.

Occupation	f
Farmer	5
Owner of small business	3
Odd-job worker	1
Clerk	4
Retired	7
Total	20

Find the index of dispersion for this distribution and interpret your answer.

3. The following shows the midterm exam scores for the female and male students taking a course in social stratification.

Females		Males	
17	6	25	22
23	30	8	6
11	29	31	24
14	26	14	25
21	24	10	15
14	15	30	40
16	21	15	16
19	23	16	42
28	25	17	17
19	29	18	19

Find the (a) range and (b) interquartile range for each group. (c) Do the groups have the same scale of measurement and mean? (d) Compare the groups by the two variation statistics. Which group is more varied? Does your conclusion change depending on the measure you use? Explain.

4. Recall the scores in Exercise 1 of Chapter 3 concerning the percent of time some social workers spent in direct contact with clients during a one-week period:

19 10 55 31 22 9 18 12 27 14 20 15 19 17 22 40 19 33 5

Find and interpret the (a) range and (b) mean deviation for these scores.

5. Use the same scores in Exercise 3 above to find the (a) mean deviation for each group and (b) descriptive population standard deviation (σ) for each group. (c) Can you directly compare the groups by either variation measure? Explain.
6. The following scores show the distribution in jail capacity for 10 small towns:

   12 17 8 10 11 9 15 5 14 19

   Find the (a) descriptive population standard deviation (σ) using the definitional formula, (b) descriptive sample standard deviation (S) using the computational formula, (c) inferential standard deviation (s) using the definitional formula, and (d) inferential standard deviation (s) using the computational formula. (e) Which of these yield equivalent results? Explain why this happens.
7. Calculate the coefficients of variation for all variables in R.E. 4.4. Order the variables according to their comparative variation.
8. Find one research article in a social science journal that uses any one of the variation measures discussed in this chapter. Explain and evaluate its use according to the *Keys to Understanding*.
9. Evaluate the usefulness of the range, mean deviation, and standard deviation if you were deciding the capacity of a shelter for abused women.
10. Consider a distribution of daily wages paid to migrant farm workers in which the range was $5. Which of the following would you know to be a true conclusion about this distribution? Explain why each of the other conclusions would not be known to be true.
    (a) The lowest wage was $5 per day.
    (b) The distribution was based on the wages of five workers.
    (c) $5 per day was the average wage.
    (d) $5 per day was the difference between the highest and lowest wages.
11. Consider a distribution of age expectancies for retired coal miners in which the standard deviation was four years. Assume that this distribution was shaped like a normal curve. Which of the following would you know to be a true conclusion about this distribution? Explain why each of the other conclusions would not be known to be true.
    (a) The range would be approximately 24 years.
    (b) The arithmetic average of the individual deviations from the mean age expectancy was four years.
    (c) Half of the retirees are not expected to live longer than 12 more years.
    (d) There was not much variation in the age expectations.

*12. The following distribution of simple frequencies describes the scores received by 25 subjects in an experiment conducted to determine rejection of homosexuals.

Rejection Score	f
8	2
7	4
6	5
5	9
4	3
3	0
2	1
1	1
	25

Find the descriptive population standard deviation (σ) using the definitional formula. (See Fig. 4.4 for the formula.)

* **13.** The following distribution of grouped frequencies describes the knowledge of ecology received by 50 respondents in a survey of adults.

Knowledge Score	f
0–8	18
9–17	16
18–26	11
27–35	5
Total	50

Find the inferential standard deviation (s) using the computational formula.

RELATED STATISTICS

Interdecile Ranges

Deciles identify the places within a distribution that have each multiple of 10% of the scores at and below them. For example, the second decile has 20% of the scores at and below it. The formula for the second decile is as follows:

$$D_2 = l_2 + \left[\frac{2N/10 - cf_b}{f_2}\right] w_2$$

where l_2 = true lower limit of the category containing D_2
cf_b = cumulative frequency below the D_2 category
f_2 = frequency within the D_2 category
w_2 = width of the D_2 category
N = number of cases

The formulas for the other deciles follow the same pattern. The interdecide ranges identify that part of the entire range which occurs between two deciles. For example, the 2nd–8th interdecile range is used to locate that part of the entire range between the second (D_2) and eight (D_8) deciles; that part excluding the extreme 20% of the scores from each end of a distribution.

Percentile Ranges

Percentile ranges identify any desired part of the entire range. For example, the 10–90 percentile range locates the part of the range that excludes the extreme 10% of the scores on each end of a distribution. The formulas for percentiles were in the section *Related Statistics* at the end of Chapter 3.

Index of Qualitative Variation (*IQV*)

The index of qualitative variation is a refined version of the index of dispersion (D). IQV is standardized for the number of categories in a measurement, unlike D. Neither IQV or D is universally superior to the other. Rather, their advantages depend on the research situation. See: Mueller et al. (1977:176–179).

Interquartile Range (*Q*) Formulas for Simple and Grouped Frequency Distributions

For simple frequency distributions:

1. Calculate a less-than cf distribution.
2. Find the position of Q_1 using the $N/4$ formula.
3. Locate the category containing this position using the cf distribution.
4. Assign to Q_1 the value of the category containing this position.

5. Repeat steps 2 to 4 for Q_3 except use the $3N/4$ formula for its position.
6. Find the interquartile range by the formula $Q = Q_3 - Q_1$.

For grouped frequency distributions:

1. Calculate a less-than *cf* distribution.
2. Locate the category containing Q_1 using the $N/4$ position formula.
3. Assign to Q_1 an estimated value by the formula

$$Q_1 = l_1 + \left[\frac{N/4 - cf_b}{f_1}\right] w_1$$

where l_1 = true lower limit of the category containing Q_1
cf_b = cumulative frequency below the Q_1 category
f_1 = frequency within the Q_1 category
w_1 = width of the Q_1 category
N = number of cases

4. Locate the category containing Q_3 using the $3N/4$ position formula.
5. Assign to Q_3 an estimated value by the formula

$$Q_3 = l_3 + \left[\frac{3N/4 - cf_b}{f_3}\right] w_3$$

where l_3 = true lower limit of the category containing Q_3
cf_b = cumulative frequency below the Q_3 category
f_3 = frequency within the Q_3 category
w_3 = width of the Q_3 category
N = number of cases

6. Find the estimated value of Q using the formula $Q = Q_3 - Q_1$.

Mean Deviation (*MD*) Formulas for Simple and Grouped Frequency Distributions

For simple frequency distributions:

$$MD = \frac{\Sigma f|x|}{N}$$

where $|x|$ = absolute value of a category's deviation score $(X - \bar{X})$
$f|x|$ = frequency of each category multiplied by the absolute value of its deviation score
Σ = summation (of the $f|x|$ products)
N = number of cases

For grouped frequency distributions:

$$MD = \frac{\Sigma f|m - \hat{\bar{X}}|}{N}$$

where $|m = \hat{\bar{X}}|$ = absolute value of the deviation of a category's midpoint from the distribution's estimated mean
$f|m - \hat{\bar{X}}|$ = frequency of each category multiplied by the absolute value of its $(m - \hat{\bar{X}})$ deviation
Σ = summation (of the $f|m - \hat{\bar{X}}|$ products)
N = number of cases

CHAPTER

DESCRIBING ASSOCIATION BETWEEN NOMINAL AND ORDINAL VARIABLES

One of the most interesting kinds of questions that researchers ask about the social world is whether characteristics are associated. Is crime associated with poverty? Is religion associated with bigotry? Is alienation associated with urbanization? Is income associated with education? Is opportunity associated with gender? Is race associated with employment? And on and on for nearly every pair of features in society. Indeed, after we ask the descriptive question of how things are distributed, it is logical to ask the association question of how those same things are connected, if at all.

At the most basic level, *association* between variables refers to the description of their joint occurrence. In the bivariate situation researchers examine data arranged in a cross-classification table to see how scores on one variable are paired or joined with scores on another variable. They seek to determine whether there is a pattern to the pairing of scores and then to describe the character of any pattern that is found. At higher analysis levels, association is examined for predictive capability, explanation of variance, and inferences. All except the last of these (inferences) are explored in this chapter and the one that follows. We look at association between nominal or ordinal variables here. We look at association between interval or ratio variables in Chapter 6. Inferences concerning association are covered in Chapter 12.

One final introductory comment must be made. Association is not the same as cause and effect. Variables can be associated without either being the cause of the other. Cause and effect is a theoretical concept, not a statistical concept. Statistical analysis may help test the reasonableness of a causal theory, but it cannot prove that variables have a cause-and-effect relationship.

It is true that association is a necessary condition for cause and effect, but it is not a sufficient condition. This means that without an association between variables there can be no causal connection. But the existence of an association is not a guarantee of a causal relationship. For example, frustration and violence are often associated; they often occur together. But frustration is not always present when violence occurs. Nor does the presence of frustration mean that violence will automatically result. One possible explanation for this is that both frustration and violence may be the result of some other factor, such as social learning, which makes them associated with each other but not causally connected. Thus we cannot say that a frustration–violence association demonstrates that frustration causes violence. It is common for people *not* to make these kinds of distinctions. Be sure that you do not read "cause and effect" when a researcher says "association."

Association at the Nominal Level

There are two very different approaches to describing association. The first is a method that focuses on the specifics of each analysis situation by devising unique techniques that will reveal associations. This method's specific focus often causes it to have limited applicability and comparability for different situations. The second approach has a much wider focus which eases diverse applications and comparisons. For nominal-level data the first approach is represented by what are called chi-square-based measures. The second is represented by what are called PRE (proportional reduction in error) measures. We will look at both types for nominal variables because they are found often in social research.

There are two main characteristics of association that are studied at the nominal level. First, researchers want to know whether an association *exists*. Second, they want to know the *strength* of any association that exists. We look at how each of the approaches to nominal association handle these two characteristics.

Basics of Association—Conditional Distributions

As noted above, association is fundamentally an examination of the joint occurrence between variables. We can see this most clearly from a bivariate distribution.

Consider the four bivariate frequency distributions shown in Table 5.1. These illustrate four possible ways in which employment status and sex could be jointly distributed. We want to know which of these shows the presence of an association between the two variables. Each subtable presents two conditional distributions. A *conditional distribution* reveals the distribution of one variable under a fixed value of another variable. If we view the four subtables from the point of view of the employment status variable, each subtable shows us how employment status was distributed separately for males and females. In other words, they tell us how employment status was distributed under the condition of being male in the first column and under the condition of being female in the second column. These columns present the two conditional distributions.

An association between two variables is said to *exist* if there are any differences in comparable conditional distributions. To make the conditional distributions comparable, they should be converted to a relative data form, such as percentages. Since we want to compare the columns to each other (males to females) for each employment status category, the percentages are based on the column totals. The column-percentaged subtables appear in Table 5.2.

Now we can compare the percentages to determine whether any differences exist. In the first distribution (a) there are no differences between the columns for each row. Sixty percent of each sex are employed. Further, 40% of each sex are not employed. Thus we can say that there is no association between sex and employment status in the (a) distribution. This means that employment status is independent of sex. It does not matter whether we look at males or females, the same percentages are employed or unemployed.

In the (b) distribution we see that there are some differences. In the first row, 80% of the males are employed but only 40% of the females are employed. A generally reversed pattern is found in the second row. There we see that only 20% of the males are not employed but 60% of the females are not employed. Thus we can say that there is some association between sex and employment status in the (b) distribution. A greater percentage of men than women are employed. A greater percentage of females than males are not employed. But the association is not without exceptions. Some women are employed and some men are not employed.

TABLE 5.1 Association for Bivariate Frequency Distributions at the Nominal Level

(a) No Association:

Employed	Sex: Male	Sex: Female	
Yes	30	30	60
No	20	20	40
	50	50	100

(b) Some Association:

Employed	Sex: Male	Sex: Female	
Yes	40	20	60
No	10	30	40
	50	50	100

(c) Perfect Association:

Employed	Sex: Male	Sex: Female	
Yes	50	0	50
No	0	50	50
	50	50	100

(d) Alternative Perfect Association:

Employed	Sex: Male	Sex: Female	
Yes	50	10	60
No	0	40	40
	50	50	100

TABLE 5.2 Conditional Distributions in Percentages for Table 5.1 Distributions

(a) No Association:

Employed	Sex: Male	Sex: Female
Yes	60	60
No	40	40
	100	100
	(50)	(50)

(b) Some Association:

Employed	Sex: Male	Sex: Female
Yes	80	40
No	20	60
	100	100
	(50)	(50)

(c) Perfect Association:

Employed	Sex: Male	Sex: Female
Yes	100	0
No	0	100
	100	100
	(50)	(50)

(d) Alternative Perfect Association:

Employed	Sex: Male	Sex: Female
Yes	100	20
No	0	80
	100	100
	(50)	(50)

In the (c) distribution another pattern emerges. Not only are there differences in the percentage of each sex employed or not employed, but one of the percentages in each row is zero. These zero percentages occur in opposite row–column combinations: employed–female and not employed–male. Thus we can say that there is a "perfect" or complete association. All cases of one sex are in one category of the employment variable, and all cases of the other sex are in the opposite employment category. There are no exceptions.

The (d) distribution shows an alternative meaning of "perfect" association. Here perfect means that all cases in one category of sex are in one category of employment status but that pattern is not mirrored for the other employment status–sex combination. Whereas all males were employed, not all females were not employed. Thus one of the conditional distributions shows a complete concurrence between sex and employment status, but the other does not.

When researchers measure association statistically, they want the measures to reflect our understanding of association as illustrated in these four distributions. Quantitatively, researchers prefer association measures that are normed. A *normed measure* has a fixed scale of values. The ideal is to have a value of 0 when the association does not exist, as in the (a) distribution. When there is a perfect association, as in (c) or (d), the measure should yield a maximum value such as 1. When there is some association that is less than perfect, the measure should have a value between 0 and 1 which proportionately reflects the extent of the association. Obviously, you need to know which definition of "perfect" is being used for each measure. There are statistics of both kinds. Each of the chi-square-based measures is only partially normed. The PRE measures are completely normed.

Chi-Square-Based Measures

A number of measures of association at the nominal level are based on Pearson's chi-square statistic. Chi-square and many other measures used for statistical description and inference were developed by Karl Pearson, an English statistician (1857–1936). Several of these measures are included in this book. See "On the Side" 6.1, p. 215, for a short biography of Pearson.

Chi-square (χ^2) is most often used as a significance test to determine whether or not sample data warrant the conclusion that an association or difference between two nominal variables exists in a population. We will save our discussion of this use of chi-square until the inference section of the book (Part III). For now we look at chi-square as a descriptive indicator of association existence and its uses with measures of association strength such as Pearson's phi (ϕ), Pearson's contingency coefficient (C) and Cramér's V. Applications of ϕ and V are illustrated in Research Examples 5.1 and 5.2. You might like to look ahead at these to see how the use of these measures commonly appears in published research.

Chi-square (χ^2)

The basic component to the computation of several measures of association at the nominal level is the chi-square statistic (χ^2). Chi-square compares the actual frequencies for each cell in a bivariate table to the frequencies that would have occurred had there been no association at all between the variables. The actual frequencies are referred to as *observed frequencies* (f_o) and the no association frequencies are referred to as *expected frequencies* (f_e). The common formula for chi-square is

$$\chi^2 = \Sigma\frac{(f_o - f_e)^2}{f_e} \qquad \text{Chi-Square Formula}$$

where f_o = observed frequency in a cell
f_e = expected frequency in a cell

The computations are performed for each cell and then summed to find chi-square for the entire table.

Let's use the Table 5.1(b) distribution as an example of finding the chi-square statistic. The observed frequencies are those listed in that subtable. The expected frequencies are determined by examining the conditional distributions assuming that employment status is independent of sex. We saw what these frequencies are in the (a) distribution of Table 5.1. It indicated that 30 of each sex were employed and 20 of each were not employed. The general logic used to determine expected frequencies is to ask how each of the row marginal totals would have been distributed if they were distributed parallel to the frequencies as shown in the column marginals. In our example, the row marginals show us that there are 60 employed and 40 unemployed persons. The column marginal shows us that there are 50 males and 50 females, half of the total in each category of sex. We first ask how the 60 employed persons would be distributed between the sexes if they were also split half and half like the column marginals. It would be 30 in each category of sex. Similarly, we ask how the 40 not employed persons would be distributed if they also were split half and half between the sexes. Obviously, it would be 20 in each category.

Because the splits are not always so simple (such as half and half), a formula can be used to determine the correct expected frequencies regardless of the marginals. It is

$$f_e = \frac{(T_r)(T_c)}{N} \qquad \text{Expected Frequency Formula}$$

where T_r = marginal total of the row for a cell
T_c = marginal total of the column for a cell
N = number of cases (the overall total)

Thus the expected frequencies for the four cells in the (b) distribution of Table 5.1 can be confirmed:

$$\text{(upper left cell) } f_e = \frac{(60)(50)}{(100)} = 30$$
$$\text{(upper right cell) } f_e = \frac{(60)(50)}{(100)} = 30$$
$$\text{(lower left cell) } f_e = \frac{(40)(50)}{(100)} = 20$$
$$\text{(lower right cell) } f_e = \frac{(40)(50)}{(100)} = 20$$

(Expected frequencies are usually shown in parentheses beside the obtained frequencies in tables displaying the calculation of chi-square.)

Chi-square proceeds to compare these no-association (expected) frequencies to the observed frequencies in the actual distribution. Following the formula above, the chi-square value for the (b) distribution in Table 5.1 is

$$\begin{aligned}
\chi^2 &= \frac{(40-30)^2}{30} + \frac{(20-30)^2}{30} + \frac{(10-20)^2}{20} + \frac{(30-20)^2}{20} \\
&= \frac{(10)^2}{30} + \frac{(-10)^2}{30} + \frac{(-10)^2}{20} + \frac{(10)^2}{20} \\
&= \frac{100}{30} + \frac{100}{30} + \frac{100}{20} + \frac{100}{20} \\
&= 3.\overline{3} + 3.\overline{3} + 5 + 5 \\
&= 16.\overline{6}
\end{aligned}$$

The chi-square computations for all four of our frequency distributions are shown in Table 5.3. The value of chi-square in the (a) distribution is 0. This will always happen when there is no difference between the observed and expected frequencies for every cell

TABLE 5.3 Finding Chi-Square for Bivariate Distributions

(a) No Association:

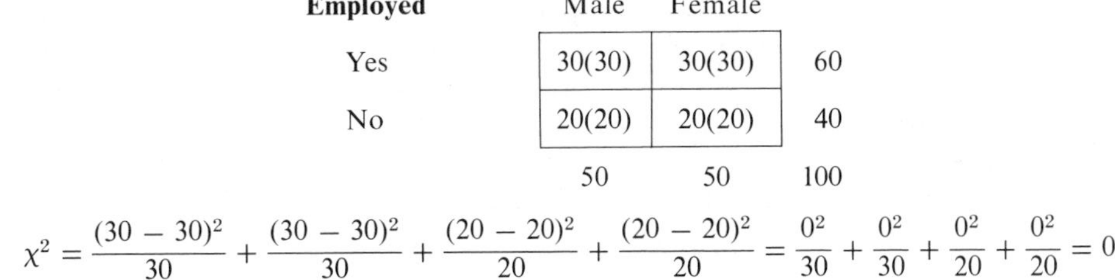

	Sex		
Employed	Male	Female	
Yes	30(30)	30(30)	60
No	20(20)	20(20)	40
	50	50	100

$$\chi^2 = \frac{(30-30)^2}{30} + \frac{(30-30)^2}{30} + \frac{(20-20)^2}{20} + \frac{(20-20)^2}{20} = \frac{0^2}{30} + \frac{0^2}{30} + \frac{0^2}{20} + \frac{0^2}{20} = 0$$

(b) Some Association:

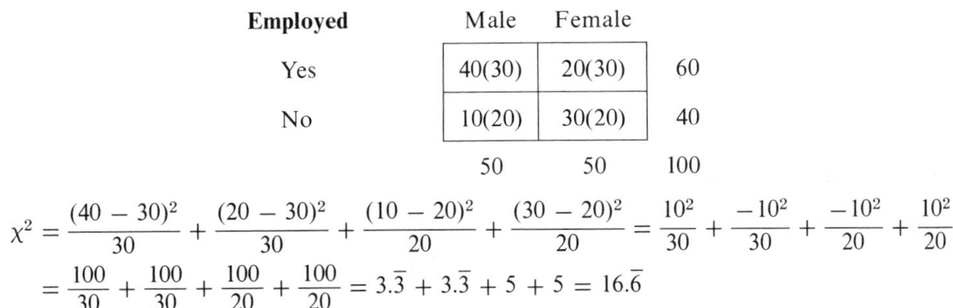

	Sex		
Employed	Male	Female	
Yes	40(30)	20(30)	60
No	10(20)	30(20)	40
	50	50	100

$$\chi^2 = \frac{(40-30)^2}{30} + \frac{(20-30)^2}{30} + \frac{(10-20)^2}{20} + \frac{(30-20)^2}{20} = \frac{10^2}{30} + \frac{-10^2}{30} + \frac{-10^2}{20} + \frac{10^2}{20}$$

$$= \frac{100}{30} + \frac{100}{30} + \frac{100}{20} + \frac{100}{20} = 3.\overline{3} + 3.\overline{3} + 5 + 5 = 16.\overline{6}$$

(c) Perfect Association:

	Sex		
Employed	Male	Female	
Yes	50(25)	0(25)	50
No	0(25)	50(25)	50
	50	50	100

$$\chi^2 = \frac{(50-25)^2}{25} + \frac{(0-25)^2}{25} + \frac{(0-25)^2}{25} + \frac{(50-25)^2}{25} = \frac{25^2}{25} + \frac{-25^2}{25} + \frac{-25^2}{25} + \frac{25^2}{25}$$

$$= \frac{625}{25} + \frac{625}{25} + \frac{625}{25} + \frac{625}{25} = 25 + 25 + 25 + 25 = 100$$

(d) Alternative Perfect Association:

	Sex		
Employed	Male	Female	
Yes	50(30)	10(30)	60
No	0(20)	40(20)	40
	50	50	100

$$\chi^2 = \frac{(50-30)^2}{30} + \frac{(10-30)^2}{30} + \frac{(0-20)^2}{20} + \frac{(40-20)^2}{20} = \frac{20^2}{30} + \frac{-20^2}{30} + \frac{-20^2}{20} + \frac{20^2}{20}$$

$$= \frac{400}{30} + \frac{400}{30} + \frac{400}{20} + \frac{400}{20} = 13.\overline{3} + 13.\overline{3} + 20 + 20 = 66.\overline{6}$$

in a table. Any calculated chi-square that is not 0 means that there is some association in a table. Thus we know that the (a) distribution has no association whatsoever, whereas the (b), (c), and (d) distributions do have some association. But we see from the (c) distribution that chi-square is not 1 when there is perfect association; its chi-square value was 100. For the alternative perfect association (d), chi-square is $66.\bar{6}$. When there is some association that is less than perfect, such as (b), the chi-square value will be between the two extremes for no association and either version of perfect association. Here it was $16.\bar{6}$. But it cannot be translated as a direct proportional indication of the strength of an association. As a result, chi-square cannot be used as a measure of association in its present form.

The maximum value of chi-square is dependent on the number of cases and the size of the table* for which it is being calculated. Its exact maximum can be found with the following formula:

$$\text{Max } \chi^2 = N(k - 1) \qquad \text{Maximum Chi-Square Formula}$$

where N = number of cases
k = number of categories in the row or column variable, whichever is less

In our example:

$$\begin{aligned}\text{Max } \chi^2 &= 100(2 - 1)\\ &= 100(1)\\ &= 100\end{aligned}$$

Chi-square will achieve this maximum only in a square table (same number of categories in each variable) with uniform distributions for both row and column marginals, such as our (c) distribution. Otherwise, it will always be less than the value found by the Max χ^2 formula. Chi-square can be made into a useful measure of association strength by comparing its actual value in a table to its maximum possible value. Each of the chi-square-based measures does this.

Pearson's Phi (ϕ)

Pearson's phi (ϕ) is a chi-square-based measure for tables with two categories in either the row or column variable. In a $2 \times k$ table (k = the number of categories in the second variable) the maximum value of chi-square is N [Max $\chi^2 = N(2 - 1) = N$]. With the addition of the square root function for other mathematical reasons, the formula for ϕ is

$$\phi = \sqrt{\frac{\chi^2}{N}} \qquad \text{Phi formula}$$

Since each of our four distributions has only two categories in either variable, phi is

*Bivariate table sizes are identified by a standardized $r \times c$ notation; r indicates the number of row categories and c indicates the number of column categories, both exclusive of any totals. Thus a 2×3 table has two row categories and three column categories.

a suitable measure of association for them. Applying its formula, we get the following results:

$$\text{(a) } \phi = \sqrt{\frac{0}{100}} = \sqrt{0} = 0$$

$$\text{(b) } \phi = \sqrt{\frac{16.\bar{6}}{100}} = \sqrt{.16\bar{6}} = .408$$

$$\text{(c) } \phi = \sqrt{\frac{100}{100}} = \sqrt{1.0} = 1.0$$

$$\text{(d) } \phi = \sqrt{\frac{66.\bar{6}}{100}} = \sqrt{.66\bar{6}} = .816$$

The values of phi are comparable for all $2 \times k$ tables. The closer it is to 1, the stronger the association is. Thus phi correctly indicates that the (b) and (d) distributions reflect a middle degree of association between (a) and (c).

Pearson's Contingency Coefficient (C)

If phi is applied to tables larger than $2 \times k$, its maximum value exceeds 1. As a result its meaning becomes ambiguous. *Pearson's contingency coefficient* (C) tries to solve this problem by constructing a measure that cannot exceed 1 for any table. Its formula is:

$$C = \sqrt{\frac{\chi^2}{\chi^2 + N}} \qquad \text{Pearson's } C \text{ Formula}$$

C will equal 0 in a table of no association and will increase in value as the association becomes stronger. But there are two difficulties with C. First, it cannot reach 1 for any table. Second, its maximum value changes for each table size. The maximum for a square table is found by the formula

$$\text{Max } C = \sqrt{\frac{k-1}{k}} \qquad \text{Maximum } C \text{ Formula for Square Tables}$$

where k = number of categories in the row or column variable of a square table

The maximum value of C cannot readily be determined for nonsquare tables. As a result, C provides a useful measure of association only for individual tables or when a researcher is comparing the same-sized tables to each other. To illustrate its computation, C is found as follows for our (b) distribution:

$$C = \sqrt{\frac{16.\bar{6}}{16.\bar{6} + 100}} = \sqrt{.142857} = .378$$

TABLE 5.4 Finding ϕ, C, and V for Bivariate Associations

(a) No Association:

$\chi^2 = 0 \qquad N = 100$

$$\phi = \sqrt{\frac{\chi^2}{N}} = \sqrt{\frac{0}{100}} = 0$$

$$C = \sqrt{\frac{\chi^2}{\chi^2 + N}} = \sqrt{\frac{0}{0 + 100}} = 0$$

$$V = \sqrt{\frac{\chi^2}{N(k - 1)}} = \sqrt{\frac{0}{100(1)}} = 0$$

(b) Some Association:

$\chi^2 = 16.\bar{6} \qquad N = 100$

$$\phi = \sqrt{\frac{16.\bar{6}}{100}} = \sqrt{.16\bar{6}} = .408$$

$$C = \sqrt{\frac{16.\bar{6}}{16.\bar{6} + 100}} = \sqrt{\frac{16.\bar{6}}{116.\bar{6}}} = \sqrt{.142857} = .378$$

$$V = \sqrt{\frac{16.\bar{6}}{100(1)}} = \sqrt{\frac{16.\bar{6}}{100}} = \sqrt{.16\bar{6}} = .408$$

(c) Perfect Association:

$\chi^2 = 100 \qquad N = 100$

$$\phi = \sqrt{\frac{100}{100}} = \sqrt{1} = 1.0$$

$$C = \sqrt{\frac{100}{100 + 100}} = \sqrt{\frac{100}{200}} = \sqrt{.5} = .707$$

$$V = \sqrt{\frac{100}{100(1)}} = \sqrt{\frac{100}{100}} = \sqrt{1.0} = 1.0$$

(d) Alternative Perfect Association:

$\chi^2 = 66.\bar{6} \qquad N = 100$

$$\phi = \sqrt{\frac{66.\bar{6}}{100}} = \sqrt{.66\bar{6}} = .816$$

$$C = \sqrt{\frac{66.\bar{6}}{100 + 66.\bar{6}}} = \sqrt{\frac{66.\bar{6}}{166.\bar{6}}} = \sqrt{.40} = .632$$

$$V = \sqrt{\frac{66.\bar{6}}{100(1)}} = \sqrt{\frac{66.\bar{6}}{100}} = \sqrt{.66\bar{6}} = .816$$

Cramér's V

Cramér's V represents an attempt to create a more universal measure of association for two nominal variables. It is found as follows:

$$V = \sqrt{\frac{\chi^2}{N(k - 1)}} \qquad \text{Cramér's } V \text{ Formula}$$

where k = number of categories in the row or column variable, whichever is less

V can be found for tables of any size, square or not. Its minimum is 0, which indicates no association. Perfect association will result in a value of 1. Values between 0 and 1 do suggest the degree of strength in an association. V for our (b) distribution would be found as follows:

$$V = \sqrt{\frac{16.\bar{6}}{100(1)}} = \sqrt{.16\bar{6}} = .408$$

Note that $V = \phi$ for a 2×2 table. The computations of ϕ, C, and V for all four distributions are shown in Table 5.4.

Interpreting Association Coefficients

Measures of association yield coefficients (indicator values) that tell you whether an association exists and how strong any existing association is. These coefficients are quantitative. As we have seen, most association measures will equal 0 when there is no association at all between two variables. Values different from 0, then, mean that there is an association. The stronger that association is, the larger the coefficient will be. The clearest way to interpret coefficients for the same measure is to compare them to each other and to their maximum possible value. Phi coefficients can be compared directly to each other for all $2 \times k$ tables. The distribution with the highest phi value is the one with the strongest association. Thus our (c) distribution had the strongest association ($\phi = 1.0$). C coefficients can be compared directly for the same-sized tables, especially square ones. Again, the largest value of C identifies the strongest association. V coefficients can be compared directly for tables of any size. Larger values mean stronger associations. One comparison we generally cannot make is between coefficients for different measures. For example, a C of .40 is not directly comparable to a V of .50.

For individual tables, the coefficients can be interpreted in reference to their maximum possible values. Since we know that the maximum value of phi in a $2 \times k$ table is 1, we can compare any computed phi to this maximum to see how close it is to the strongest possible association. Values of V can also be compared to 1. A C value must be compared to the exact maximum value that could occur for the specific size of the table for which it is being computed.

It is common for researchers to go beyond this quantitative meaning of association strength by attaching such verbal phrases as "weak," "moderately strong," "very strong," and so on. This is more "an exercise in art than science." The verbal phrases are somewhat arbitrary if they are not grounded in the context of the specific research being conducted. In one context a researcher may be very surprised to find much association at all between two variables, and therefore would be inclined to call a V of .250 a moderate association. In another context a researcher may be disappointed to find a V of "only" .250 and call it a very weak association.

The verbal interpretations may also be influenced by the strength of other associations in the same area of research. If most of the associations are .750 and above, any that are much below this will seem modest, if not weak, in comparison. On the other hand, if most of the association coefficients are running .330 or less, a few that are .450 or better will seem strong to the researcher. So you see that the verbal meanings you find in research reports may be quite subjective.

Figure 5.1 presents an approximate guide to use for giving verbal interpretations of

FIGURE 5.1 Approximate Guide to Verbal Interpretations of Strength for Association Measures Ranging from 0 to (±)1

Value of Measure	Verbal Interpretation
0	No association
.01–.25	Weak association
.26–.55	Moderate association
.56–.75	Strong association
.76–.99	Very strong association
1	Perfect association

strength to association coefficients. This guide is no less arbitrary than the varied methods referred to above. However, it does match with most common practice in social research and should be known in a general way for this reason. The cautionary note still holds: When you read research reports, remember that the verbal interpretations are flexible. They are an imprecise auxiliary to the quantitative coefficients that have been carefully found by statistical analysis. Only those quantitative values have constant meanings.

Research Example 5.1 shows the use of the phi measure for two samples of unwed mothers in North Carolina (whites and blacks). The study examined stereotyped notions of unwed mothers as deviants. In the first table presented we see that most unwed mothers did live in the same town or place as their sex partners. After calculating chi-square, the phi coefficient was found to be .20 for this table. This means that there was only a weak association between race and place where the unwed father lived compared to the unwed mother. The second table shows that most of the families of the unwed mothers did know the unwed fathers. Again, phi indicates only a weak association between race and family acquaintance with the unwed fathers ($\phi = .22$.) Both tables show some rejection of common wisdom about illegitimacy.

RESEARCH EXAMPLE 5.1 Using Phi to Measure Association

From Hallowell Pope, "Unwed Mothers and Their Sex Partners," Journal of Marriage and the Family, 29 *(3), August 1967:555ff. Reprinted by permission.*

[Previous perspectives] encouraged the view that it takes a very deprived, a very immoral, a very stupid, a very deviant, or perhaps a very exploited girl to make such a disastrous mistake as to have an illegitimate child . . . [But the] probability of a girl having a birth out of wedlock is related to many interconnected factors, among them: the composition of her field of eligibles, the nature of her heterosexual partnerships, frequency of premarital intercourse, her fecundity, her knowledge and use of contraception; and, if she becomes premaritally pregnant, her attitudes toward and the availability of abortion, her possibilities of getting married before giving birth, etc. There is no reason to believe that a "normal" person in a "normal" relationship might not become premaritally pregnant *if we avoid the tautology* of arguing that a relationship is abnormal or deviant *because* it results in a premarital pregnancy for the girl. The premarital pregnancy may indicate a prior deviant act as defined by traditional moral standards, but it is a mistake nowadays to assume that such an act must have been committed by an abnormal person, within a deviant relationship, and in an unusual social setting. We also must avoid the mistake of using single-cause theories and commonsense research categories in trying to account for illegitimacy. . . .

The universe sampled consisted of all those women in selected counties of North Carolina who were recorded on birth certificates as mothers of illegitimate children during 1960 and 1961—North Carolina had about 1,700 white and 8,200 nonwhite, live illegitimate births for each of these years. (The illegitimacy ratio for 1960 was 9.0 percent overall: 2.2 percent for whites and 23.8 percent for nonwhites.) Counties in each of the state's major socioeconomic regions were chosen so that interviews in them might complete a sample that would be reasonably representative for the state. The goal was to contact all officially recorded white unwed mothers and a one-third sample of Negro unwed mothers in each county selected.

The completed sample included over 1,000 interviews, including 939 with women who had never before been married. The completed cases represent 32 percent of the white cases in sampled counties and 42 percent of those cases actually sought by interviewers. . . . For Negroes, 65 percent of the cases in the sampled counties brought completed interviews or 67 percent of the respondents actually sought. . . .

What was the social context within which the fathers and these primiparous unwed mothers associated? Two-thirds of the white and over four-fifths of the Negro unwed mothers had sex partners who were living in the same localities as they did (same town or "place"—see Table 1 for question used). In addition, in Table 2 over one-half of the white and three-quarters of the Negro unwed mothers' families knew the alleged father well; on the other hand, one-fifth of the white and one-tenth of the Negro families of the unwed mothers did not know the father at the time the woman became pregnant. Among the younger girls (aged 16–20) of both races, the proportion of parents who did not know the father before the pregnancy was even less. Parents of both races apparently had an opportunity to become acquainted with their daughters' sex partners, and, consequently, it may be assumed that they had an opportunity to exercise control over the relationship. . . .

TABLE 1. Comparison of the Place Where the Father and Unwed Mother Lived When She Met Him, by Race[a]

	Place Where Father Lived[b]		
Race	Same As Unwed Mother	Different from Unwed Mother	Total
White	68%	32%	100%
	(212)	(101)	(313)
Negro	85%	15%	100%
	(216)	(38)	(254)

Chi-square = 22.8 Phi = .20

[a]Adapted from Table 1 in the original.

[b]The question used: "Where was he living when you met him—in the same town or place you were, or did he usually live in (come from) a different place?"

TABLE 2. Degree to Which the Unwed Mother's Family Knew the Father at the Time She Became Pregnant, by Race[a]

	Degree to Which the Unwed Mother's Family Knew the Father[b]			
Race	Knew Him and Had Seen Him a Lot	Knew Him but Had Seen Him No More Than Now and Then	Did Not Know Him	Total
White	55%	27%	18%	100%
	(172)	(83)	(57)	(312)
Negro	77%	15%	9%	101%
	(193)	(37)	(22)	(252)

Chi-square = 28.2 Phi = .22

[a]Adapted from Table 2 in the original.
[b]The question used: "Did your family know him?" (If yes) "How well did they know him—had they seen him very much?"

Research Example 5.2 shows Cramér's V being used to measure association between race and confidence in the U.S. Supreme Court. As the excerpt notes, polls in the 1960s found a large difference in levels of confidence between whites and blacks. But in the mid-1970s that difference had disappeared almost completely. A Cramér's V of only .029 confirmed this impression for the more recent data.

The chi-square-based measures have an important limitation. None of them is capable of direct ratio comparisons in the strength of associations. For example, a Cramér's V of .50 does not represent an association that is twice as strong as one with a V of .25. This is one of the major reasons these measures are considered to be unsatisfactory to many researchers. The PRE measures to follow do not have this limitation.

RESEARCH EXAMPLE 5.2 Using Cramér's V to Measure Association

From Lee Sigelman, "Black–White Differences in Attitudes Toward the Supreme Court: A Replication in the 1970s," Social Science Quarterly, *60 (1), June 1979:113ff. Reprinted by permission.*

Do black and white Americans hold dissimilar attitudes toward the Supreme Court? In a study published in 1968, Hirsch and Donohew found that blacks were much more positively oriented toward the Court than were whites; favorable attitudes toward the Court were expressed by less than 30% of the white sample, but by almost 72% of the blacks. This racial difference withstood controls for several other factors, including education, region, income, party identification and sense of political efficacy. . . .

Has support for the Court continued at mid-1960s levels despite the turbulent events of the late 1960s and early 1970s? Or has the pervasive political malaise of the 1970s undermined confidence in the Court? If so, has this impact been felt more or less equally by blacks and whites, so that Court support, although perhaps lower for both groups than it was in the 1960s, continues to be higher among blacks? Or, to take another possibility, have the changes in the Court eroded confidence among blacks in particular, to the point that racial differences in Court support have diminished? . . .

In the yearly General Social Surveys it has conducted since 1973, the National Opinion Research Center (NORC) has asked random samples of Americans about their confidence in a number of key social, cultural, economic and political institutions. NORC's confidence question is the same one that the Harris Poll introduced during the 1960s:

> I am going to name some institutions in this country. As far as the people running these institutions are concerned, would you say you have a great deal of confidence, only some confidence, or hardly any confidence at all in them?

[The U.S. Supreme Court is included in the list.] Each yearly NORC sample numbers approximately 1,500 respondents, of whom an average of between 11% and 12% are black. Pooling the yearly samples over the entire 1973–77 period produces an omnibus sample of 6,644 whites and 900 blacks, although missing data reduce these numbers to approximately 6,300 and 770, respectively. . . .

[T]he first, and in many respects most important, conclusion to be drawn from the five years of NORC surveys is simply that by the mid-1970s there was no longer any substantial black-white difference in confidence in the Court. As [the table] reveals, during this period the very same proportion of blacks and whites (15.6%) expressed "hardly any confidence at all" in the Court and blacks and whites differed very little from one another (by only 4.2%) in the tendency to express "a great deal" or "only some" confidence in the Court. As a reflection of these minimal differences, Cramer's V for the relationship between race and Court support between 1973 and 1977 is only .029. . . .

Black and White Levels of Confidence in the Supreme Court, 1973–1977 (%)

	Race	
Confidence	White (N = 6,334)	Black (N = 776)
A great deal	35.3	31.1
Only some	49.1	53.4
Hardly any	15.6	15.6

$V = .029$

PRE Measures

PRE (proportional reduction in error) measures are more versatile and more informative than are the chi-square-based measures. All PRE measures are normed; they use a standardized scale where the value 0 means there is no association and 1 means there is perfect association. Any value between these extremes indicates the relative degree of association in a ratio comparison sense. The number of cases, the table size, and the variables being measured do not interfere with the interpretation that can be given to them. A PRE measure with a value of .50 does represent an association that is twice as strong as one that has a PRE value of .25.

PRE measures examine our ability to predict the individual scores for one or the other of the variables in a bivariate distribution. The variable that is being predicted is called the dependent variable. Recall from Chapter 2 that bivariate tables can be viewed as showing the joint distribution of one independent (X) and one dependent (Y) variable. The dependent variable is the one that may be explained by the independent variable. In our earlier examples of association between sex and employment status it would be natural for us to think of sex as the independent variable which explains (in part, at least) differences in employment status, the dependent variable.

PRE measures recast our earlier ideas of association as simple joint occurrence into terms of explanation and prediction. For example, we now think of our ability to predict people's employment status on the basis of knowing their sex. If sex if truly associated with employment, we should be able to predict their employment status by knowing their sex. The stronger the association is, the more accurate our prediction will be. But if there is no association, we will not be able to predict anyone's employment status correctly solely on the basis of knowing his or her sex.

PRE measures are based on a comparison of the errors produced by two methods of prediction. In the first method we predict individual scores on the dependent variable using only information from that same variable's univariate distribution. We calculate our number of errors by comparing the predicted score on the dependent variable for every case in the distribution to its actual score.

In the second method we predict each score on the dependent variable using information in the bivariate distribution of the dependent and the independent variables. Again, we calculate the number of errors we make by comparing our predicted scores to the actual scores for every case.

A PRE measure computes a ratio of the reduced number of errors between the two prediction methods to the number of errors made by the first method alone. If the second method improves our prediction ability, it will have fewer errors than the first method and the PRE measure will show the proportional reduction in errors that occurs. The improved prediction will be a consequence of the association between the two variables. Therefore, we can reason that PRE values directly measure the degree of association that exists in a bivariate distribution.

The basic PRE formula for measures of association is

$$\text{PRE value} = \frac{E_1 - E_2}{E_1} \qquad \text{Basic PRE Formula}$$

where E_1 = number of errors made by the first prediction method

E_2 = number of errors made by the second prediction method

Lambda (L)

Lambda (L) is a commonly used PRE measure of association for nominal variables. Let's look at a new example to see how lambda works. Suppose that we were examining a possible association between a person's view toward the need for greater social legislation and race. Table 5.5 shows three possible distributions for such an association based on 200

TABLE 5.5 Three Possible Bivariate Distributions at the Nominal Level

(a) No Association:

(Y) **Need More Social Legislation**	**(X) Race** White	Nonwhite	
Yes	50	50	100
No	50	50	100

(b) Some Association:

(Y) **Need More Social Legislation**	**(X) Race** White	Nonwhite	
Yes	30	80	110
No	70	20	90

(c) Perfect Association:

(Y) **Need More Social Legislation**	**(X) Race** White	Nonwhite	
Yes	0	100	100
No	100	0	100

cases. These show no association (a), some association (b), and perfect association (c). If lambda performs as we expect, it will be proportionally larger as the association becomes stronger and we will have a fully comparable statistic as a descriptive measure.

Lambda uses the basic PRE formula given above. Like all PRE measures, lambda's first prediction method examines the univariate distribution of the dependent variable. In our present example we will consider the social legislation attitude to be the dependent variable that potentially is explained or predicted by race, the independent variable. Let's look at the (b) distribution from Table 5.5, which we believe has some association between the variables. Follow the computations as they are shown in Table 5.6 while we discuss them.

E_1 for lambda is found by determining the mode for the dependent variable and using this as the predicted Y score for every case. We use the mode because we know that more cases have that score than any other single score. We see that the mode was "yes" for the univariate distribution of the need for social legislation. Therefore, we predict that every case said "yes." We are correct for 110 cases, but we are in error for 90 cases. $E_1 = 90$.

The second prediction method examines the bivariate distribution of the two variables. E_2 for lambda is found by determining the dependent variable's mode within each category of the independent variable, white and nonwhite. In our example the modal attitude response for the whites was "no." We now predict that all whites said "no." We are correct for 70 whites, but in error for 30. The mode for the nonwhites was "yes." We now predict that all nonwhites said "yes." We are correct for 80 nonwhites and in error for 20. Together we make 30 + 20 errors with our second prediction method. $E_2 = 50$.

Lambda is found by substituting our values for E_1 and E_2 into the basic PRE formula on page 162.

TABLE 5.6 Finding Lambda for the Table 5.5(b) Bivariate Distribution of Nominal Data

(Y) Need More Social Legislation	(X) Race: White	Nonwhite	
Yes	30	80	110
No	70	20	90

E_1 = The mode of the dependent variable is "yes," so we predict every case to say "yes." We are correct for 110 cases and in error for 90 cases.
= 90

E_2 = The mode for whites is "no," so we predict every white to say "no." We are correct for 70 cases and in error for 30 cases. The mode for nonwhites is "yes," so we predict every nonwhite to say "yes." We are correct for 80 cases and in error for 20 cases.
= 30 + 20 = 50

$$\text{lambda} = \frac{E_1 - E_2}{E_1} = \frac{90 - 50}{90} = \frac{40}{90} = .444$$

$$L = \frac{E_1 - E_2}{E_1} \qquad \text{Lambda Formula}$$

where E_1 = number of errors made by predicting all cases to have the same Y score as the overall Y mode

E_2 = number of errors made by predicting all cases in each X category to have the same Y score as the Y mode within that category

In our example we make 40 fewer errors with the second prediction method than the first ($E_1 - E_2 = 90 - 50 = 40$). The proportionate reduction in error is $40/90 = .444$. This is the value for lambda here.

PRE coefficients often are interpreted in one of two ways. The first way is a general statement that can be applied to all PRE measures regardless of the specific methods of prediction used. This general interpretation simply speaks in terms of "explaining or predicting differences" in the dependent variable by the independent variable. In our present example a researcher might say, "44.4% of the differences in people's attitude toward the need for social legislation can be explained (or predicted) by knowledge of their race." You may find other researchers giving a second type of interpretation which is much more specifically tied to the mechanics used to arrive at the E_1 and E_2 values for the exact measure employed. For our same example such as interpretaton might be: "Modal knowledge of the bivariate distribution of race with the attitude toward the need for social legislation improves our explanation (or prediction) of that attitude by .444 (44.4%)." The second interpretation refers to the use of modes reflecting the specific techniques of lambda. Both types of interpretations are correct.

A lambda value of .444 is approximately half of the maximum possible strength (1.0) and usually would be considered a moderate association. It is directly comparable to the lambda value for any other bivariate table of nominal data regardless of the table size, number of cases, or context of the research. The table with the higher lambda has the stronger association. Further, the lambdas can be compared in a ratio sense. If the second distribution has a lambda of .222, its association is only half the strength of ours.

The computations of lambda for the (a) and (c) distributions from Table 5.5 are shown in Table 5.7. We see that in the no-association distribution (a) there was no reduction in prediction errors by the second method over the first. Thus lambda was 0. In the perfect association distribution (c) there was a complete elimination of errors by the second prediction method. Thus lambda was 1.

As stated above, lambda can be calculated and compared for tables of any size. One limitation to remember is that lambda examines data at the nominal level. It will not be sensitive to any ordering or scaling of categories for variables at higher levels of measurement.

A note of caution is in order about lambda and many other measures of association, as well. Although the values of 0 and 1 are always possible, this does not mean that a value of 0 always means that there is no association *of any kind* and that 1 means there is perfect

TABLE 5.7 Finding Lambda for the Table 5.5(a) and (c) Bivariate Distributions

(a) No-Association Distribution:

(Y) Need More Social Legislation	(X) Race White	Nonwhite	
Yes	50	50	100
No	50	50	100

E_1 = There is no mode on the dependent variable, so we can predict either all "yes" or all "no" responses. Either way we are correct for 100 cases, and in error for 100 cases.
= 100

E_2 = There is no mode for whites, so can predict either all whites say "yes" or all say "no." Either way we are correct for 50 cases and in error for 50 cases. There is no mode for nonwhites, so we can predict either all nonwhites say "yes" or all say "no." Either way we are correct for 50 cases and in error for 50 cases.
= 50 + 50 = 100

$$\text{lambda} = \frac{E_1 - E_2}{E_1} = \frac{100 - 100}{100} = \frac{0}{100} = 0$$

(c) Perfect Association Distribution:

(Y) Need More Social Legislation	(X) Race White	Nonwhite	
Yes	0	100	100
No	100	0	100

E_1 = There is no mode on the dependent variable, so we can predict either all "yes" or all "no" responses. Either way we are correct for 100 cases and in error for 100 cases.
= 100

E_2 = The mode for whites is "no," so we predict all whites to say "no." We are correct for all 100 cases and in error for none. The mode for nonwhites is "yes," so we predict all nonwhites to say "yes." We are correct for all 100 cases and in error for none.
= 0 + 0 = 0

$$\text{lambda} = \frac{E_1 - E_2}{E_1} = \frac{100 - 0}{100} = \frac{100}{100} = 1.0$$

association *of every kind* in a distribution. For example, lambda only examines a distribution in terms of its modes (univariate and bivariate) and not in any other way. Thus a lambda of 0 means that there is no reduction in predictive errors when modes within the conditional distributions are used for predictions as compared to using the dependent variable's overall mode for predictions. This will happen whenever within-category modes of the independent variable all occur in the same category of the dependent variable. But there may be some association in the sense of percentage differences between conditional distributions. Similarly, a lambda of 1 means that there is complete reduction in predictive errors using the same information about modes, but not necessarily using some other information. Other measures have their own methods of prediction and exact meanings for a coefficient of 0 and 1. The general point to remember is that each measure incorporates its own definition of association and it is therefore limited to that definition.

Statisticians call lambda an *asymmetrical* measure. This means that it distinguishes between an independent and dependent variable. In our example, we predicted the social legislation attitude from race and not the reverse. Had we done the reverse we would be finding a different lambda. Every bivariate table has two asymmetric lambdas (often with different values) depending on which variable is put in the dependent role. There is a third lambda, which gives a weighted average of the other two lambdas. This third one is a *symmetrical* measure. The chi-square-based measures of association above were also symmetrical. Symmetrical measures do not distinguish between the independent and dependent variables. There is only one phi, *C*, or *V* for a table. Researchers who use lambda or other asymmetric measures will tell you which variable is in each role by the way they subscript the statistical symbol. L_{yx} means that the *Y* variable (generally the row variable) is dependent and being predicted by the *X* variable (generally the column variable). L_{xy} means that the *X* variable is dependent and being predicted by the *Y* variable.

There are other PRE measures of association for nominal data beyond lambda. One of these is phi-squared (ϕ^2). Our chi-square-based measure phi can be converted to a PRE measure for a 2×2 table (only) simply by squaring its calculated value. In our earlier example we found that ϕ was .408 in the Table 5.1 (b) distribution of employment status and sex. Therefore, ϕ^2 is $(.408)^2 = .166$ in this table. This is interpreted to mean that knowledge of sex allows us to reduce by .166 (16.6%) the errors in predicting employment status. Because phi-squared is symmetrical, we can also say that knowledge of employment status allows us to reduce by .166 the errors in predicting sex.

* Goodman and Kruskal's Tau-*y* (T_y)

Another popular PRE measure for nominal variables is Goodman and Kruskal's tau, sometimes called tau-*y*. Like lambda it can be found for tables of any size and is an asymmetric measure. For 2×2 tables tau-*y* is equal to ϕ^2. Like lambda (and ϕ), its possible range is from 0 to 1. But the 0 value occurs only when there is truly no association—no percentage differences between conditional distributions. The 1 value occurs only when there is perfect association under the condition that the number of categories in the independent variable is equal to or greater than the number of categories in the dependent variable.

Tau-*y* achieves these more admirable characteristics by using a different and more complicated procedure for making predictions than lambda. Whereas lambda makes predictions based on modes, tau-*y* makes predictions based on the proportional distribution of the dependent variable with the independent variable. Because of this method, it is often preferred over lambda by social researchers.

Tau-*y* can be found by the usual form of a PRE statistic's formula:

$$T_y = \frac{E_1 - E_2}{E_1} \qquad \text{Tau-}y\text{ Formula}$$

where E_1 = sum for all Y categories of the product: Y proportion of all cases that are *not* in a category multiplied by that category's frequency

E_2 = sum for all X-Y category cells of the product: proportion of cases *not* in an X category multiplied by that cell's frequency

Table 5.8 presents a bivariate distribution that we can use to demonstrate the computation of tau-*y*. Our dependent variable is regular attendance at religious meetings (church, synagogue, mosque, and so on) and the independent variable is marital status. Faced with such a table we would want to know how useful marital status is in order to predict whether a case does or does not attend such meetings.

TABLE 5.8 Finding Tau-*y* for a Bivariate Distribution of Nominal Data

Regular Attendance at Religious Meetings	Marital Status: Married	Single	Other	
Yes	35	30	10	75
No	15	10	20	45
	50	40	30	120

E_1 = for "yes": $\frac{120-75}{120}(75) = \frac{45}{120}(75) = .375(75) = 28.125$

for "no": $\frac{120-45}{120}(45) = \frac{75}{120}(45) = .625(45) = 28.125$

= 56.25

E_2 = for "married-yes": $\frac{50-35}{50}(35) = \frac{15}{50}(35) = .30(35) = 10.5$

for "single-yes": $\frac{40-30}{40}(30) = \frac{10}{40}(30) = .25(30) = 7.5$

for "other-yes": $\frac{30-10}{30}(10) = \frac{20}{30}(10) = .6\bar{6}(10) = 6.7$

for "married-no": $\frac{50-15}{50}(15) = \frac{35}{50}(15) = .70(15) = 10.5$

for "single-no": $\frac{40-10}{40}(10) = \frac{30}{40}(10) = .75(10) = 7.5$

for "other-no": $\frac{30-20}{30}(20) = \frac{10}{30}(20) = .3\bar{3}(20) = 6.7$

= 49.4

$$T_y = \frac{E_1 - E_2}{E_1} = \frac{56.25 - 49.4}{56.25} = \frac{6.85}{56.25} = .122$$

The first prediction method corresponds to assigning randomly each of the 120 cases to one of the two categories of the dependent variable in the same proportions as now exist. Thus we first want to know how many errors in prediction we would make if 75 cases at random were predicted not to be regular attenders ("no"). This number of errors is found as follows:

$$E_1 = \Sigma\left[\frac{N - F_y}{N}(F_y)\right] \qquad \text{Computation of } E_1 \text{ for Tau-}y$$

where F_y = frequency in each category of the dependent variable (Y)
N = total sample size

The E_1 computation is made separately for each category of the dependent variable and the resulting subtotals are then summed. This work is shown in Table 5.8 below the distribution. We see that we would make 28.125 errors for each category, and thus $E_1 = 56.25$.

The second prediction method corresponds to assigning randomly the cases in each category of the independent variable to one of the two categories of the dependent variable, again preserving the existing proportions. The number of errors that result is found as follows:

$$E_2 = \Sigma\left[\frac{F_x - f}{F_x}(f)\right] \qquad \text{Computation of } E_2 \text{ for Tau-}y$$

where F_x = frequency in each category of the independent variable (X)
f = cell frequency

The E_2 computation is made separately for each cell in the table and the resulting subtotals are then summed. Again this work is shown in Table 5.8. We see that 10.5 errors would be made in randomly assigning 35 of the 50 married cases to the "married–yes" cell, for example. The total for E_2 is 49.4. Thus

$$T_y = \frac{56.25 - 49.4}{56.25} = \frac{6.85}{56.25} = .122$$

We interpret this to mean that .122 (12.2%) of the errors in predicting whether or not the cases are regular attenders of religious meetings can be reduced by knowing their marital statuses. This is a low PRE value, indicative of a weak association between the two variables in this distribution. Marital status has only limited value in predicting this dependent variable.

Association at the Ordinal Level

Association between ordinal variables involves one new characteristic that the researcher wants to identify. In addition to the existence and strength of a relationship, the researcher wishes to know its *direction.* If a bivariate relationship shows a pattern of having the higher X scores occurring jointly with higher Y scores, the direction is *positive.*

If the pattern has higher X scores occurring with lower Y scores, the direction is *negative.* If these positive or negative patterns are consistent throughout the distribution, they are said to be *monotonic.* If the pattern reverses partway through the distribution, it is called *polytonic.*

Study the three distributions in Table 5.9 by looking for the highest frequencies. Education and income have historically had a positive association (a). People with higher levels of education tend to have higher incomes. Education and prejudice have had a negative association (b). Those with higher levels of education tend to have lower levels of prejudice. Income and family size have had a polytonic association (c). The association is negative from low- to middle-income levels as family size decreases. But then it reverses from middle- to high-income levels as family size increases. Most measures of association describe only consistently positive or negative directions. They do not look for reversals such as the (c) distribution. None of those discussed here are appropriate for polytonic distributions.

Measures of association for ordinal variables are developed for two basic types of ordinal measurement. One type is the situation that typically has categories of "low," "middle," and "high," like those in Table 5.9. We can call these *data grouped in ordered categories* because many cases often share the same classification. The second type is typified by ranking the cases from the first to the last according to an ordered dimension. They are called *data in ranked order.* The order of finish in an athletic contest or on a test are examples. We will look at representative measures for each of these types.

TABLE 5.9 Three Possible Bivariate Distributions at the Ordinal Level

(a) Positive Association:

	Education		
Income	High	Middle	Low
High	80	15	15
Middle	30	90	35
Low	10	15	70

(b) Negative Association:

	Education		
Prejudice	High	Middle	Low
High	15	20	75
Middle	30	80	35
Low	75	20	10

(c) Polytonic Association:

	Education		
Family Size	High	Middle	Low
Large	65	10	60
Medium	35	35	40
Small	20	75	20

The normed scale for ordinal measures ranges from -1 to $+1$. These extremes represent perfect association. No association is represented by a value of 0 and the strength of association is indicated by its value between 0 and either extreme. The sign indicates whether the association is negative ($-$) or positive ($+$). Interpreting the strength in verbal terms is as arbitrary for ordinal variables as it is for nominal variables. In the absence of any substantive information, the same approximate guidelines are often used, as indicated in Figure 5.1. Merely ignore the sign for an ordinal measure when using that guide.

Data Grouped in Ordered Categories

The analysis strategy used for association measures with data grouped in ordered categories is based on the idea of looking at the cases in pairs. We examined this idea briefly in Chapter 4 for the index of dispersion. The total number of unique pairs that can be created in a distribution is determined as follows:

$$P_t = \frac{N(N-1)}{2} \qquad \text{Total Number of Unique Pairs Formula}$$

where P_t = total number of unique pairs
N = number of cases

Table 5.10 shows a bivariate distribution for the ordinal variables of alienation and urbanization. Each variable is trichotomized into high, middle, and low categories. Since $N = 200$ in this table, the total number of unique pairs is

$$P_t = \frac{200(199)}{2} = \frac{39{,}800}{2} = 19{,}900$$

The total number of unique pairs for bivariate ordinal distributions can be divided into five types of pairs:

1. Agreement pairs (P_a).
2. Disagreement pairs (P_d).
3. Tied-on-X pairs (P_x).
4. Tied-on-Y pairs (P_y).
5. Tied-on-both-X-and-Y pairs (P_{xy}).

FIGURE 5.2 Examples of Types of Pairs Formed from Table 5.10

Cells	X	Y	Type of Pair
a–e	High–Middle	High–Middle	Agreement
a–h	High–Middle	High–Low	Agreement
c–e	Low–Middle	High–Middle	Disagreement
c–h	Low–Middle	High–Low	Disagreement
f–i	Low–Low	Middle–Low	Tied on X
g–h	High–Middle	Low–Low	Tied on Y
f–f	Low–Low	Middle–Middle	Tied on both X and Y

Suppose that we look at examples of some pairs of cases from the cells of Table 5.10, as shown in Figure 5.2. The first pair is formed by choosing one case from **cell a** and another case from **cell e.** Focusing first on the *X* variable (urbanization), we see that the **cell a** case has a score of high and the **cell e** case has a score of middle. The *X* pattern is for the **cell a** case to have a higher score than the **cell e** case. Looking at the *Y* variable (alienation), the **cell a** case has a score of high and the **cell e** case has a score of middle. Again, the **cell a** case has a higher score. Because the *X* and *Y* patterns are alike, we call this an agreement pair.

The second example shows a pair formed by putting one case from **cell a** with one case from **cell h.** On *X* the **cell a** case has a score of high and the **cell h** case has a score of middle. Just like the *X* pattern for the **a–e** pair above, in this *X* pattern the **cell a** case is higher. On *Y* the **cell a** case is high and the **cell h** case is low. Even though this *Y* pattern is not exactly like the *X* pattern, the **cell a** case is still higher than the **cell h** case. This pair is also called an agreement pair.

In the third example one case from **cell c** is paired with one from **cell e.** On *X* the **cell c** case has a lower score than the **cell e** case, low versus middle. But on *Y* the **cell c** case has a higher score than the **cell e** case, high versus middle. Because these two cases show an opposite pattern for the *X* and *Y* variables, they form a disagreement pair.

The fourth example pairs one case from **cell c** with one from **cell h.** This time the *X* pattern has the **cell c** case with a lower score than the **cell h** case, low versus middle. On *Y* the **cell c** case has a higher score than the **cell h** case, high versus low. This reversal identifies a disagreement pair even though the patterns on *X* and *Y* are not exactly opposite.

The fifth, sixth, and seventh examples show pairs with ties. Tied pairs are found by matching cases from the same rows and/or columns of the table. A pair from **cells f** and **i** have the same *X* scores (low) but different *Y* scores (middle versus low). It is called a tied-on-*X* pair. In contrast, a pair from **cells g** and **h** are a tied-on-*Y* pair. They have the same *Y* scores (low) but different *X* scores (high versus middle). Pairs of cases drawn from the same cell are the tied-on-both-*X*-and-*Y* type. Two cases from **cell f** would have the same *X* scores (low) and the same *Y* scores (middle).

There are several ordinal measures of association that use this information about the five types of pairs. In general, they determine the predominance of some types over others. Of these measures, we will only discuss Goodman and Kruskal's gamma. Some of the others are listed in the section *Related Statistics* at the end of the chapter.

TABLE 5.10 Bivariate Distribution of Data Grouped in Ordered Categories

	(*X*) Urbanization			
(*Y*) Alienation	High	Middle	Low	
High	a 50	b 20	c 10	
Middle	d 20	e 40	f 15	
Low	g 10	h 5	i 30	
				200

Goodman and Kruskal's Gamma (G)

The most commonly used measure of association for data grouped in ordered categories is Goodman and Kruskal's *gamma* (G). Gamma can also be applied to ungrouped (ranked) ordinal data, but this use is not common and it requires a different procedure than the one to be shown here. Research Example 5.3 illustrates the ordinary use of gamma in a study of Chicano attitudes toward police and civil liberties. Three tables of data grouped in ordered categories are presented. Both chi-square and gamma are reported for these three distributions. We will return to this example after we have discussed the computation and interpretation of gamma.

Like all PRE measures, gamma compares two methods of prediction. The first method is based on randomly predicting all untied pairs of scores to be either in agreement or disagreement. The second method predicts all untied pairs to be of the same type. Whether that is agreement or disagreement is determined by the direction of the bivariate distribution. In a table with a prevailing positive pattern to the association, we would predict all untied pairs to be in agreement. In a table with a prevailing negative pattern, we would predict all untied pairs to be in disagreement.

The number of prediction errors made by each method could be compared by the basic PRE formula as usual to find the value of gamma for a table. However, there is a more efficient method to reach the same result. This is achieved by using the following formula to compute gamma:

$$G = \frac{P_a - P_d}{P_a + P_d} \qquad \text{Gamma Formula}$$

where P_a = number of agreement pairs
P_d = number of disagreement pairs

As the formula indicates, gamma uses only the agreement and disagreement pairs in its measure of association. The three types of tied pairs are ignored. The ordinal measures listed in the section *Related Statistics* at the end of the chapter do use some of these tied pairs in their computations.

Agreement pairs identify the extent of a positive direction in a distribution. Disagreement pairs identify the extent of a negative direction in a distribution. Thus gamma determines whether the distribution is predominately positive or negative. The extent of the dominance of either type is reflected in the strength of gamma.

Table 5.11 shows the computation of gamma for our bivariate distribution of alienation and urbanization. Note that the table is arranged so that the high–high combination on X and Y is in the upper left corner. If you arrange other tables in this way, you can follow the same steps we are using here to compute gamma.

We find the number of agreement pairs (P_a) by multiplying the frequency for each cell in the table by the sum of the frequencies from all cells that are both to the *right* and *below* it. We begin with the upper left cell **(a)**, which identifies the positive diagonal in the table. Scores in this diagonal represent exact (perfect) positive association. Scores just off this diagonal represent inexact (imperfect) positive association. Both types are being considered in our tabulation of agreement pairs. The frequency in **cell a** (50) is multiplied by the sum of 40 + 15 + 5 + 30 (90) to equal 4500 pairs. The four numbers making up the sum (90) represent all of the cells in the table that are to the right and below **cell a.**

Then we move to the next cell **(b)** in the top row. Its frequency (20) is multiplied by

TABLE 5.11 Finding Gamma for a Bivariate Distribution of Data Grouped in Ordered Categories

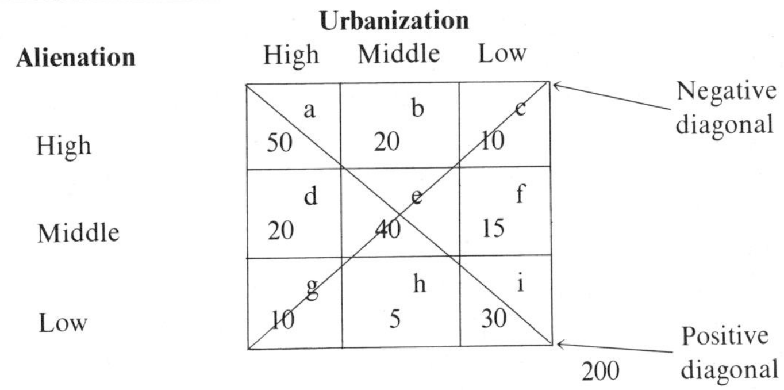

Agreement Pairs:

$$
\begin{aligned}
a \times (e + f + h + i) &= 50 \times (40 + 15 + 5 + 30) = 50(90) = 4500 \\
b \times (f + i) &= 20 \times (15 + 30) = 20(45) = 900 \\
d \times (h + i) &= 20 \times (5 + 30) = 20(35) = 700 \\
e \times (i) &= 40 \times (30) = 1200 \\
P_a &= 7300
\end{aligned}
$$

Disagreement Pairs:

$$
\begin{aligned}
c \times (e + d + h + g) &= 10 \times (40 + 20 + 5 + 10) = 10(75) = 750 \\
b \times (d + g) &= 20 \times (20 + 10) = 20(30) = 600 \\
f \times (h + g) &= 15 \times (5 + 10) = 15(15) = 225 \\
e \times (g) &= 40(10) = 400 \\
P_d &= 1975
\end{aligned}
$$

$$G = \frac{P_a - P_d}{P_a + P_d} = \frac{7300 - 1975}{7300 + 1975} = \frac{5325}{9275} = +.574$$

15 + 30 (45) to equal 900 pairs. The next cell **(c)** has no cells to its right, so it is not included in the calculation of agreement pairs. Next we go to the second row and repeat the process for **cells d** and **e.** This yields 700 and 1200 agreement pairs, respectively. **Cell f** has no cells to its right, and none of the cells in the third row have cells below them, so they do not contribute further to the computation. The total number of agreement pairs in Table 5.11 is 7300.

The number of disagreement pairs (P_d) is found by multiplying the frequency for each cell in the table by the sum of the frequencies from all cells that are both to the *left* and *below* it. We begin with the upper right cell **(c)**, which identifies the negative diagonal in the table. Again, exact and inexact negative association are included in the computation. The frequency of **cell c** (10) is multiplied by the sum of 40 + 20 + 5 + 10 (75) to equal 750 pairs. This process is repeated for all cells that have other cells to their left and below them. The total number of disagreement pairs in Table 5.11 is 1975.

The P_a and P_d totals are substituted into the gamma formula to find its value. Gamma is +.574 for Table 5.11. This means that there is a positive association of moderate to strong strength between urbanization and alienation. Because gamma is a PRE measure, we can add the specialized interpretations that are a part of all such measures. In a general way we can say that knowledge of urbanization allows us to explain .574 (57.4%) of the differences in alienation. Specifically, we can say that .574 (57.4%) of the errors made in randomly predicting agreement or disagreement between alienation and urbanization are reduced if we predict only agreement.

Gamma is a symmetrical measure; it does not make a distinction between an independent and dependent variable. Thus our PRE interpretation above applies whether we are thinking of explaining alienation from urbanization or the reverse. In Table 5.11, 57.4% of the differences in urbanization are explained by alienation.

Like other PRE measures, gamma values can be compared as ratios between distributions. If gamma was .191 in another distribution, we could correctly conclude that Table 5.11 shows an association which has three times the strength of the second distribution, but both were positive in direction.

Also like some other measures, the meaning of 1 (+ or −) for gamma is not always obvious. A perfect association will produce a gamma of 1. But so will some imperfect distributions. This limitation is controlled if gamma is used where only a small proportion of the pairs of scores are of the tied-on-X or tied-on-Y types. Perfect association can be verified by examining the table visually. If all the scores are in one diagonal, you can be sure the association is really perfect.

Look closely at R.E. 5.3. In Table 1 we see that gamma is .35 for a distribution of the two variables of support for police and fear of attack. This gamma represents a moderate association in the positive direction. Greater support for increasing police power tends to occur jointly with greater fear of attack. In Table 2 gamma is .27 for a relationship between support for civil liberties and fear of police. The association is weak to moderate in strength but positive like the first. Greater support of increasing civil liberties occurs jointly with greater fear of police. In Table 3 gamma is .72 for a relationship between support for police and fear of police. The association is strong to very strong and presented as being positive in sign. The author notes that in this association greater support for limiting police power jointly occurs with greater fear of police. All other things being equal, we can conclude that the third relationship is the strongest of the three.

The three tables in R.E. 5.3 illustrate the importance of carefully reading tabular distributions as well as their summarizing statistics. All three tables are reported as having positive associations. However, it may not be readily apparent from the tables that this is true. The difficulty arises from two sources. The author presents the row variable in Table 2 in a low-to-high order while the format in Tables 1 and 3 is high-to-low. Further, the author sometimes expresses a variable's name in two opposite ways; "fear of police" or "police respect." These combine to make verifying the interpretations of direction challenging. The author's interpretations are correct, but the example shows how you must be very thoughtful in comparing the reported direction of an association statistic with the tabular presentations.

Data in Ranked Order

Ordinal data are not always grouped in ordered categories with several cases sharing the same score values. Some ordinal variables make distinctions between all or nearly all of the cases. This is the situation when cases are ranked. It is an "improved" measurement over data grouped in ordered categories. It can also be viewed as an inadequate approximation of interval measurement where distinctions of known quantitative differences are made. Data in ranked order do not have known quantitative differences; they have only rank differences. Thus this type of measurement fits between the usual ordinal and interval levels.

RESEARCH EXAMPLE 5.3 Using Gamma to Measure Association

From Alfredo Mirandé, "The Chicano and the Law," Pacific Sociological Review, *24 (1), January 1981:65ff. Reprinted by permission.*

The rising rate of crime and urban unrest in the late 1960s led to an overwhelming concern not only with issues of crime and civil disorders but with official violence and the violation of civil liberties. Of particular interest was whether fear of crime increased support for police power and decreased support for civil liberties. . . .

The U.S. Commission on Civil Rights, in perhaps the most systematic and far-reaching study of Chicanos and the legal and judicial system, found strong evidence of a pattern of systematic harassment and abuse of Chicanos by the police. . . . There is a need for research that examines the attitudes of barrio residents toward law enforcement not only during major incidents but in their day-to-day contact with police. This study attempts to add to our understanding of Chicano-police relations by presenting an overview of conflict with police in an urban barrio, testing several hypotheses concerning fear of the police, attitudes toward increasing or curtailing police power, support of civil liberties, and fear of crime. . . .

The setting for this case study of Chicano-police conflict is a barrio in a Southern California community of 150,000 inhabitants. The barrio was selected not only because it has been the site of recent civil disorders which have gained national attention, but because it appeared representative of other barrios in the Southwest in being isolated from Anglo society and having a long history of conflict with the police. Like most barrios, it is a distinctive community within the city, extending over an area that is approximately one square mile and includes about 3000 persons within its boundaries. . . . A random sample of households yielded 170 completed interviews. . . .

The first hypothesis posits that fear of crime among barrio residents is translated into a greater willingness to support increases in police power. . . . From Table 1 it is clear that the hypothesis is supported in this sample. Those barrio residents who say street attack is very likely are more likely to favor increases in police power. Eighteen percent of the respondents who believe attack is very likely support increasing police power, whereas only 4% of those who feel it is very unlikely do so. Chicanos who see attack as very unlikely, on the other hand, are much more inclined to want to curtail police power (54%) than those who see attack as very likely (26%). . . .

TABLE 1 Support for the Police and Fear of Attack[a] (%)

	Fear of Attack			
Support Police	Very Likely	Somewhat Likely	Somewhat Unlikely	Very Unlikely
More	18	18	19	4
Enough	55	69	45	41
Curtail	26	12	35	54
N	(38)	(49)	(31)	(46)
		Gamma = .35	Chi-square = 22.42	

[a]Adapted from Table 1 in the original.

The second major hypothesis concerns the effects of fear of the police on support for civil liberties among barrio residents. My hypothesis is that those Chicanos who are most fearful of police tend to be more supportive of civil guarantees, since such guarantees are designed to limit police abuses. Table 2 shows that the relationship is in the predicted direction but [weakly moderate]. . . .

TABLE 2 Support for Civil Liberties and Fear of the Police[a] (%)

Support Liberties	Police Respect: Very Good	Pretty Good	Not So Good
Less support	50	38	30
Full support	50	62	71
N	(34)	(60)	(61)
	Gamma = .27	Chi-square = 3.95	

[a]Adapted from Table 2 in the original.

The preceding hypotheses have been based on the assumption that much of the willingness of barrio residents to increase police power and limit civil guarantees is grounded in fear, fear both of crime and of the police. Fear of crime is likely to lead some residents of the barrio to support increasing the power of the police. Fear of the police, on the other hand, intensifies the support of civil liberties, thereby limiting police abuses. Just as fear of the police should lead to greater support for civil liberties so should it increase the desire to limit the power of police in the barrio, a hypothesis which will now be examined.

From Table 3 it is clear that the relationship between fear of the police and support for the police is strong. . . . As predicted, those Chicanos who fear the police most are least likely to support increases in police power. Of the respondents who fear the police, 63% (i.e., who see police respect toward people like themselves as "not so good") desire to curtail or to limit police power, whereas only 9% of those who do not fear the police desire to curtail it. . . .

TABLE 3 Support for the Police and Fear of the Police[a] (%)

Support Police	Police Respect: Very Good	Pretty Good	Not so Good
More	32	20	2
Enough	59	67	35
Curtail	9	13	63
N	(34)	(61)	(60)
	Gamma = .72	Chi-square = 51.84	

[a]Adapted from Table 3 in the original.

Spearman's Coefficient of Correlation (r_s)

Spearman's Coefficient of Correlation (r_s) is a measure of association specifically designed for rank-ordered data. It is also known as Spearman's rank-difference correlation coefficient or Spearman rho. It should not be used for data grouped in ordered categories.

Spearman's r_s is based on a comparison of the squared differences in ranks between two variables in an actual distribution to the maximum possible value of those squared differences. In this way it technically requires that the ranks represent equal interval differences. But you will frequently find it used on ranked data with unknown intervals. It uses the following formula:

$$r_s = 1 - \left[\frac{6\,\Sigma D^2}{N(N^2 - 1)}\right] \qquad \text{Spearman's } r_s \text{ Formula}$$

where D = differences between X and Y ranks for each case
ΣD^2 = sum of the individual squared differences
N = number of pairs of ranks

If the ranks for two variables are identical for every case, r_s will be $+1$, perfect positive association. If the ranks are exactly opposite for the two variables, r_s will be -1, perfect negative association. If there is no pattern to the pairs of ranks, r_s will be 0, no association. r_s will take on a value between these to indicate the strength of any existing association.

Table 5.12 presents an example illustrating r_s's application in a hypothetical study of small-group dynamics. Such studies typically have group members identify two types of

TABLE 5.12 Finding Spearman's r_s for a Bivariate Distribution of Data in Ranked Order

Group Member	Rank in Task Leadership	Rank in Social-Emotional Leadership	D	D^2
Pat	5	6	−1	1
Sue	2	2	0	0
Carmella	9	5	4	16
Mike	4	10	−6	36
Linda	1	4	−3	9
Teresa	8	7	1	1
Carla	3	1	2	4
Tom	7	8	−1	1
Hal	10	9	1	1
David	6	3	3	9
			$\Sigma D = 0$	$\Sigma D^2 = 78$

$$\begin{aligned} r_s &= 1 - \left[\frac{6\,\Sigma D^2}{N(N^2 - 1)}\right] \\ &= 1 - \left[\frac{6(78)}{10(10^2 - 1)}\right] = 1 - \left[\frac{468}{10(99)}\right] = 1 - \left[\frac{468}{990}\right] \\ &= 1 - .4727 = +.5273 \\ &= +.527 \end{aligned}$$

leaders: the leader who helps the group achieve a task and the leader who helps the group maintain its social and emotional ties. Suppose that a small group of 10 adolescents has been studied in this way. Each group member's rank in the peer ratings for the two types of leadership are shown in the table. From a visual inspection of the table it appears there is an imperfect positive pattern to the rankings. Spearman's r_s will tell us whether that is true.

The difference (D) between the two ranks is computed for each case. For example, Pat's difference in ranks is -1. Notice that the sum of differences (ΣD) is 0. It will be 0 for every distribution. The individual rank differences are squared and then summed (ΣD^2). This sum (78) and N (10) are substituted in the r_s formula as shown below the table. The calculated value is .527. Since it is not 0, we can conclude that an association does exist in this bivariate distribution of ranks. The size of the coefficient tells us its strength. The usual verbal interpretation would be that this association is moderate. Values of r_s are directly comparable across distributions of ranked scores. Larger values indicate greater strength. The sign of the coefficient is positive here, meaning that the association tends to be one in which higher rank in task leadership occurs jointly with higher rank in social-emotional leadership.

r_s can be given a PRE interpretation if it is squared first. In our same example, the PRE value would be .278 ($.527^2$). At the general level this means that knowledge of task leadership explains 27.8% of the differences in social-emotional leadership. At the more specific level this means that errors in predicting rank in social-emotional leadership without knowledge of rank in task leadership can be reduced by .278 (27.8%) if the second ranked variable is used to preduct the first. This statement can be reversed to note that the reduction in errors if you predict task leadership from social-emotional leadership is also .278. That should tell you that r_s is a symmetrical measure.

Many researchers would consider this same r_s value of .527 to represent a moderate-to-weak association, not a moderate one as noted above. They say this because the PRE value of r_s reveals that the proportion of one variable's differences that has been explained by the other is only .278 ($.527^2$), not .527 itself. This second interpretation of strength has considerable appeal if one is focusing on the power of one variable to predict the other. However, many researchers do not focus on this; rather, they think in the more general terms of joint occurrence. You should expect to find both types of interpretations in research reports. The second one actually is more accurate from a strictly statistical viewpoint.

More on Ranking Scores

Not all variables analyzed by Spearman's r_s are measured directly as ranks. r_s is often used when one or both variables are measured first as scores at the interval or ratio level. These scores must be changed to ranks before r_s can be applied. Ranks may be assigned to scores in one of two ways. Either the highest score or the lowest score may be given rank = 1. Although it does not matter to r_s which method is used, it does matter to the researcher and you who must interpret the result.

The ranks should be assigned consistently for the two variables being examined so that the direction of the association will not be confused. The key is to decide which end of the scale represents the "high" end. Rank = 1 can then be assigned alike to either the "high" or "low" end of both variables. Usually, the scores that are largest quantitatively reflect the high end of the scale conceptually and are assigned rank = 1. For example, older ages and larger test scores indicate the high ends of age and test scales.

But sometimes the lowest scores might seem to be the high end of a scale. Suppose that a researcher was measuring error rates for people who filled our bureaucratic forms. Should rank = 1 be assigned to the highest error rate (the lowest performance) or to the lowest error rate (the highest performance)? If an association between test scores and error rates were found to have an r_s value of +.75, what would you think that meant? If rank = 1 had been assigned to the highest (or lowest) quantitative scores for both variables, the r_s value would mean that there was a strong positive relationship in which higher test scores occurred with higher error rates. If rank = 1 had been assigned to the highest test score and the lowest error rate, for example, the r_s value would mean that this was a strong positive association in which higher test scores occurred with lower error rates. Most researchers would use the first of these ranking methods. Clearly, you must know the method of ranking in order to interpret the sign of the r_s statistic.

Another common problem in ranking is the issue of tied scores. What rank should be given to scores that are the same? This is handled by observing the rank positions occupied by the tied scores, finding the arithmetic mean of those positions, and assigning this mean rank to those tied scores.

Look at Table 5.13 as an example. Therein interval scores on social class and illness are presented for 15 persons. For each variable, we assign rank = 1 to the highest score. Note that persons CX and DW have the same social class scores. Since there are three scores higher than them, these two cases occupy the fourth and fifth rank positions. We find the mean of these two ranks [(4 + 5)/2] and assign the rank 4.5 to both cases. The next lowest social class score is given rank = 6. Note that cases FU, GT, and LO share the same illness scores. Again their rank positions are averaged [(11 + 12 + 13)/3] and they are each assigned the mean rank (12). The next lowest illness score is given rank = 14. Once the ranks have been determined, Spearman's r_s can be calculated as before.

Do not let this discussion of tied ranks make you forget the introductory comment concerning Spearman's r_s. It should not be used with data that are grouped in ordered categories, that is, data that are often tied. If too many scores are ties, gamma is a more

TABLE 5.13 Assigning Ranks to Tied Scores

Case	Score on Social Class	Rank on Social Class	Score on Illness	Rank on Illness
A Z	12	14	80	1
B Y	27	8	70	3
C X	33	4.5	27	10
D W	33	4.5	41	8
E V	15	13	53	6
F U	31	6	20	12
G T	40	1	20	12
H S	17	12	50	7
I R	20	11	75	2
J Q	39	2	16	14
K P	30	7	35	9
L O	25	9	20	12
M N	35	3	14	15
N M	21	10	62	5
O L	10	15	66	4

appropriate measure of association than r_s. The exact limit on tied ranks for r_s is debated. Many researchers will not calculate it if more than 20% of the scores represent ties.

Elaborating a Bivariate Association

Before ending our discussion of bivariate association for variables at the nominal and ordinal levels, a brief introduction to the topic of elaborating an association is appropriate. Often, this topic is not covered in introductory statistics texts, but it is so important for understanding social statistics that you should be introduced to it. So far we have been talking about association as though it were possible to examine two variables in isolation from all the other variables that are operating in any real social situation. Usually, the association between two variables is the product of several variables influencing each other. To view the relationship of only two variables alone is a fiction. The basic statistical technique to overcome this at the nominal and ordinal levels is to *elaborate* the bivariate relationship. This means that we look at the relationship under other conditions representing the categories of a third variable.

Elaboration could be applied usefully to every association we have examined in this chapter, hypothetical and real. Do you think there are any conditions under which the association between sex and employment status would change? Social class or marital status might affect this bivariate relationship. How about race and attitude toward the need for social legislation? Income and occupation may affect this association. What about urbanization and alienation? Economic diversity and racial segregation might influence their relationship. Do you think education or region might affect the association between

RESEARCH EXAMPLE 5.4 Elaborating on Association Using Gamma

From Laurily Keir Epstein, "Individual and Contextual Effects on Partisanship," Social Science Quarterly, *60 (2), September 1979:314ff. Reprinted by permission.*

To what extent is the distribution of partisanship structured by the environment in which it is located? That is, in addition to the effects on partisan behavior of well-defined individual properties such as income, education and occupation, can we also expect that individuals will be affected by the patterns of interaction they experience? While it is acknowledged generally that partisan identity tends to be ordered by individual-level social-class characteristics, relatively few studies in recent years have examined the extent to which various contextual configurations modify otherwise similar social characteristics. The research presented here demonstrates that both individual and contextual variables shape or determine patterns of partisan affiliation. . . .

The data presented below are drawn from a series of six surveys of New Jersey residents conducted by the Eagleton Poll of New Jersey from September 1976 to April 1978. The data from the six polls are collapsed into one sample in an effort to eliminate the two major sources of sampling error: short-run time effects and sample size.

The individual-level data are coded trichotomously. . . . Ideology is divided into the conventional liberal, moderate and conservative categories.

Coding for the contextual variables is relatively straightforward. There are 40 legislative districts in New Jersey, each of which elects one state senator and two members of the Assembly. Since U.S. Census, voting and New Jersey Poll data can be aggregated by legislative district, the districts provide the boundaries for contextual analysis.

Income density is measured by coding the 40 legislative districts. The 10 districts with the highest proportion of low-income residents are coded as low-income areas. Low-income areas are ranked by percent of families under the poverty line in 1970. High-income areas are the 10 districts with the highest proportion of high-income families, i.e., families with income of more than $15,000 at the time of the 1970 Census. The remaining 20 districts are defined as medium-income areas. . . .

To what extent does ideology affect the individual's partisan choice? The expectation generally is that liberals will be more likely to favor the Democratic party and conservatives more likely to think of themselves as Republicans. The data displayed in Table [1] bear this out. Liberals are, in fact, more likely to be Democrats and conservatives are more likely to be Independent or Republicans. Moderates are more likely to exhibit Independent status than express a partisan preference.

In Table [2] we see the effects that income density has on the distribution of partisanship by ideological preference. Once again, both individual and contextual differences emerge and persist. Liberals remain more predisposed to the Democratic party than do conservatives, regardless of contextual dimension. However, the tendency of liberals to favor the Democrats diminishes as the individual moves from a low-income area to a high-income neighborhood. Within the low-income context, there is a 13.3% difference between liberal and conservative preference for the Democratic party. There is a 6.4% difference between liberal affiliation and a 21.9% difference between conservative affiliation for the Democratic party as the individual moves from a low-income to a high-income neighborhood. Conservative adherence to the Republican party also undergoes considerable change depending upon income context. There is a 24.8% shift from low-income to high-income areas. Social context, then, once again modifies the individual characteristic of ideology. . . .

My data show that while at the gross level the relationship between party preference and ideology is not overwhelming, the association (as measured by the gamma) grows stronger as the income context increases. Thus, while party and ideology may not manifest a strong relationship at the national level, in New Jersey (a context in itself) partisanship and ideology do indeed show a correspondence. . . .

TABLE 1 Partisanship by Ideology[a] (%)

	Liberal	Moderate	Conservative	**Total**
Democratic	47.6	32.4	25.0	34.2 (2049)
Independent	41.2	51.7	42.2	46.9 (2809)
Republican	11.2	16.0	32.8	18.8 (1127)
Total	23.6 (1415)	52.6 (3148)	23.8 (1422)	(5985)
		Gamma = .30		

[a]Reprinted from Table 3 in the original.

TABLE 2 Partisanship by Ideology, Controlling for Income Context[a] (%)

	Low-Income Areas			Medium-Income Areas			High-Income Areas		
	Liberal	Moderate	Conservative	Liberal	Moderate	Conservative	Liberal	Moderate	Conservative
Democratic	51.0	42.3	37.7	48.4	32.4	25.4	44.6	23.7	15.8
Independent	38.2	44.9	42.9	41.1	52.9	43.6	42.7	35.5	40.0
Republican	10.8	12.8	19.4	10.5	14.7	31.0	12.6	20.8	44.2
Total	100.0	100.0	100.0	100.0	100.0	100.0	99.9	100.0	100.0
		Gamma = .16			Gamma = .31			Gamma = .42	

[a]Reprinted from Table 4 in the original.

race and confidence in the U.S. Supreme Court? The author in R.E. 5.2 looked at this and other conditions and found that none of them had any appreciable influence. Do you think that age, length of residence, and sex could affect the associations concerning support for police, fear of attack, and support for civil liberties in R.E. 5.3? That example's author looked at these and other third variables and found that many of them did influence the basic bivariate relationships.

Let us look at a new example to illustrate the logic and procedures for elaborating a bivariate association. Research Example 5.4 presents an examination of the association between political partisanship and ideology. In Table 1 we see the bivariate relationship between these two variables. Gamma was .30 for this relationship, indicating a positive association of moderate strength. There was a tendency for the more liberal respondents to be more Democratic. (Note that the partisanship variable is treated as an ordinal variable, with the "high" end being Democratic and the "low" end being Republican. In the ideology variable liberal was the "high" end and conservative was the "low.") Income context was one of the third variables on which the author focused. This was a trichotomized variable representing low-income, medium-income, and high-income areas of New Jersey. The author wanted to know whether the basic association between partisanship and ideology was influenced by the income level of the area where the respondent lived. Table 2 in R.E. 5.4 contains the answer.

In low-income areas the association had a gamma of only .16, still positive but weak. In medium-income areas the association had a gamma of .31, virtually unchanged from the basic association. In high-income areas the association had a gamma of .42, still positive but somewhat stronger than the basic association. Because the value of gamma changed in these three conditions of income context, we can conclude that income context does make a difference (in a descriptive, although not necessarily an inferential, sense) in the association between partisanship and ideology. We see that as level of income context increases, the partisanship–ideology relationship becomes stronger.

As this example shows, the way we elaborate a bivariate relationship for nominal and ordinal variables is to look at that relationship within each category of a third variable. Since everyone in the low-income area subtable has the same income context, we know that income context cannot account for the relationship we see there between partisanship and ideology. The same is true for each of the other two subtables.

A Short Vocabulary Lesson

Researchers use a special vocabulary to refer to the process and outcomes of elaborating an association. When they examine a bivariate relationship within the conditions of a third variable, that third variable is said to have been *controlled.* The subtables are referred to as *conditional* or *partial tables* which show *conditional* or *partial associations.* The statistics calculated in these tables are called *conditional* or *partial measures* (e.g., conditional gammas for Table 2 in R.E. 5.4).

The basic association without controls is called the *zero-order* or *simple association.* When we control for only one variable as we did in R.E. 5.4, the conditional associations are also called *first-order associations.* In addition, we can look at the basic association while we control two variables at once. This would be done by pairing each category of one control variable with every category of a second control variable and then looking at the relationship between the two main variables within these combinations of conditions. In R.E. 5.4 we could examine the relationship between partisanship and ideology while controlling both income context and individual income level. These kinds of double controls give us *second-order associations.* Although it is possible to go on with the process to third- or fourth-order examinations, most social research stops with the first or second order. It requires a rather large number of cases to accomplish elaboration beyond this point with nominal or ordinal variables. Notice the large sample size in R.E. 5.4 (N = 5985).

When researchers elaborate a relationship, there are basically four possible outcomes.

1. The association's strength may increase. If so, the association is said to have been *suppressed* by the control variable. The association is actually stronger than it appeared to be in the zero-order examination.
2. The association's strength may be decreased. If so, the association is said to have been *partly explained* by the control variable. The association was actually weaker than it first appeared to be.
3. The association may disappear completely. If so, the association is said to have been *spurious.* It was completely explained away by the control variable. Its prior appearance was not genuine.
4. The association may not change. If so, the control variable is said to be *unrelated* or *extraneous* to that association.

You are likely to encounter most of these terms if you read much social research.

This quick look at elaborating a relationship is only the beginning of a rather large topic that can become quite complex. But it will suffice to open your eyes to an important idea and set of techniques in social statistics which attempt to accommodate the complexities of the social world we wish to study.

Keys to Understanding

The keys to understanding association between nominal and ordinal variables are organized into three general points. First, be familiar with the data that are being analyzed. Second, know the meaning of association that is being used. Third, distinguish between simple association and more sophisticated examinations of association.

Key 5.1. Become familiar with the data the researcher is analyzing. When we first discussed how distributions are tabulated in Chapter 2, several specific keys were suggested for their understanding. We can apply some of these same ideas to our present interest in association. Read the tables and surrounding text to identify the names of the variables, their methods of measurement, and their levels of measurement. Verify that association was a goal of the research by locating statements of purpose or objectives. Note whether any hypotheses mention association or relationship. Visually inspect the distributions shown and form some tentative descriptions about their associations in your own words. What strength and direction do they seem to have? Determine whether the measure of association used is appropriate for the level of measurement in the data. Finally, you can attempt to verify the calculations for the reported association measures. This will sensitize you to the data that are reported.

Key 5.2. Know the meaning of association that the researcher is using. There are several specific things for you to think about here. First, you want to know the meaning of any existing association in view of the data and the measure being used. You should know how the data would be distributed both if there were no association and if there were perfect association between the variables. You should know whether the measure used would yield a value of 0 if the no-association distribution occurred and 1 if the perfect association distribution occurred. Ask yourself how the measure would reflect the strength of an imperfect association. You must know what kind of comparisons can be made with each statistic used. Is the measured normed or not? That is, can you compare calculated values only to the theoretical extremes (e.g., 0 and 1) for the measure, to calculated values with the same measure for other distributions in a relative sense, or can PRE-type comparisons be made? For ordinal association, you should know what both a positive and a negative direction would mean for the data. This can be facilitated if you identify the positive and negative diagonals in the table showing the distribution. You should also look at tables to discover any distributions that may have reversals in the direction of their association. These will require special analysis.

You should also remind yourself of the exact statistical definition of association that each measure uses. The chi-square-based measures (ϕ, C, and V) define association by comparing the calculated chi-square value to its maximum possible value in a table with a specific size and number of cases. Lambda defines association in terms of the usefulness of knowing the dependent variable's modes within each of the categories of the independent variable (X) when we try to predict the dependent variable (Y) scores for each case. Tau-y* defines association as the usefulness of knowing the proportional distribution of the dependent variable within each category of the independent variable when predicting scores on the dependent variable. Gamma defines association as the utility of knowing how well the numbers of agreement and disagreement pairs in an ordinal distribution allow us to reduce errors when we predict the direction for all untied pairs in a distribution. Spearman's r_s defines association as the ability of ranks for one variable to predict the ranks for another variable. These definitions may be somewhat cumbersome to non-researchers, but they do tell you exactly what a researcher means when a particular association measure is used.

Key 5.3. Distinguish between simple association and the more sophisticated analyses of association. Most research begins with an examination of association between two variables without consideration of any other factors. But this look at simple association is usually followed by more complex analyses. With nominal and ordinal variables researchers often use the technique of elaboration to determine whether the bivariate relationship they first observed holds under various relevant conditions. The tabular displays of these conditional associations may not be shown, but you can still read the narrative to determine whether the researcher considered the influence of third variables. If not, a justification should be given for keeping the analysis at the most basic level.

As in Chapter 1, we can add the caution that not all the information you need to evaluate measures of association may be immediately available. You may need to read more than one report by one researcher to get a complete picture. Remember, too, that we are discussing only the introduction to social statistics. Some of your difficulties in understanding published research may be related to your limited exposure to it. The only cure for that is more study. The next chapter will point the way to some of the more complex and useful analysis procedures now being used in social research. Although we will continue to stay on the beginning level, a mastery of this material will be extremely helpful for consuming a large part of contemporary social data analysis.

Figure 5.3 shows some guidelines about the use of various measures of association at

*Covered in an optional * section of the chapter.

FIGURE 5.3 Characteristics of Common Measures of Association for Nominal and Ordinal Variables

Level of Measurement	Measure	PRE	Symmetrical/Asymmetrical	
Nominal	ϕ	Yes (if squared in 2×2 table)	×	
	C	No	×	
	V	No	×	
	T[a]	No	×	
	Lambda	Yes		×
	Tau-y	Yes		×
Ordinal	Gamma	Yes	×	
	Q[a]	Yes	×	
	Tau-a[a]	Yes	×	
	Tau-b[a]	Yes	×	
	Tau-c[a]	Yes	×	
	Sommer's d[a]	Yes		×
	Spearman's r_s	Yes (if squared)	×	
One nominal and one ordinal variable	θ[a]	Yes	×	

[a]Listed in the section *Related Statistics.*

the nominal and ordinal levels. This chart includes the measures that we have been discussing in this chapter and those listed in the section *Related Statistics* on page 189. It should help you to get organized.

KEY TERMS

Association
Conditional Distribution
No Association
Perfect Association
Normed Measure
Chi-Square (χ^2)
Observed Frequencies
Expected Frequencies
Pearson's Phi (ϕ)
Pearson's Contingency Coefficient (C)
Cramér's V

PRE Measure
Lambda (L)
Asymmetrical Measure
Symmetrical Measure
Association Existence
Association Strength
Association Direction
- Positive
- Negative
- Monotonic
- Polytonic

*Goodman and Kruskal's Tau-y (T_y)
Goodman and Kruskal's Gamma (G)
Agreement Pairs
Disagreement Pairs
Spearman's r_s
Elaborating on Association
Zero-Order Association
Controlling a Variable
Conditional Association
First-Order Conditional Association
Second-Order Conditional Association
Suppressed Association
Partly Explained Association
Spurious Association

EXERCISES

1. For the following distribution: (a) Compute chi-square. (b) Compute phi and interpret it in terms of existence and strength of an association for these data. (c) Square the value of phi and give it a PRE interpretation for these variables.

Presence of Stigma Feelings	Source of Public Aid Received: Local	Federal
Yes	45	25
No	10	20

2. For the following distribution: (a) Compute chi-square. (b) Compute C and V and interpret each in terms of existence and strength of an association for these data. (c) Which is easier to interpret, and why?

Fuel Source	Ownership of Utility Company: Private	Public	Cooperative
Coal	11	10	19
Petroleum	15	29	11
Nuclear	21	12	7

3. Verify the calculation of phi in Table 1 of R.E. 5.1 by doing your own computation.
4. Compute phi for the race–social legislation (b) distribution in Table 5.5. Compare your answer to the lambda for that table. What can you say about how and why they compare as they do?
5. Verify the calculation of V in the table shown in R.E. 5.2 by doing your own computation. Note that you must first convert the percentages to the nearest whole frequencies.
6. Go back to R.E. 2.8 and compute (a) phi, (b) C, and (c) V. (d) Compare the three results.
7. For the following distribution: (a) Compute chi-square. (b) Compute V and interpret it in terms of existence and strength of an association for these data.

Perceived Sex of Silhouette	Sex of Viewer: Male	Female
Male	52	23
Female	10	30
Uncertain	13	22

8. For the same data as Exercise 7: (a) Compute lambda by predicting perceived sex. (b) Interpret your result in terms of existence and strength of an association for these data.

9. For the following distribution: (a) Compute lambda by predicting style of supervision. (b) Interpret your result in terms of existence and strength of an association for these data. (c) Give a PRE interpretation of lambda for these variables. *(d) Repeat parts (a)–(c) using tau-y.

Style of Supervision	Type of Firm: Industrial	Investment	Retail	
Authoritarian	8	12	40	60
Paternal	10	36	14	60
Production-Centered	42	12	6	60
	60	60	60	

10. For the sex–employment status distribution in part b of Table 5.1: (a) Compute lambda by predicting employment status. (b) Interpret your result in terms of existence and strength of an association. (c) Give a PRE interpretation of lambda. *(d) Repeat parts (a)–(c) using tau-y.
11. For the following distribution: (a) Compute gamma. (b) Interpret your result for existence, strength, and direction of an association for these data. (c) Give a PRE interpretation of gamma for these variables.

Length of Sentence	Severity of Crime: High	Medium	Low
10+ Years	15	8	2
3–9 Years	8	15	10
0–2 Years	5	7	10

12. For the following distribution: (a) Compute gamma. (b) Interpret your result for existence, strength, and direction of association for these data. (c) Give a PRE interpretation of gamma for these variables.

Behavior Toward Ending Air Pollution	Attitude Toward Ending Air Pollution: Should Be Stopped	Should Be Reduced	Should Do Nothing
Works in an Activist Organization	6	4	0
Signs Petitions	34	51	5

(d) Would you say that the people represented by these data show consistency between what they feel (attitude) and what they do (behavior)?

13. Verify the calculation of gamma in Table 1 of R.E. 5.3 by doing your own computation. First, convert the percentages to the nearest whole frequencies.
14. Verify the calculation of gamma in Table 2 of R.E. 5.3 by doing your own computation. First, convert the percentages to the nearest whole frequencies.
15. Go back to R.E. 2.9: (a) Compute gamma. First, convert the percentages to the nearest whole frequencies and arrange the table properly. (b) Interpret your result in terms of existence, strength, and direction of association for these data.
16. A group of 12 preschool children were ranked according to their verbal abilities and frequency of physical conflict during play. Rank = 1 was assigned to the highest ability and highest frequency, respectively, in the following distribution. (a) Compute Spearman's r_s. (b) Interpret your result for existence and direction of an association for these data. (c) Square the value of r_s, interpret the strength of the association, and give a PRE interpretation for these variables.

Child	Rank in Verbal Ability	Rank in Frequency of Conflict
A	5	7
B	12	2
C	6	8
D	8	4
E	4	9
F	3	11
G	10	3
H	9	5
I	7	6
J	11	1
K	2	12
L	1	10

17. In the following distribution, nine cities have been ranked according to the density of their central city's population. Rank = 1 was assigned to the highest density. The percentage of the city's revenue that comes from local taxation is also listed. (a) Convert the revenue data into ranks by assigning rank = 1 to the highest percentage. (b) Compute Spearman's r_s. (c) Interpret your result in terms of existence and direction of association for these data. (d) Interpret the strength of the association.

City	Rank in Population Density	Percent of Revenue from Local Taxes
A	5	41
B	2	21
C	9	31
D	1	15
E	4	20
F	6	34
G	3	25
H	7	36
I	8	38

(e) Now rerank the revenue variable by assigning rank = 1 to the lowest percentage. Recompute Spearman's r_s, interpret it, and compare your result to the first value and interpretation.

18. For each of the following decide what measure(s) of association would be optimum. Choose from those discussed in the body of the chapter, excluding the additional ones listed in the section *Related Statistics.*

Predominant Reaction	Reaction to Former Mental Patient	Reaction to Former Convict
Fear	16	18
Disgust	4	14
Sympathy	14	2

Desire to Change Residence	Number of Friends: Many	Some	Few	None
Great	1	3	7	11
Some	5	10	11	5
Little	10	6	3	2

Perceived Health Status	Sex: Male	Female
Good	114	131
Not-So-Good	86	69

Religiosity	Age: Young	Middle-Aged	Old
High	24	12	52
Medium	31	21	29
Low	45	62	19

Country	Rank in GNP	Rank in Size of Military
A	6	5
B	2	2
C	1	1
D	3	4
E	4	3
F	5	6

19. Suppose that a researcher has measured the following three variables in the categories listed. (a) Sketch a table from which we could calculate gamma for the association between self-esteem and analytical ability with a hypothetical distribution. (b) Sketch the conditional tables from which we could calculate conditional gammas for the same association while controlling for ethnicity.

Self-esteem (high, medium, low).
Analytical ability (high, medium, low).
Ethnicity (white, black, Native American).

20. Find one example of (a) phi, *C*, or *V* and of (b) gamma being used in the professional social science literature. Write a short narrative evaluating each measure's application using the *Keys to Understanding* in this chapter.

Tschuprow's *T*

T is a chi-square-based measure of association for nominal data arranged in a bivariate table of any size. It ranges from 0 to 1; however, it can achieve 1 only in a square table. It was designed as an improvement over Pearson's *C*.

$$T = \sqrt{\frac{\chi^2}{N(r-1)(c-1)}}$$

where r = number of categories in the row variable
c = number of categories in the column variable

See: Blalock (1979:304).

Yule's *Q*

Q is a PRE measure for 2 × 2 tables of nominal or ordinal data. It is a special case of gamma. It ranges from 0 to (±)1; however, it can achieve 1 both whenever one or two (diagonal) cells have a frequency of zero. That is, it uses the alternative definition of a perfect association.

$$Q = \frac{ad - cd}{ad + cd}$$

where a = frequency of upper left cell
b = frequency of upper right cell
c = frequency of lower left cell
d = frequency of lower right cell

See: Mueller et al. (1977:219ff).

Kendall's Tau-*a*

Tau-*a* is a symmetrical PRE measure for ordinal data in a bivariate table of any size. It ranges from 0 to (±)1; however, it cannot achieve 1 if there are any tied pairs, which there often are.

$$T_a = \frac{P_a - P_d}{P_t}$$

where P_a = agreement pairs
P_d = disagreement pairs
P_t = total number of unique pairs

See: Loether and McTavish (1980:234ff).

Kendall's Tau-*b*

Tau-*b* is a symmetrical PRE measure for ordinal data in a bivariate table of any size. It ranges from 0 to (±)1; however, it cannot achieve 1 unless the table is square. Therefore, it is used with square tables when the researcher wishes to include pairs that are tied on *X* or *Y*.

$$T_b = \frac{P_a - P_d}{\sqrt{(P_a + P_d + P_y)(P_a + P_d + P_x)}}$$

where P_a = agreement pairs

P_d = disagreement pairs
P_y = pairs tied on Y
P_x = pairs tied on X

See: Loether and McTavish (1980:236ff).

Kendall's Tau-*c*

Tau-*c* is a symmetrical PRE measure for ordinal data in a bivariate table of any size. It ranges from 0 to (±)1 regardless of table size.

$$T_c = \frac{2k(P_a - P_d)}{N^2(k - 1)}$$

where P_a = agreement pairs
P_d = disagreement pairs
N = number of cases
k = number of categories in the row or column variable, whichever is less

See: Loether and McTavish (1980:236).

Somer's *d*

Somer's *d* is an asymmetrical PRE measure for ordinal data in a bivariate table of any size. It ranges from 0 to (±)1. It is used for evaluating the predictive ability of Y by X (or the reverse) including ties on the predicted variable.

$$d_{yx} = \frac{P_a - P_d}{P_a + P_d + P_y} \quad \text{or} \quad d_{xy} = \frac{P_a - P_d}{P_a + P_d + P_x}$$

where P_a = agreement pairs
P_d = disagreement pairs
P_y = pairs tied on Y
P_x = pairs tied on X

See: Loether and McTavish (1980:235ff).

Wilcoxon's Theta (θ)

Theta is a special-purpose measure of association for one nominal and one ordinal variable in a bivariate distribution. It ranges from 0 to 1 and provides a PRE interpretation.

$$\theta = \frac{\Sigma D}{T_2}$$

where ΣD = sum of the absolute differences between the total number of frequencies above and below each ordinal category for pairs of nominal categories
T_2 = total frequency of each nominal category multiplied by the total frequency of every other nominal category

See: Champion (1981:341ff).

CHAPTER 

DESCRIBING ASSOCIATION BETWEEN INTERVAL AND RATIO VARIABLES

This chapter introduces the basic ideas and techniques used by social researchers to describe association between interval and ratio variables. These ideas and techniques are an extension of those used with nominal and ordinal variables. Mathematics play a more dominant part in the relevant statistics because of the higher level of measurement with interval and ratio variables. This makes the statistics we are going to discuss both more useful and more complex to understand. The social science literature is filled with the types of analysis that we are exploring. The four research examples in this chapter demonstrate some of these. Skim over these now to see what we are aiming to understand.

Correlation is the statistical term commonly used for association at the interval and ratio levels. Some use it for association at the lower measurement levels, too. It represents the idea of mutual or "co-relation" between two or more variables. At the interval and ratio levels there are three kinds of correlation to consider. *Simple or zero-order bivariate correlation* is the association between two variables while ignoring any other variables. *Partial correlation* is the association between two variables while controlling one or more other variables. *Multiple correlation* is the simultaneous association between three or more variables. We look at each of these in this chapter, although the concentration will be on the first one. The measurement of correlation between interval and ratio variables is based on a technique called regression, with which we will begin.

Regression

Regression is a special PRE (proportional reduction in error) method for predicting the scores on one variable from knowledge of the scores on another variable using statistics of variation. The accuracy of the predictions is a direct consequence of the association between the variables. The more highly related the variables are, the more accurate the regression predictions will be. Some scatter diagrams can illustrate this technique.

Scatter Diagrams of Association

Figure 6.1 shows scatter diagrams of four possible bivariate distributions for scores on an application test and beginning hourly wages. Let's examine each of these with an eye toward guessing the strength and direction of the association. For direction we scan across the graph to see how the data points follow either a pattern in which the scores increase on both variables (positive) or in which the scores increase on one variable while they decrease on the other (negative). For strength we look to see how consistently concentrated the data points are in any pattern.

In the (a) graph the plotted points show a pattern of higher test scores occurring with higher beginning wages. Thus the association has a positive direction. The pattern is entirely consistent across all data points. This means that the strength of the association is

FIGURE 6.1 Scatter Diagrams of Various Types of Association for Interval/Ratio Data

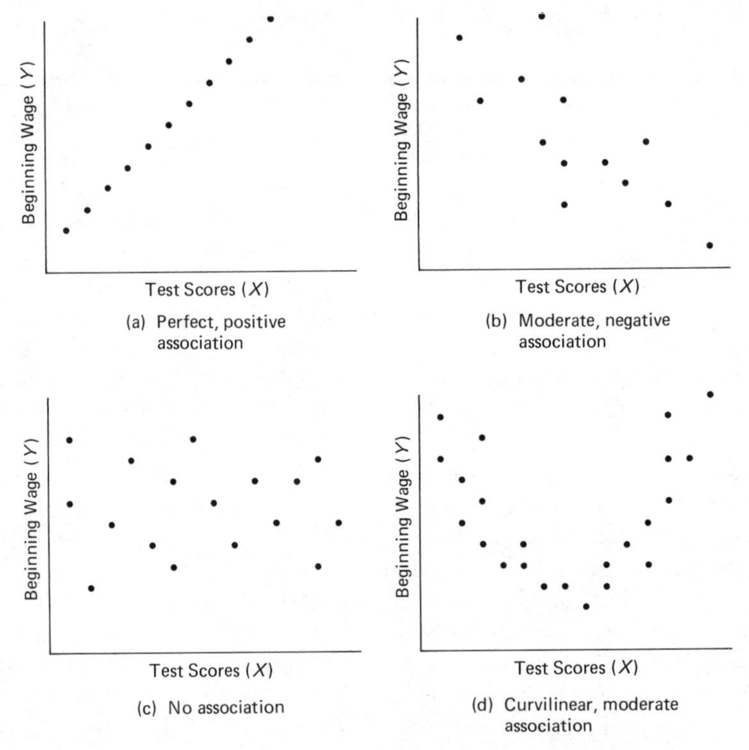

perfect or at the maximum possible level. In the (b) graph the points are arranged so that higher test scores usually occur with lower beginning wages. Thus the direction is negative. The points are not very consistently scattered, though. Sometimes a small increase in test scores occurs with a large decrease in wage and sometimes with a small decrease in wage. The strength is only moderate. The (c) graph has no apparent direction. Test scores of most every level are paired with wages of most every degree. The inconsistency is so widespread that we say that there is no visible association between the two variables.

In the (d) graph the association follows a rather consistent pattern which means that it is of moderate strength. But the direction changes from negative in the first half of the graph to positive in the second half. This directional reversal is called *curvilinear.* It is similar to the polytonic pattern noted for ordinal variables in Chapter 5. Statisticians give it a new name because interval and ratio variables have continuous distributions across several measurement categories, whereas ordinal variables have discrete distributions restricted to scores at only a small number of categories.

Our examination of association will focus on the consistently positive or negative patterns. These are referred to as *linear associations,* for which we will discuss *linear regressions* and *linear correlations.* It is important for you to realize that the application of linear statistics to curvilinear distributions is a mistake. For example, suppose that you calculated a measure of linear correlation for Figure 6.1(d). This would lead you to conclude that these data had no association because their pattern does not have a consistent direction. In fact, there is a moderate curvilinear association in these data. To avoid this error, researchers frequently use scatter diagrams to check for linearity before deciding on the analysis procedures to follow.

The Regression Line

Our interpretations of the distributions in Figure 6.1 were based on our ability to describe visually the patterns graphed. You probably imagined drawing a straight line through the center of each pattern to help you describe it. This is not always an easy or reliable technique when done by imagination alone. Methods of linear regression find the one line that truly best describes the pattern in the data. It is the *line of best fit* in the statistician's vocabulary. Exactly how this line is found can be demonstrated with a specific example beginning with the bivariate frequency distribution shown in Table 6.1.

Suppose that the data in this table represent five new employees at the High Caloric Donut Factory. The management at this establishment has been using a special application test to help evaluate potential employees. One of the new employees has discovered

TABLE 6.1 Bivariate Distribution of Beginning Hourly Wages and Application Test Scores for Five New Employees at the High Caloric Donut Factory

Employee	(*X*) Test Score	(*Y*) Beginning Wage ($)
A	4	9
B	5	6
C	3	4
D	2	2
E	1	4

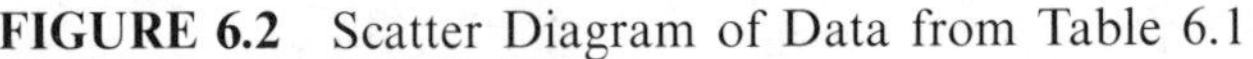

FIGURE 6.2 Scatter Diagram of Data from Table 6.1

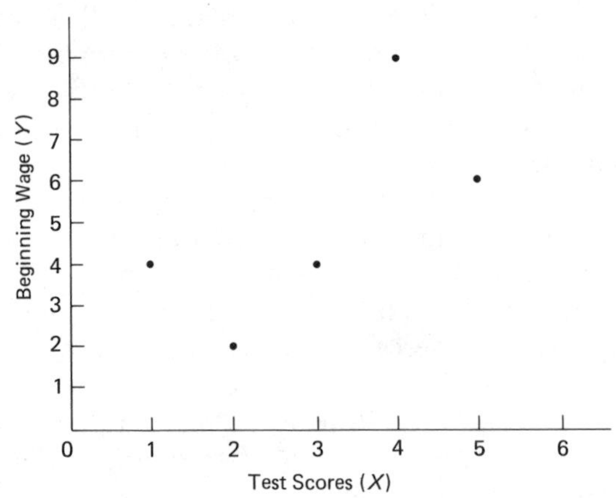

that his wage is below that of several others and has decided to engage in some investigative research to determine whether the test scores truly are related to beginning wages. The application test scores and the beginning wages (in dollars) are displayed in the table for each of the five employees. Our employee-researcher is case D. We will follow this example throughout this entire chapter to give you a stable reference point for the various issues to be discussed.

Our first step is to graph these data in a scatter diagram. Figure 6.2 shows this. Notice that we have put the beginning wages on the Y-axis as the dependent variable. The application test scores are put on the X-axis as the independent variable which may help explain the differences in beginning wages. If you have temporarily forgotten the procedures for graphing data in scatter diagrams, you may want to reread the discussion in Chapter 2. Note the correspondence between each employee's pair of scores from Table 6.1 and the data point in the scatter diagram that represents that employee. The data points do seem to follow a linear, rather than a curvilinear, pattern, so we can proceed to determine their linear regression.

Any straight line, including a linear regression line, can be described by an equation of the following basic type:

$$Y = a + bX \qquad \text{Equation for Any Straight Line}$$

where Y = scores on the dependent variable
X = scores on the independent variable
a = Y-axis intercept of the line
b = slope of the line

Let's review the basic mathematics behind such an equation. Suppose we know that the equation for a straight line is $Y = 4 + 2X$. This means that the line would cross (intersect) the vertical Y-axis at the value 4 ($a = 4$). See Figure 6.3(a). The line would have a positive slope, going across the graph from the lower left to the upper right. We know that this is the direction because the sign of the b value is positive. The exact degree of slope for this line is such that it increases by two units on Y for every one unit increase on X. This is indicated by the value of b, 2. That b value is understood to be a ratio of change on the Y-axis per one unit increase on the X-axis. Thus a b value of $+2$ represents the fraction $+2/+1$. This ratio is illustrated with a small triangle drawn below the line with a base equal to a one-unit increase in X. The associated change in Y is a two-unit increase.

Consider another equation for a straight line: $Y = 18 - .7X$. This means that the line crosses the Y-axis at the value 18 ($a = 18$). The line has a negative slope going from the upper left to the lower right, as indicated by the sign of b. The precise slope of that line is such that it decreases by seven-tenths (.7) of a unit on Y for every one-unit increase on X. The slope ($-.7$) is understood to represent the fraction $-.7/+1$. This line is shown in Figure 6.3(b).

The basic equation for a straight line is modified slightly for our linear regression line to accommodate the reality that we are predicting values for scores on the dependent variable, not defining them as they actually are case by case. To make this obvious, we use a special symbol for the predicted values, $\widehat{Y}$ ("Y hat"). The circumflex over the Y means that the scores are estimates based on the regression prediction. Our linear regression equation is thus

$$\widehat{Y} = a + bX \qquad \text{Linear Regression Equation}$$

where $\widehat{Y}$ = predicted scores on the dependent variable
X = actual scores on the independent variable
a = Y-axis intercept of the regression line
b = slope of the regression line

Table 6.1 provided the values for the independent variable, test scores, needed to calculate our regression equation. But where do we get the values for the a and b constants in that equation? Each of these has its own formula based on the X and Y scores in the bivariate distribution.

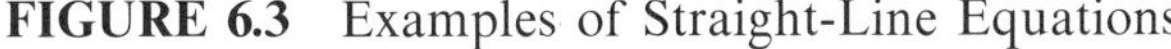

FIGURE 6.3 Examples of Straight-Line Equations

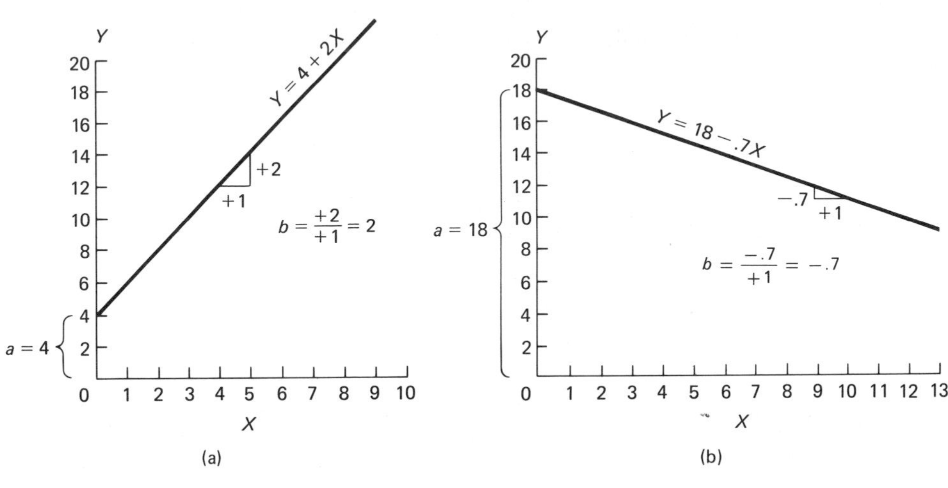

The slope (b) expresses the degree of association between the X and Y scores in terms of deviations. Its definitional formula demonstrates this:

$$b = \frac{\Sigma (X - \overline{X})(Y - \overline{Y})}{\Sigma (X - \overline{X})^2} \quad \text{or} \quad \frac{\Sigma\, xy}{\Sigma\, x^2}$$

Definitional Formula of Slope (b)

The numerator is the sum of the product of deviations from the X and Y means for each case. It is also known as the *covariation,* the extent to which the X and Y scores vary together. The numerator determines the sign of the slope. Thus, if the covariation is positive, so is the slope, and likewise for the negative direction. The denominator expresses the sum of squared deviations in the independent variable, the total variation in that variable. Therefore, the slope is a ratio of covariation between X and Y to the total variation in X alone.

Commonly used computational formulas for b and a are

$$b = \frac{N(\Sigma XY) - (\Sigma X)(\Sigma Y)}{N(\Sigma X^2) - (\Sigma X)^2}$$

Computational Formulas for: Slope (b)

$$a = \overline{Y} - b\overline{X}$$

Y-axis Intercept (a)

where N = number of cases (pairs of scores)
ΣXY = sum of all X times Y products
ΣX = sum of all actual X scores
ΣY = sum of all actual Y scores
ΣX^2 = sum of the squared X scores
$(\Sigma X)^2$ = square of the ΣX quantity
$\overline{X}$ = mean of the X scores
$\overline{Y}$ = mean of the Y scores

Table 6.2 shows the preliminary computations necessary to calculate the a and b values for the regression line that best fits our data. These are substituted into the formulas as follows:

$$b = \frac{N(\Sigma XY) - (\Sigma X)(\Sigma Y)}{N(\Sigma X^2) - (\Sigma X)^2} = \frac{5(86) - (15)(25)}{5(55) - (15)^2}$$
$$= \frac{430 - 375}{275 - 225} = \frac{55}{50} = 1.1$$

$$a = \overline{Y} - b\overline{X} = 5 - (1.1)(3)$$
$$= 5 - 3.3 = 1.7$$

Substituting these values into our linear regression equation, we find that

$$\widehat{Y} = a + bX$$
$$= 1.7 + 1.1X$$

TABLE 6.2 Computations Necessary for Calculating Regression Equation

X	Y	X^2	XY
4	9	16	36
5	6	25	30
3	4	9	12
2	2	4	4
1	4	1	4
15	25	55	86

$\bar{X} = 15/5 = 3$ $\bar{Y} = 25/5 = 5$

This equation means that the regression line will cross the Y-axis at \$1.70 (a). Its slope (b) means that this line will incline by \$1.10 along the Y-axis for every one point of increase in test score along the X-axis. Both the a and b constants were positive in our example, but they would not always be this direction. If the intercept had been negative, it would mean the regression line crosses the Y-axis below the point of origin. If the slope had been negative, it would mean the beginning wage decreased with every one-point increase in test score.

We can plot the regression line on our scatter diagram by finding any two X and Y coordinates (pairs of scores) that satisfy the equation. These can then be plotted and connected with a straight line. For example:

$$\text{Let } X = 0. \text{ Then } \hat{Y} = 1.7 + 1.1(0)$$
$$= 1.7 + 0 = 1.7$$
$$\text{Let } X = 5. \text{ Then } \hat{Y} = 1.7 + 1.1(5)$$
$$= 1.7 + 5.5 = 7.2$$

These two coordinates (0, 1.7 and 5, 7.2) have been plotted in Figure 6.4 and connected with a straight line. This line is the linear regression line for predicting beginning wage from test score. Notice that the line does cross the Y-axis at the value of a (1.7). By drawing a small triangle below the regression line with a base equal to a one-point increase in X (text scores), we see that the line slopes upward such that the change in Y (beginning wage) is +1.1, which is the value of b.

We can now use this same regression equation to make predictions about the beginning wage that an employee would receive at any relevant test score under the assumption that these two variables were perfectly correlated. This technique is exactly like that just above when we found two coordinates to use for plotting the regression line on the graph. We simply choose some test score of interest, substitute it for X in the equation, and solve the equation for $\hat{Y}$. What would we predict the wage to be for someone who scored 6 on the test?

$$\hat{Y} = 1.7 + 1.1(6)$$
$$= 1.7 + 6.6 = 8.3$$

What would we predict the wage to be for our employee-researcher who scored a 2 on the test?

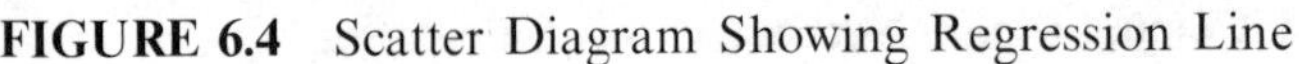

FIGURE 6.4 Scatter Diagram Showing Regression Line

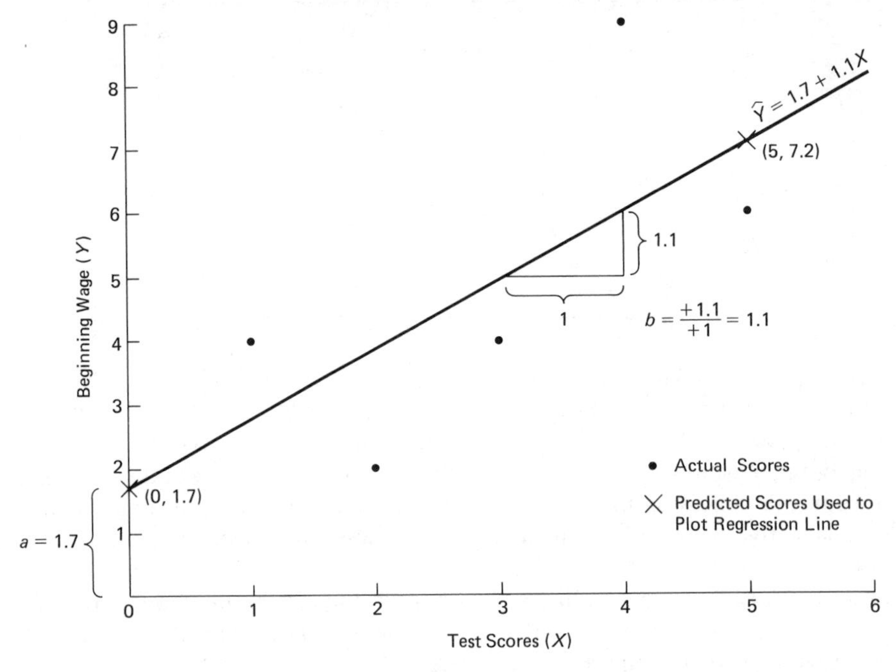

$$\begin{aligned}\hat{Y} &= 1.7 + 1.1(2)\\ &= 1.7 + 2.2 = 3.9\end{aligned}$$

Our regression prediction of his wage is \$3.90. Actually, he received \$2.00. One conclusion we can reach from this is that test score was not the only thing being used to set this person's beginning wage. Otherwise, the regression prediction would have been perfectly accurate. Does this mean that test scores are not related at all to beginning wages? Not necessarily. It may be related but less strongly than a perfect association. We will pursue this possibility later.

There are three important restrictions on the use of regression equations for making predictions. First, we must confine ourselves to the general range of independent scores represented in the actual distribution. We are not at liberty to speculate on the wages for persons who might have test scores at such extremely high levels as 25 or 50, for example. That would imply that the two variables have the same relationship at all imaginable values, which may not be true. It is entirely possible that people with exceptionally high test scores might receive wages that were much below the values predicted by the regression equation because the employer was unwilling to pay anyone above a set amount. In other situations, we can imagine scores on an independent variable much below those that were a part of the original data set. Again, these theoretical values might have a very different association with the dependent variable if they actually occurred compared to the X and Y scores in an original distribution.

The second restriction is to remember that we are discussing the technique for predicting scores on the dependent variable from actual scores on the independent variable and not the reverse. That is to say that regression is an asymmetrical technique. In our example the formulas would need to be adjusted if we wanted to predict test scores from knowledge of beginning wages. We are "regressing beginning wages on test scores," in the parlance of statistics. The reversed formulas are presented in the section *Related Statistics* at the end of the chapter.

The third restriction is to know that most regression prediction is different from a forecast.* We are not predicting future scores that may occur with new observations that may be collected. We are simply predicting existing scores based on limited knowledge of the cases we have already studied. This is the same meaning of prediction that we used in Chapter 5 concerning PRE measures of nominal and ordinal association.

Interpreting Regression Coefficients

One additional piece of information is necessary to interpret many published reports of bivariate linear regression. The slope of the regression line (b) is also called the *unstandardized regression coefficient.* As mentioned above, this coefficient tells you the predicted change in the Y variable for every one-unit change in the X variable. A second kind of closely related value is often reported in the literature on social research. It is the *standardized regression coefficient* (b^*). It is also known as *beta* or the *beta weight* (β). It, too, tells you about the predicted change in the Y variable from one-unit changes in the X variable. But whereas the unstandardized coefficient does this in terms of the original units of measurement for those two variables, the standardized coefficients express this in a way that makes all regression coefficients directly comparable. The formula for the standardized coefficient is as follows:

$$b^* = b\left(\frac{S_x}{S_y}\right)$$ Standardized Regression Coefficient Formula

where b^* = standardized regression coefficient
b = unstandardized regression coefficient
S_x = standard deviation of the independent variable (X)
S_y = standard deviation of the dependent variable (Y)

As an illustration, let's consider an extension to our example. We know from our data in Table 6.1 that the unstandardized regression coefficient is 1.1. As already explained, this means that for every one-point increase in test score, we predict that the employee will receive \$1.10 more in beginning wage. But suppose that you wanted to compare this regression coefficient with another from a separate regression equation. We might want to regress beginning wages on another variable, say, years of previous work experience. We would follow the same procedure as above and find the new regression equation. Suppose that it was

$$\hat{Y} = 2.1 + .8X$$

The unstandardized regression coefficient in this equation is .8. This means that for every one-year increase in work experience, we predict beginning wage to increase by

*Some researchers do use regression to predict future scores. But they can do so only under the assumption that the future association of X and Y is exactly the same as that for the existing data.

\$.80. So how do the two regression coefficients compare? Which variable, test scores or work experience, results in the prediction of a greater increase in beginning wage? We cannot compare these directly because one point on the application test is not equivalent to one year of work experience. But the standardized versions of these same coefficients would not have these original units of measurement and thereby they would be directly comparable. For our Table 6.1 data, the standard deviation of X is 1.414 and of Y is 2.366. The standardized regression coefficient is thus

$$b^* = b\left(\frac{S_x}{S_y}\right) = 1.1.\left(\frac{1.414}{2.366}\right) = 1.1(.5976) = .657$$

Suppose that the standard deviation for our work experience variable was 2.199. The standardized regression coefficient for the second regression would be

$$b^* = .8\left(\frac{2.199}{2.366}\right) = .8(.9294) = .744$$

Our comparison of these two standardized coefficients shows that one standardized unit increase in work experience would result in a greater increase in predicted beginning wage than would one standardized unit increase in test score (\$.744 versus \$.657, respectively). Most published research reports these kinds of standardized regression coefficients because of their obvious advantage for comparative interpretations. They may also report the unstandardized regression coefficients for their descriptive merits.

Research Example 6.1 illustrates how the regression line can be helpful in understanding association between interval or ratio variables. The authors of this example examined the claims of extreme old age in Vilcabamba, Ecuador, a place noted for these claims. To evaluate the assertions, they collected independent information about peoples' ages (records of birth, marriage, and death) for comparison with the stated ages of village residents. A scatter diagram of these two age variables is shown in the figure of the research example.

RESEARCH EXAMPLE 6.1 Using the Regression Line to Examine Association

From Richard B. Mazess and Sylvia H. Forman, "Longevity and Age Exaggeration in Vilcabamba, Ecuador," Journal of Gerontology, *34 (1), January 1979:94ff. Reprinted by permission.*

The population of Vilcabamba in southern Ecuador is one of three populations in the world noted for extreme longevity and for the rarity of chronic diseases which usually afflict the elderly. . . . Documentation of great longevity in Vilcabamba, however, has been viewed as difficult because of the destruction of early baptismal records in a church fire. Controversy has developed as a result of both uncritical acceptance and uncritical rejection of the extreme ages reported to investigators by the inhabitants. We have attempted to evaluate the extent of age exaggeration in this population by comparing reported ages with those derived from birth records and from other records of the age of the inhabitants when they were young adults (age 20 to 50 years).

Age exaggeration appears to be a common finding in the extreme elderly throughout the world and appears associated with illiteracy and absence of actual documentation. . . . Illiteracy has been common in Vilcabamba and there has been little attempt at documentation of ages until recently. Although there are few social and economic benefits associated with extreme age in this region, there does appear to be high prestige accorded the aged. Consequently some overstatement of ages might be expected. Fortunately birth, marriage, death, and other records have been kept over the past century and these permit objective documentation of actual ages. . . .

Linear regressions were calculated between the stated age (or the reported age at death) and the age based on the actual or estimated birthdate. There was no difference between the regression based on actual birthdates and those based on the estimated year of birth. There was only a slight difference between the regression for living subjects and that for the recently deceased. [The figure] shows the stated age (or age at death) plotted against the age provided by the records. There was little age exaggeration evident up to 60 or 70 years, but after this systematic age exaggeration was evident. A regression was calculated for the 110 cases above "70" years. At a stated age of 80 the estimated age was 77, while at "100" and "130" years the estimates were 84 and 95 years, respectively. The high correlation ($r = 0.75$) and small standard error of estimate (4.8 years) suggested that this regression would be useful in interpreting grouped data (e.g., from census or death records) from this population, and for correcting the systematic age exaggeration. Such a regression cannot be used for individual cases, particularly those beyond "110" years, because of the large confidence limits. . . .

Stated Age vs. Actual Age in Vilcabamba. Age exaggeration is described by the regression line for cases above 70. [Figure 1 in the original.]

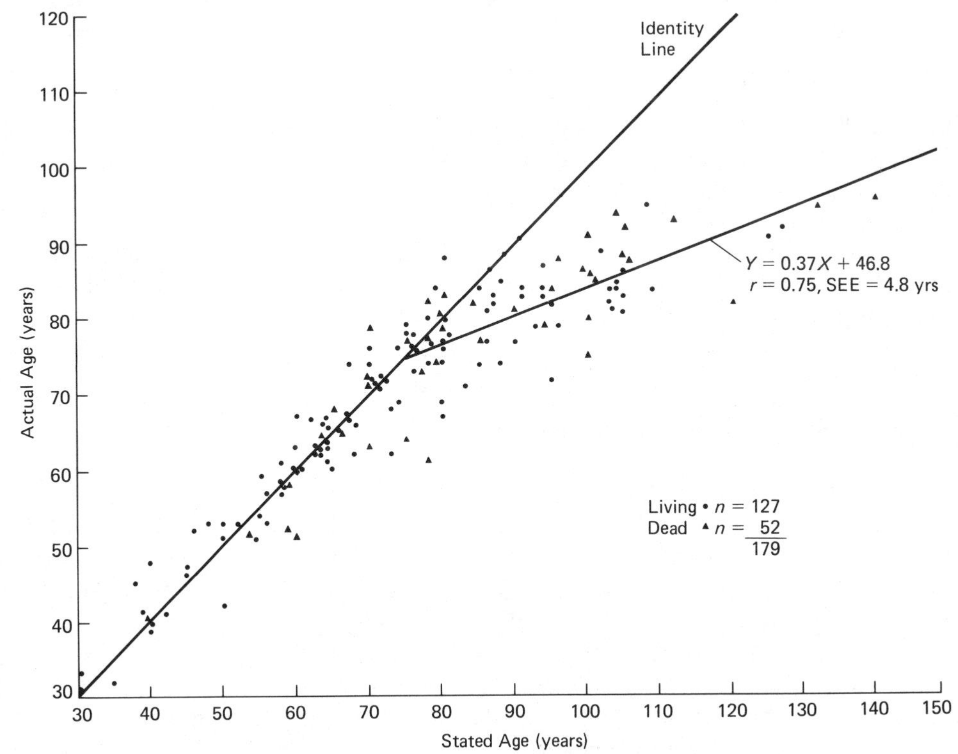

It has not been possible to completely resolve the controversy concerning longevity in Vilcabamba, but we were able, using the above approach, to show that the elderly had overstated their ages. This was expected, since it has been so often the case in other populations. In fact, none of the 23 investigated cases of living or recently deceased "centenarians" had survived over 100 years. The average age of these reputed centenarians was 86 (SD $\pm$ 5) years with a range of from 75 to 96 years. Among nonagenarians (n = 15) the mean age was actually 81.5 (SD $\pm$ 4) years with a range of from 72 to 88 years. These results, together with the age regression, suggested that reported ages over 90 years must be viewed with extreme caution. Such ages were typically overstated by at least a decade, and in many cases by as much as two decades. The reasons for such age exaggeration were unclear, but some investigators have suggested that the tendency has been worsened by recent scientific and tourist incursions. Examination of the death records from the first half of this century, however, showed many deaths occurring at above 90 or 100 years of age. This suggested that the exaggeration has been traditional, and that it was so while the area was still isolated from surrounding zones. . . .

Notice that the data follow one linear trend up to the age of about 70 and another after that. The under-70 trend is well represented by an "Identity Line" which shows perfect correspondence between the stated age and the actual age as confirmed by records. The after-70 trend is represented by a bivariate regression line having the equation $Y = .37X + 46.8$. Put in the form we have been discussing, this is $\hat{Y} = 46.8 + .37X$, where $\hat{Y}$ is the predicted actual age and X is the stated age. The unstandardized regression coefficient means that for every one-year increase in stated age the estimated actual age increases only .37 year. Roughly speaking, it suggests that the over-70 residents were exaggerating their age by three times (on the average) once they passed their 70th birthdays ($1/.37 \simeq 3$). This regression is a good illustration of the error that would be involved if predictions were attempted for ages much below those on which the regression was based. It would be meaningless to substitute age values such as 10 or 20 for X into the equation since it was intended only to characterize those aged 70 or older.

The accuracy of the regression predictions in R.E. 6.1 is noted by the "$r = .75$" and "SEE = 4.8 years" statistics shown below the equation in the figure. The first of these is the correlation coefficient, which is discussed later in this chapter. The second is the standard error of the estimate, which is included in the section *Related Statistics* at the end of the chapter.

Least Squares Principle for Regression

We have described the linear regression line as a line of best fit for data. No other straight line would describe the pattern in a data set with more accuracy. The accuracy to which we refer is based on the mathematical *principle of least squares.* This principle will explain to you not only the meaning of best fit which is operating but also the PRE meaning of regression.

One of the characteristics of the mean for a distribution is that the deviations of the individual scores from it will be a minimum if squared and then summed for all cases. For example, when we find the deviations around the five beginning wages in Table 6.1, square these, and sum them, we get the total 28. If you were to find the deviations of these same

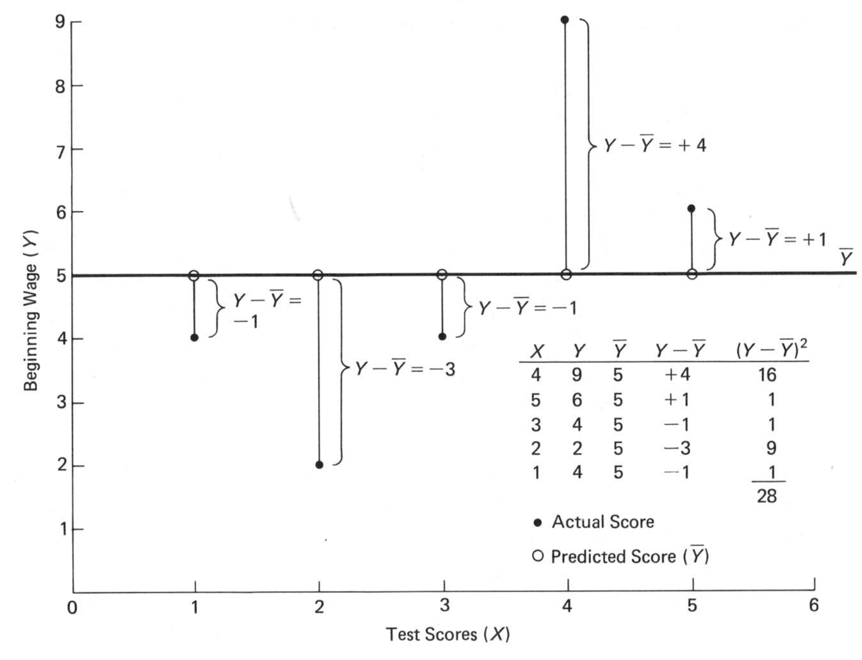

FIGURE 6.5 Scatter Diagram Showing Total Variation Around $\overline{Y}$

five scores around any value other than the mean beginning wage ($5), square these, and sum them, you would get a sum larger than 28. The same pattern is true for any data set: The smallest (minimum) sum of squared deviations are those around a data set's mean.

One use of this characteristic of the mean for regression is that it provides us with a first prediction method for a PRE statistic of association for interval or ratio variables. This is shown graphically in Figure 6.5. The mean beginning wage is shown as a level line across the graph at that value on the *Y*-axis ($5). Using the mean as a prediction of individual beginning wages, we predict each of them to be $5. The errors that we make in this prediction are represented by the vertical lines drawn from the mean line to each of the plotted points. These errors (deviations) are labeled $(Y - \overline{Y})$ because they are the difference between each case's actual beginning wage (Y) and the mean wage $(\overline{Y})$. If these values are each squared and then summed, we have $\Sigma\,(Y - \overline{Y})^2$. This is the total variation in the *Y* variable. It constitutes the value of E_1 for our PRE-type formula of association. In our example E_1 is 28.0.

As noted above, this sum is the smallest sum of squared deviations from any one value for these scores. That is one reason we use the mean as the basis of this prediction effort. It gives us the least error when using one value as a predictor. Recall from Chapter 5 that we are restricted to using only information about the dependent variable itself for making E_1-type predictions about association for PRE measures.

Our second prediction method is based on the regression equation. This is shown graphically in Figure 6.6. Now vertical lines are drawn between each actual beginning wage and the regression line. These lines represent the errors that would be made if we predicted each employee's beginning wage to be the exact value determined by the regression equation (i.e., to be on that line). These errors are appropriately labeled $(Y - \widehat{Y})$ and are known as *residuals.* Notice that three of these five errors are smaller than the errors when the mean beginning wage was used for the predictions. This does not always happen, but usually many of the regression errors are smaller than the errors from the mean. If, in aggregate, the residuals are smaller than the deviations from the mean, it means that there is some association between test scores and wages. To make these second errors comparable to the first set, we again square and sum them. This sum is labeled $\Sigma\,(Y - \hat{Y})^2$, is known as the *unexplained* or *error variation,* and constitutes the value of E_2 for our PRE formula. In our example the error variation is 15.90. This represents the smallest sum of squared differences around any one straight line that could be drawn through the same data. This (regression) line is the *line of least squares.*

The two methods of predictions are illustrated on the same graph in Figure 6.7. This graph shows that the difference between the errors in the two predictions for each case leave a component we can label $(\hat{Y} - \bar{Y})$. This is shown for two of the points as illustrations in the figure. If squared and summed for all cases, this yields the quantity $\Sigma\,(\hat{Y} - \bar{Y})^2$, which is known as the *explained* (*by regression*) *variation.* In our example the explained variation is 12.10. Notice that if you add the error variation and the explained variation you have the total variation: 15.90 + 12.10 = 28.00.

Conceptually, what this means is that the regression method of prediction shows the total variation in beginning wages to be composed of two parts. One component is explained or accounted for by the regression equation. It describes the amount of errors that is *not* being made by the regression prediction compared to using the mean as the prediction. The other component is the amount of errors still remaining from the regression prediction. If these two components are substituted into a regular PRE formula, we will

FIGURE 6.6 Scatter Diagram Showing Error Variation Around $\hat{Y}$

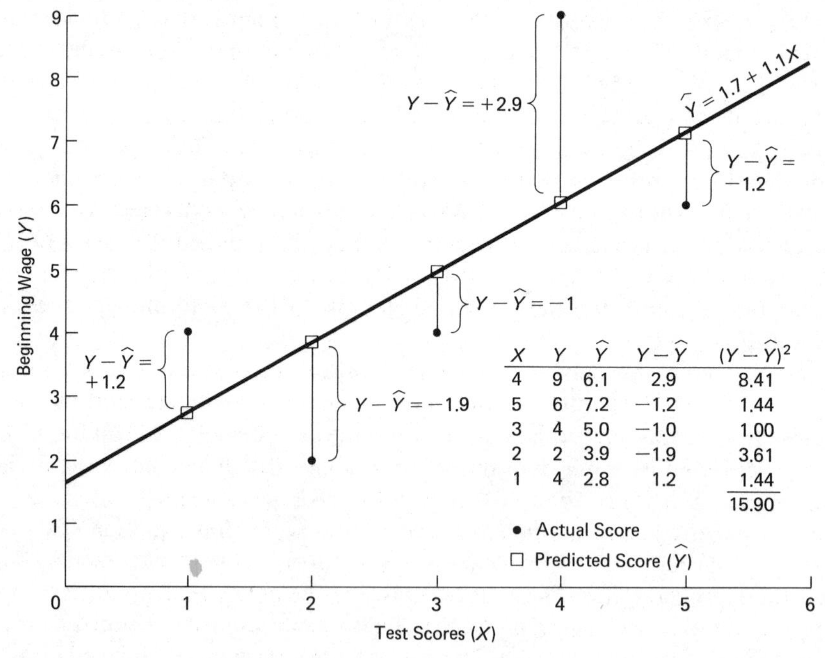

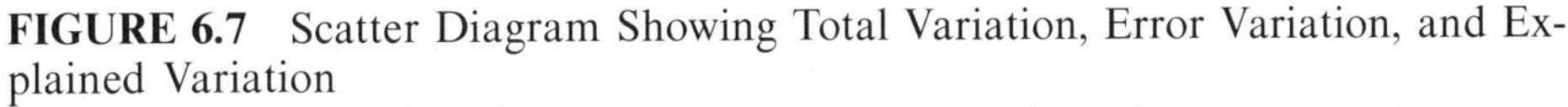

FIGURE 6.7 Scatter Diagram Showing Total Variation, Error Variation, and Explained Variation

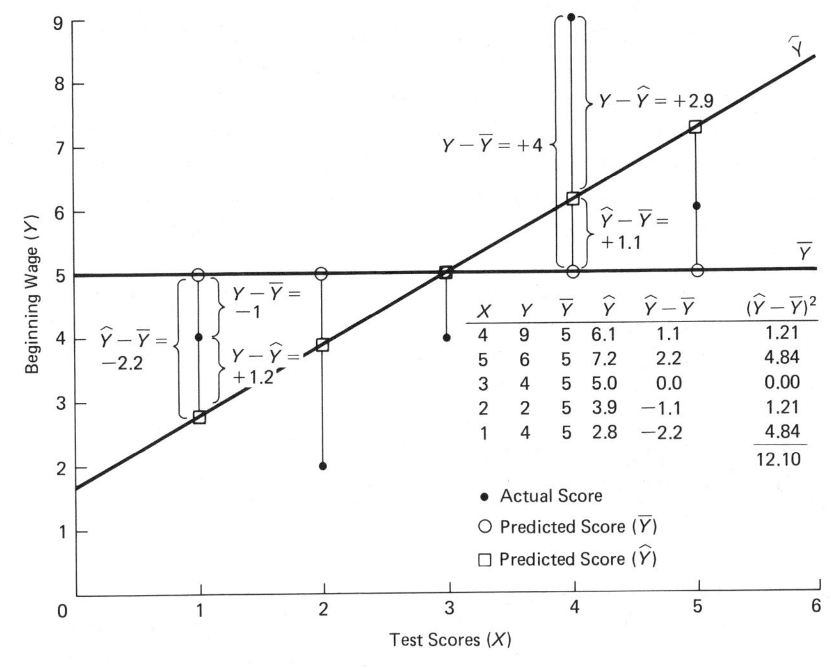

know how helpful the independent variable (test scores) is in predicting beginning wages:

$$
\begin{aligned}
\text{PRE} &= \frac{E_1 - E_2}{E_1} \\
&= \frac{\text{total variation} - \text{error variation}}{\text{total variation}} = \frac{\text{explained variation}}{\text{total variation}} \\
&= \frac{28.00 - 15.90}{28.00} = \frac{12.10}{28.00} \\
&= .432 = .432
\end{aligned}
$$

As we learned in Chapter 5, there are two levels at which such a PRE statistic can be interpreted. At a general level, we can say that .432 (or 43.2%) of the differences in beginning wages is explained or predicted by test scores. At a more specific level, we can say that regressing beginning wages on application test scores explains .432 (43.2%) of the total variation in beginning wages.

This is a perfectly legitimate method to arrive at a normed statistic of correlation for interval and ratio variables. It provides us with a measure that varies from 0 to 1. Zero means there is no reduction in predictive errors using the independent variable to predict the dependent variable. The value 1 means there is complete reduction in such errors. Values between these represent ratio differences in the reduction and are directly comparable. However, this is not the method most often used to determine an index of correlation. An equivalent but much simpler method is used instead.

Correlation

Pearson's Product-Moment Coefficient of Correlation (*r*)

Whereas regression tells us the equation for the line that best fits a data set, correlation tells us how well that line fits the data. The discussion of the least squares principle demonstrated how the total variation around a dependent variable's mean may be reduced by fitting a regression line to those same data. We referred to any improvement in predicting Y scores that results from this procedure as explained or regression variation. In effect, this quantity tells us how much of the total variation (around the Y mean) is explained by a relationship between the independent and dependent variables. That is, it is a measure of association.

The PRE formula above shows how the ratio of regression variation to total variation provides a direct index of correlation for interval and ratio data. The square root of this index is known as *Pearson's product-moment coefficient of correlation* (r). It is more commonly defined by the following formula:

$$r = \frac{\Sigma (X - \bar{X})(Y - \bar{Y})}{\sqrt{[\Sigma (X - \bar{X})^2][\Sigma (Y - \bar{Y})^2]}} \quad \text{or} \quad \frac{\Sigma xy}{\sqrt{(\Sigma x^2)(\Sigma y^2)}}$$

Definitional Formula of Pearson's Coefficient of Correlation

Notice that the numerator is the covariation between X and Y, just as with the regression coefficient, b. The denominator is the square root of the product of the total variations in X and in Y. This denominator calculates the maximum possible value that the covariation could become. Thus r expresses the ratio of actual covariation between X and Y to their maximum possible covariation.

When there is no association between X and Y, their covariation will be zero and thus r will equal zero. When there is a perfect positive association, the actual covariation will equal the maximum possible covariation and r will be $+1$. When there is perfect negative association, the actual covariation will be equal in value to the maximum possible covariation but negative in sign, so r will be -1. Further, the r statistic is independent of the original units of measurement for X and Y. They cancel in the formula, appearing in both numerator and denominator.

The perfect association values (± 1) occur when every data point falls exactly on the regression line. This means that there is an exact linear fit to the data so that we could predict without any error every case's score on the dependent variable by knowing its score on the independent variable. The no-association value (0) occurs when the regression line coincides with the mean line. This means that the data fit the mean line better than any other straight line that can be drawn through the plotted points. In this situation the independent variable's scores would be of no help in predicting the dependent variable's scores.

Although r could be found through the regression-based PRE formula above or by the definitional formula just given (as well as some other formulas), it usually is not. A much easier computing formula is used when working from the original data:

$$r = \frac{N(\Sigma XY) - (\Sigma X)(\Sigma Y)}{\sqrt{[N(\Sigma X^2) - (\Sigma X)^2][N(\Sigma Y^2) - (\Sigma Y)^2]}}$$

Computational Formula for Pearson's Coefficient of Correlation

where N = number of cases (pairs of scores)
ΣXY = sum of all X times Y products
ΣX = sum of all X scores
ΣY = sum of all Y scores
ΣX^2 = sum of all squared X scores
$(\Sigma X)^2$ = square of the ΣX quantity
ΣY^2 = sum of all squared Y scores
$(\Sigma Y)^2$ = square of the ΣY quantity

The computational formula for r can be applied to the data for our example of beginning wages and test scores. Table 6.3 shows the preliminary calculations needed. Notice that the only new calculations from the raw data required over what we used with the formula for the regression coefficient (b) is the ΣY^2 quantity:

$$r = \frac{5(86) - (15)(25)}{\sqrt{[5(55) - (15)^2][5(153) - (25)^2]}}$$
$$= \frac{430 - 375}{\sqrt{[275 - 225][765 - 625]}} = \frac{55}{\sqrt{[50][140]}}$$
$$= \frac{55}{\sqrt{7000}} = \frac{55}{83.6660} = .65738 = +.657$$

Pearson's r is $+.657$ for our data. The sign of this coefficient means that higher test scores tend to occur with higher beginning wages. A negative sign would mean that higher test scores occurred with lower wages. The interpretation of the strength of this coefficient is not so obvious.

You may be tempted to interpret the strength of correlation values between 0 and ± 1 according to the guidelines given in Chapter 5 for such coefficients. There are two ways in which Pearson's r's are interpreted that parallel the practice for Spearman's coefficient, r_s. Verbal interpretations are often applied directly to the coefficients themselves, as suggested by the Figure 5.1 guide. The r coefficient for our example (.657) would be called strong. Pearson's r's of .40 would be called moderate and those of .15 would be called weak, for example.

This practice of interpreting Pearson's r's is acceptable if done carefully. But like Spearman's r_s values, Pearson's r's are not directly comparable in a ratio sense. A Pearson's r of .6 does not represent an association that has twice the strength of a Pearson's r

TABLE 6.3 Computations Necessary for Calculating Pearson r

X	Y	X^2	Y^2	XY
4	9	16	81	36
5	6	25	36	30
3	4	9	16	12
2	2	4	4	4
1	4	1	16	4
15	25	55	153	86

of .3. A ratio-type comparison is possible only if the values are first squared. In our example, $r^2 = (.657)^2 = .432$. This means that 43.2% of the total variation in beginning wages can be explained by test scores. Sound familiar? That is the same interpretation presented earlier when the PRE meaning of regression was being discussed. Indeed, the PRE value shown there for this same example was .432. This squared correlation value is known as the *coefficient of determination* and does use r^2 as its symbol. It tells you the proportion of variation in one variable that is explained or determined by the second variable. Presumably, you will agree that finding this value is much simpler when we use the computational formula for r and then square it rather than trudging through the least squares formulas. However, the least squares formulas show you the PRE meaning behind the coefficient of determination, whereas the definitional and computational formulas for r do not.

We can now return to the concern of our employee-researcher who wanted to know whether the test scores were being used to set beginning wages. This employee's regression-predicted wage was \$3.90, compared to his actual wage of \$2.00. In spite of the error in this individual prediction, we have determined that there is a strong (although, not perfect) positive association between the two variables. It is true that, in general, those with higher test scores also have higher wages. But we can say nothing about the test scores being used to set those beginning wages. This issue is one of cause and effect. Pearson's r, like all other measures of association, can only tell you if variables are related, not whether they are causally connected. One might say that we have shown "coincidental evidence" that the test scores were being used in part to set beginning wages but we have not proved it. Later, this chapter looks at partial and multiple correlation. These techniques can help us to evaluate the persuasiveness of the coincidental evidence that we have for a causal explanation. But even these will not convert association to causation. That remains beyond the limits of our statistics.

There is one important difference between regression and correlation other than those already mentioned. Pearson's r is a symmetrical measure whereas regression is asymmetrical, as noted earlier. We would get the same calculated value of r for our beginning-wages example if we reversed the labeling of the two variables so that we considered the test scores to be the dependent variable (Y) and the wages to be the independent variable (X). We could correctly conclude that beginning wages explain 43.2% (r^2) of the total variation in test scores.

Research Example 6.2 shows Pearson's r being reported from four early studies of the relationship between body type and intelligence. The article notes the long history of interest in the possibility of an association between these two variables. But most of the studies show little evidence to support the idea. A majority of the correlations are either negligible or very weak. The only suggestion of a consistent association was for a small and positive relationship between verbal ability and ectomorphy, the linear and fragile body type. You can find these correlations in the last column of the reprinted table.

Intercorrelation Matrix

Researchers frequently calculate several correlation values in their analysis of social data. Sometimes these will show how one variable of interest is associated with a list of other variables. Research Example 6.2 demonstrated this by showing how body type was (or was not) associated with several measures of intelligence. Other times researchers will report how all possible pairs of variables are related to each other. These are sometimes referred to as *intercorrelations.* A convenient form to present a set of intercorrelations is a

RESEARCH EXAMPLE 6.2 Using Pearson *r* to Measure Association

From Raymond Montemayor, "Men and Their Bodies: The Relationship Between Body Type and Behavior," Journal of Social Issues, *34 (1), Winter 1978:48ff. Reprinted by permission.*

For at least two thousand years people have believed that a man's character is revealed by his body. As early as the fifth century B.C. the Greek physician Hippocrates divided men into two physical types, *phthisic habitus* (long and thin) and *apoplectic habitus* (short and thick), and suggested that men with the phthisic physique are particularly susceptible to tuberculosis while apoplectics are predisposed to diseases of the vascular system. Although less systematic in his observations than Hippocrates, Shakespeare's more memorable descriptions of Cassius as "lean and hungry" and Falstaff as indolent and gluttonous are meant to reveal something more profound about these men than their caloric intake. More recently Nichols (1975) suggests that men with muscular bodies have a distinct social advantage over thin or fat men since people have a more positive stereotype of muscular men than of non-muscular men. . . .

Ever since Sir Francis Galton observed that men of genius tend to be above average in height and weight (Galton, 1896/1962), researchers have sought to establish the true relationship between men's physical traits and their intelligence. Galton's observation has not been verified by subsequent research which has suggested that when such factors as social class, family size, and birth order are controlled for, any relationship between height and weight and intelligence disappears. . . .

Researchers who had been defeated in their quest to find a simple relationship between height or weight and intelligence moved to the more complex hypothesis that a man's physique, the shape of his body, might be related to his intelligence and to other aspects of his behavior. A variety of typologies have been proposed to

Correlations Between Physique and Measures of Intellectual Abilities[a]

Study[b]	(*N*)	Endomorphy	Mesomorphy	Ectomorphy
Sheldon (1942)				
IQ	(110)	.07	−.07	.19
Child and Sheldon (1941)				
SAT verbal	(518)	.00	.01	.09
SAT math	(518)	−.04	−.10	.08
Smith (1949)				
Thurstone verbal	(105)	.05	−.32	.25
Thurstone reasoning	(87)	−.02	.01	.05
Thurstone space	(85)	−.11	.06	.03
Thurstone memory	(90)	.04	.06	.01
Davidson, McInnes, and Parnell (1957)				
Stanford Binet	(50)	.31	.01	.19
WISC non-verbal	(50)	−.03	.15	—
WISC verbal	(50)	−.06	−.35	.34

[a]Table 1 in the original.
[b]The Davidson et al. study measured 7-year-old boys; the other three were college populations.

classify physique, the most notable and widely researched being William Sheldon's (1940, 1942, 1949, 1954). Sheldon and his co-workers developed a highly objective and reliable technique for classifying men's physiques. Sheldon argues that a man's physique can be described in terms of the relative presence of three physical components: *endomorphy*—soft and round with a highly developed digestive viscera, *mesomorphy*—hard and rectangular with a predominance of muscle and bone, and *ectomorphy*—linear and fragile with a flat and delicate body. Based on information obtained from standardized photographs, a man's physique can be assigned a somatotype, which is the degree of presence of each of the three components of physique and is expressed as a three digit number. . . .

A number of studies have examined the relationship between intellectual ability and endomorphy, mesomorphy, and ectomorphy. The results of these studies [see the table] generally suggest that there is little if any consistent relationship between intellectual ability and somatotype. The most that can be said from these data is that a small positive relationship may exist between verbal ability and ectomorphy, although the size of that relationship is not very large. . . .

matrix. The matrix allows you to review and compare easily a large number of correlations in a small space.

Table 2 of R.E. 6.3 shows a matrix of the Pearson *r*'s among a set of variables concerning the level of state maximum monthly AFDC payments (Aid to Families with Dependent Children), labeled X_1. At a glance we can determine that these correlations range from virtually 0 to a high of −.64. By examining the matrix one variable at a time we can begin to construct generalizations about the patterns of intercorrelation. For example, X_1, X_2, X_3, and X_4 seem to form a mutually related subset. The same is true for X_5, X_6, X_7, X_8, and X_9. But these two subsets do not correlate well with each other. This is one of the common uses of correlation matrices. Notice the diagonal of coefficients equal to 1.00. These are the correlations of each variable with itself, which is perfect, of course. Many matrices omit this diagonal because it is always filled 1.00's.

RESEARCH EXAMPLE 6.3 Using the Multiple Regression Equation and Correlation Matrix to Examine Association

From Martha N. Ozawa, "An Exploration into States' Commitment to AFDC," Journal of Social Service Research, *1 (3), Spring 1978:245ff. Reprinted by permission.*

. . . The present study attempts to investigate the degree to which each of the selected economic, sociological, political, and demographic variables predicts the level of state maximum monthly AFDC payments, controlling for the rest of the independent variables in the multiple regression model. For the purpose of this study, state maximum payments for a family of two are used as the dependent variable. . . .

A stepwise multiple regression analysis . . . was performed to investigate the relative strength of each independent variable in predicting the level of state maximum monthly AFDC payments, controlling for the rest of the independent variables in the multiple regression model.

TABLE 1 Multiple Regression Analysis of State Maximum Monthly AFDC Payments for Family of Two in 1976[a]

Independent Variables	*B* Coefficient (Slope)	Standardized Beta Weight
Percentage of nonwhite population	−3.52077	−.43181
Per capita personal income	.04112	.42171
Tax effort	1.40000	.25606
Percentage of population in metropolitan areas	−.35005	−.12609
Density of population	.03332	.10365
AFDC child recipient rate	−.83002	−.03878
Unemployment rate	−1.14722	−.03535
Percentage of children and the elderly	−.62662	−.01278
Constant	−80.93435	

[a]Adapted from Table 1.

As seen in Table 1 . . . the independent variable most strongly associated with state maximum monthly AFDC payments is the percentage of nonwhite population (Standardized Beta weight = −.43181 . . .). The direction of the regression coefficient is negative, indicating that a state with a higher percentage of nonwhites tends to have a low level of maximum payment. Controlling for other variables, one percentage point increase in nonwhite population is associated with a decrease of state maximum monthly AFDC payments by $3.52. To illustrate the strong predictive power of this variable, the maximum payments in a state with a nonwhite population of 20% (such as North Carolina) are expected to be $70.40 less than in a state with 0% (such as Vermont), other things being equal.

The state per capita personal income was found to be the second most statistically significant independent variable in predicting the level of state maximum monthly AFDC payments (Standardized Beta weight = .42171 . . .). Controlling for other variables, a $1 increase in per capita personal income in the state increased by $.04 the state maximum monthly AFDC payments. For example, a rich state with $6,800 per capita personal income (such as Connecticut) is expected to pay $112 more in maximum payments than a poor state with $4,000 per capita personal income (such as Mississippi), other things being equal.

Tax effort is the third strongest independent variable in predicting the level of state maximum monthly AFDC payments (Standardized Beta weight = .25606 . . .). The regression coefficient is 1.4, which means that each time there is an increase of $1 in per capita taxes for each $1,000 per capita personal income the maximum monthly payment is expected to go up by $1.40, other things being equal.

All other variables—state unemployment rate, the state combined percentage of children and elderly, the AFDC child recipient rate, urbanization, and density of population—are insignificantly related to the level of state maximum monthly AFDC payments, given the multiple regression model. Also, the zero-order correlation coefficients between these variables and the dependent variables are all low: .27 or less [See Table 2.] These variables do not contribute much toward explaining the variance in the dependent variable. Nor do they contribute to the predictive power

TABLE 2 Zero-Order Correlation Matrix[a]

	1	2	3	4	5	6	7	8	9
1	1.00	.57	.44	−.64	.02	−.23	−.10	.10	.27
2		1.00	.22	−.29	.18	−.54	.16	.60	.49
3			1.00	−.20	.24	−.07	.22	.01	.19
4				1.00	.13	.03	.38	.13	−.03
5					1.00	−.27	.51	.49	.56
6						1.00	−.12	−.49	−.39
7							1.00	.41	.47
8								1.00	.56
9									1.00

X_1 = state's maximum monthly AFDC payments.
X_2 = state per capita personal income.
X_3 = tax effort in the state.
X_4 = nonwhite percentage of state population.
X_5 = state unemployment rate.
X_6 = combined state percentage of children ages 1–17 and of the elderly aged 65 and over.
X_7 = AFDC child recipient rate in the state.
X_8 = urbanization in the state.
X_9 = density of population in the state.

[a]Reprinted from Table 2 in the original.

of the multiple regression equation. All this means that these variables are not directly associated with the level of state maximum monthly AFDC payments, when all other variables are held constant. . . .

* Partial and Multiple Regression and Correlation

Partial Associations

Examining the association between only two interval or ratio variables often represents an oversimplification of social reality. Our example of beginning wages and test scores may be an illustration of this. In a short time we could list several other variables that might be associated with one or both of these. For example, work experience, education, and age are possibilities. If some of these were considered together with our basic wage-test score association, we might find that one of these actually explained beginning wage better than did test scores. Or we might discover that they did not alter our previous finding. There are other possible outcomes, too.

We looked at this kind of issue at the end of Chapter 5 with nominal and ordinal variables under the heading of elaborating an association. There the technique was to split the cases into subcategories according to a third variable and then calculate separate measures of association for each of the resulting subtables. The same procedure can be used with interval and ratio variables. For example, we might look at the wage-test score association controlling for work experience. We could divide the cases into those with no previous experience and those with some experience. Separate regressions and Pearson correlations could be calculated for the two subsets and compared. If, for instance, the *partial r's* for both of these subsets were substantially less than our earlier zero-order bivariate correlation r of .657, we would conclude that test scores were not as strongly related to wages as we thought.

There are two disadvantages of this technique. First, we must reduce all control variables to categorical data by discarding the interval or ratio quality that they may have. This loss of measurement information lessens our understanding of the relationships we are studying. No such loss would be involved if the control variable was only nominal at the start. But these can be handled alternatively by giving them binary (dichotomous) coding (0, 1) in the procedure known as dummy variable coding, which has been mentioned before.

The second disadvantage to splitting the cases into subcategories is that a large number of cases are required to achieve meaningful subsets. In our example with only five employees we have no margin for this kind of manipulation. Even with larger data sets of moderate size (e.g., 100 cases) only a few subsets can be formed and the possibility of second- or third-order elaborations is limited.

The more common method to calculate partial regression and correlation values uses some extended mathematical manipulations. In effect, these methods control for third (or higher) order variables by holding them constant statistically rather than by actually sorting the cases into subsets. As a result the number of cases does not diminish as the controls are considered. Nor is the level of measurement reduced for interval and ratio variables. There are some consequences of these techniques for drawing inferences from sample to population data, though. But these are not germane to our current discussion of descriptive association.

Partial Correlation

Let us see how statistical controls work with our example of wages and test scores. We want to find the *first-order partial correlation.* This is the correlation between two variables (e.g., wages and test scores) while controlling for differences in one other variable (e.g., work experience of the employees). This can be determined by using the following formula:

$$r_{y1.2} = \frac{r_{y1} - r_{y2}r_{12}}{\sqrt{(1 - r_{y2}^2)(1 - r_{12}^2)}}$$

First-Order Partial Correlation Coefficient

where $r_{y1.2}$ = partial correlation between the dependent variable (Y) and the first independent variable (X_1) while controlling the second independent variable (X_2)

r_{y1} = zero-order bivariate correlation between Y and X_1

r_{y2} = zero-order bivariate correlation between Y and X_2

r_{12} = zero-order bivariate correlation between X_1 and X_2

As you can tell from this formula, the partial correlation coefficient when controlling for one independent variable is found by a combination of zero-order bivariate correlations between the three variables involved. This same model can be extended for finding partial correlations of the second order (controlling two variables) and other more complex problems.

Imagine that we had calculated the three zero-order bivariate correlations for our example to find that: wages and test scores: $r_{y1} = .657$; wages and experience: $r_{y2} = .800$; test scores and experience: $r_{12} = .500$. The partial correlation between wages and test scores controlling for work experience would be:

$$r_{y1.2} = \frac{.657 - (.800)(.500)}{\sqrt{[1 - (.800)^2][1 - (.500)^2]}}$$

$$= \frac{.657 - .400}{\sqrt{1 - .640][1 - .250]}} = \frac{.257}{\sqrt{[.360][.750]}}$$

$$= \frac{.257}{\sqrt{.270}} = \frac{.257}{.5196} = .495$$

Our partial correlation between wages and test scores when we controlled for work experience is somewhat less than their zero-order correlation. This means that the wage-test score association is actually weaker (although still substantial) when differences in work experience are removed. This would dilute the weight of the coincidental evidence that could be cited to claim that test scores were playing an important role in setting beginning wages. Regardless, it would still be only suggestive evidence, not conclusive.

Multiple Associations

Multiple associations are closely related to the topic of partial associations. In multiple association analysis we determine the simultaneous association between three or more variables. Researchers using this concept want to know how several variables are related at once. But this is not a matter of simply adding together the separate bivariate zero-order correlations of each independent variable with the dependent variable. These variables are usually related to each other in such a way that the unique association between each independent variable and the dependent variable while holding all the others constant (controlling them) must be found. Then these unique associations can be combined to find the *coefficient of multiple correlation* (R).*

Multiple Regression

The interconnection between multiple and partial correlation can be seen in the multiple regression equation. This equation tells us the best fit between a set of two or more independent variables and one dependent variable. The basic formula for that equation is a straightforward extension of the bivariate equation:

$$\hat{Y} = a + b_1X_1 + b_2X_2 + \cdots + b_kX_k$$ Multiple Regression Equation

where $\hat{Y}$ = predicted scores on the dependent variable
a = predicted Y score when all independent variables equal 0

*A formula expressing this process with two independent variables for R^2 is $r_{y1}^2 + r_{y2.1}^2(1 - r_{y1}^2)$.

ON THE SIDE 6.1 Karl Pearson

Karl Pearson (1857–1936), a product of the Victorian age, embodied the emerging freedom that science experienced as the traditions of orthodox religious control over thought and life fell away. He was born in London, where he spent nearly all his life. At age 9 he began his long association with University College when he was sent to its school for children. After a stint with a private tutor during a time of ill health, he studied at King's College, Cambridge, in mathematics and later in Germany at Heidelberg and Berlin. He had developed a great interest in history and literature and revealed an antiauthority spirit which at one point led him to defy successfully the rule of compulsory attendance at divinity lectures at King's College. He struggled for many years with the tension between the old religious ideas and the ways of science. He was greatly influenced by the philosopher Spinoza and later published several works on religious topics, including *The Trinity; a Nineteenth Century Passion Play* (1882) and *The Ethic of Freethought* (1901), and articles on German social thought and Martin Luther. His interest in social history and welfare led him to become a socialist as a young scholar, even contributing to the Socialist Song Book. He had also changed the spelling of his first name to the German "K" from the English "C," for reasons not entirely clear from the record. Perhaps it was merely reflective of his homage to German culture.

Many of Pearson's early works reveal a continuing characteristic of his thought and personality. While espousing the virtues of rejecting traditional dogma, he expressed his own views with such vigor that they seemed to be dogmatic to others. It was this blend of the rejection of the old myths with a self-assured assertiveness that made him revered by many but slow to accept legitimate criticisms. Many of his emerging ideas about the nature of truth and the role of science in its pursuit were presented in his *The Grammar of Science* (1892).

In 1884, Pearson received an appointment to the Chair of Applied Mathematics and Mechanics at University College. This position required him to teach mathematics to engineering students, which he did with great success by the use of graphical techniques over pure mathematics and the engaging strength of his character. During the next few years he published several books on topics in mechanics while continuing to give lectures on freethought and social problems.

In 1890, Pearson's life took an important turn. He received a lectureship in geometry at Gresham College. This gave him the freedom to lecture on topics of his choosing while continuing his teaching at University College. He first chose the philosophy of science and later statistics and probability. When the dual work load became too much, he gave up the lectureship at Gresham (in 1894). Also in 1890, W. F. R. Weldon, a biologist, was appointed to the chair in zoology at University College and came to exert a great influence on Pearson through their long collaboration. Weldon was intent on demonstrating Darwinian evolution via mathematics. He turned to Pearson for the help he needed with the technical matters. Pearson's ideas were also profoundly influenced by Francis Galton's work on heredity, in which he had developed the early foundations of statistical correlation and its coefficient, *r*. Pearson poured himself into the analytical problems, producing over 35 relevant papers between 1893 and 1901. Pearson clarified some of Galton's ideas by developing the theory of multiple correlation and worked at problems of fitting a curve to the distributions of natural observations using the "method of moments." The method of moments had great utility for practical problems and led Pearson to one of his greatest contributions, the chi-square goodness-of-fit test. This provided a simple means of determining the probability that a set of observations could be described by a hypothetical distribution.

Weldon and Pearson's interest in hered-

ity and evolution took them to the development of the new field of biometry, the use of mathematical methods in the study of biology. At first biometry research was to be guided by a committee of the Royal Society headed by Galton. This structure proved impractical because of the vigorous objections of older biologists such as William Bateson, who rejected the use of mathematics and instead held to the theories of Gregor J. Mendel (the noted botanist). Consequently, the committee was disbanded. A new journal, *Biometrika*, was begun by Galton, Weldon, and Pearson to publish biometrical research which the Royal Society had decided to refuse in order to avoid controversy. The conflict between the sides became so heated that Pearson never became a member of the Royal Statistical Society.

In 1903, Pearson's department at University College received the first of 30 annual grants from the Worshipful Company of Drapers. The grants were used to further research through the Biometric (and Eugenics) Laboratory established in the department. In 1906 Weldon died suddenly of pneunomia at the age of 46, a crushing blow to Pearson, for they had been the best of friends. Weldon had gone to Oxford, where he too was developing a group of young biometricians working to establish the new field. With Weldon's death, the old-line critics became even stronger in their attacks on Pearson and successfully painted a division between themselves and biometrics, which Pearson claimed never to have espoused. He continued to see his ideas as a statistical description of heredity, not a theoretical attack on Mendelian explanations.

In 1911, Pearson's laboratory became the basis of the new Department of Applied Statistics, the first of its kind. Galton had died that year and his estate provided an endowment for a chair in eugenics, which was occupied by Pearson. Pearson could now give up teaching engineers, and he used the new freedom to study biometrics, eugenics, and mathematical statistics more exclusively. The department and *Biometrika* produced a series of tables describing several theoretical distributions of statistical relationships which became the grounds for many subsequent developments in modern statistics. Pearson also produced a series of papers combining biometrical and historical research. He conducted studies of craniometry, anthropometry, and dog breeding, as well. His output was slowed by World War I and by his work on a three-volume biography of Galton, finished in 1930. The book demonstrated the continuing themes of eugenics by tracing Galton's ancestory back through his cousin, Charles Darwin, and on to the Emperors of Byzantium.

Pearson resigned from his professorship in 1933. To his disappointment the College decided to divide his department into two units; a Department of Eugenics and a Department of Statistics. R. A. Fisher (Pearson's long-time acquaintance and rival) became head of Eugenics and Pearson's son, E. S. Pearson, became head of Statistics. Karl Pearson was provided an office at the College where he continued to edit *Biometrika* and do his own research until his death in 1936.

Based on E. S. Pearson, "Karl Pearson, An Appreciation of Some Aspects of His Life and Work," *Biometrika*, 28, December 1936:193–257, and 29, February 1937:161–248.

X_1 = actual scores on the first independent variable
X_2 = actual scores on the second independent variable
⋮
X_k = actual scores on the last independent variable
b_1 = partial regression coefficient for X_1 with all other independent variables being controlled

b_2 = partial regression coefficient for X_2 with all other independent variables being controlled

$\vdots \quad \vdots$

b_k = partial regression coefficient for X_k with all other independent variables being controlled

The partial regression coefficients could be more explicitly labeled by expanding their subscripts to show the controlled variables. For example, the b_1 value in the multiple regression equation could be written as $b_{y_{1.2} \cdots k}$. We will use the shortened version here, but you should remind yourself that it is a partial coefficient, not a zero-order coefficient.

The computing formulas for the a and b constants in the multiple regression equation are too complex for our consideration here. They are presented in the section *Related Statistics* at the end of the chapter for your convenience if you wish to examine them. An intuitive understanding of these is more useful to our discussion now. We continue to use our familiar beginning wages example to demonstrate this interpretation.

We could consider the multiple regression of beginning wages with two independent variables: test scores (X_1) and work experience (X_2). Our unsolved equation would look like this:

$$\widehat{Y} = a + b_1X_1 + b_2X_2$$

Suppose that this had been found to be: $\widehat{Y} = .5 + .7X_1 + .6X_2$. The constant a in this equation represents the predicted beginning wage when both of the independent variables equal zero (0 on the test and no work experience). Here that predicted wage is \$.50. The first regression coefficient (b_1) represents the increase in predicted beginning wage (\$.70) for every one-point increase on the application test while controlling for work experience. The second regression coefficient (b_2) tells us that beginning wage is predicted to increase by \$.60 for every one-year increase in work experience while controlling for test scores.

Does this mean that test scores have the greater influence on beginning wages? Not necessarily. The regression coefficients produced by the multiple regression equation above yield unstandardized coefficients. To make them directly comparable they need to be transformed to standardized coefficients as we did earlier with zero-order bivariate regression coefficients. Because of their complexity the formulas for this are in the section *Related Statistics* at the end of this chapter. But you can correctly assume that they have gone through the same kind of mathematical manipulations involving standard deviations as was shown before for the zero-order coefficients.

Standardized regression coefficients allow you to compare the relative influence of the independent variables on the dependent variable. Suppose that our standardized coefficients were as follows:

$$b_1^* = .4$$
$$b_2^* = .7$$

This would mean that the second independent variable (work experience) did have the greater influence on beginning wages of the two independent variables considered. Its influence was 75% greater than the first independent variable, test scores $[(.7 - .4)/.4 = .75]$.

Table 1 of R.E. 6.3 shows a multiple regression equation predicting the level of state maximum monthly AFDC payment. We looked at this same research example above to illustrate an intercorrelation matrix. The exact multiple regression technique used was "stepwise," which, in short, arranges the independent variables in descending order according to their power to predict the dependent variable. The "standardized beta weight" (standardized regression coefficient) column demonstrates this. As you read down this column note that every successive entry is smaller. The unstandardized (regression) coefficients and constant (*a*) in the first column can be used to write the multiple regression equation. Try it. Note that the labeling of the variables is reported in Table 2 of the example. It assigns X_1 to the dependent variable, rather than *Y*, as we have been doing. Using the example's labeling of the variables our equation would be

$$\widehat{X}_1 = -80.9435 + .04112X_2 + 1.40000X_3 - 3.52077X_4 - 1.14722X_5 - .62662X_6 - .83002X_7 - .35005X_8 + .03332X_9$$

As the author notes, the first of these unstandardized coefficients as listed in Table 1 (-3.52077) means that every 1% increase in a state's nonwhite population (X_4) results in a predicted decrease of \$3.52 in its monthly AFDC payment. Remember that this is an average increment and it assumes that all other variables are being held constant.

Multiple Correlation

The multiple regression equation provides a basis for determining a measure of the total association between a set of independent variables and one dependent variable. Such a measure is called the *coefficient of multiple correlation, R,* as noted above. If squared, this gives us the *coefficient of multiple determination,* R^2. R^2 is a PRE measure of multiple association which varies from 0 to $+1$, as did r^2. R^2 tells us the proportion of total variation in the dependent variable that is explained by (or related to) the collective set of independent variables being considered simultaneously.

In our continuing example we might find R^2 for the association of beginning wages with the same two independent variables of test scores and work experience. The relevant formulas are given in the section *Related Statistics* at the end of the chapter. Suppose that this R^2 was equal to .727. This strong coefficient means that 72.7% of the total variation in beginning wages is explained by these two variables. Such a finding would be rather astounding in social research because most explanations of variation are far less successful. R^2 values on the order of .6 or .5 are usually the best that can be achieved in any real comparable situations of social research. Many are far less than this.

Multiple regression and correlation have become very popular statistical techniques. Social researchers often use multiple regression to find the relative power of a number of independent variables to explain a dependent variable. They use multiple correlation to find the collective relationship between a set of variables. These two goals are at the heart of much of the research that is currently being published.

Research Example 6.4 demonstrates the combined use of Pearson *r* and multiple correlation, *R* and R^2. Like the previous research example, this one concerns welfare payments to the poor. Research Example 6.4 is an attitude study in which correlates of opposition to hypothetical increases in welfare payments were sought. Four categories of independent variables were considered: factual beliefs about welfare recipients, social class of respondents, ideology of respondents, and beliefs about the work motivation of the poor. The association between individual measures within each of these four categories and opposition to increased welfare were measured with zero-order Pearson *r*. You can see from the first column of the example's table that most of these were very low for factual beliefs and social class variables. But they were moderate or nearly so in strength for the ideology and work motivation variables.

RESEARCH EXAMPLE 6.4 Using Pearson *r* and Multiple *R* to Measure Association

From John B. Williamson, "Beliefs about the Welfare Poor," Sociology and Social Research, *58 (2), January 1974:163ff. Reprinted by permission.*

. . . This study is based on interviews which were conducted during the Spring of 1972 with 375 men and women in the Boston area. For the main sample we used area probability sampling with substitutions when necessary after three call backs. The main sample includes 50 white women from each of six sampling areas: two upper-income, two middle-income, and two lower-income. The sample design was constructed in such a way as to assure over-representation for both the upper and the lower extremes of the income distribution. Most of the data presented in this article is based on this "main sample" of 300 white female respondents. . . .

There are a number of grounds on which a case can be made for rectifying misconceptions that are commonly held about welfare recipients. Of crucial importance is the relationship between these misconceptions and the extent of support for or opposition to increasing welfare payment levels. A major hypothesis which this study has been designed to test is that those who have anti-welfare biases in their beliefs about welfare recipients with respect to idleness, dishonesty, fertility, illegitimacy, and current benefit levels would tend to oppose the idea of increasing welfare benefits.

The measure of opposition to increased welfare payments used in this study is an index based on the responses to the following three questions:

> Do you think payments to welfare recipients in Massachusetts should be increased, decreased, or kept about the same as they are now?
>
> How about for the nation as a whole? Do you think that welfare payments should be increased, decreased, or kept about the same as they are now?
>
> In Massachusetts what do you think should be the average monthly AFDC payment to a family of four with no other source of income?

Factual beliefs. The correlations between the factual beliefs and support for increased welfare payments are presented in [the table]. The only correlations that account for more than 1 percent of the variance are those for the question on the percent of welfare recipients that are dishonest and the question on the number of children in the average AFDC family. Those who overestimate the percent of welfare recipients who are dishonest the most, tend to show less support for increased welfare payments; while this relationship is not strong it is in the predicted direction. . . . Our original hypothesis was that the general public would use the presumed large families of welfare recipients as grounds for opposing increased welfare payments (lest they feel able to support still more children). An alternative interpretation is that the children are viewed as a deserving category of the welfare population; those who believe that a particularly high proportion of AFDC recipients are children would for this reason be likely to support greater welfare payments. This interpretation is supported by the positive correlation (.15) between a question asking, "What percent of the poor are children?" and support for increased welfare payments.

The most striking aspect of the correlations between the factual beliefs about

Opposition to Increased Welfare Payments As Predicted by Factual Beliefs About Welfare Recipients, Social Class, Ideology, and Beliefs About the Work Motivation of the Poor[a]

Predictor	Pearson Correlation (r)	Multiple Correlation (R)	Variance Explained ($R^2 \times 100$)
Factual beliefs about welfare recipients			
Percent dishonest	.13		
Percent idle	.04		
Children in average AFDC family	.13		
Birth rate	.03		
Illegitimacy rate	.00		
AFDC payments in Massachusetts	.09		
AFDC payments in US	.03		
All seven factual belief predictors		.21	4%
Social class			
Total family income	.07		
Per capita family income	.05		
Income of residential area	.16		
Education of respondent	.01		
Education of family head	.03		
Occupation of family head	.10		
All six social class predictors		.22	5%
Ideology			
Liberalism	.52		
Work ethic	.42		
Both ideology predictors		.59	35%
Beliefs about the work motivation of the poor			
Poor want to get ahead	.31		
Poor try harder than everyone else	.30		
Circumstances	.18		
Blacks try harder	.22		
Poor try harder than wealthy	.37		
All five motivation predictors		.43	19%
Twenty predictors		.66	43%

[a]Adapted from Table 4 in the original.

welfare recipients and extent of opposition to increased welfare payments is that they are so low. The data for the set of seven factual belief items are summarized in the multiple regression results. The multiple correlation is only .21 or, in other words, this entire set of predictors explains only 4 percent of the variance in opposition to increased welfare payments. . . .

Social class. The correlations between several indicators of social class and opposition to increased welfare payments are also presented in [the table]. Again, the relationships are on the whole quite weak. The set of all six social class predictors together explain 5 percent of the variance in opposition to increased welfare payments. The argument can be made that the items assessing factual beliefs about welfare recipients are almost as important as social class in predicting opposition to increased welfare payments; but the more salient result is that neither of these sets of predictors accounts for much of the variance. . . .

Ideology. We asked, "In terms of your outlook on welfare issues, do you think of yourself as being very conservative, moderately conservative, moderately liberal, or very liberal?" In spite of difficulty on the part of some low-education respondents in understanding the question, it turned out to have a high correlation (−.52) with opposition to increased welfare payments. Those who consider themselves the least liberal (most conservative) show the most opposition to increased welfare payments. . . .

There is a high correlation (.42) between work ethic and opposition to increased welfare payments, indicating that those who most strongly support the work ethic show the most opposition to increased welfare payments.

Together the two ideology predictors explain 35 percent of the variance in support for increased welfare payments. This is a very respectable portion, particularly when compared to the amount of variance that is accounted for by social class and factual beliefs about welfare recipients.

Beliefs about the work motivation of the poor. While it is clear that the factual beliefs considered are only weak predictors of opposition to increased welfare payments, this leaves open the possibility that more subjective beliefs, particularly those relating to the motivation of the poor, might prove to be better predictors. . . .

As is indicated in [the table], the correlations for these items are all relatively high ranging from −.18 to −.31. Those who believe that the poor are highly motivated tend to show the most support for and least opposition to increased welfare payments. Together these items have a multiple correlation of .43 and account for 19 percent of the variance in opposition to increased welfare payments; this is considerably above the 4 percent accounted for by the set of seven questions assessing the more factual beliefs about welfare recipients. . . .

Multiple R and R^2 were calculated for the variables collectively from each of the four categories. Notice that the author speaks of "variance explained" rather than "variation explained," as we have been doing. They are interchangeable phrases for our purposes here. The second and third columns of the table show that factual beliefs explained only 4% of the total variation in the dependent variable. The highest explanation was achieved by the two ideology variables (35%).

All variables combined to explain 43% of the variation in opposition to increased welfare payments. This is a respectable showing, although it does leave more variation unexplained (57%) than has been accounted for. There are probably both conceptual and methodological reasons why this is a common outcome in social research. Both our theoretical understanding and measuring techniques need improvement. Many people advocate that we try to work on these two goals together rather than adhering to a strict division of labor whereby some people are only theorists and others are only researchers. It seems like a good idea. You should think about it for yourself.

Keys to Understanding

There are two major keys to understanding measures of association between interval and ratio variables. The first is to know whether the data meet the assumptions of measurement required by the statistics. The second is to know the exact meaning of association that is being examined.

Data Assumptions

Key 6.1. Determine whether the data meet the assumptions of measurement. At the descriptive level of analysis there is only one major data assumption required for the statistics discussed in this chapter. There are some additional assumptions when these same statistics are used for inferences. The one pertinent assumption is that the data have at least interval quality of measurement. The elaborate mathematical manipulations of scores that we have seen require that interval-level differences between those scores be known. The more exacting characteristic of a true zero point found with ratio measurement is not required, however.

Although it may now seem fairly simple to you to determine if a variable has interval measurement quality, it can actually be rather difficult to verify. A great effort has been made in social research to design measurements that truly have interval differences. Much of this effort has been aimed at social attitudes and behaviors. The controversy over interval measurement was mentioned beginning in Chapter 1 and has been repeated in some of the subsequent chapters. This controversy is highly relevant to the regression and correlation measures of this chapter. Let us emphasize it again.

One of the problems with attitude and behavior measurement has been to establish acceptable reliability for the measures used. Do you know that someone who gave a certain response to an attitude question today would answer it the same way tomorrow? If we asked the question in a slightly different way, but with the same meaning (to us), would our respondent give us the same answer? If we changed the wording of our response categories, would the respondent still give us essentially the same reply? These are the kinds of troubling questions that researchers must consider.

Some rather involved techniques have been devised to increase the reliability of social measurements that we want to have interval quality. These fall under the general heading of scale analysis. There are two main ideas to creating a reliable measurement at the interval level. First, we should use a series of questions to measure one concept. We do not rely on only one question for each idea. This allows us to look at the overall pattern of several responses to see whether a consistent message is coming through. We are not dependent on the whims of a respondent's mood, interest, or ability to read or listen at the one moment in which a single question is asked. This technique is known as using multiple measures.

The second idea is to construct a single index that summarizes responses to the series of questions we have decided successfully measure one concept. This often involves looking at the intercorrelation matrix from a larger set of responses and deciding which questions fit together meaningfully. We did that briefly when talking about R.E. 6.3. Determining a summary index can be very complex. Some researchers use a highly mathematical technique called factor analysis. This identifies the homogeneous subsets from a large number of questions and tells one how to put the responses to those questions together into the desired index number. It is based on statistics of correlation like those we have been discussing. It is much too mathematical for our consideration here. (See Kerlinger, 1964, and Kim and Mueller, 1978.) But you will notice in many research reports a mention of factor scores and index construction. You can take these as cues that the researchers were attempting to meet the goal of finding reliable measures with interval quality. They may not have succeeded, of course. But you know they were trying. Research Examples 6.2 and 6.4 did use these kinds of methods.

As a first key to understanding the statistics covered in this chapter you should look for indications that the researchers have satisfied the interval measurement requirement.

There are some measures that may achieve this without a technique such as factor analysis, but you should be looking for persuasive evidence that the requirement has been approximated, at least.

Meaning of Association

Key 6.2. Know the meaning of association that is being examined. Having nearly completed two chapters on the topic of association, it should be obvious to you that there are many specific ways in which the term "association" can be interpreted. You need to know exactly what meaning of association for interval and ratio variables a researcher is using in order to understand the analyses and conclusions that you find. Perhaps the simplest way to do this is to identify the research according to a series of five distinctions.

First, is the researcher interested in finding an index of association, or in predicting specific scores for one variable from knowledge of a second variable? If the first, correlation should be the technique. If the second, regression should be the technique.

Second, how many variables are being examined? If there are only two, then zero-order bivariate relationships should be the focus. If there are two variables while controlling for a third (or more), partial relationships should be the focus. If there are several variables being examined at once, a multiple relationship should be examined.

Third, are unstandardized or standardized measures being used? If unstandardized, one can interpret results in their original units of measurement but cannot compare them directly. If standardized, direct comparisons can be made.

Fourth, are linear or nonlinear statistics being used? If linear, we can interpret results to mean that a constant incremental change in one variable occurs with a constant incremental change in another. If nonlinear, the increments of mutual change are varied, not constant.

Fifth, is the researcher's goal to explore association or causation? If association, the techniques we have been discussing are appropriate. If causation, these techniques are inadequate. Some additional ones plus a close interaction with theoretical reasoning will be required.

If you examine social research with these ideas in mind, you will be giving attention to those issues that are at the basis of association statistics. As before, you should not be discouraged if you cannot answer immediately every question that has been suggested about a piece of research. You must continue to read research articles and reports and to practice your statistical consuming skills to become more proficient. If you compare your current understanding of social statistics with your understanding when we began this book, you should be able to give yourself a well-deserved note of congratulations. There's more to follow, but you must have made great progress already.

KEY TERMS

Correlation
Simple (Zero-Order) Bivariate Correlation
Bivariate Correlation
Partial Correlation
Multiple Correlation
Regression
Linear
Curvilinear
Regression Line
Straight-Line Equation
Regression-Line Equation
Y-Intercept
Slope
Unstandardized Regression Coefficient (*b*)

Standardized Regression Coefficient (b^*)
Principle of Least Squares
Total Variation
Residual
Unexplained (Error) Variation
Explained (by Regression) Variation
Pearson's Coefficient of Correlation (r)
Coefficient of Determination (r^2)
Intercorrelation Matrix
*Partial Correlation
*First-Order Partial Correlation
*Multiple Regression
*Coefficient of Multiple Correlation (R)
*Coefficient of Multiple Determination (R^2)

EXERCISES

1. The following shows the bivariate distribution of years of work experience and days of vacation earned at a print shop. (a) Plot these scores in a scatter diagram. (b) Compute the regression equation values a and b and write the regression equation. (c) Plot the regression equation on your scatter diagram.

Employee	(X) Years of Work Experience	(Y) Days of Vacation
A	2	7
B	1	5
C	5	21
D	3	14
E	4	14
F	7	23

2. The following shows the bivariate distribution of literacy scores and number of felony convictions for some prison inmates. (a) Plot these scores in a scatter diagram. (b) Compute the regression equation values a and b and write the regression equation. (c) Plot the regression equation on your scatter diagram.

Inmate	(X) Literacy Score	(Y) Number of Felony Convictions
716309	1	12
222101	7	3
933998	3	5
468264	5	4
375991	4	6

3. Suppose that the regression equation for predicting social class on a 0–100 scale from income (in \$1000 units) was $\widehat{Y} = 10 + 2X$. What is the predicted value of social class when income is (a) 1; (b) 15; (c) 40?
4. Suppose that the regression equation for predicting age from height in inches for a set of elementary school students was $\widehat{Y} = 2.7 + .11X$. What is the predicted age for a student whose height is (a) 42 inches; (b) 48 inches; (c) 55 inches?
5. Imagine that the regression equation for predicting mental age from physical age was $\widehat{Y} = .1 + 1.1X$. (a) What is the predicted value of mental age at which the plotted regression line would intersect with the Y-axis? (b) What is the predicted change in mental age for every one-year increase in physical age?

6. Suppose that the regression equation for predicting the crude birth rate (per 1000) from city population (also in 1000s) for a set of medium-sized cities was $\hat{Y} = 40 - .10X$. (a) What is the predicted value of the birthrate at which the plotted regression line would intersect with the Y-axis? (b) What is the predicted changes in birthrate for each 1000 increment of increase in city population?
7. Suppose that separate bivariate regressions predicting subjects' lung capacity in cubic centimeters (Y) for a group of joggers found the following unstandardized regression coefficients: $b = .55$ with mean weekly miles run (X_1) and $b = .38$ with mother's lung capacity in cubic centimeters (X_2). (a) Interpret each of these unstandardized coefficients. (b) Assume that the standard deviations were as follows: S for subject's lung capacity $=5$, S for miles run $= 1.5$ and S for mother's lung capacity $= 4$. Find the standardized regression coefficients. (c) Is running or inheritance more important to increased lung capacity from these data? (d) Would you be justified in concluding from these data that running is not related to increased lung capacity? Explain.
8. The following shows the bivariate distribution of reading and mathematical performance scores for 10 junior high students. (a) Compute the Pearson correlation coefficient (r) for these data. (b) Interpret this coefficient's direction and strength.

Student	(X) Reading Score	(Y) Mathematical Score
Steve	7	7
Paul	8	7
Bill	6	6
Judy	5	8
Ralph	3	4
Jill	9	5
Joe	4	6
Marjorie	1	2
Dorothy	7	9
Al	2	3

9. The following shows the bivariate distribution of marital satisfaction scores and IQ scores for seven participants in a weekend conference on sexual awareness. (a) Compute the Pearson r for these data. (b) Interpret this coefficient's direction and strength.

Conference Participant	(X) Marital Satisfaction Score	(Y) IQ Score
HGB	10	100
KAM	5	120
BBC	9	88
QLJ	7	115
WOC	2	90
PKU	3	102
GGG	8	95

10. Suppose that a bivariate regression analysis of job satisfaction and salary has found that the total variation was 100 and the unexplained (error) variation was 25. (a) Find the explained variation. (b) Find the coefficient of determination. (c) Give a PRE interpretation of this coefficient assuming job satisfaction was the dependent variable. (d) Find the Pearson's r value.
11. Imagine that a bivariate regression analysis of prejudice scores and number of driving viola-

tions has found that the total variation was 200 and the explained variation was 30. (a) Find the unexplained variation. (b) Find the coefficient determination. (c) Give a PRE interpretation of this coefficient assuming that prejudice score was the independent variable. (d) Find the Pearson r value.

12. Suppose that two measures of self-worth have a Pearson's r value of .60. (a) What percentage of shared variation do these measures have? (b) Would you consider either to be a reliable measure of the other? Why?

13. Suppose that two measures of institutional discriminaton have a Pearson's r value of .95. (a) What percentage of shared variation do these measures have? (b) Would you consider either to be a reliable measure of the other? Why?

*** 14.** Imagine that the multiple regression equation for predicting income in dollars (Y) from education in years (X_1) and age in years (X_2) was $\hat{Y} = 500 + 600X_1 + 300X_2$. (a) What is the predicted income when both education and age are zero? (b) What is the predicted change in income when education is increased one year while age is controlled? (c) What is the predicted change in income when age is increased one year while education is controlled? (d) What is the predicted income when education is 16 and age is 40?

*** 15.** Imagine that the multiple regression equation for predicting total agency budget in dollars (Y) from staff size (X_1) and client size (X_2) is $\hat{Y} = 20{,}000 + 5750X_1 + 660X_2$. (a) What is the predicted budget when staff size is 10 and client size is 50? (b) What is the predicted budget when staff size is 15 and client size is 200? (c) What is the predicted budget when staff size is 25 and client size is 150?

16. Determine what type of correlation statistic [zero-order bivariate, partial or multiple] should be used to answer each of the following questions concerning the relationships between scores on a quality of life index (Y) with job satisfaction, housing conditions, health, and leisure activity. (a) What is the association between quality of life and each one of the other four variables considered singly? (b) What is the association between quality of life and all four of the other variables considered simultaneously? (c) Which one of the four variables is most strongly associated with quality of life while controlling the influence of the others?

17. Imagine that three separate zero-order bivariate correlations with voting frequency over a 10-year period (Y) have been found as follows: $r = .62$ with income (X_1), $r = -.15$ with family size (X_2), and $r = .47$ with years of education (X_3). (a) Does this mean that voting frequency is most highly related to income if both family size and education are ignored? (b) Does this mean that voting frequency is least related to family size if both income and education are controlled? (c) What correlation statistic should be used to determine which one of the three independent variables is most highly related to voting frequency while controlling the influence of the other two?

*** 18.** Suppose that the multiple regression equation for Exercise 17 was $\hat{Y} = 1.2 + 2.3X_1 - .1X_2 + .9X_3$. (a) Interpret the "1.2" value. (b) Interpret the "2.3" value. (c) Does the equation mean that income has the greatest influence on voting frequency? Explain.

*** 19.** Suppose that standardized partial correlations of age at death (Y) with a health index (X_1) and occupational risk score (X_2) have been found as follows: $r_{y1.2} = .65$ and $r_{y2.1} = .32$. Is health or occupation the greater contributor to explaining age at death? Explain.

20. Imagine that someone wanted to investigate the association between individual feelings of social alienation and population density for a set of communities. Name three variables that you would suggest should be controlled to help determine the true magnitude of the correlation between these two main variables.

21. Suppose that you were studying the variation in college grade-point averages (GPAs) for students. List five independent variables you think would be important in finding a high R^2 value for their simultaneous association with GPA.

22. Find one actual research example in the social science literature of any kind of correlation or regression covered in this chapter. Summarize and critique its descriptive use of the statistical technique shown according to the *Keys to Understanding* for association between interval and ratio variables.

Regression of X on Y

The roles of the two variables X and Y can be reversed in regression so that we predict X from Y. This will tell us how helpful knowledge of the Y scores is for predicting individual X scores. The formula needed for this regression is as follows:

$$\hat{X} = a + bY$$
$$\text{where } b = \frac{N(\Sigma XY) - (\Sigma X)(\Sigma Y)}{N(\Sigma Y^2) - (\Sigma Y)^2}$$
$$a = \bar{X} - b\bar{Y}$$

Correlation Ratio, *Eta*

The correlation ratio is a PRE measure of association used in two situations: (1) one interval-level dependent variable with one nominal-level independent variable and (2) one interval-level dependent variable with one interval-level independent variable when their relationship is curvilinear rather than linear. Eta uses the following computational formula:

$$\text{Eta}^2 = 1 - \frac{\Sigma Y^2 - \Sigma n_k \bar{Y}_k^2}{\Sigma Y^2 - N\bar{Y}^2}$$

where $n_k \bar{Y}_k^2$ = number of cases times the squared mean of the kth subcategory of the independent variable

See: Loether and McTavish (1980:261–265).

Multiple Regression Coefficients

Formulas for the multiple regression coefficients can be illustrated as follows for the situation in which there are two independent variables, X_1 and X_2, being used to predict the dependent variable, Y.

$$\hat{Y} = a + b_1X_1 + b_2X_2$$
$$\text{where } a = \bar{Y} - b_1\bar{X}_1 - b_2\bar{X}_2$$

unstandardized coefficients:

$$b_1 = \frac{S_y}{S_1}\left(\frac{r_{y1} - r_{y2}r_{12}}{1 - r_{12}^2}\right) \qquad b_2 = \frac{S_y}{S_2}\left(\frac{r_{y2} - r_{y1}r_{12}}{1 - r_{12}^2}\right)$$

standardized coefficients:

$$b_1^* = b_1\frac{S_1}{S_y} \quad \text{or} \quad = \frac{r_{y1} - r_{y2}r_{12}}{1 - r_{12}^2}$$
$$b_2^* = b_2\frac{S_2}{S_y} \quad \text{or} \quad = \frac{r_{y2} - r_{y1}r_{12}}{1 - r_{12}^2}$$

where S_y = standard deviation of Y
S_1 = standard deviation of X_1
S_2 = standard deviation of X_2
r_{y1} = zero-order r between Y and X_1
r_{y2} = zero-order r between Y and X_2
r_{12} = zero-order r between X_1 and X_2

See: Ott et al. (1978:420–421) and Mueller et al. (1977:299–300).

Multiple Correlation Coefficient, R^2

Formulas for the multiple correlation coefficient, R^2, can be illustrated as follows for the situation in which there are two independent variables, X_1 and X_2, and one dependent variable, Y.

$$R^2 = b^*_{y1.2}r_{y1} + b^*_{y2.1}r_{y2}$$

where $b^*_{y1.2}$ = standardized regression coefficient for X_1
$b^*_{y2.1}$ = standardized regression coefficient for X_2
r_{y1} = zero-order r between Y and X_1
r_{y2} = zero-order r between Y and X_2

See: Mueller et al. (1977:301–302).

Standard Error of the Estimate, *SEE*

The standard error of the estimate, *SEE*, is a measure of variation for the accuracy of regression predictions. Both a descriptive and inferential form of this measure are shown below. The inferential measure is the more commonly reported one.

$$\text{Descriptive:}\quad SEE = \sqrt{\frac{\Sigma\,(Y - \hat{Y})^2}{N}}$$

$$\text{Inferential:}\quad SEE = \sqrt{\frac{\Sigma\,(Y - \hat{Y})^2}{N - 2}}$$

See: Loether and McTavish (1980:247).

Path Analysis

Path analysis is an application of multiple regression to the examination of causal ordering among a set of variables. Theory is used to suggest a path diagram (model) that illustrates the ordering of the variables using arrows to represent the presumed direction of causal influence (the "paths"). Separate multiple regression equations are determined for each variable that is viewed to be the effect (ultimate or intermediate to the model) of some other variable(s). The nature of the presumed links is described by a series of path coefficients which are the standardized regression coefficients resulting from the multiple regression equations. Using inference tests, the theoretical path diagram is revised by dropping paths with insignificant path coefficients. The model is analyzed for direct and indirect effects of causal flow and this helps the researcher to evaluate theoretical notions about the interrelationships between the variables.

See: Bouden (1965), Duncan (1966), and Hadden and DeWalt (1974).

Other Special Measures of Correlation

Biserial correlation (r_b): The biserial correlation coefficient measures linear association between two continuous variables, one of which has been recoded to a dichotomy and which was normally distributed.

Point biserial correlation (r_{pb}): The point biserial correlation coefficient measures association between one continuous variable and one truly dichotomous variable which is discrete.

Tetrachloric correlation (r_{tet}): The tetrachloric correlation coefficient measures linear association between two continuous variables, both of which have been recoded to a dichotomy and which were normally distributed.

See: Minium (1978:156).

PART II

TRANSITION FROM DESCRIPTION TO INFERENCE

CHAPTER 7

INTRODUCTION TO PROBABILITY AND SAMPLING

In the first section of this text we looked at various ways in which researchers describe social data. The data that were being summarized constituted the entire set under consideration. The major issues were deciding which features of that data to highlight and how it should be done. This was referred to as descriptive statistics. Now our attention turns to the ways in which researchers describe social data when those data at hand represent only a part of the entire set being considered. In these situations the data set at hand is called a *sample* and the entire set of interest is called the *population.* Constructing a description of the population on the basis of a sample is the work of inferential statistics. A single statistical description based on sample data is called a *statistic.* Its counterpart in a population is called a *parameter.* The mean from sample data is then a statistic. The mean for an entire population is a parameter. Inferential statistics are concerned with finding statistics and using them to estimate parameters. An *inference* is an estimated description of a population (an estimated parameter) based on a description of a sample (a statistic).

Because an inference is an estimate we are never sure that it is correct. However, statistical theory tells the researcher how reasonable inferences are under certain conditions. The accuracy of an inference is defined by probability. This chapter is devoted to explaining the basics of probability and sampling relevant to inferential statistics. In the next chapter we focus on the logic used to make inferences when specific hypotheses are being tested. Together, these two chapters will give you the foundation for understanding the inferential statistics presented in the third section of the text that follows.

Probability

Probability is the likelihood that certain outcomes will occur in a situation over the long run. Any situation is thought of as a collection of possible events that may occur, "outcomes." For example, when we roll a common die there are six possible outcomes, one represented by each face on the die. When a court case is decided by a judge there are only two possible outcomes, guilty or not guilty. Probability is a mathematical description of the likelihood that each of the possible outcomes will actually happen over a series of "trials." The probability of an outcome can be expressed as a proportion (f/N) which compares the expected frequency of occurrence to all possible outcomes combined. In a formula it looks like this:

$$\Pr(A) = \text{expected}\left(\frac{f}{N}\right)$$

Probability of an Outcome (A)

where f = frequency of outcome A's occurrence
N = frequency of occurrence of all possible outcomes

In our six-sided die example, N equals 6. Each of the possible outcomes has only one face representing it, and all faces are equally likely to occur. Thus the expected probability for any one of the outcomes, say, a "3," is

$$\Pr(\text{“3”}) = 1/6 = .1\overline{6}$$

Or, suppose that 25 people each puts his or her name on a slip of paper and tosses it in a hat. What is the expected probability that one particular name will be drawn, say "Willie"?

$$\Pr(\text{Willie}) = 1/25 = .04$$

Notice that the probabilities are usually expressed as decimals, like all proportions. Probabilities will vary from 0, meaning no likelihood of occurrence, to 1, meaning certainty of occurrence.

The method used in the examples above to find the f and N values for calculating probability is the *theoretical method.* Theoretical probability is based on mathematical principles of what should happen in the long run. Since a die has six equal sides, 1/6 was the expected fraction to work with to find probability. Since there were 25 equal slips of paper in the hat, an honest draw would create an expected probability of 1 in 25 (.04) for any one of the names to occur. These are only two simple examples of many in which probabilities are found theoretically. We will be looking at similar and more complex examples later.

A second method for finding probabilities is the *empirical method.* Empirical probability is based on observing the actual number of times certain outcomes occur over a number of trials. This observation can take place in experiments, natural settings, surveys, or any number of ways. This may be the only method to find probabilities for situations that have unknown or indefinite theoretical properties. For example, how would we find the probability that a name picked at random from a student telephone book would be that of a female when we do not know the sex distribution of those students? One workable method would be to draw a random sample of a fair size and record the number of female names that occurred. Suppose this distribution showed that there were 15 females and 10 males in a sample of 25. The expected probability of randomly choosing a female's name could be set at $15/25 = .60$. In effect, we have empirically estimated the actual distribution of sex in the student population to answer our question.

What is the probability that an office worker randomly chosen for an errand would be under 25 years old? We can answer by determining the age distribution of the office workers in a random sample. If six of 30 in the sample are under 25, the expected probability is $6/30 = .20$.

The earlier examples of tossing the die and drawing names from a hat could be examined by the empirical method as well as the theoretical. We could roll the die a number of times, say 20, and record the outcome for each trial. From this a frequency distribution could be constructed and we could use the information to calculate the expected probabilities. If a "3" occurred four times, the expected probability for that face to occur in a future roll could be set at

$$\Pr(\text{“3”}) = 4/20 = .20$$

With names in a hat, we could also make several draws and record the results. The name drawn each time would be replaced in the hat for the next draw. Suppose that the name Willie was drawn twice in 35 trials. The expected probability of drawing that same name in a future draw could be set at

$$\Pr(\text{Willie}) = 2/35 = .057$$

These probabilities were determined by observation, not theory. Social researchers use the empirical method to construct inferences in many situations. However, a theoretical method is preferred, if it is available. It is preferred because it avoids the practical difficulties of conducting a large number of absolutely fair trials. Further, we can never be sure that an empirical distribution fairly represents the theoretical possibilities. However, we do know that the more of these trials we have, the closer our empirical distribution will come to the actual theoretical distribution. This is known as the *Law of Large Numbers.*

One distinction of great importance to probability is between situations whose outcomes have a discrete distribution and those having a continuous distribution. Each of the examples above was presented as having a discrete distribution. Only a limited number of outcomes were possible. The die could have only the "1," "2," "3," "4," "5," or "6" face turn up. Nothing else in between these or beyond was possible. The same pattern was true for the other examples. We call probability with discrete distributions *simple probability.* We can continue to look at this kind of situation to explain more of the basics of probability. Probability with continuous distributions is an extension of these basics and it guides most statistical inference. We will look at it later.

Simple Probability

There are two basic properties of probability that can be illustrated easily with a discrete distribution example. These are the addition and multiplication rules of probability. Suppose that we have a mint-condition coin with one side defined as a head and the other as a tail. Theoretically, we know that an honest flip of this coin will result in either side having an equal probability of landing face up. Put in terms of our formula, there are two outcomes ($N = 2$) and the expected probability of each accounts for half of this total:

$$\Pr(\text{head}) = 1/2 = .50$$
$$\Pr(\text{tail}) = 1/2 = .50$$

What is the expected probability that either a head or a tail will result from one honest flip of this coin? The key to the answer is to add the separate probabilities of these two outcomes:

$$\Pr(\text{head or tail}) = \Pr(\text{head}) + \Pr(\text{tail}) = .5 + .5 = 1.0$$

The *addition rule* tells us to add the probabilities of *individual outcomes* when those outcomes are mutually exclusive and we want to know the likelihood that any one of them will occur within one trial. Mutually exclusive outcomes are those that cannot occur simultaneously. In our example, the head and tail outcomes cannot both occur at once. In this instance, our calculated probability was certainty (1) because we were asking about the likelihood that either one of the two possible outcomes would occur. An honest flip is certain to result in either a head or a tail because there are no other possibilities. Notice, too, that asking about the probability of any one of several outcomes is mathematically equivalent to asking for the sum of their probabilities; "or" means "add" in these situations.

Consider a different question. What is the probability that in a first honest flip of the coin the result will be a head and in a succeeding second honest flip the result will be a tail? We are now asking about a series of outcomes over two trials, not about two alternatives within the same trial. Probabilities for independent outcomes over a series of trials use the *multiplication rule.* Independent outcomes are not contingent on each other. What happens in the first trial does not influence what happens in the second trial. Our question fits this description; therefore, we multiply the individual probabilities for the series of trials:

$$\Pr(\text{head, tail}) = \Pr(\text{head}) \times \Pr(\text{tail}) = (.5)(.5) = .25$$

What about a series of three tails in a row?

$$\Pr(\text{tail, tail, tail}) = (.5)(.5)(.5) = .125$$

Suppose that we have the empirical distribution shown in Table 7.1 for the predominant racial background of 20 persons. What is the probability of randomly selecting someone from this group who is white? Since the frequency of whites in the distribution is 6, the probability is found as follows:

$$\Pr(\text{white}) = 6/20 = .30$$

What is the probability of randomly selecting someone who is black? There are 10 blacks in the distribution, so

$$\Pr(\text{black}) = 10/20 = .50$$

What is the probability of randomly selecting someone who is Hispanic?

TABLE 7.1 Predominant Racial Background of 20 Persons

Predominant Racial Background	f
White	6
Black	10
Hispanic	4
Total	20

$$\Pr(\text{Hispanic}) = 4/20 = .20$$

How about the probability of randomly selecting either a person who is black or Hispanic? This problem uses the addition rule because we are asking about mutually exclusive outcomes in one trial:

$$\Pr(\text{black or Hispanic}) = 10/20 + 4/20 = 14/20 = .70$$

What is the probability of selecting at random either a person who is white or black?

$$\Pr(\text{white or black}) = 6/20 + 10/20 = 16/20 = .80$$

What about the probability of randomly selecting the sequence of a person who is black followed by one who is Hispanic?

$$\Pr(\text{black, Hispanic}) = (.50)(.20) = .10$$

The separate probabilities are multiplied because we are asking about a series of independent outcomes over two trials.

What is the probability of randomly selecting three whites in a row?

$$\Pr(\text{white, white, white}) = (.30)(.30)(.30) = .027$$

Again, we multiply because we are considering a series of independent outcomes.

How would you find the probability that a first random draw would produce either a white or black person and a second random draw would produce either a black or Hispanic person?

$$\Pr[(\text{white or black}), (\text{black or Hispanic})] = (.30 + .50)(.50 + .20) = (.80)(.70) = .56$$

Notice that this problem combines the addition and multiplication rules.

What about a three-draw sequence in which you are interested in selecting either a white or a Hispanic person each time?

$$\Pr[(\text{white or Hispanic}), (\text{white or Hispanic}), (\text{white or Hispanic})]$$
$$= (.30 + .20)(.30 + .20)(.30 + .20) = (.50)(.50)(.50) = .125$$

Be sure to study these simple examples so that you can quickly distinguish between questions about mutually exclusive outcomes in one trial and independent outcomes over a series of trials. Many of the probabilities underlying tests of statistical inference are based on these two rules.

Binomial Sampling Distribution

So far we have looked at each probability example as a unique problem. Actually, many examples conform to one of a few common types of distributions. Once we know which common distribution a problem belongs to we can solve the probability questions in a more standardized way. These distributions are known as sampling distributions. A *sampling distribution* is a theoretical probability distribution describing all possible sample outcomes for a sample statistic (e.g., a frequency, proportion, or mean). You must be very careful to distinguish a sampling distribution from both a sample's distribution and a population's distribution. A sample's distribution shows the distribution of scores for those individual cases contained in the sample. For example, the sex distribution of males and females in one sample of 25 students is a sample's distribution. A population's distribution shows the distribution of all individual scores contained in the population. For instance, the entire sex distribution of males and females for all students in a college of 5000 students is a population's distribution. In contrast to both of these, a sampling distribution for the same situation might show the distribution of proportions of females (or males) for every possible sample of a set size that could be drawn from that population. To continue our example, we might think of every possible sample of 100 students from a population of 5000. This would be a very large number of samples, about 54×10^{137}! For each there is a proportion of the cases that are females. The first sample may have .60 females. The second sample may have .50 females. A third may have .57 females, and so on. We can construct a frequency distribution of the proportions of females for all these samples. This would be the sampling distribution of proportions for this situation.

There are several sampling distributions in common use. One of the simplest and most common is the binomial. The *binomial sampling distribution* is a theoretical probability distribution for dichotomized variables that requires only nominal measurement. In these situations we divide all possible outcomes into two nominal categories.

The coin-tossing example above conformed to a binomial distribution. We have only two categories of outcomes, head and tail, and they represent only nominal measurement. Suppose that we theoretically toss our honest coin four times and consider all the possible sequences of outcomes that are possible. To make things more interesting, answer the following question before you continue. Are you willing to bet that half of the time there will be just as many heads as tails occurring from our honest coin tosses? The possible results for the four tosses are as follows:

	Toss			
Sequence	**1**	**2**	**3**	**4**
1	H	H	H	H
2	H	H	H	T
3	H	H	T	H
4	H	T	H	H
5	T	H	H	H
6	H	H	T	T
7	H	T	H	T
8	H	T	T	H

	Toss			
Sequence	**1**	**2**	**3**	**4**
9	T	H	T	H
10	T	H	H	T
11	T	T	H	H
12	T	T	T	H
13	T	T	H	T
14	T	H	T	T
15	H	T	T	T
16	T	T	T	T

We see that there are 16 possible sequences for the four tosses. There is only one sequence in which all results are heads (number 1) and only one in which all are tails (16). There are four sequences in which there are three heads and one tail (2–5) and another four with three tails and one head (12–15). There are six sequences in which there are two heads and two tails (6–11).

The probability that any one of these sequences will occur is the same as any other as long as a head and a tail are equally likely on each toss (Pr = .50). What is the probability of one of these sequences, say, four heads in a row?

$$\Pr(\text{H, H, H, H}) = (.50)(.50)(.50)(.50) = .0625$$

Similarly for another sequence, such as alternating heads and tails:

$$\Pr(\text{H, T, H, T}) = (.50)(.50)(.50)(.50) = .0625$$

But what is the probability of getting two heads and two tails, regardless of the order? That was the question you were asked to bet on. Since there are six possible sequences in which the same number of heads and tails occur, we must find each of their separate probabilities and then add them. As indicated, the probabilities within each sequence are multiplied [(.50)(.50)(.50)(.50) = .0625] because it is a series of independent outcomes. The probabilities for the six sequences now are added because we are asking for the likelihood that any one of the mutually exclusive series will occur. The entire solution looks like this:

$$\Pr(\text{sequence 6 or 7 or 8 or 9 or 10 or 11}) = \\ .0625 + .0625 + .0625 + .0625 + .0625 + .0625 = .375$$

If you bet that half (.50) of all results would have as many heads as tails in four tosses, you were wrong. Only three-eighths (.375) of the possible results are like this, 6 of 16. The entrepreneurs among you should see some immediate utility of this information for redistributing the wealth among your uninformed friends.

The entire sampling distribution of this four-tosses-of-a-coin example is shown in Table 7.2. Notice the probabilities total 1.0 for all 16 sequences. They must since there are no other sequences possible.

Recognition of this example as a member of the binomial distribution allows you to shorten the procedures in its solution. The binomial's standard reference is to one category of outcome(s) as *success* (labeled p) and the other as *failure* (labeled q). Let's arbitrarily have a head be success (p) and a tail be failure (q). We can find the probability for any number of successes and failures in any number of trials (cases) using the formula

$$\Pr(r \text{ successes}) = (C_r^N)(p^r)(q^{N-r})$$

Binomial Probability of Success Formula

where r = number of successes

N = number of trials (cases)

C = number of combinations or sequences

$$C_r^N = \frac{N!}{r!(N-r)!}$$

TABLE 7.2 Theoretical Sampling Distribution for Tossing an Honest Coin Four Times When Pr(Head) = .5 and Pr(Tail) = .5

Sequence	Toss 1	Toss 2	Toss 3	Toss 4	Probability				Result Summary
1	H	H	H	H	(.5)(.5)(.5)(.5)	=	.0625		4 Heads
2	H	H	H	T	(.5)(.5)(.5)(.5)	=	.0625	.25	3 Heads 1 Tail
3	H	H	T	H	(.5)(.5)(.5)(.5)	=	.0625		
4	H	T	H	H	(.5)(.5)(.5)(.5)	=	.0625		
5	T	H	H	H	(.5)(.5)(.5)(.5)	=	.0625		
6	H	H	T	T	(.5)(.5)(.5)(.5)	=	.0625	.375	2 Heads 2 Tails
7	H	T	H	T	(.5)(.5)(.5)(.5)	=	.0625		
8	H	T	T	H	(.5)(.5)(.5)(.5)	=	.0625		
9	T	H	T	H	(.5)(.5)(.5)(.5)	=	.0625		
10	T	H	H	T	(.5)(.5)(.5)(.5)	=	.0625		
11	T	T	H	H	(.5)(.5)(.5)(.5)	=	.0625		
12	T	T	T	H	(.5)(.5)(.5)(.5)	=	.0625	.25	3 Tails 1 Head
13	T	T	H	T	(.5)(.5)(.5)(.5)	=	.0625		
14	T	H	T	T	(.5)(.5)(.5)(.5)	=	.0625		
15	H	T	T	T	(.5)(.5)(.5)(.5)	=	.0625		
16	T	T	T	T	(.5)(.5)(.5)(.5)	=	.0625		4 Tails
					Total	=	1.000		

The factorial symbol (!) in the formula for combinations means that we multiply a number by each successively smaller whole number until we reach the number 1. 3! means (3)(2)(1), for instance. The entire probability of successes formula works as follows for our problem of having exactly two heads occur in four tosses of a coin:

$$\begin{aligned}
\Pr(2 \text{ successes}) &= (C_2^4)(p^2)(q^{4-2}); \quad \text{since } p \text{ and } q \text{ have probabilities of .50:} \\
&= \frac{4!}{2!(4-2)!}(.50^2)(.50^2) \\
&= \frac{4\cdot 3\cdot \not{2}\cdot \not{1}}{2\cdot 1(\not{2}\cdot \not{1})}(.25)(.25) \\
&= \frac{4\cdot 3}{2\cdot 1}(.0625) \\
&= \frac{12}{2}(.0625) \\
&= (6)(.0625) \\
&= .375
\end{aligned}$$

Again, we find that the expected probability of getting two heads and two tails in any possible sequence from four tosses is .375. It is the same answer as before and we did not have to construct a complete theoretical distribution to find it.

The same procedure can be used with other binomial questions. For instance, what is the probability of six pregnant women giving birth to four female and two male children? Let the probability of a female child be .60 and a male child be .40. A female birth can be success (p) and a male birth be failure (q). Further, we will restrict the situation to only single births for each woman.

$$\begin{aligned}
\text{Pr(4 successes)} &= (C_4^6)(.60^4)(.40^2) \\
&= \frac{6!}{4!2!}(.1296)(.16) \\
&= \frac{6\cdot5\cdot\cancel{4}\cdot\cancel{3}\cdot\cancel{2}\cdot\cancel{1}}{(\cancel{4}\cdot\cancel{3}\cdot\cancel{2}\cdot\cancel{1})(2\cdot1)}(.020736) \\
&= \frac{6\cdot5}{2\cdot1}(.020736) \\
&= \frac{30}{2}(.020736) \\
&= 15\,(.020736) \\
&= .31104
\end{aligned}$$

These computations tell us that there are 15 possible sequences of four females and two males (e.g., female, female, female, male, female, female, female, male) from the six pregnancy results and that each of them has a separate probability of .020736. The probability that any one of them will occur is thus 15 times the single sequence probability, or approximately .31.

This method of finding probabilities is obviously less cumbersome than determining the complete theoretical distribution. It is easy to omit some of the possible sequences if we try to list them all, even with as few as five trials. Inasmuch as five trials are analogous to having a sample of only five cases, this is not encouraging. Obviously, many research problems are much larger. The second method makes them more manageable.

Many research problems conform to the binomial distribution. Some do it naturally, like a coin toss or the sex of a child. Others can be converted to a dichotomy, like our earlier illustration of office workers who were either under 25 years of age or not. As long as a situation can be usefully thought of as a distribution of a variable with two nominal categories, the binomial should be considered as a method for finding probability. It does not always apply but it often does.

One further feature of probability can be illustrated with a binomial problem. So far we have been considering questions of expected probability for specific outcomes: three whites in a row, two heads in four tosses, and so forth. We call this *point probability*. We can also be interested in finding something called *cumulative probability*. For example, what is the probability of having *at least* two heads occur in four tosses? Success would be two, three, or four heads resulting.

This problem is solved by finding the probabilities for each of the separate set of outcomes that satisfy our definition of success and then adding them when our variable has a discrete distribution. It is equivalent to asking about the probability of getting either two, three, or four heads in four tosses. We already know that Pr(2 heads) = .375. Table 7.2 tells us that the probability for three heads and one tail is .25 and for four heads it is .0625. When we combine terms we see that

$$\begin{aligned}
\text{Pr(2 or 3 or 4 heads)} &= .375 + .25 + .0625 \\
&= .6875
\end{aligned}$$

FIGURE 7.1 Theoretical Distributions for Coin Tosses When $N = 4$, 10, and 20 and Pr(Success) = .5

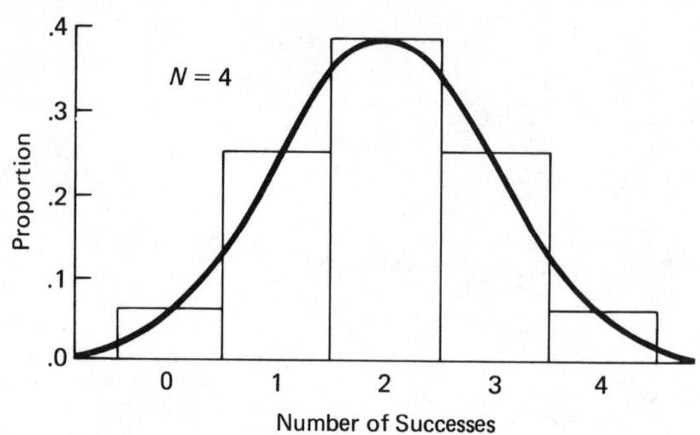

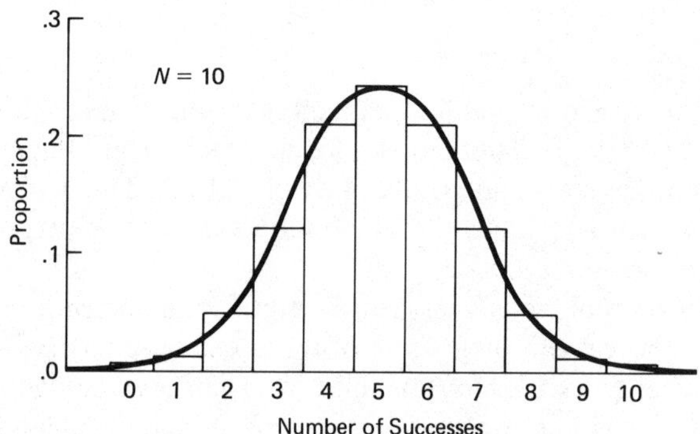

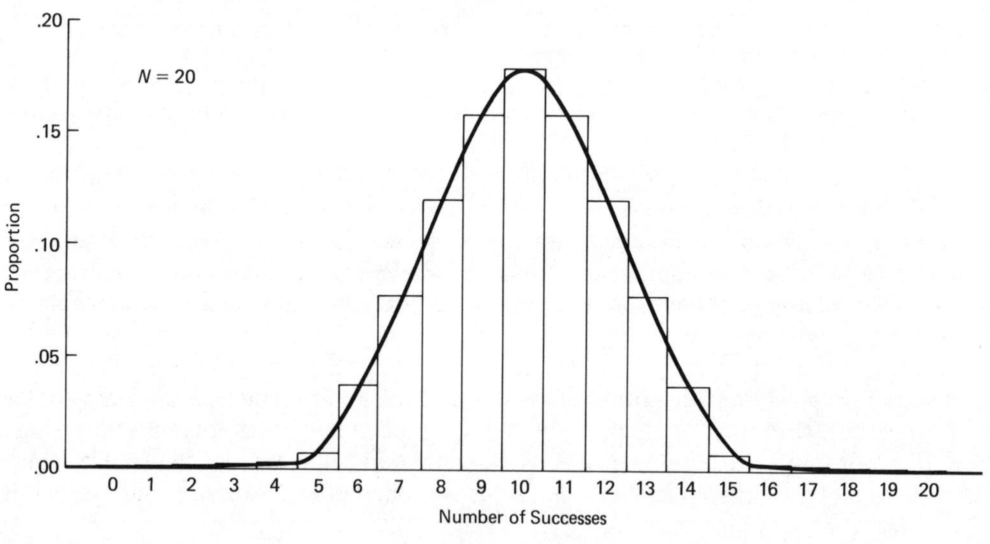

This combination of outcomes illustrates our meaning of cumulative probability. We think in terms of the accumulation of a number of separate outcomes within a set range. Our example's range was from two to four heads. Note that the binomial used for this problem still required only a nominal variable (heads versus tails), but the accumulation of a number of successes (2, 3, or 4) is, of course, a ratio phenomenon.

Simple probability can be concerned with either point or cumulative probabilities. In contrast, probability with continuous distributions is nearly always of the cumulative type. We will now look at this kind of problem.

Probability for Continuous Distributions

Not all situations usefully conform to a binomial distribution. Either there are more than two categories of important outcomes or there are several categories that we want to keep at a level of measurement higher than the nominal. One major classification of these more complex distributions is a collection of distributions for variables with continuous, not discrete, forms. We can sometimes approximate these distributions by examining a binomial distribution over a large number of trials.

If we theoretically toss our still honest coin 4, 10, and 20 times and graph the frequency of successes (heads) for each conceptual experiment, we begin to see the shape of a cumulative distribution in approximation. This is shown in Figure 7.1. There the expected frequencies have been plotted as histograms. Only the actual plots at each number of successes conforms to the theoretical, discrete distributions for our coin tosses. But a smoothed line connecting those plots suggests the results if our variable had been continuous instead. It approximates what would happen if we were looking at the distribution of a truly continuous variable. In these illustrations the approximated continuous distributions become more and more like a normal curve as N increases. Not all continuous variables take on this shape. Other common ones are J-shaped or shaped like one-half of a normal curve. But the normal distribution is the most useful of the continuous distributions for us to study as we lay the foundation for inferential statistics.

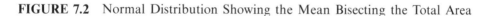

FIGURE 7.2 Normal Distribution Showing the Mean Bisecting the Total Area

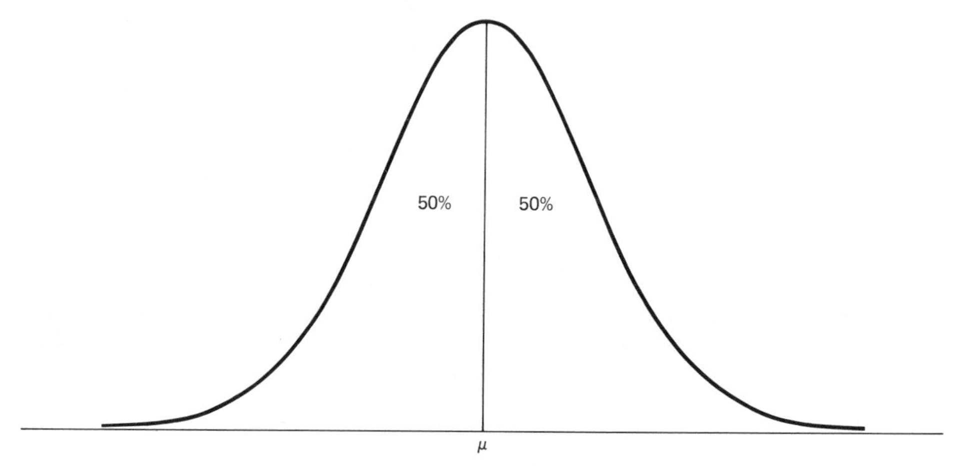

The Normal Distribution

The normal distribution is a bell-shaped, symmetrical distribution based on theory. We first encountered it in Chapter 4 when we discussed the special meaning of the standard deviation. It was developed by Carl F. Gauss around 1800 to describe random errors made in astronomy observations, so some call it the Gaussian distribution. The name "normal" refers to the common occurrence of empirical distributions of the same form. Such widely divergent variables as errors of astronomic observations, human heights, and IQs all correspond approximately to the normal distribution.

Let us review and elaborate some of the features of a normal distribution. The theoretical normal distribution is perfectly symmetrical, unimodal, and continuous. Therefore, its mean equals both its median and mode. According to the theory, the curve never quite reaches the X-axis; it goes to infinity at either end. The area between the curve and the X-axis contains 100% of the cases represented in the distribution. Fifty percent of this area is on either side of a line drawn from the mean to the curve, since the mean exactly bisects the distribution. (See Figure 7.2.)

All the areas under the normal curve are known by formula. Figure 7.3 shows some of these areas in percentage. Notice that 34.13% of the total area is between the mean and the mean plus one standard deviation. Between the mean and two standard deviations from the mean is 47.72%. The area from the mean to three standard deviations from it contains 49.87%, or nearly all of the 50% on one side of the distribution. The same percentages hold for parallel areas below the mean because the curve is symmetrical.

Figure 7.4 shows the areas between one, two, and three standard deviations on both sides of the mean. Notice that a little more than two-thirds of the area (68.26%) is between one standard deviation above and below the mean. About 95% (actually, 95.44%) is between two standard deviations above and below the mean. And nearly all of the area

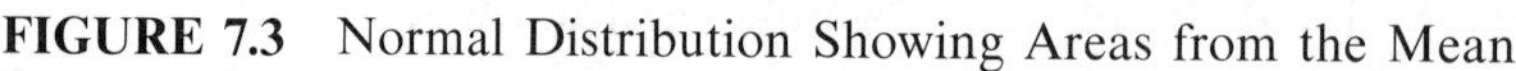

FIGURE 7.3 Normal Distribution Showing Areas from the Mean

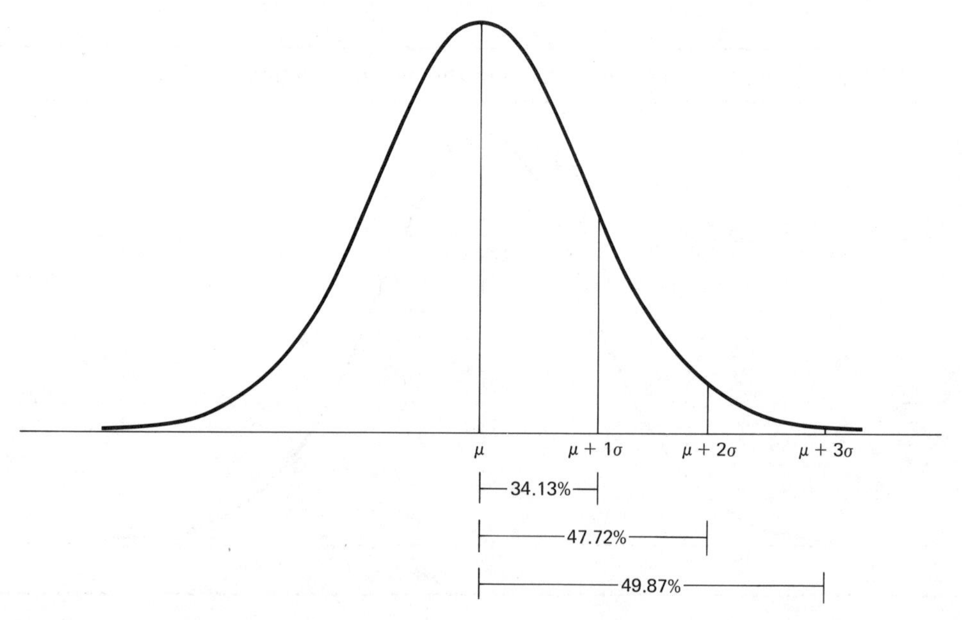

FIGURE 7.4 Normal Distribution Showing Areas Between Standard Deviations

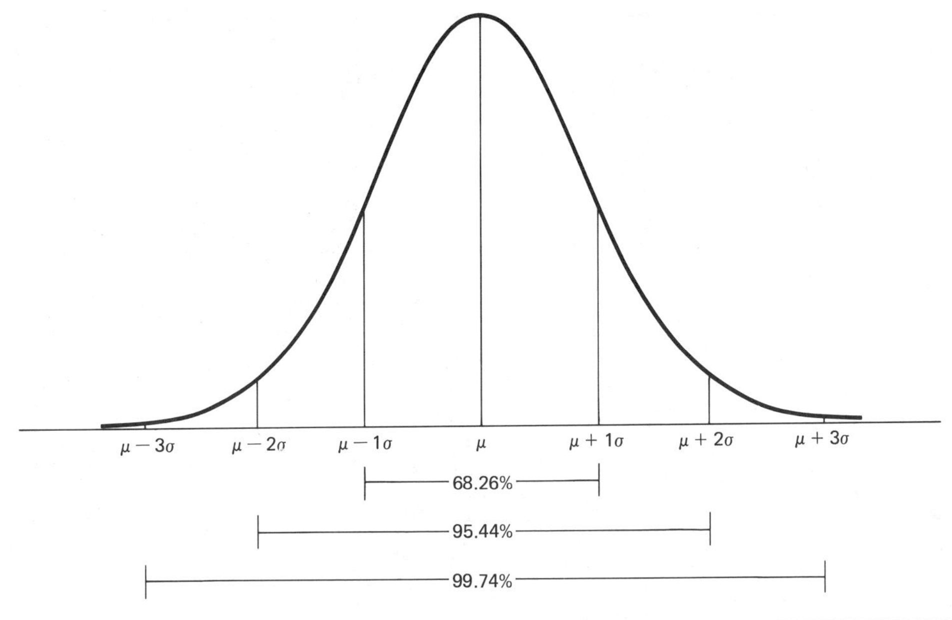

(99.74%) is between three standard deviations on each side of the mean. (Remember from Chapter 4 that $R \simeq 6\sigma$ in a normal distribution.)

These areas are determined by formula and reported in various tables. One of these is Table D2 in Appendix D. It lists the areas in percentages between the mean and distances above the mean for the *standard normal distribution.* This is a normal distribution having a mean of 0 and a standard deviation of 1. We can transform any actual distribution that is normally shaped to this standard normal distribution in order to take advantage of our knowledge of it. Since the distribution is symmetrical only areas on the positive side (above) of the mean need to be shown in Table D2. Those for the negative side (below) are identical. Further, because so little area is beyond ± 3 standard deviations from the mean, only a few tabular entries are given after this point.

To use Table D2 with actual data, the raw scores must be transformed from their original units of measurement (inches, dollars, and so forth) to standard deviation units in what are called *z scores* or *z standard scores.* The *z* standard scores use a distribution's own standard deviation as their unit of measurement. This transformation is accomplished by the following formula:

$$z = \frac{X - \bar{X}}{S} \qquad z \text{ Standard Score Formula}$$

where X = raw score
$\bar{X}$ = mean score
S = descriptive standard deviation

Figure 7.5 shows three empirical normal distributions. The (a) graph shows the distribution of heights for a group of 50 people, (b) shows the IQs for a group of 75 people,

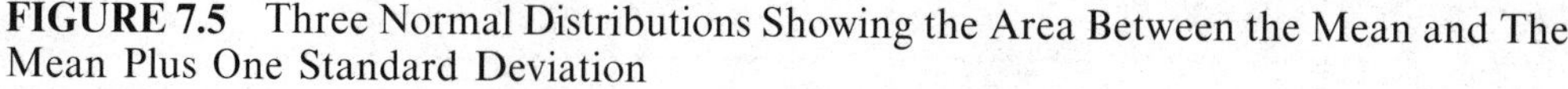

FIGURE 7.5 Three Normal Distributions Showing the Area Between the Mean and The Mean Plus One Standard Deviation

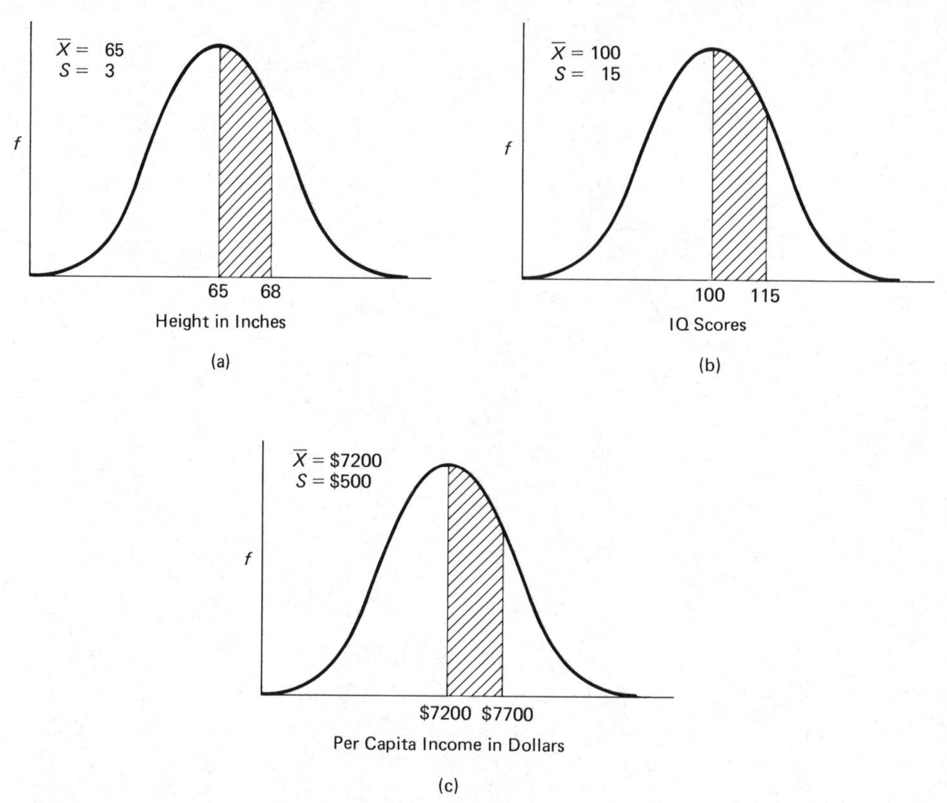

and (c) shows the per capita income for 100 cities. The means and standard deviations for each distribution are noted in the figure.

Let's transform the distribution shown in Figure 7.5(a) to the scale of a standard normal distribution. We can do this by changing each of several possible raw scores (e.g., 65, 68, 71, 74, 62, 59, 56) to z scores as follows:

$$\text{For } X = 65: \quad z = \frac{65 - 65}{3} = \frac{0}{3} = 0$$

$$\text{For } X = 68: \quad z = \frac{68 - 65}{3} = \frac{3}{3} = 1$$

$$\text{For } X = 71: \quad z = \frac{71 - 65}{3} = \frac{6}{3} = 2$$

$$\text{For } X = 74: \quad z = \frac{74 - 65}{3} = \frac{9}{3} = 3$$

$$\text{For } X = 62: \quad z = \frac{62 - 65}{3} = \frac{-3}{3} = -1$$

$$\text{For } X = 59: \quad z = \frac{59 - 65}{3} = \frac{-6}{3} = -2$$

$$\text{For } X = 56: \quad z = \frac{56 - 65}{3} = \frac{-9}{3} = -3$$

The new transformed distribution is shown in Figure 7.6. The z scores tell you how far a raw score is from its mean in standard deviations. For example, a height of 68 inches is one standard deviation above the mean height in that distribution ($z = 1$). A height of 59 inches is two standard deviations below that mean ($z = -2$).

In the same way, a raw IQ score of 90 from the distribution in Figure 7.5(b) is equivalent to a z standard score of $-.\overline{66}$:

$$z = \frac{90 - 100}{15} = \frac{-10}{15} = -.\overline{66}$$

This IQ is two-thirds of a standard deviation below the mean IQ of that distribution.

A city with a per capita income of \$8005 from the distribution in Figure 7.5(c) converts to a z standard score of 1.61:

$$z = \frac{\$8005 - \$7200}{\$500} = \frac{\$805}{\$500} = 1.61$$

Once empirical normal distributions have been converted to standard normal distributions, we can use Table D2 to answer probability questions about them. For instance, what is the probability of randomly selecting someone from our distribution of heights [Figures 7.5(a)] who is between 65 and 68 inches tall? The raw scores of 65 and 68 are equivalent to z scores of 0 and 1, respectively. We use Table D2 to find the area between them. A sketch like Figure 7.6 with the area in question shaded helps clarify the problem.

Table D2 is used by finding the row and column intersection that best approximates our standard scores. The z score of 0 is superfluous to the problem because we know that there is no area between the mean and a z of 0. Our z score of 1.00 is found in the row labeled 1.0 opposite the first column (.00). The number shown in the body of the table for this z score is 34.13. This is the percentage of the total area under the curve that is between

FIGURE 7.6 Standard Normal Distribution for Figure 7.5(a)

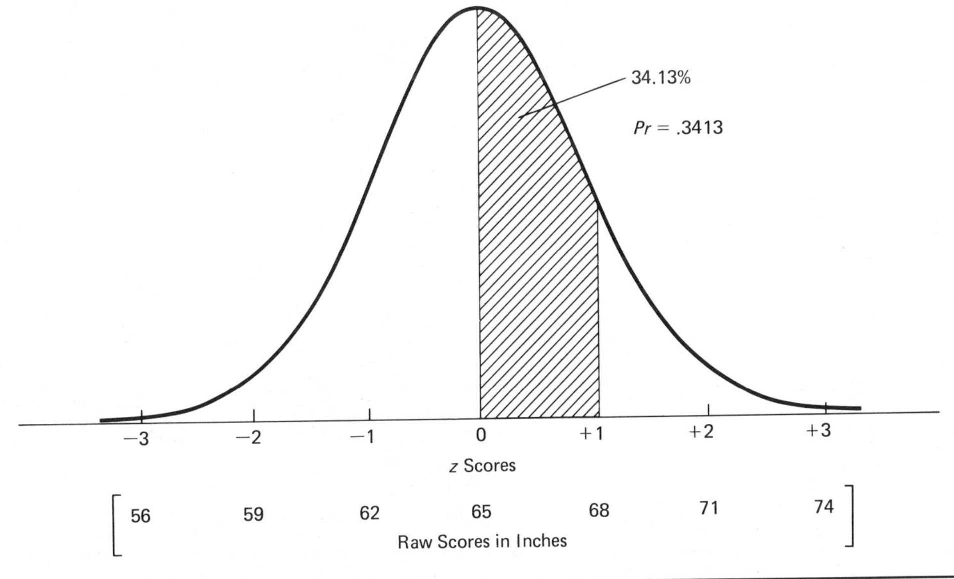

the mean and one standard deviation from the mean, to the nearest one-hundredth of a percent. Note that this is the same percentage given to you in Figure 7.3 for this area.

The percentage can be translated into probability rather easily. We are asking for the probability of randomly drawing someone from the distribution whose height is between 65 and 68 inches. Since 34.13% of these people have such heights, we need only convert our percentage to a proportion, the form of all probabilities. Stated as a probability, 34.13% is .3413. Simply move the decimal point two places to the left. Note that we state and solve the problem in terms of cumulative, not point, probability.

Now let us find the probability of randomly picking someone from the distribution in Figure 7.5(b) who has an IQ between 90 and 100. Again the raw scores are converted to z scores and a sketch is drawn. Our relevant z score is $-.6\bar{6}$. Remembering that we ignore the minus sign, we look in Table D2 for the nearest entry. It is .67. The table shows that 24.86% of the area under the curve is between the mean and this standard score. Therefore, the probability of randomly selecting someone who has an IQ between 90 and 100 from this distribution is .2486. (See Figure 7.7.)

What is the probability of randomly selecting a city from the distribution in Figure 7.5(c) with a per capita income between \$8005 and \$7200? We are seeking the area in Table D2 associated with a standard score of 1.61. The tabular entry opposite this z score is 44.63%. Thus the probability of the selection occurring is .4463. (See Figure 7.8.) Of the three examples, we are most likely to see the third one happen (it has the highest probability).

We can expand our examples to show the addition and multiplication rules applied to a continuous distribution such as the standard normal. What is the probability of drawing two people from the heights distribution who are between 65 and 68 inches tall? Like simple probability, we find the probabilities for each trial (draw) and then multiply them because this is a problem of a series of independent outcomes. We already know

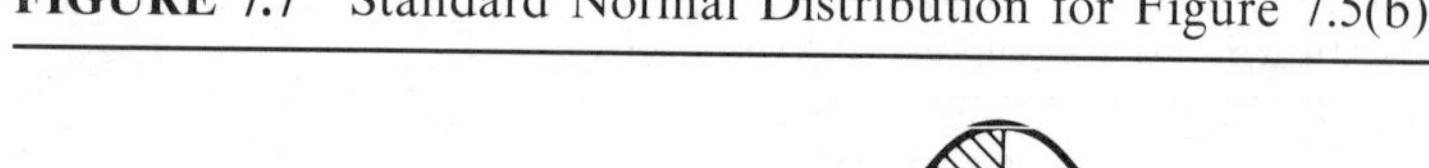

FIGURE 7.7 Standard Normal Distribution for Figure 7.5(b)

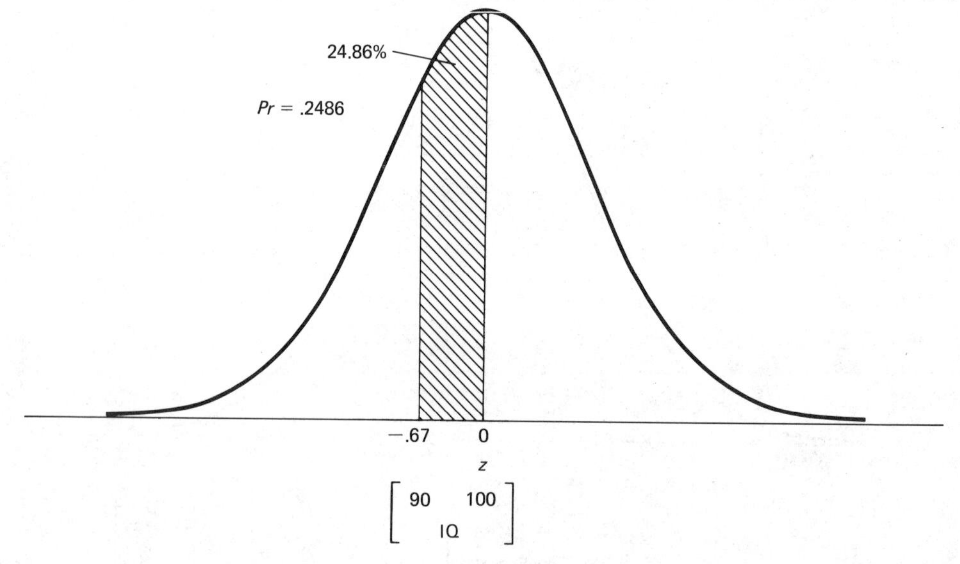

FIGURE 7.8 Standard Normal Distribution for Figure 7.5(c)

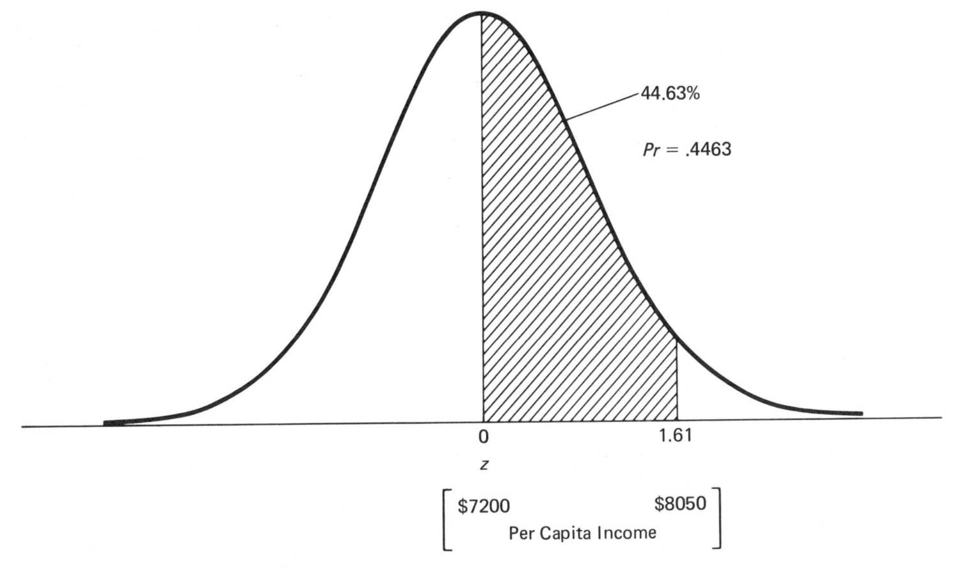

that the probability for the first draw is .3413. Because the second draw would be completely independent of the first (it is not influenced by what happens the first time), its probability would also be .3413. The series of the two would have a probability of (.3413)(.3413) = .1165.

What would be the probability of selecting from the IQ distribution either someone between 90 and 100 or someone between 100 and 110? This is a problem of mutually exclusive outcomes in one trial, so the addition rule is used. We already know that the probability for selecting someone between 90 and 100 is .2486. We find the probability for selecting someone between 100 and 110 in the same way. First it is transformed to a z score. Then the area in percentage form is found from Table D2. That percentage is converted to a probability as always.

$$z = \frac{110 - 100}{15} = \frac{10}{15} = .6\overline{6}$$

Again, the percentage is 24.86 and the probability is .2486. The probability that either of these two outcomes will happen is .2486 + .2486 = .4972. Figure 7.9 illustrates the solution of the problem.

What is the probability of randomly selecting a city from the distribution in Figure 7.5(c) with a per capita income of $8200 or more? Figure 7.10 shows the area under the standard normal curve that we are seeking. An extra step is required to solve this problem. As usual we begin by transforming the raw score to a standard score:

$$z = \frac{\$8200 - \$7200}{\$500} = \frac{\$1000}{\$500} = 2.00$$

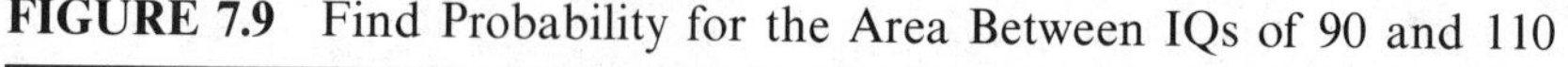

FIGURE 7.9 Find Probability for the Area Between IQs of 90 and 110

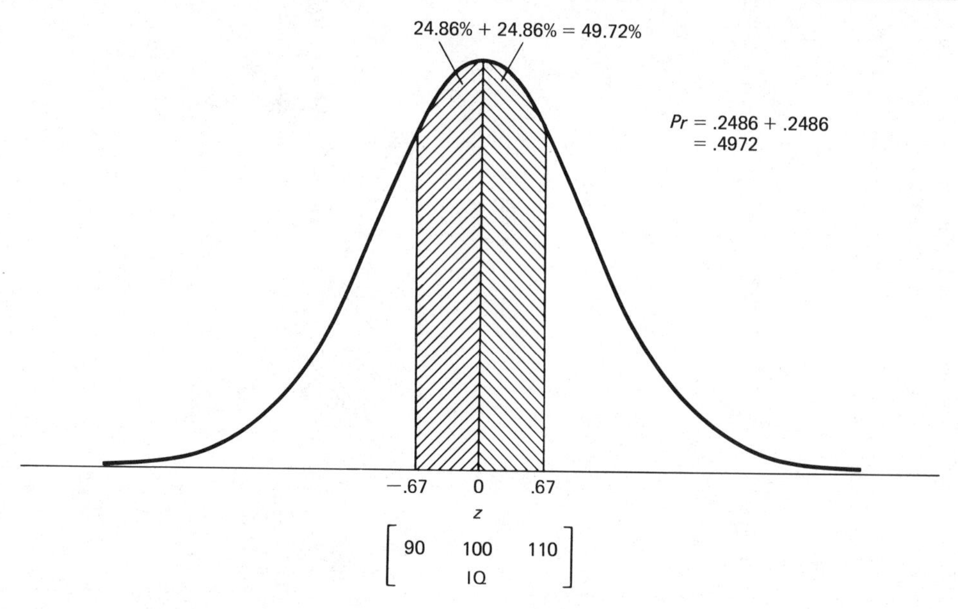

We find that the area between this z score and the mean in Table D2 is 47.72%. The probability of selecting at random a city with a per capita income between \$8200 and \$7200 (the mean) would be .4772. But that was not our question. To find the percentage of the area (and its associated probability) for \$8200 *and above,* we take advantage of our knowledge that 50% of the area is on each side of the mean. Since 50% of the area is between \$7200 and positive infinity (theoretically) and 47.72% is between \$7200 and \$8200, we know that subtraction is the key. Subtracting these percentages we find that $50.00 - 47.72 = 2.28$. This is the percentage of cities with incomes of \$8200 and more. The probability of randomly selecting one of these high income cities is therefore .0228, not a very high likelihood.

As can be shown from these examples, a sketch is very useful in solving the problems. It helps to clarify which area is to be found and whether addition, subtraction, and/or multiplication are needed. The reader should practice finding areas in every part of a standard normal distribution and solving probability problems using both the addition and multiplication rules. This is a skill that is relevant to several parts of the text that follow.

Many social research problems use the standard normal distribution as their sampling distribution when doing inference. Under the right conditions, the collection of outcomes from all possible samples for certain statistics take on the shape of the standard normal distribution. Just what those conditions and statistics are will have to wait for later chapters. But for now, we may note that there are research situations in which the probabilities from the standard normal distribution will be used to draw inferences from samples to populations. This will be done using a z-score transformation similar to what we have just been examining.

FIGURE 7.10 Finding Probability for the Area of $8200 and Above

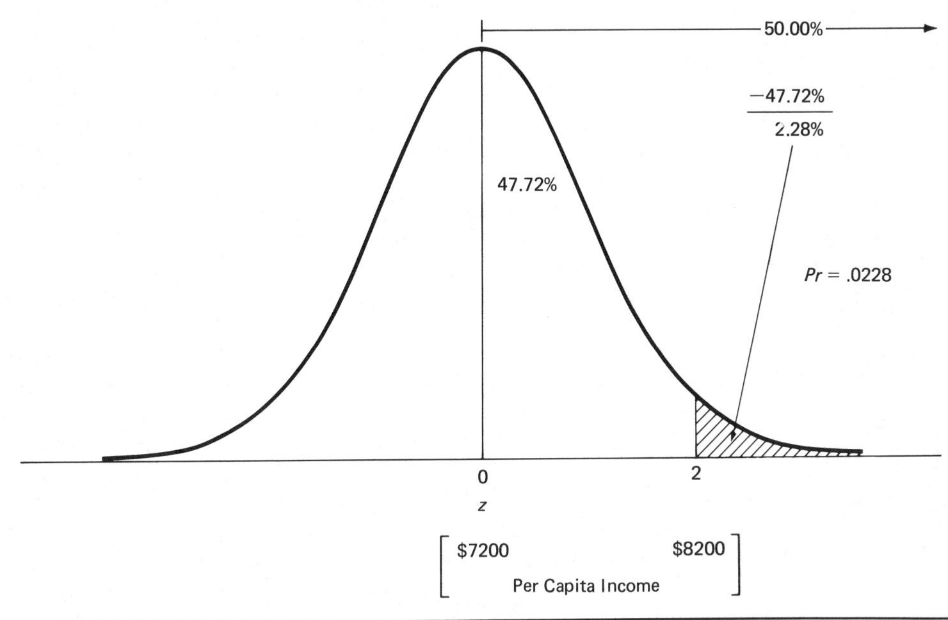

ON THE SIDE 7.1 Carl Friedrich Gauss

Carl Friedrich Gauss (1777–1855) was the dominant figure in Western science in the nineteenth century. His only equals in intellectual history are said to be Archimedes and Newton. He was a German child prodigy who reputedly could count before he could talk. At the age of 10 he astounded his teacher by solving an arithmetic twister using his knowledge about the symmetry of arithmetic progressions. By the age of 11 he had enrolled in a gymnasium and was told by his mathematics professor that it was unnecessary for one as gifted as he to attend lectures. Through the intervention of a professor's assistant, the reigning Duke Carl Wilhelm Ferdinand of Braunschweig became his patron. The Duke supported him through the remainder of his education and early years until the Duke's death in 1806 at the battle of Auerstadt, lost to Napoleon. When 19, Gauss published his first scientific paper on the construction of a regular 17-sided polygon, previously thought impossible. This discovery was based on Gauss's extension and revision of earlier ideas on prime, natural, and imaginary numbers. The paper made him instantly famous among mathematicians. He quickly went on to develop many ideas in number theory, which were later collected in his *Disquisitiones arithmeticae* (Arithmetical investigations). This, in itself, had made a reputation unequaled by his contemporaries. But it was only a beginning. He went on to work in algebra, function theory, differential geometry, probability theory, astronomy, mechanics, geodesy, hydrostatics, electrostatics, magnetism, and optics.

In 1804, Gauss was appointed director of a new observatory to be built in Göttingen. When it was sufficiently completed in 1807, Gauss moved there and remained for the rest of his life. This move coincided with his

interest and enormous regard within astronomy. Astronomers had begun to observe new objects in the sky in 1801, but they puzzled over their nature and orbits. Gauss successfully predicted the times and places of reappearance for these asteroids. His methods were published in 1809 in a work still studied as a classic in theoretical astronomy (*Theoria motus corporum coelestium in sectionibus conics solem ambientium*—Theory of the motion of the heavenly bodies moving about the sun in conic sections.)

The *Theoria motus* contained Gauss' first publication on the method of least squares used in making his predictions about orbits. This caused a tempest of sorts because another mathematician (Legendre) had already published the ideas in 1806. However, records indicate that Gauss had made the discovery first (in 1802) and that the men had worked independently. Gauss had delayed publishing his ideas in concert with his aversion to claiming priority rights through the device of publishing an advance notice. Instead, he held to a principle of thoroughly developing and polishing an idea before publication. This practice caused other such priority disputes throughout his life.

The method of least squares was developed to describe the expected distribution of measurement errors when making the astronomical observations of the newly discovered asteroids. Gauss used the method to determine their orbits based on the scattered reports of their sightings. Gauss derived the "error curve" by determining the equation that best corresponded to the distribution of measurement errors around the arithmetic average of the observations. Working between a set of measurement and mathematical assumptions, he was able to develop an ideal "normal density curve" which he could use to assess the probable error in a set of observations. By this means he was able to show that the average error of the arithmetic average declines with the square root of the number of observations.

These ideas are central to the methods of regression and hypothesis testing as applied to social statistics. Gauss's personal application of the ideas (other than to orbits) was in saving the pension fund for professors' widows at Göttingen by correcting the actuarial basis for computing the annuities. Gauss went on to many other topics, such as the invention of the first electromagnetic telegraph, in his usual mix of the pure and applied sciences. He lived and died in an environment alienated from most of his relatives and contemporaries because of his genius and fame.

Based on Tord Hall, *Carl Friedrich Gauss, a Biography*, translated by Albert Froderberg. Cambridge, Mass.: The MIT Press, 1970.

Sampling Methods

We have been speaking casually about "randomly" selecting cases from distributions as though we knew what that meant. If the term conveyed a sense of impartiality, that was sufficient for our earlier purposes. But now we need to be more exact by defining sampling methods, random and otherwise. Some actual examples of sampling methods from social research are presented in Research Examples 7.1 to 7.7. Some have very short descriptions, others quite lengthy. Can you tell which of these are truly random methods? At the end of this section you will better understand what these researchers were doing.

Why Sample?

Researchers use a sample rather than a complete enumeration (*census*) of a population for three basic reasons. First, sampling is an *efficient* method to gather information. A census of every case in a population would usually be very costly, time-consuming, and wasteful of human effort. Imagine the cost of doing even a small census, such as asking every student in a college a few questions. Even if we used the telephone, which is usually the least expensive personal method, the task and cost could be very large.

The U.S. Bureau of the Census conducts a complete population study only once every 10 years, and then chiefly because it is mandated by law. The 1980 Census of the Population cost over $1 billion and most people were contacted by mail, not in person. The Bureau uses samples for nearly all the other surveys it conducts because they are so much more efficient. The same dynamics of cost, time, and staff demands lead most researchers to select a sample of a population, not a census.

Second, sampling is a *practical* method. Even if money and time were not limiting factors, it is often not feasible to take a census. When we want to study individuals it is seldom possible to find every person who belongs to a population. Could we really locate every individual with a residential address in a community, every worker listed as employed in a large factory, every student officially enrolled at a university, every person ever arrested in a jurisdiction, and so on? Probably not. People are too mobile for this effort.

It is sometimes easier to locate all cases when we study larger social units, such as cities, schools, police departments, and formal organizations. These are more stable than individuals. Yet we may have difficulty with definitions (What is a city?) and with gaining access (How do you get every business to cooperate?). Larger social units tend to have well-developed barriers to access by outsiders. It is often more practical to concentrate our efforts on locating and gathering information from a smaller number of cases drawn from a population of individuals or larger social groupings.

Third, sampling is *effective*. Through probability theory researchers know how to draw representative samples and make reasonable inferences to populations. It is not a matter of luck. There may still be error, but it is predictable and therefore manageable. With samples the prevailing issue is *sampling error*. This is the error that results from drawing any sample from a population. Even a perfectly drawn sample may misrepresent the population to some degree. One sample may estimate a population's mean age to be 27, another may be 29, and still another may be 28. The variation between a sample's result and its population is sampling error. However, we know that this type of variation is not systematically biased and we know what can be expected in a probability sense *if the sample is drawn randomly*.

On the other hand, the errors associated with attempting to conduct a census of a large population are likely to have systematic biases. We are more likely to miss more of the very rich and very poor, the most mobile and reclusive, those on the fringes of society and its laws, those in minority groups, and others of the more extreme characteristics than those in the middle of the ranges. These kinds of bias are not well described by probability theory, and thus adjustments cannot readily be made for them. Although the same kinds of problems affect samples, they are exaggerated with attempts to take censuses.

There is no need to fear sampling. We take samples every day. We do not sniff all the air before we breathe, tally every risk before we travel, give all our blood to the doctor for a test, learn everything about someone before we form a judgment, study a political candidate's every word and act before deciding how to vote, or learn everything about statistics before deciding if it is a useful subject. Sampling is a very common activity. But as these examples suggest, not all samples are equally commendable. Learning to evalute sampling methods is important.

How to Sample?

The goal of sampling is to obtain a smaller representation of a population. An accurate representation is best achieved by a probability sampling design. Sampling designs are divided into two major types, probability and nonprobability. In *probability sampling* the researcher knows the probability that each case in a population may be included in the sample. In *nonprobability sampling* these case probabilities of inclusion are not known. If the probabilities are not known, inferences cannot be made with known degrees of assurance and the appropriate population to which the inferences can be extended remains unclear. Only with probability sampling is a researcher able to assess the likely extent to which sampling error is at work.

Probability sampling uses random methods of case selection. We are going to discuss some random and nonrandom sampling methods to illustrate exactly what a random sample is and is not.

Random Methods

A *simple random* method of sampling gives every case in a population an equal and known probability of being selected. No preferential treatment is given to any case or set of cases. No researcher, interviewer, subject, respondent, or other individual is allowed to alter the equal probability for inclusion of any case. The only operating factor in selecting cases for the sample is chance. Simple random selection is not haphazard or accidental. These terms reflect biases that will be detailed later.

A simple random method of sampling chooses cases from a population in a single selection process using a series of randomized numbers. The process requires one first to attain a list of all the cases in the population. Every student in a university, every adult in a community, every defendant in a court, every business in an association, and so on, must be identified and listed. The cases are assigned numbers from 1 to N. The researcher then uses a table of random numbers (or other random-number-generating process, such as a computer program) to select the cases numbers (and thus the cases) for inclusion until the desired sample size is reached.

Suppose that we want to draw a simple random sample of 25 individuals from a professional association having 305 members. Our first step is to get a complete membership list from the association. Next we number the names on the list from 001 to 305. Now we consult a table of random numbers to draw our 25 cases for the sample. A table of random numbers is presented as Table D1 in Appendix D. A short version of such a table is reprinted here as Table 7.3. Let us use this one to draw our sample.

A table of random numbers has no prescribed order. Any digit (0 to 9) is equally likely to be found in any part of the table. In using such a table we first determine how many digits we need to use at once. The population size tells us this. If we have 500 cases in our population list, we need to read the table in sets of three digits (001 to 500). If our list has 5000 cases, we need to read in four digits (0001 to 5000). Since our list has 305 names on it, we use three-digit numbers.

Next we determine a starting point in the table. This is best done by using two arbitrary figures. We might use the next-to-the-last digits in a baseball team's winning

TABLE 7.3 Short Version of a Table of Random Numbers

8	5	3	9	9	7	1	4	5	0	8	5	1	9	0	8	5	9	7	6
7	0	6	1	8	4	9	3	7	8	9	5	9	3	5	2	0	5	2	3
7	8	5	6	1	2	7	4	0	0	2	6	8	9	2	3	1	9	5	4
9	0	1	0	8	9	0	9	8	4	5	1	6	6	7	7	0	0	0	3
2	3	2	8	2	3	5	3	8	2	1	8	3	6	7	7	5	4	5	6
6	5	3	3	5	0	7	5	4	9	4	0	0	3	9	2	1	4	8	2
0	5	5	2	0	8	0	0	5	6	4	2	2	9	7	8	0	2	7	8
2	5	7	0	3	6	7	6	2	8	5	7	7	3	0	6	0	7	4	0
9	2	9	2	9	1	8	8	2	9	6	4	0	4	6	7	9	8	8	9
3	3	3	1	5	8	0	1	4	3	8	3	4	8	5	9	9	8	9	1
4	2	5	8	4	5	3	0	9	2	9	6	6	7	3	3	6	5	1	5
2	6	0	5	4	3	2	4	4	9	4	3	6	4	2	9	9	3	0	8
2	7	0	9	3	0	1	8	9	4	6	5	3	0	3	5	0	3	6	7
1	3	7	7	3	6	9	2	5	4	2	4	1	2	0	0	6	0	4	5
0	2	7	0	4	6	2	1	4	4	8	2	3	9	7	7	6	2	3	3
2	8	7	0	7	0	2	6	9	8	9	1	1	4	5	3	6	5	7	2
8	5	2	7	3	2	9	3	1	0	1	1	6	3	4	0	3	8	7	5
1	5	3	1	1	8	8	2	0	9	2	6	2	9	6	8	9	8	7	0
2	0	6	4	2	1	3	8	0	9	6	1	9	0	1	6	6	6	9	1
1	4	3	2	4	0	2	8	8	6	5	4	6	2	5	7	0	7	8	9

percentage and in the daily average from a stock exchange, both as reported for the previous day. Use one of these digits to pick the row and the other for the column at which to enter the table. Suppose that our numbers were three and five. We enter our table at the third row and fifth column.

Beginning at the entry point we record numerals in the number of digits already determined. Our first number is 127. Thus, case 127 is in our sample. We then proceed through the table in any direction, either by columns or rows. Let's go down the columns working from left to right.

The next number we find is 890. This number is not within our range (001 to 305), so we ignore it. Next comes 235. This is in our range, so case 235 is in the sample. We continue the process until a sample size of 25 is reached. The cases selected for our sample are as follows:

127 235 080 301 188 213 056 143 092 254 144 209 268
183 241 116 262 035 200 016 257 052 195 148 027

The simple random method of sampling uses replacement. That is, once a case number is selected it is not removed from further consideration. To do otherwise would alter the equal probability requirement for complete randomness. In our example, we had 305 cases in the population. Each of these must have a 1 in 305 (Pr = .00328) chance of being included at every selection. If the first case selected was removed from consideration after it was drawn, the second selection would give the other cases a 1 in 304 (Pr = .00329) chance. Similarly, without replacement the chances for inclusion in the sample would inch upward for the remaining cases with each selection. Of course, the larger the population and the smaller the sample, the more minute these differences in probabilities would be. But strictly speaking, selection without replacement is not simple random sampling.

This said, however, sampling without replacement is tolerated in social research. It is

RESEARCH EXAMPLE 7.1 Systematic and Cluster Sampling Methods

From Harold C. Meier and William Orzen, "Student Legitimation of Campus Activism: Some Survey Findings," Social Problems, *19 (2), Fall 1971:182.*

. . . The findings reported here are based on an aggregate of four systematic samples of University of New Mexico undergraduates surveyed during the Spring and Fall Semesters of 1969 and 1970. The samples were drawn by an *N*th interval procedure from the computerized university enrollment records. Thus, for all practical purposes, the samples may be considered simple probability samples of the undergraduate population. . . .

done both because the differences in probabilities are usually insignificant and because replacement is often impractical. How do you convince someone to be in the same experiment twice or to answer a questionnaire three times! Not only would one have to agree to do this, but for many studies they would have to behave as though the redundant experiment or questionnaire was completely new to them. Some analyses require that the observations be independent from one case to another, which is impossible when the same person represents more than one case. For these and similar reasons, you seldom find a social researcher using a simple random sampling method including replacement. None of the research examples in this chapter use this method.

A *systematic random* method of sampling selects cases using a set interval from a population list. This method may select every 20th case from a list of agency clients, every 100th name from a telephone book, or every 5th year from an almanac of production figures. The method is only quasi-random in design but it can produce random-like results.

Once the starting point is set for a population list and a selection interval decided, the remainder of the sample has been determined. Every case thereafter is either certain to be included or excluded. Hence it is not strictly random. Systematic methods will seriously misrepresent a population when there is some meaningful order to the population list. Clients at the top of the list may be different from those below them if the list reflects some agency screening or other procedures. Production figures may follow a time-series cycle retained in a time-oriented report list. Some researchers use the systematic method when they have an existing population list and they do not inspect it for natural order. The consequences can be disastrous.

However, systematic methods can produce representative samples little different from simple random methods when the population list itself is essentially randomized. Taking names at a set interval from a telephone directory will usually do this. One does need to be careful about such things as multiple lines to one household and no lines to a few people, however. A randomized population list can be sampled systematically with reliable probability results and is, therefore, an accepted procedure in social research.

Research Example 7.1 shows the use of a systematic random method to sample college students. The university's enrollment records provided the population list. Then an unspecified interval was used to select students from the fall and spring semesters of two school years. Those four semesters constitute four strata in which the systematic method was used. Thus the example illustrates both systematic and stratified techniques (to be discussed next).

A *stratified random* method of sampling selects cases randomly but separately from subpopulations (strata) of the overall population. It is used when we want to ensure sufficient representation of important subgroups for statistical analysis reasons. We may want to have representatives of all ethnic groups, all religions, selected income levels, certain educational backgrounds, or any specified categories of other crucial variables to a study. These subgroups are the strata from which we sample, not the entire population. A sample size is determined for each of the strata as though it were a separate population.

Sometimes the strata can be identified geographically. For instance, ethnic groupings are often physically segregated. Other times a researcher must screen selected sampling units to determine if cases represent the desired strata before proceeding with data collection. Educational strata might require this technique.

Once the strata are identified they are treated as independent populations and cases are selected randomly. Imagine that we wanted to select a stratified random sample of social agencies in a state on the basis of control source (public versus private) and staff size (under 21, 21–50, and over 50). First we would precisely define "social agency" and then consult a directory to amass our population list. Next we would divide the list into the unique combinations of the stratifying criteria. Our example would have six such strata:

Public:	under 21	Private:	under 21
Public:	21–50	Private:	21–50
Public:	over 50	Private:	over 50

Then we would select the agencies from each strata using one of the random methods.

Stratifying a sampling procedure is only appropriate when the stratification characteristics are related to the variables under study. We stratified by control source and staff size because we thought these described important differences between social agencies and we did not want to risk omitting representative numbers of these strata in our sample. We did not stratify by sex of agency director or type of heating source used in the agency's building because these seemed irrelevant.

Research Example 7.2 illustrates a stratified random sample. The author was studying federal support of social research. The sample was stratified by academic discipline of the questioned respondents. Those strata were anthropology, economics, political science,

RESEARCH EXAMPLE 7.2 Stratified Sampling Method

From Michael Useem, "State Production of Social Knowledge: Patterns in Government Financing of Academic Social Research," American Sociological Review, *41 (4), August 1976:615.*

. . . The primary data of the present study are from a questionnaire survey of 1,079 social scientists in the disciplines of anthropology, economics, political science and psychology. The population consisted of the academic members listed in the most recent directory of the major professional associations representing the four disciplines. . . . A random sample of 500 was drawn from each of the discipline directories, and an initial 12-page form was sent in December, 1973. Four months later, after a second mailing of the questionnaire and a final follow-up letter, 1,079 usable responses had been obtained, for a return rate of 54.0 percent, a fairly common response level for college faculty. . . .

RESEARCH EXAMPLE 7.3 Stratified Sampling Method

From Thomas Carothers and Howard Gardner, "When Children's Drawings Become Art: The Emergence of Aesthetic Production," Developmental Psychology, *15 (5), September 1979:572.*

. . . Subjects were selected at random from an elementary school having a predominantly middle-class population. The group consisted of 11 males and 11 females at each of the grade levels of first, fourth, and sixth. . . .

and psychology. Random samples of 500 were selected separately from the lists of professionals in each of these four disciplines.

Another illustration of stratifying a sampling plan is R.E. 7.3. In this study of elementary school children the strata were the three grade levels of first, fourth, and sixth combined with the sex of the student.

These two research examples are like many in social research. Because the researchers want to compare various subgroups, they use a stratified sampling procedure to ensure the inclusion of relevant strata. Without the stratification, they might end up with a sample that omits some important groups. This would compromise their planned analysis.

A *cluster random* method of sampling is a multistage random selection process for choosing cases from a population. It is often used to sample large and dispersed populations. Most national opinion polls and other large-scale surveys by reputable researchers use the cluster method. In most of these situations it is not possible to construct a list of all population members. Further, the cases are so widely spread that sending interviewers (if that is the technique used) to randomly selected persons would be a huge expenditure of travel time and money. Instead, geographic clusters of individuals are selected.

Suppose that we ambitiously decide to personally interview a sample of adults in a major urbanized area. First, we divide the entire area into smaller areas for which we know the approximate number of residents. Since the Bureau of the Census already has done this for the entire country, we can use their same subareas, called census tracts. These constitute our *primary sampling units* or PSUs. The tracts consist of a number of city blocks

RESEARCH EXAMPLE 7.4 Cluster Sampling Method

From Mokhtar M. Abd-Ella, Eric O. Hoilberg, and Richard D. Warren, "Adoption Behavior in Family Farm Systems: An Iowa Study," Rural Sociology, *46 (1), Spring 1981:47.*

. . . The population of this study consists of all Iowa farms operated by families or family corporations, which had gross sales of agricultural products of at least $2,500 in 1976, and have a married couple living on the farm at the time of data collection. A self-weighting multistage cluster sample was drawn, and both the farm operator and his spouse on each sample farm were interviewed. Data about the farm enterprise as well as personal information about the operator were obtained from the operator. Data about the family, household, and personal information were obtained from the spouse. Although information was collected for 933 farm units, the present analysis will be based on the 844 farm units with both the operator and spouse present. . . .

RESEARCH EXAMPLE 7.5 Cluster and Systematic Sampling Methods

From Nan Lin, John C. Vaughn, and Walter M. Ensel, "Social Resources and Occupational Status Attainment," Social Forces, *59 (4), June 1981:1163–1181.*

. . . The data . . . were collected in the Spring of 1975 with a modified random sample of males, aged 20 to 64 years, residing in the Albany–Schenectady–Troy, New York Area. All respondents were presently or had been members of the United States civilian labor force. Dwelling units were selected by a systematic cluster-sampling plan. Qualified males in three preselected contiguous dwelling units were interviewed; 399 out of a total of 440 cooperated, resulting in a 91 percent completion rate. Similarities between the distribution of frequencies in the U.S. Bureau of the Census occupational categories and those of our sample increased our confidence that the sample was representative of the male labor force in this tri-city area. . . .

that are contiguous. In the first stage of the sampling method we can use a table of random numbers to select some of these tracts.

Next we enumerate the dwelling units in the selected clusters. This is a process of identifying the number and location of each separate housing unit in the selected clusters. We can use whatever definition of a housing unit is appropriate for our study. If we use the Bureau of the Census' definition, it will consist of a house or apartment that has separate cooking and sleeping quarters. These housing units are our *secondary sampling units.* In the second stage of the sampling process we can make a simple random selection of some of these housing units.

In the third stage of the procedure we randomly select adults from each housing unit to interview. The adults are our *tertiary sampling units.* In this example they are also our final or *ultimate sampling units.* Other cluster methods may have a different number of stages. Prior to their selection, the interviewer enumerates the selected housing units to determine the number and ages of the residents. Then a simple random selection from the adults in those housing units can be made.

In our example a simple random selection process was used at each stage of the cluster method. In many cluster selection methods this is compromised at one or more stages. For instance, we might decide to cut further travel costs by grouping the housing units selected. We might randomly pick a starting point in a tract and then instruct the interviewer to choose nearby housing units in a systematic way. As a result the interviewer works in only one section of each tract. Any cluster method is not completely random. Differences in sizes of tracts and housing units will cause some ultimate sampling units to have different probabilities for inclusion than others. At each stage the procedure causes some individuals to have no further chance at all of inclusion in the sample. However, the results of carefully executed cluster sampling plans have been remarkable in their accuracy. Most preelection polls based on cluster samples are very close to the election results, for example.

Cluster random methods tend to produce samples that are less variable than simple random methods. People who live near each other tend to be more alike than those who are separated. Cluster methods have larger sampling errors. Thus a cluster sample will ordinarily be less representative of a population than a simple random sample. But this must be weighed against its cost and practical advantages. Many populations simply could not be sampled at all if the simple random method were the only possibility.

Research Examples 7.4 and 7.5 illustrate the cluster random method. Family farm residences are the main clusters in R.E. 7.4, urban dwellings in R.E. 7.5. Like most published reports using this method, few details are given about the actual procedures followed. Note that R.E. 7.5 combines the cluster technique with the systematic.

Nonrandom Methods

A nonrandom method of sampling selects cases from a population using a method having unknown inclusion probabilities. Cases may or may not have equal selection probabilities; we have no way to know. Nonrandom methods usually produce samples without a basis for evaluating statistical inferences. These methods generally should not be used in social research that is designed to draw inferences. They are sometimes used regardless because of cost constraints or problems in executing a sampling design. But the results can be useless or misleading for making inferences. Thus their cost, although reduced, may produce ambiguous findings. Our discussion should help you recognize the four main types of nonrandom methods and their products.

A *quota nonrandom* method of sampling selects cases nonrandomly from the strata of a population. Quota sampling is the nonrandom counterpart to stratified random sampling. A sample size (the "quota") is determined for each stratum and then cases are selected in any convenient way for all strata. Market research and amateur opinion polls typically reflect this method.

RESEARCH EXAMPLE 7.6 Quota, Cluster, and Systematic Sampling Methods

From Louis A. Penner and Tran Anh, "A Comparison of American and Vietnamese Value Systems," Journal of Social Psychology, *101 (2), 1977:191–192.*

. . . A quota sampling technique was used. The demographic characteristics upon which the quotas were set were gender, occupation (farmer, civil servant, other), religion (Buddhist, Catholic, other), and place of residence (rural, urban). The quota percentages were the following: (a) gender—50% male, 50% female, (b) occupation—36.5% farmer, 36.5% civil servant, 27.0% other, (c) religion—45.0% Buddhist, 27% Catholic, 27% other, and (d) place of residence—60% rural, 40% urban. The rationale behind the setting of quota percentages was not that they match the population percentages on these characteristics. Rather, the goal was to insure that the various sub-groups would have sufficient representation to allow statistical analyses of them. Although the authors desired to obtain 550 interviews, the quota percentages were based on the assumption that only 275 interviews might be conducted. Within the rural area, two villages in each province district were selected by chance. Since all areas of the province were secure at the time of the study, all districts were included in the pool from which the villages were drawn. Within the large city, two of the 10 precincts were selected by chance. . . .

Each interviewer was given a quota to fill and a geographic location in which to fill it. Once in geographic location, an interviewer was to employ the following procedure: A house from a row of houses was to be selected by chance and the first person in that household to fit the characteristics prescribed by the interviewer's quota was given the interview. Thereafter, every third house was to be approached and the same procedure followed until the interviewer's quota was filled. . . .

Research Example 7.6 refers to a quota method of sampling individuals for an interview about values. Allowing the interviewer to select the first person in a household meeting the researcher's criteria precluded any random selection at this point in the procedure. Note that the sampling plan also combined cluster and systematic methods with the quota technique.

When you read research reports of sampling methods you must be careful to distinguish between quota and stratified. Many researchers do not make a distinction in their vocabularies, so that they may say "quota" but mean "stratified." The authors in R.E. 7.6 seem to mean "quota" when they say it.

A *judgment nonrandom* method of sampling (also known as a *purposive* method) selects cases based on a researcher's or interviewer's opinion of their representativeness. An interviewer may be instructed to talk to "typical" shoppers. A researcher may choose "average" students to study. A reporter may do a feature story on an "ordinary" citizen. Pollsters may watch "bellwether" precincts. Seldom can the expert or the untrained observer avoid personal biases in making such choices. The impartial chances of random selection are not allowed to operate. Judgment sampling can be useful when one is exploring a new field prior to making a well-defined study. It is also appropriate for a case study when one purposely wants to examine one social situation with specific characteristics. But a judgment method remains nonrandom for probability purposes.

A *convenience nonrandom* method of sampling selects cases based on their ease of inclusion. A pollster goes to the handiest spot to find potential voters (a shopping mall) or a teacher asks her students to complete a questionnaire. This method is more haphazard than the judgment method. There is no attempt to find the "typical" or "average." It corresponds to much of the public's erroneous understanding of "random." But the difference between the common definition and the statistical definition of random is one of nearly opposites. Convenience samples are the least random of all methods listed here.

A *volunteer nonrandom* method of sampling selects cases based on their willingness to be chosen. This is a very popular sampling technique. Magazines use it to study their readers. Businesses use it to study their customers. Social researchers may use it to study controversial issues. Examples you may have seen include printing a questionnaire in a magazine and publishing the results in a later issue, sending a questionnaire to a politician's constituents and then reporting the findings from the returned forms, and asking for volunteers in a study of sexual attitudes and practices. The historic Kinsey Report used this method as have others.

The results from volunteer samples may be revealing and interesting but they are not based on a random sample. Obviously, the very act of volunteering makes the people who are in a sample different from those who are not. Volunteer samples seldom represent any general population because of the self-selected bias produced by the procedure. But it may be the only way to get any information at all on some topics.

Research Example 7.7 illustrates a volunteer component to a sampling plan in a study of family status. The researcher used volunteers from a set of organizations, which had also volunteered to participate in the research. The sample included some college undergraduates who may have been selected by the volunteer or the convenience method. The focus of the study was on a research technique (the vignette), not on making inferences to a general population. Thus a volunteer sampling plan is not as serious an issue as it might otherwise be.

As the foregoing discussion of sampling methods and examples shows, many studies use a combination of sampling techniques. Stratified-cluster, simple random following a judgment preselection of a setting, and voluntary-quota are common blends. The more

RESEARCH EXAMPLE 7.7 Voluntary Sampling Method

From T. A. Nosanchuk, "The Vignette as an Experimental Approach to the Study of Social Status: An Exploratory Study," Social Science Research, *1 (1), April 1972:107, 112.*

. . . This exploratory study reports a first attempt to measure general family status in an urban setting, and further, to do this in an experimental way. . . . The respondents were mainly members of consenting voluntary organizations in the Boston area, with the questionnaire being filled out either at some point during a meeting, often in groups of approximately 20, or taken home and returned later. The remainder of the respondent set were undergraduates from Tufts University with a variety of academic backgrounds. . . . It was our intention to obtain our sample from outside the university but time-pressure forced some compromise. . . .

complex the sampling method becomes, the harder it is to evaluate. Worse, complete description of the sampling method used is not universal in the published reports of research. Although the emphasis in most research is properly on the substance of the study and not its methods, the two are intertwined. Often, more complete information about the sample selected for study is needed in social research.

Sample Representativeness

We can never be certain that our sample represents the intended population. Nonrandom methods seldom do this, although it can happen. Even random methods do not ensure accurate representation. One check is possible in studies of the general population. We can compare our sample's demographic characteristics against those reported in the official census data. Do we have the same proportions of men and women, whites and nonwhites, young and old, and so on? A standard part of many technical reports of social research is this comparison to help you evaluate the sample's representativeness. We do know the likelihood that random methods will produce representative samples; that is why they are preferred over nonrandom methods.

The quality of random sampling methods can be undermined by problems in executing a sampling plan of any type. Interviewers may make "convenient" substitutions when a case is not located. A significant proportion of persons selected may not respond to a mailed questionnaire and destroy the randomness of the sample that remains. Increasing rates of refusals are an important problem to survey researchers. They are finding the public to be ever more reluctant to cooperate. Some of this is due to the large number of surveys, which saturate the population and abuse public goodwill. But this is also compounded by growing numbers of sales and political groups impersonating legitimate researchers and creating greater distrust by potential respondents. Refusals would be no special problem if they were randomly distributed among the population. But they usually represent people with extreme political attitudes, personal problems, or marginal status.

These kinds of failures and imperfections are the "Achilles' heel" of social research based on samples. Regardless of the elegance of the problem statement and analysis, if the sample was improperly drawn, the entire project may be in question. Watch research reports for response rates. Rates as low as 40% are not uncommon. But rates below about 80% warrant special attention. Beyond the overall response rate, some individual questions may evoke an unusually high number of refusals, which further attack the randomness of the data. An 80% response to an item by a sample having an overall response rate of 80% gives us only a 64% rate on that item. Watch for question-by-question response rates as well as the overall rate.

What Sample Size?

Perhaps the most often asked question by neophyte researchers is: "What sample size should I use?" and then, "Does it depend on the population size?" These are good questions but they have very complex answers. Beginners often feel as though the statisticians they ask are protecting some very secret information when all they seem to get is a complicated reply. Unfortunately, the answer really is complicated. Let us outline the relevant issues.

There are three main factors to be considered (aside from practical ones like cost and time): confidence, precision, and population variation. When you ask a statistician how big a sample you should use, he or she will ask you to specify how sure you want to be of your results (confidence), how close you want to be to the true population values (precision), and how diverse the population is on the characteristics that are important to your study (population variation). *Confidence* is thought of in percentaged probability terms (e.g., 95% confidence) and it is discussed more beginning in the next chapter. *Precision* refers to such statements as being within two or three percentage points on a poll. It is calculated from the data and is difficult to control independent of confidence. We will be looking at it in Chapter 9 and beyond for each kind of statistical inference we cover. *Population variation* is a situation-by-situation phenomena. The irony is that you must know something about the population you want to study *before* you draw your sample to study it. This requires some advanced scouting and comparison with similar studies already completed. Stratified sampling is a method that specifically reflects this kind of prior knowledge. The substance of the research problem contributes heavily to this consideration.

You notice that none of these three factors was the population size. In general, the proportion of the population that is found in the sample is of only negligible impact. Samples of the same size tend to represent populations of different size equally well. Most samples are a very small proportion of their populations and therefore the proportion is trivial.

The size of the sample alone does have an important bearing on our inferences. Larger samples tend to give us more precision and confidence with equally variable populations. Sampling errors do go down as sample size goes up. However, increasing the sample size has diminishing returns. The advantage of a sample size of 1200 over one of 1000 is much less than is one of 300 over one of 100. This is demonstrated with specific statistical tests of inference in Part III later.

As mentioned earlier, the question of sample size is complex. We will consider some of the related issues as we go along. But we will not return directly to the question. A more complete explanation is beyond the scope of this text. There are some specialized sources which show you how to choose a sample size once you understand the problems involved. (See Kish, 1965 and Mueller et al., 1977:407–411.)

Keys to Understanding

There is one major key to understanding probability, regardless of how complex it may appear in some research. *Key 7.1.* Remember that probability can be calculated whenever we know or can assume a sampling distribution. Probability is a proportion. It merely compares the expected frequency of some outcome(s)'s occurrence to the total frequency for all possible outcomes in a situation. Therefore, if you have the sampling distribution, you can calculate probabilities. The sampling distribution can be based on either empirical experience or theory. One major theoretical basis for discrete, dichotomous variables is the binomial distribution. Continuous variables often rely on the standard normal distribution for their theoretical backing.

There are two major keys to understanding sampling. *Key 7.2.* Remember that the probabilities for inferences can be found only for random sampling designs. Case inclusion probabilities are known with random methods, but unknown with nonrandom methods. A simple random method of sampling gives every case in a population an equal chance of being selected. This method can seldom be used perfectly in most social research situations. But it is approximated by such techniques as systematic, stratified, and cluster methods. Nonprobability (nonrandom) methods, such as judgment, quota, convenience, and voluntary, do not produce samples with known probabilities. Hence they ordinarily should not be used for inferential statistics.

Key 7.3. Remember that all sampling methods, even simple random, are subject to sampling error. Samples are drawn in hopes of representing their parent populations. But all sampling plans are open to chance variations which threaten their representativeness (i.e., sampling error). However, the probabilities of sampling error can be determined for random samples.

Keep these few points in mind as you consume social research. They will focus your attention on some of the key ideas that provide the basis for inferential statistics.

KEY TERMS

Sample
Population
Statistic
Parameter
Inference
Probability
- Theoretical
- Empirical
- Simple
- Point
- Cumulative

Law of Large Numbers
Addition Rule
Multiplication Rule
Sampling Distribution
Binomial Sampling Distribution
Success (p)
Failure (q)
Binomial Probability of Success
Normal Distribution
Standard Normal Distribution
z Standard Score
Census
Sampling Error
Probability Sampling
Nonprobability Sampling
Simple Random
Systematic Random
Stratified Random
Cluster Random
Primary Sampling Units
Secondary Sampling Units
Ultimate Sampling Units
Quota Nonrandom
Judgment Nonrandom
Convenience Nonrandom
Volunteer Nonrandom

EXERCISES

1.

Household Size	f
1	11
2	25
3	38
4	48
5	43
Total	165

Determine the probability of randomly selecting the following (with replacement): (a) a household of size 2 in one draw; (b) a household of size 4 in one draw; (c) a household of size 2 or 4 in one draw; (d) the series of a household of size 3 in the first draw and size 1 in the second draw; (e) the series of either a household of size 1 or 2 in the first draw and either 3 or 4 in the second draw; (f) a household of size 3 or more in one draw.

2.

Type of Industry	f
Manufacturing	12
Mining	4
Construction	6
Transportation	8
Total	30

Determine the probability of randomly selecting (with replacement) the following: (a) a mining industry in one draw; (b) one manufacturing industry in one draw; (c) either one mining or one manufacturing industry in one draw; (d) the series of a construction industry in the first draw followed by a transportation industry in a second draw; (e) the series of either one mining or manufacturing industry in the first draw, either one construction or transportation industry in the second draw, and either one mining or construction industry in the third draw.

3. Assuming that the data represent a sample of observations, (a) are the problems in Exercise 1 based on theoretical or empirical probability? (b) Are the problems in Exercise 2 based on theoretical or empirical probability?
4. Find the probabilities for the following considering a regular six-sided die: (a) one toss yielding a 2; (b) one toss yielding either a 2 or 3; (c) one toss yielding a 4, 5, or 6; (d) the series of the first toss yielding a 2 and the second toss yielding a 4; (e) the series of one toss yielding either a 1 or 2, a second toss yielding either a 3 or 4, and a third toss yielding either a 5 or 6; (f) a three-toss series yielding the outcomes 1, 2, and 3, in that order; (g) a three-toss series yielding all 3's.
5. Suppose that we put 10 red balls, 5 blue balls, and 15 green balls in a jar and mix them thoroughly. Find the following probabilities for random selection (with replacement): (a) one red ball in one draw; (b) one blue ball in one draw; (c) either one red or one blue ball in one draw; (d) either one blue or one green ball in one draw; (e) the series of one red ball in the first draw and one blue ball in the second draw; (f) the series of three draws each producing a red ball; (g) the series of either one red or blue ball in the first draw, either one blue or green ball in the second draw, and one red or green ball in the third draw; (h) the series of five draws each producing a blue ball.
6. (a) Are the problems in Exercise 4 based on theoretical or empirical probability? (b) What about those in Exercise 5?

7. Suppose that we make random selections with replacement from an ordinary deck of 52 playing cards having four suits of 13 cards each. Determine the probability for each of the following: (a) getting the ace of spades in one draw; (b) getting any king in one draw; (c) getting any club in one draw; (d) getting any ace or king in one draw; (e) getting the series of four aces in four draws; (f) getting the series of a 2 or 3 in the first draw, a 7 or 8 in the second draw, and jack or queen in the third draw.
8. Imagine that we flip a *dishonest* coin which has been weighted so that heads have a probability of .6 and tails of .4. Determine the probabilities for the following: (a) obtaining a head in one toss; (b) obtaining a tail in one toss; (c) obtaining either a head or tail in one toss; (d) obtaining two heads in a row from two tosses; (e) obtaining three tails in a row from three tosses.
9. Imagine that we flip the same dishonest coin from Exercise 8 three times. (a) List all the head–tail sequences for the three tosses. (b) How many sequences are there for getting 3 heads. (c) For 2 heads and 1 tail? (d) For 1 head and 2 tails? (e) For 3 tails? (f) How many sequences in total are there? (g) What is the probability of getting 3 heads? (h) What is the probability of getting 2 heads and 1 tail? (i) What is the probability of getting 1 head and 2 tails? (j) Compare and explain your answers to parts (h) and (i).
10. Suppose that we flip an honest coin [Pr(heads) = .5] five times. Use the probability of success formula to find the following probabilities: (a) 5 heads in a row; (b) 3 heads and 2 tails; (c) 1 head and 4 tails; (d) 4 or more heads.
11. Given a normal distribution of children's ages in a preschool educational program where the mean is 4 years and the standard deviation is 1 year, determine the following: (a) the z standard score for an age of 5; (b) the percentage of ages between 4 and 5; (c) the probability of randomly selecting an age between 4 and 5; (d) the percentage of ages 5 and over; (e) the probability of randomly selecting an age 5 and over; (f) the z standard score for an age of 2; (g) the percentage of ages between 2 and 4; (h) the probability of randomly selecting an age between 2 and 4; (i) the percentage of ages 2 and under; (j) the probability of randomly selecting an age 2 and under; (k) the probability of randomly selecting (with replacement) two ages between 2 and 4 in two draws.
12. Suppose that a researcher has a normal distribution of voter turnout over 10 years of electións in a precinct with a mean of 250 voters and a standard deviation of 50. Find the following: (a) the z standard score for an election in which 250 people voted; (b) the z standard score for a voter turnout of 325; (c) the z standard score for a voter turnout of 110; (d) the number of voters equivalent to a z standard score 2.2; (e) the number of voters equivalent to a z standard score of -1.6; (f) the probability of randomly selecting a turnout between 250 and 305; (g) the probability of randomly selecting a turnout between 250 and 215; (h) the probability of randomly selecting the series of two turnouts of 150 or under.
13. A researcher reports a normal distribution of days hospitalized on a psychiatric ward has a mean of 12 and standard deviation of 2.2. Find the following: (a) the z standard score for a hospitalization of 11 days; (b) the percentage of hospitalizations between 12 and 11 days; (c) the probability of randomly selecting a hospitalization score between 12 and 11 days; (d) the z standard score for a hospitalization of 18 days; (e) the percentage of scores 18 days and longer; (f) the z standard score for a hospitalization of 8 days; (g) the percentage of scores 8 days and shorter; (h) the probability of randomly selecting either a hospitalization of 18 days and longer or 8 days and shorter.
14. A distribution of number of "serious dating partners" before marriage is reported to have a mean of 5 and standard deviation of 1.5. The researcher reports that approximately 80% of the study's respondents had between 6.9 and 3.1 partners. Are these findings consistent with a claim that the distribution was normal? Explain.
15. A distribution of number of criminal convictions is reported to have a mean of 8.1 and a standard deviation of 2.3. The researcher reports that approximately 72% of the study's subjects had between 2 and 14 convictions. Are these findings consistent with a claim that the distribution was normal? Explain.
16. Determine whether the following illustrate random or nonrandom techniques of sampling: (a)

instructing an interviewer to talk with any three adults and three children waiting in line for a movie; (b) telling a research assistant to collect genealogies from the first five natives who agree to be informants; (c) asking readers who have never before been in a survey to return a mailed questionnaire; (d) using randomized numbers to select 10 case histories from a computerized list of clients.

17. Name the specific method of sampling used in each of the following: (a) Every fifth resident of a retirement home is interviewed. (b) Random members of farm households are interviewed in randomly chosen households from randomly selected 5-square-mile segments of a rural township. (c) A television station watches for early poll returns from community A because of its record in predicting statewide election results. (d) A student is instructed to gather information on the sleeping habits of three close friends.

RELATED STATISTICS

Bernoulli Trials

A binomial distribution can be established by empirical methods as well as the theoretical ones illustrated in the chapter. Rather than relying on the theory of what should happen when we toss a coin, we can actually conduct a number of trials and record the results. These empirical tests of binomial situations are called Bernoulli trials after James Bernoulli, a mathematician from Switzerland. Bernoulli devised the basis for these trials in the later part of the seventeenth century.

See: Ott et al. (1978:149).

Poisson Distribution

In binomial situations where the probability of outcomes is extremely small and the sample size is quite large, the results do not conform to the theoretical binomial distribution. Instead, they follow the Poisson distribution. The probabilities can be found by formula or table. Studies of very rare phenomenon (e.g., suicide) require this distribution.

See: Neter et al. (1978:139–142) and Blalock (1980:172–173).

Hypergeometric Distribution

The distributions in this chapter assumed that sampling occurred with replacement. Sometimes this is not the case. When sampling is done without replacement and the population is relatively small, probabilities of outcomes can be found with the hypergeometric distribution.

See: Neter et al. (1978:220–222) and Blalock (1980:171–172).

CHAPTER 8

INTRODUCTION TO HYPOTHESIS TESTING

In Chapter 7 we discussed some of the basic ideas about probability and sampling. Three points were emphasized. One, probability can be calculated whenever we know the distribution of a situation's possible sample outcomes, its sampling distribution. Two, probabilities can be found for randomly drawn samples. Three, even random samples are subject to sampling error. In this chapter we want to show how these ideas help researchers to make decisions about competing hypotheses of the proper inference that should be drawn from sample results to a population.

The type of hypothesis testing discussed in this text is known as the Neyman-Pearson method. Jerzy Neyman and E. S. Pearson developed its logic and techniques in the early 1900s. Today, it is the predominant approach to inferences in social research. However, not all statistical inference is based on the Neyman-Pearson method. One alternative is a set of parameter estimating methods that do not require a priori hypotheses. These methods are introduced later in this chapter and demonstrated in Chapter 9. A second alternative is Bayesian statistics, which use information from sources beyond the data being analyzed, unlike the Neyman-Pearson approach. (See Hays, 1973:809–860.) Mostly, our attention will be given to the Neyman-Pearson method of hypothesis testing.

The Logic of Hypothesis Testing

Hypotheses

Consider the three following examples of research situations in which inferences are to be made. First, a politician claims that .80 (80%) of his constituents support his plan to alter taxes. An independent poll is conducted which finds that the proportion of adults in a random sample from the district in support of the plan is only .40 (40%). We do not know how the politician got his figures. We do know how the independent poll got its figures. There are two possible explanations for the difference between the politician's claim and the poll's results.

1. Sampling error caused the independent poll to misrepresent the district's sentiment and the true magnitude of support for the plan really is .80.
2. The independent poll is a fair representation of the district's sentiments and the true magnitude of support for the politician's plan is not .80.

Second, a researcher is interested in studying individual feelings of powerlessness using a standardized attitude scale. The mean powerlessness score for a sample of urban residents is 15, whereas the mean for a sample of rural residents is 10. There are two possible meanings for the findings.

1. Sampling error caused each of the two samples to misrepresent its intended population (urban or rural residents) and those populations really do not differ in mean powerlessness.
2. Each of the two samples is a fair representation of its intended population and these populations really do differ in mean powerlessness.

Third, a local citizen pens a Letter to the Editor in which he argues that declining reading ability is responsible for increased juvenile delinquency. A stratified random sample of delinquent and nondelinquent juveniles are tested for reading ability and questioned about their acts of delinquency over the past two years. The study finds that there is a positive correlation (r) of .60 between reading ability scores and number of delinquent acts committed. The findings can have two possible interpretations.

1. Sampling error caused the sample to misrepresent the community's juveniles and there truly is not an association between reading ability and acts of delinquency.
2. The sample is a fair representation of the community's juveniles and there truly is an association between reading ability and acts of delinquency.

These three examples illustrate how social scientists think about research findings when drawing inferences. For each finding there are two opposite explanations. These are called the null and alternative statistical hypotheses. The *null hypothesis* (H_0) is a possible description of a population that we wish to test for likely accuracy using sample data. It proposes a value for a population parameter and attributes any difference between that proposed population parameter and the sample result to sampling error. The *alternative hypothesis* (H_1) is a possible description of a population that is opposite to the null hypothesis. It attributes any difference between the population parameter proposed by the null hypothesis and the sample result to a faulty null hypothesis, not to sampling error.

In each of our three examples the first-listed explanation above corresponds to the null hypothesis and the second explanation is the alternative hypothesis. Each of these can be rewritten in the formal style of social research.

For the politician's tax plan:

H_0: $P = .80$ The proportion of constituents in the population who support the politician's tax plan is .80. A random sample may differ from this because of sampling error.

H_1: $P \neq .80$ The proportion of constituents in the population who support the politician's tax plan is not .80. Sampling error would not be the cause of a random sample differing from this.

For the powerlessness study:

H_0: $\mu_1 = \mu_2$ The populations of urban and rural residents have the same mean powerlessness scores. Two random samples may differ from this because of sampling error.

H_1: $\mu_1 \neq \mu_2$ The populations of urban and rural residents do not have the same mean powerlessness scores. Sampling error would not be the cause of random samples differing from this.

For the citizen's idea about delinquency:

H_0: $\rho = 0$ The population of juveniles is one in which there is no correlation between reading ability and acts of delinquency. A random sample may differ from this because of sampling error.

H_1: $\rho \neq 0$ The population of juveniles is one in which there is a correlation between reading ability and acts of delinquency. Sampling error would not be the cause of a random sample differing from this.

Note that the hypotheses are focused mainly on the characteristics of populations, not samples. This is why Greek rather than Roman symbols are used to represent the hypotheses, at least when a common distinction is made. The symbol μ refers to a population's mean and $\bar{X}$ is the sample's mean. The symbol ρ refers to a population's Pearson correlation coefficient, compared to r for a sample. The symbol P is ordinarily used for the population proportion and p for the sample proportion, an exception to the pattern.

The hypotheses are focused on population parameters because the goal of hypothesis testing is to draw an inference about a population. There is no need to speculate about the sample's characteristics; we already know them from the data collected. Descriptive statistics are sufficient for the sample. We do not know about the population. Thus we need inferential statistics.

You may be curious about the meaning of the names for the hypotheses. There are two meanings of "null." First, the null hypothesis suggests that a sample result will not differ from a proposed population parameter. Here null refers to this idea of no difference. Second, null comes from the word nullify. The null hypothesis is a possible explanation that the researcher tests by determining whether it can be shown to be improbable, that is, whether it can be nullified. Typically, but not always, it is a statement of no difference between groups or no association between variables. This happens because most research is guided by theoretical ideas suggesting that groups do differ or that there is some association between characteristics of interest. Hence the alternative hypothesis commonly reflects the researcher's expectation (insofar as the theory is concerned). By default, the null hypothesis states the opposite possibility.

Decisions About Hypotheses

We have noted that the researcher tests the null, not the alternative, hypothesis. Keep in mind that the null hypothesis is one in which the researcher often has no faith. Regardless, the researcher does this rather than try to verify the alternative hypothesis. Why? The reason is somewhat involved, but we can sketch its basic logic.

Most social science theory suggests rather inexact hypotheses about the empirical world. Political opinion theory may be sufficient for a researcher to predict a broad range of values in which public sentiment will fall for something like a tax proposal. But it will not usually be able to specify an exact value such as .80. Social psychology may be developed enough to warrant the prediction that urban versus rural place of residence is an important factor affecting feelings of powerlessness, but not so good as to predict the exact direction or magnitude of the difference on an attitude scale. Theory of juvenile delinquency may be sophisticated enough to suggest that acts of delinquency and reading ability are related in some way. But it may not be able to say exactly how. As a consequence of this state of social theory, alternative hypotheses must be somewhat imprecise. Their logical opposites, null hypotheses, therefore reflect exact possibilities.

Further, we usually must advance our understanding of phenomena by discarding those possible explanations of it that can be shown to be unlikely to be true. This is due to the nature of "proof." Ordinarily, it is not possible to prove or disprove any idea with certainty. There are too many circumstances and outside influences to make it practical to set upon such a course. It is too big a task. We can only determine what is probable. We do this most efficiently by searching for contradictory evidence rather than for confirmatory evidence. The more of those explanations that we can show to be improbable, the more support we are amassing for their alternatives. This suits the social science field well since our null hypotheses tend to be very specific explanations which can be more easily tested than the broad statements of our alternative hypotheses.

You can see this with our three earlier examples. Rather than try to show that the politician's tax plan is at an inexact level of support, we try to show that it is unlikely to be .80. Rather than try to prove that urban and rural residents have some unspecified mean powerlessness scores that differ, we seek to find evidence that they do not have the same exact means. We do not put our effort into showing definitively that reading ability and acts of delinquency have a certain correlation range. Instead, we see if it is really probable that the two variables have no correlation at all.

A final consideration lies with our knowledge of sampling errors. We noted from our three examples above that sampling error was the critical difference in interpreting the meaning of study results. The null hypothesis stated that our sample result may differ from the proposed population value because of sampling error. The alternative hypothesis said that sampling error would not be the reason for such a difference. We can only decide between these by assessing the probability that sampling error influenced our sample results. Our strategy is to assume that the null hypothesis is true, find the probability that sampling error is the cause of our sample's deviation from the proposed population value, and thereby decide the likelihood that the null hypothesis should continue to be assumed to be true.

If our null hypothesis is correct, we expect our analysis to show that there is a rather high probability of sampling error being the reason for getting a sample result that differs from the population value proposed by the null hypothesis. If this happens, we draw the inference that the null hypothesis should not be discarded. This leaves the alternative hypothesis in greater doubt as a result of the study. If our null hypothesis is incorrect, we expect our analysis to show that there is a rather low likelihood that sampling error was causing the difference between the sample result and proposed population value. If this happens, we draw the inference that the null hypothesis is wrong. That leaves the alternative hypothesis as an idea which has gained credibility because of the study.

What does this mean for our examples? Take the first null hypothesis, that $P = .80$. We begin by assuming that this hypothesis is true. Then we evaluate the actual sample result and compare it to that hypothesis. Suppose that the sample found that .40 of the people support the tax plan. Now we determine the probability that this discrepancy could be due to sampling error. We ask ourselves: How probable is it that we have drawn a random sample from a population in which .80 support the tax plan but we find only .40 supporting it in the sample? If we find that there is a high probability that sampling error could lead to a difference such as that which we found, we conclude that this is what may have happened. Thus we retain H_0 as a viable explanation. If we find that there is a low probability that sampling error could lead to a difference such as that which we found, we conclude that sampling error is not the reason for the difference. Since the probability was calculated on the assumption that the null hypothesis was true, we now decide that this assumption is suspect. We conclude that H_0 should be discarded.

In the second example, the null hypothesis said that $\mu_1 = \mu_2$. We first assume that this is true. We now compare the sample results (suppose that $\bar{X}_1 = 15$ and $\bar{X}_2 = 10$) to that hypothesis. There is an apparent discrepancy between them because the sample means were not equal. But could they differ by this much (5 points) because of sampling error? We evaluate this difference (5 points versus the null hypothesis prediction of 0 difference) by determining the probability that two samples randomly drawn from populations with identical means could produce what we observed. If that probability is rather high, we decide that H_0 is a reasonable explanation of what happened and we retain it. If that probability is rather low, we decide that H_0 is not a very good explanation for what happened and we reject it.

Our logic is the same with the third example. The null hypothesis was that $\rho = 0$. For purposes of our analysis we initially assume that this is true. Now we look at the actual sample result, e.g., $r = .60$. Then we find the probability that we may have drawn a random sample in which $r = .60$ while the population had no correlation at all. If that probability is rather high, we must admit that the null hypothesis provides a likely explanation for the discrepancy. Thus H_0 should be kept. However, if the probability is rather low, we decide that the null hypothesis is an unlikely explanation for what we saw. In this case, we conclude that H_0 should not be kept.

In sum, this means that there are two possibilities when we make a decision about a null hypothesis. Either we *reject* it or we *fail to reject* it. There is a preference to state the second possibility as "fail to reject" rather than "accept" to avoid the permanency attached to the latter term. We have conducted only one study at a time and, given the weight of theory often behind an alternative hypothesis compared to a null hypothesis, we are in no rush to make the definitive conclusion to accept the null hypothesis. The very next study may find the contradictory evidence we may have expected. The language reflects our desire to be cautious and to hedge our conclusion when we reach such a result.

FIGURE 8.1 Making a Decision About a Null Hypothesis

If the probability for a sample result	$\leq$	level of significance,	then reject H_0
If the probability for a sample result	$>$	level of significance,	then fail to reject H_0

Level of Significance

Whether we decide to reject or fail to reject a null hypothesis is determined by a value known as the *level of significance* or *alpha level* (α). The level of significance is a probability value that is used to define that set of sample outcomes which are so unusual (from the point of view of the null hypothesis) as to warrant rejection of the null hypothesis, if any occur. Before the probability has been found for a sample outcome, we must decide what an "unusual" level would be. A value such as .8 or .9 seems obviously to identify a usual event, so we would naturally fail to reject H_0 if it occurred. A value such as .0001 or .00004 seems obviously unusual, so it would lead us to reject H_0. But where do we draw the line? The level of significance provides us with a firm, although arbitrary, value against which to make the judgment.

There are only three alpha levels in common usage in social science; .10, .05, and .01. The .10 level is restricted to use in very exploratory studies which have a high degree of uncertainty surrounding their theory and methods. The .01 level is ordinarily used in the more sophisticated studies where little uncertainty prevails. The vast majority of studies fall in between and thus use the .05 level.*

Using the .05 level of significance means that we only reject null hypotheses when we get sample results whose sampling error probabilities are as low or lower than .05. Otherwise, we fail to reject the null hypothesis. This gives the null hypothesis a real opportunity to be kept, even though we may not have much faith in it. If we are successful in rejecting it anyway, we can have considerable confidence in that decision. This lends greater credibility to our alternative hypothesis as a consequence. The general rule for making a decision about a null hypothesis is shown in Figure 8.1.

Suppose that we are evaluating the null hypothesis that a coin to be used in a gambling house is honest, with Pr(head) = Pr(tail) = .5. We suspect that the coin's owner is using a dishonest coin to win our money. To evaluate the hypothesis we ask that we be allowed to sample its performance by having it flipped five times in a row so that we can determine the probability for whatever occurs.

Imagine that the first flip yields a head. For analysis purposes we assume that the coin is honest (i.e., assume H_0 is true). Thus the probability of getting a head on the first flip is .5. This does not cause us to doubt the honesty hypothesis. Imagine that the second flip is also a head. Getting two heads in two trials has a probability of (.5)(.5) = .25. We still do not doubt the coin's honesty because such events happen one-fourth of the time by chance alone. Imagine that the third flip is another head: Pr = (.25)(.5) = .125. This is still an ordinary probability, but we are beginning to wonder about the honesty claim.

Imagine that the fourth flip is yet another head: Pr = (.125)(.5) = .0625. Now we may be getting nervous. Four heads in four trials from an honest coin should happen only

*For a criticism of the use of the .05 level of significance, see Skipper et al. (1967).

a little more than six times in a hundred, but that is exactly what happened. Is this one of the six-in-a-hundred events, quite rare but possible? Or is the coin not really honest? The .05 level of significance tells us to keep faith in the honesty hypothesis because .0625 is not sufficiently rare to call the coin dishonest.

The fifth flip will tell the story. It is a head! The probability of getting five heads in five trials from an honest coin is (.0625)(.5) = .03125. We have had enough. Things that happen with only about a three-in-a-hundred chance are too rare to be believed by our .05 alpha criterion. We decide to reject the null hypothesis. It is true that what we witnessed could legitimately have been one of the three in a hundred possibilities. Sampling error could have caused us to see a series of five unrepresentative trials from an honest coin. But we doubt it. We decide instead that the assumption under which the probabilities for these sample results were calculated (an honest coin) is wrong. We will keep our money and not risk any of it on the possibility that the coin really is honest. We decide that there is a "significant difference" between our sample result and the honesty proposal contained in the null hypothesis.

The same procedure is followed in all hypothesis testing. This can be seen again with any of our earlier examples. Take the first one; H_0: $P = .80$. Let us use the .01 alpha level, for illustration purposes. We find the probability of randomly drawing a sample of constituents in which .40 support the tax plan when the population is assumed to have .80 supporting it. If that probability is .01 or less, we would reject H_0. If it is greater than .01, even .02 or .03, we would fail to reject H_0.

These examples illustrate the general process for testing hypotheses. We set up our null and alternative hypotheses. We choose a level of significance. We draw a random sample. We evaluate the sample's result by calculating its probability of occurrence assuming the null hypothesis is true. We compare that probability to the level of significance. We then make a decision about the null hypothesis. If we reject the null hypothesis, we draw the inference implied in our alternative hypothesis to the population. If we fail to reject the null hypothesis, we draw the inference contained in the null hypothesis to our population.

Errors in Decisions

Regardless of the decision we make about a null hypothesis, we must constantly remind ourselves that it can be wrong. This is shown in the fourfold table of Figure 8.2. There you see that each decision has two possibilities when compared to the truth of a null hypothesis. If we decide to reject the null hypothesis (as we often hope to do), we may be correct because it is, in fact, false. Or we may be mistaken because that hypothesis is

FIGURE 8.2 Possibilities When Making a Decision About a Null Hypothesis

Decision About H_0	**H_0 Is Actually:** False	True
Reject	Correct decision	Type I error (α)
Fail to Reject	Type II error (β)	Correct decision

actually a true statement. This is called a *Type I* or *alpha* (α) *error.* If we fail to reject the null hypothesis, we may be correct because it is a true statement. Or we may be mistaken because that hypothesis was false and deserved to be rejected. This is a *Type II* or *beta* (β) *error.*

Let's consider a new example. A social agency believes that it should try an experimental program to help child abusers. There is some theory and prior research suggesting that the program would succeed. An alternative hypothesis is established which states that the program will be effective. The null hypothesis becomes the idea that the program will not be effective. (See Figure 8.3.) Suppose that the program is tried and a study is conducted to evaluate it. If we decide that the study findings warrant the decision that the null hypothesis should be rejected, there are two possibilities. First, we may be correct; the program really works and the null hypothesis of ineffectiveness should be rejected. Second, we may be mistaken; the program really is ineffective, but sampling error caused us to get study results that suggested it was effective instead. We made a Type I error; we rejected a true null hypothesis. If we decide that the findings mean that we should fail to reject the null hypothesis, there are also two possibilities. First, we may be correct; the program really was not effective and the null hypothesis of ineffectiveness should not have been rejected. Second, we may be mistaken; the program really was effective, but sampling error caused us to get study results which suggested that it was ineffective instead. We made a Type II error; we failed to reject a false null hypothesis.

Can you apply this to our tax plan example? The null hypothesis is that $P = .80$. If we decide that P is not .80, but it truly is, we make a Type I error. If we decide that P is .80, but it truly is not, we make a Type II error. (See Figure 8.4.)

You may think that making a Type I or Type II error is equally as grevious. Often it is not. The consequences in the real world for one kind of mistake can be quite different from the other. Our child abuse program example shows this. If we make a Type I error, we will have decided that an ineffective program is effective. That means that we will continue to use it and may be wasting our money and effort. If we make a Type II error, we will have decided that an effective program is ineffective. Then we would be ending a useful program and start searching for a needless replacement. We would need to weigh the potential harm that could result from deceiving ourselves and clients about a failing program against the harm of withholding a successful program that would prevent some future child abuse. Which would cause less harm?

FIGURE 8.3 Possibilities for H_0: Child Abuse Program Is Ineffective

Decision About H_0	**Program Is Actually:** Effective	Ineffective
Reject (Program Is Not Ineffective)	Correct decision	Type I error
Fail to Reject (Program Is Ineffective)	Type II error	Correct decision

FIGURE 8.4 Possibilities for H_0: $P = .80$

Decision About H_0	Proportion Is Actually: Not .80	.80
Reject ($P \neq .80$)	Correct decision	Type I error
Fail to Reject ($P = .80$)	Type II error	Correct decision

In our tax plan example, a Type I error may cause us to decide that the politician was engaged in self-serving propaganda. This might encourage others to defeat a plan that is supported by a high proportion of the public. A Type II error may cause us to decide that the politician's plan suits the public when it really does not. As a result some may work for a plan that the public truly does not support so highly.

These are the same kind of dilemmas faced by physicians when they consider the risks of using or not using a particular treatment. They must ask themselves whether trying something that may not work is worse than not using something that may succeed. The common use of X-rays is such an example. (See Figure 8.5.)

Suppose that our null hypothesis is that the X-ray will not be useful. A Type I error is deciding the X-ray will be useful when it will not be. We needlessly risk tissue damage for a useless picture. A Type II error is deciding that the X-ray will not be useful when it would have been. We do not damage any tissue, but we sacrifice a useful picture that might have helped the patient. Physicians typically decide the possible knowledge gained by an X-ray is more important than the possible tissue damage caused by the X-ray itself. Thus they see a Type II error as the worse possibility.

Another example is provided by the US justice system. Is it worse to decide that a guilty person is innocent or that an innocent person is guilty? We have taken the position that it is better to let some of the guilty go, than to imprison the innocent. (See Figure 8.6.) Here a Type I error is the greater mistake, assuming the null hypothesis is that the person is not guilty. There are many other important real-life examples as well. Type I and II errors are not wholly theoretical issues.

FIGURE 8.5 Possibilities for H_0: X-Ray Will Not Be Useful

Decision About H_0	X-Ray Is Actually: Useful	Useless
Reject (X-Ray Is Not Useless)	Correct decision	Type I error
Fail to Reject (X-Ray Is Not Useful)	Type II error	Correct decision

FIGURE 8.6 Possibilities for H_0: Person Is Not Guilty

Decision About H_0	Person Is Actually: Guilty	Not Guilty
Reject (Guilty)	Correct decision	Type I error
Fail to Reject (Not Guilty)	Type II error	Correct decision

Controlling Decision Errors

Fortunately, we have some control over the chances that we will make decision errors in our hypothesis testing. This is especially true for Type I errors. A researcher can determine the maximum level of risk he or she is willing to take of making a Type I error before making a decision. This is set by the same level of significance which we use to make that decision about a null hypothesis.

Any alpha level establishes the maximum probability that a researcher will make a Type I error if the null hypothesis is rejected. Hence the .05 level means that the researcher is willing to test a null hypothesis knowing that if he or she later decides to reject that hypothesis, the probability that this decision is wrong will be no more than .05. Similarly, the .01 level means that the rejection of a null hypothesis will have a probability of being wrong of no more than .01.

Why take any risk of making a Type I error? Why not set alpha at the lowest acceptable level? The researcher must take some risk of committing this error or there is no possibility of ever rejecting any null hypothesis. In fact, the *lower* the risk is set, the *harder* it is to make the rejection decision. Remember that the level of significance is the criterion against which a sample result's probability is compared to decide whether the null hypothesis is to be believed. It is only when that sample probability is equal to or lower than alpha that we can reject the null hypothesis. The lower we set alpha, the more difficult it is to get a sample result's probability that matches it or is below it. The .05 level means we can only reject a null hypothesis when the probability of the sample result falls to .05 or less. If we lower the alpha level to .01, we must get a sample result probability of .01 or less in order to reject the same null hypothesis. Often the researcher's goal is to nullify (reject) the null hypothesis because it is believed to be a false statement. This means that we would like to make the rejection decision whenever it is possible. So you see that the desire to make the rejection decision works in the opposite way of the desire to avoid Type I errors in setting an alpha level.

Whether or not the researcher believes in the null hypothesis, he or she must overcome any inclination to adjust the alpha level in a way that makes it easier to reach a desired conclusion. An honest researcher is interested in conducting a fair test of H_0, not in coaxing a particular decision about it. The conventional use of an arbitrary alpha level such as .05 helps to lessen the temptation to manipulate the inference warranted by the research.

The possibility of a Type II (β) error also enters the picture. Unlike Type I errors, we have no simple mechanism such as the alpha level to control the risk of making a Type II error. We cannot arbitrarily set a beta level. It depends on the value of the population parameter being studied and the sample size. Since that parameter is usually unknown, we cannot predetermine a beta level.

The alpha level does have some indirect influence, however. The lower the alpha level becomes, the more likely you are to be unable to reject a null hypothesis, even if it is actually false. That means that a very low alpha level may cause you to fail to reject the null hypothesis when you should have rejected it—a Type II error. Thus as the alpha level is increased, it becomes easier to reject a null hypothesis, less likely that you will make a Type II error, but more likely that you will make a Type I error. On the other hand, as the alpha level is decreased, it becomes harder to reject a null hypothesis, more likely to make a Type II error, but less likely to make a Type I error.

All of this can be illustrated with a ridiculous example. Suppose that you are employed by an eccentric professor. You are assigned to work for him 20 hours a week. However, the professor offers you a proposition each Monday morning. You are given the opportunity to guess which of two classes taught by the professor is doing better. The professor will let you draw a sample of five scores from one of the grade books each time to help you guess. He will tell you the mean score for the previous week in each of the classes, but not tell you which class has each mean. If you guess correctly which class is doing better, you will get paid without having to work the entire week. If you guess that class A is doing better, but it is not, you must spend all week counting paper clips. If you guess that class B is doing better, but it is not, you must spend all week cleaning erasers. This is diagrammed in Figure 8.7.

So each week you draw a sample of five scores, compare its mean to each of the class means, and find the probability that such a sample result could have come from each class. Suppose you let H_0 be that Class A is doing better. If you set alpha extremely high, you increase the chance of deciding that class B is doing better than class A (rejecting H_0). This means that you will reduce your chance of having to count paper clips all week (Type II error), but you increase your chance of having to clean erasers (Type I error). If you set alpha extremely low, you increase your chance of deciding that class A is doing better than class B (fail to reject H_0). This means that you will reduce your chance of having to clean erasers all week (Type I error), but increase your chance of counting paper clips (Type II error). There is no way both to increase the chance of getting off work and to avoid a penalty of doing a meaningless task. This is what we mean about being unable to set alpha at such a point where you can maximize the possibility to reject a null hypothesis and avoid both Type I and II errors. You can only find a compromise between these.

FIGURE 8.7 Possibilities for H_0: Class A Is Doing Better

	Class A Is Actually:	
Decision About H_0	Better	Worse
Reject (Class B Is Better)	Correct decision (no work)	Type I error (clean erasers)
Fail to Reject (Class A Is Better)	Type II error (count paper clips)	Correct decision (no work)

One of the related major concerns of researchers is with a concept known as the *power of a test.* This is the probability that a test will cause the researcher to reject a null hypothesis that is actually false. Of course, a researcher hopes to reject false null hypotheses and hence wants a procedure that maximizes the chances of this happening. The probability of power is equal to $1 - \beta$, one minus the probability of a beta error. But as noted, beta error and, therefore, power cannot be set a priori at some maximum level as it was possible to do for alpha error. Beta error and power are determined by a combination of factors which are only partially controlled by the researcher. Among these factors are the actual value of the population parameter being studied, the standard deviation around that population parameter, the level of significance chosen (α), the sample size, and the use of directional versus nondirectional hypotheses (discussed below). As you may suspect, this is a rather complex topic and is better left to more advanced texts for a complete explanation. (See Minium, 1978:364–373.)

In brief, the researcher has the most influence over the last three of these factors. You already know that Type I and II errors are inversely related. As one goes down the other goes up. Thus raising the level of significance (α) to reduce beta error would increase the power of a test. However, it may be a shortsighted strategy, as demonstrated above, because alpha error would occur more often.

One way a researcher can reduce both Type I and II errors is to increase the sample size. The increased stability in a statistic based on a larger sample will make the entire decision-making process more trustworthy. We will be more likely both to reject false null hypotheses (greater power) and to fail to reject true null hypotheses. As a consequence, alpha *and* beta errors decrease and power increases.

A related strategy used by researchers is to replicate a study. Suppose that we reject a null hypothesis in one study using the .05 level of significance. The risk of having made a Type I error in this study is no more than .05. Suppose that we do a second independent study replicating the first one. Again assume that we reject the same null hypothesis at the .05 level. The risk of having made a Type I error is no more than .05 in the second study. But considered as a series of tests, the probability that both studies erroneously rejected the same null hypothesis is $(.05)(.05) = .0025$. The probability of H_0 being falsely rejected is only .0025 now.

The more replications we make, the lower the probability of a Type I error becomes for the series of studies. Further, the replications do not increase the probability of making Type II errors. Like increasing sample size, the replications may actually decrease it.

The selection of directional rather than nondirectional hypotheses also reduces Type II errors. This distinction is important in its own right and is discussed next.

Directional and Nondirectional Tests

So far we have skirted an important distinction in hypothesis testing. This concerns the nature of the difference between the null and alternative hypotheses. Some alternative hypotheses are a simple contradiction of a null hypothesis, $P \neq .80$, for example. Others may indicate that an expected value is over or under a proposed population parameter, $P < .80$, for example. The first of these is called a *nondirectional* hypothesis because it does not specify whether it expects the parameter to be over or under a certain value. The second is called a *directional* hypothesis because it does specify the direction of an expected parameter from a value proposed in a null hypothesis. Directional tests tend to have greater power than their nondirectional counterparts.

Nondirectional hypotheses are sometimes known as *two-tailed tests* because they establish two "rejection regions" in some sampling distributions. Directional hypotheses are sometimes known as *one-tailed tests* because they establish only one rejection region in some sampling distributions. A *rejection region* describes all those sample results, in one or two extreme ends of a sampling distribution, whose cumulative probability is equal to the level of significance chosen for a hypothesis test. In those problems using the .05 level of significance, the rejection region(s) contain those extreme sample results whose combined probability is .05. A directional test may commit all of this alpha probability to one rejection region in one extreme end ("tail") of a sampling distribution. A nondirectional test may divide the alpha probability into two equal parts such that one part is in the extreme right tail and the other part is in the extreme left tail of a sampling distribution.

If an actual sample result falls into a rejection region, it means that its probability is equal to or less than the level of significance. Thus, as before, it would lead a researcher to reject the null hypothesis. This is why it is called a rejection region. If an actual sample result falls outside a rejection region, its probability is greater than the level of significance. Thus it would lead to a decision of fail to reject.

We can illustrate these points with our tax plan example from earlier in this chapter. Let us assume that this problem conforms to a normal distribution. H_0 was that $P = .80$; H_1 was the $P \neq .80$. It was, thus, nondirectional. The alternative hypothesis contained two possible options to the null hypothesis. Either P was over .80 or it was under it; $P > .80$ or $P < .80$. When we set a level of significance we must cover all possibilities in opposition to H_0. In our case it must cover both of the options implied by H_1. As you will see in the next chapter, the appropriate sampling distribution for this problem has two rejection regions. Thus, if alpha is .05, it must be split in half with each part (.025) covering one of the options. This creates a two-tailed test of H_0. This is illustrated in Figure 8.8.

The sketch shows that there are two rejection regions. One is the extreme .025 of the area at the right end of the sampling distribution. This describes those sample results far

FIGURE 8.8 Rejection Regions in a Two-Tailed Test

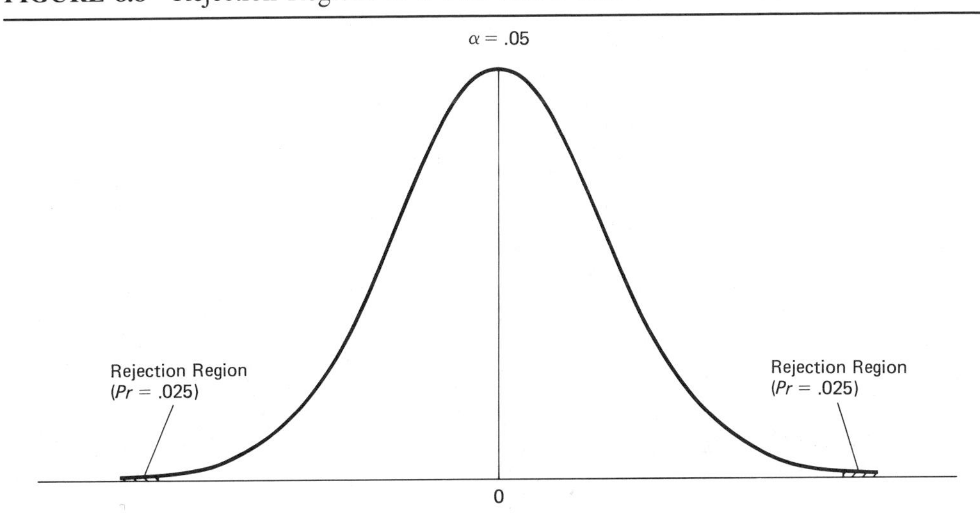

above .80. The second region is the extreme .025 of the area at the left end. This describes those sample results far below .80. After we transform our sample result to a z score (much as we did in Chapter 7), we would locate it on our sketch. We can then determine if it falls in either of the rejection regions. If it does, we know that it could be the effect of sampling error only 5 times in 100, at the most. We would reject H_0. If the result does not fall in either rejection region, we know that sampling error could be its cause more than 5 times in 100. We would fail to reject H_0. This graphic method of making a decision is a common technique used in hypothesis tests with continuous sampling distributions such as the standard normal.

One must be cautious, though. Not all sampling distributions have two rejection regions for a nondirectional test. Some have only one. In those situations the rejection area is double what is used for a directional test. Thus a .05 nondirectional test will have twice the area in the rejection region (of such a sampling distribution) that a .05 directional test has. This will become clearer to you through some examples in the chapters to follow.

Nondirectional tests are commonly used in social research when our theory is sufficient only to state a nondirectional ($\neq$) alternative hypothesis. That was what we did in our tax plan example. We only speculated that the proportion of public support was not .80. We did not say whether it would be greater or less than .80.

In other situations we can make a directional alternative hypothesis and thereby use a one-tailed test. We could alter our tax plan example to show this. Suppose our political theory suggested to us that tax proposals are not the kind of thing that elicits an extremely high level of support. Thus we might say that we expect the proportion of support to be below .80. Symbolically, H_1: $P < .80$. By default, the null hypothesis becomes both of the other possibilities. Either the proportion of support is equal to .80 or it is greater. Symbolically, H_0: $P \geq .80$. If we keep the level of significance at .05, we have only one rejection region in our sampling distribution sketch. The entire .05 probability is committed to this area. This is shown in Figure 8.9.

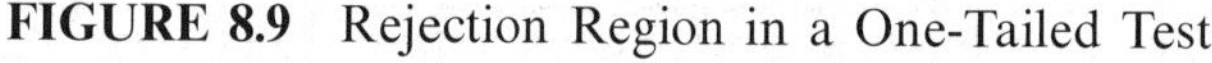

FIGURE 8.9 Rejection Region in a One-Tailed Test

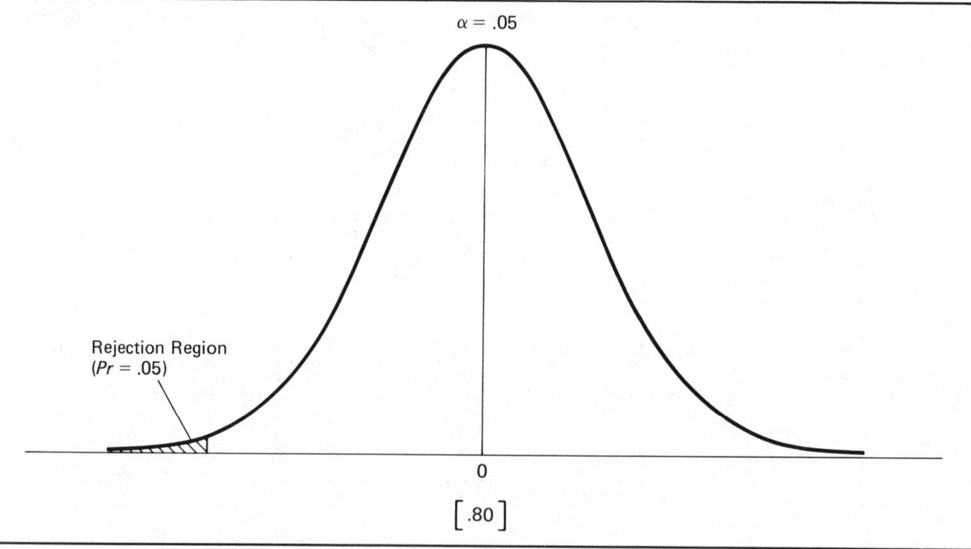

This sketch shows how the entire probability of our alpha (.05) is put in the extreme left rejection region. It is in that end because we expect the sample result to be below the assumed population value (.80). The consequence of using a directional test is that it will be relatively easier to reject H_0 than if we had used a nondirectional test when H_0 is actually false. There is more rejection area in the one end where we expect to find our sample result. Thus Type II errors will be reduced and power increased, while Type I error goes unchanged. However, it is completely impossible to reject H_0 if our sample result is above .80. No matter how extreme it might be, a sample result above .80 will cause us to fail to reject H_0. Thus we must be quite certain that we want a directional test before we devise one. Again, this will become more evident to you after you have seen some examples in the coming chapters.

Parametric and Nonparametric Tests

There are some important divisions in the types of hypothesis tests that are useful to our discussion about the basics of statistical inference. One of these is the difference between parametric and nonparametric tests. All the tests we will be looking at have some requirements for their use. One universal requirement is that we have data from a random method of sampling. Some of the tests will require us to know certain population parameters. For example, these may require the population of scores to have a known standard deviation or to have a normal distribution. The tests that have requirements about population characteristics are called *parametric tests.*

Other tests will not require anything of the population being studied. We can use the test regardless of how the population's distribution is shaped or whether its standard deviation is known. Because these tests do not require such restrictive population parameters, they are called *nonparametric tests.* Some also refer to these as *distribution-free tests,* although this is a less encompassing phrase.

Throughout Part III we will be examining various kinds of research problems; comparing one sample to one population, two samples to two populations, three or more samples to three or more populations, and others. In each situation a researcher may face a choice between a parametric and nonparametric test. It should be obvious that it will always be easier to defend a nonparametric test's requirements because they are less restrictive. But a researcher will want to use the parametric alternative, if possible. The reason for this is that parametric tests have greater power. They are more likely than their nonparametric alternatives to lead to a decision to reject a null hypothesis when it should be rejected. This gives us one more potential means to boost the power of a hypothesis test; use a parametric test. But we must meet the parametric requirements or the meaning of the test becomes ambiguous.

Not all situations can meet parametric requirements. For example, the population may not be distributed normally. Or its standard deviation may be completely unknown. Some researchers believe that the parametric requirements are so difficult to meet that they recommend that most tests be nonparametric, even those that nearly meet the requirements. Others are willing to use parametric tests when the requirements have been only approximated. The issue cannot be ignored, for it finds its way into many aspects of research. We can anticipate seeing one researcher using a parametric test and another using a nonparametric test for the same type of data.

Tests of Significant Differences and of Association

It is convenient to differentiate between two types of tests in another way. Some look for *significant differences* and others look for *significant associations.* The differences tests examine such features of sample data as central tendency, variability, and form for single-group, two-group, and multiple-group comparisons. The first three chapters of Part III cover tests of significant differences. The association tests examine some measure of association (e.g., gamma or Pearson's *r*) for sample data to determine what inferences can be drawn to populations about relationships between variables. The fourth chapter of Part III covers tests of significant associations.

Hypothesis Testing and Parameter Estimation

We have concentrated exclusively on hypothesis-testing procedures in this chapter. Although this is the main type of statistical inference, it is not the only type. Hypothesis testing begins with a hypothesis about a parameter in one or more populations. We then test this hypothesis with a sample statistic. We end with an inference about the population parameter(s). A second type of inference problem is called *parameter estimation.* It does not begin with any hypothesis about a population. Rather, it simply uses a sample statistic to construct an estimate of an unknown population parameter. You might think of the difference diagrammatically as shown in Figure 8.10.

The most common techniques of parameter estimation are the point estimate and the confidence interval. A point estimate gives a single estimated value of an unknown parameter. A confidence interval provides a range of values centered on a sample statistic. The interval gives us a hedged inference about a population's unknown parameter to which a certain degree of confidence can be attached. For example, a "95% confidence interval around a sample mean" is a usual form of estimation. We will look at point estimates at various places in Part III and at some confidence intervals together with their related hypothesis tests in Chapter 9.

Keys to Understanding

Our keys to understanding can be explained by looking at the general outline of the hypothesis-testing process. That outline can be summarized to consist of the following five steps.

FIGURE 8.10 Hypothesis Testing Versus Parameter Estimation

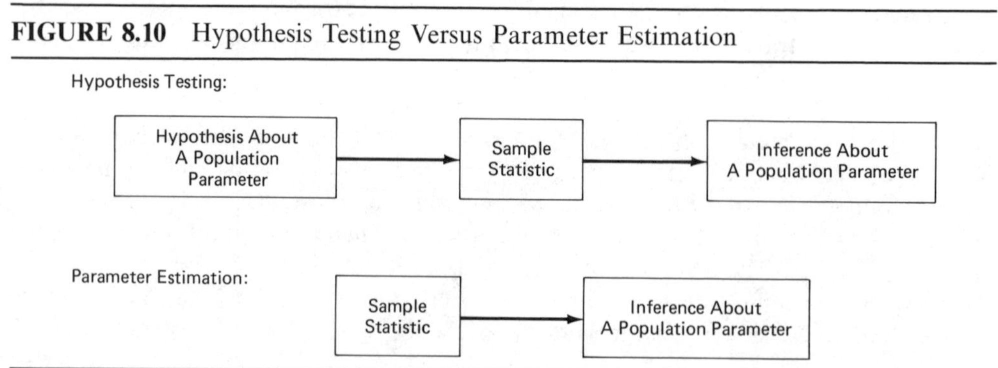

Key 8.1. First, state the null and alternative hypotheses. These should be presented both verbally and symbolically, as demonstrated earlier in the chapter. The focus should be on the characteristics of the population, not the sample. This is reflected both in the verbal and symbolic forms of the hypotheses. A choice must be made between using directional (one-tailed) or nondirectional (two-tailed) hypotheses.

Key 8.2. Second, determine the criteria for rejecting the null hypothesis. This includes choosing a level of significance, identifying an appropriate sampling distribution, and finding the rejection region(s) of that sampling distribution. If these things are done at this early point in the process, we can enhance the test's credibility because the rejection criteria are not being swayed by the actual study findings.

Key 8.3. Third, evaluate the sample results. This usually requires the greatest computational effort. There can be several substeps, depending on the type of analysis being done. The end result will reveal the probability that the sample data could be the product of sampling error under the assumption that the null hypothesis was true.

Key 8.4. Fourth, make a decision about the null hypothesis. This is the statistical goal of hypothesis testing. The result from step 3 is compared to the rejection criteria from step 2. If the sample result meets the criteria for rejection, the null hypothesis is rejected. If the sample result does not meet those criteria, we fail to reject the null hypothesis.

Key 8.5. Fifth, reach a substantive conclusion. Once a statistical decision is made about the null hypothesis, the researcher considers the meaning of the test for the subject matter under study. This may include some discussion of the theory that guided the research. Is it still useful? Does it need modification? What needs further study? What seems fairly certain? The substantive conclusion may also include some practical considerations. What policy implications are there from the test? What program changes may be needed? What does the test mean for the respondents, subjects, officials, communities, or other units included in the research problem? We will not always be able to say much about this final step in a test here because it is directly related to the subject matter of each particular study. Our focus is on the methods instead. But as a consumer of research you will be looking for the substantive conclusions. You need to be evaluating the fit between the methods (either descriptive or inferential) and the conclusions. Our overall goal remains to make you more able to do that evaluation on your own. We are trying to plug any gap in statistical literacy that otherwise prevents attaining this goal.

Of necessity, this chapter has been somewhat more abstract than most of the others. The logic of hypothesis testing is a highly rationalized and formal process. The remainder of the book will show how this logic is put to use in many situations commonly faced by social researchers.

KEY TERMS

Null Hypothesis (H_0)
Alternative Hypothesis (H_1)
Decision About H_0
  Reject
  Fail to Reject
Level of Significance (α)
Type I (Alpha, α) Error
Type II (Beta, β) Error
Power of a test ($1 - \beta$)
Directional Test
Nondirectional Test
One-Tailed Test
Two-Tailed Test
Rejection Region
Parametric Test
Nonparametric Test
Significant Difference
Significant Association
Hypothesis Testing
Parameter Estimation

EXERCISES

For each situation in Exercises 1 to 6, state the usual null and alternative hypotheses, both verbally and symbolically.

1. A local government official announces that a new intercity freeway is needed and desired by the public. Given information from officials in other similar cities, the local official says that public support should be about 60% ($P = .60$). A social impact study will be made of the project, including a survey of public opinion to determine the actual level of support.

2. A youth offender agency is interested in distinctive characteristics of its clients. The mean age of all teenage youth in the community is 16.0 years. A review of the agency's records is undertaken to see if their clients have a different mean age.

3. Achievement levels at two junior highs are to be compared. The mean scores for two random samples of seventh-grade students on a state-administered test will be used to make the comparison. Those doing the study expect the schools to differ.

4. The fuel economy of two lines of cars is to be compared. A random sample of 10 cars from each line will be chosen for testing. The mean MPG (miles per gallon) over a 100-mile course will be the statistical measure of fuel economy. A difference is expected.

5. An anthropologist plans to investigate altitude and life expectancy. A set of otherwise similar communities is chosen for a study in which the two variables will be measured. Pearson's r is expected to find an association between them.

6. A researcher is interested in stratification and institutionalized mental illness. A random sample of records from a county mental health hospital will be drawn. Information about social class and duration of hospitalization will be gleaned from the records. Gamma will be used to measure association, which is expected to be found.

For each situation in Exercises 7 to 12, decide whether the decision about the null hypothesis should be to reject or fail to reject it.

7. H_0: $\mu = 100$ on a test of mechanical aptitude. Level of significance = .05. A random sample of test scores has a mean of 80. The probability of this happening assuming that H_0 is true is .15.

8. H_0: $P = .50$ concerning the proportion of females on county governing boards. Level of significance = .05. A random sample of boards has .10 females. The probability of this happening assuming that H_0 is true is .03.

9. H_0: $P = .50$ concerning the proportion of working teenagers. Level of significance = .01. A random sample of teenagers shows that .20 have jobs. The probability of this happening assuming that H_0 is true is .03.

10. H_0: $\mu_1 = \mu_2$ for the mean incomes of whites and nonwhites in a state. Level of significance = .01. A random sample of whites shows a mean income of $15,000. A random sample of nonwhites shows a mean income of $12,000. The probability of this happening assuming that H_0 is true is .04.

11. Same as Exercise 10, except that the level of significance = .05.

12. H_0: $\rho = 0$ for the association between yearly frequency of wearing a suit and annual income. Level of significance = .10. A random sample finds that $r = .20$. The probability of this happening assuming that H_0 is true is .25.

13–18. For each situation in Exercises 7 to 12, describe the Type I and Type II errors that could occur.

For each situation in Exercises 19 to 22, decide if a directional or nondirectional test should be used and state the null and research hypotheses symbolically.

19. A consumer advocate believes that an advertiser is exaggerating when he claims that .75 of the public have used a certain soap at least once.

20. A welfare rights worker doubts that a new federal law concerning energy prices will affect equally those above and below the official poverty line. The worker believes that those below will be affected more. The mean change in part of family income going to energy costs will be compared in two random samples of households.

21. A researcher believes that people from rural places feel more social isolation than people from urban places. A standardized scale will be used to establish mean social isolation scores from random samples of rural and urban adults.

22. A teacher is uncertain how boys and girls will compare in performance on an experimental math learning module. The instructor materials provided with the module make no mention of sex differences in expected achievement. Mean scores on a quiz at the completion of the module will be used to test for any sex differences in the teacher's class.

PART III

INFERENTIAL STATISTICS

CHAPTER 9

INFERENCES FROM SINGLE SAMPLES

Part II introduced the fundamental ideas that researchers use to draw inferences from samples to populations. At least in principle, these ideas explain that this task is accomplished by replacing the uncertainty that inevitably surrounds inference work with probability. Now we are ready to see how this is done in specific research problems. This chapter looks at some of the many types of inference problems in which the result from a single sample is used to make a generalization about a population value. We examine how this is done for sample proportions, means, and frequencies in hypothesis testing and parameter estimation. Both parametric and nonparametric techniques will be included.

A Single-Sample Proportion Test

We have previewed some of the considerations for a *single-sample proportion test* in Chapter 7. There we learned how to answer simple probability questions involving proportions by using the binomial sampling distribution. For example, we wanted to know the probability of getting two heads and two tails in four tosses of a coin or the probability of six pregnancies resulting in four female and two male children. The same basic procedure used to solve these problems could be applied to inference questions concerning entire sample proportions. The single-sample proportion example of the tax proposal in Chapter 8 could be solved using a binomial distribution.

However, the binomial solution is seldom used to draw inferences from a sample to a population, for two related reasons.

1. The binomial probability of success formula becomes very burdensome when the sample size is much over 25 or so. But most social research uses samples this large and larger.
2. The implicit question being asked in most inference problems is one that requires the computation of cumulative probabilities, not exact point probabilities. We do not simply want to know the probability that a particular sample result would occur while assuming that a null hypothesis is true. Rather, we usually want to know the probability that a result as different *or more different* from that predicted by the null hypothesis would occur. This is an accumulation of the probabilities for the observed result plus all results more extreme than it. The binomial probability of success formula would need to be used many times to solve a single inference problem. This is seldom practical.

As a consequence, the binomial sampling distribution is ordinarily used only for inferences about an exact proportion from a very small sample. With larger samples, the more common solution is to use the standard normal distribution as an approximation. This is considered acceptable to researchers whenever three conditions can be satisfied: (1) independent observations, (2) a random sample, and (3) $NP \geq 5$ and $NQ \geq 5$.

We have independent observations whenever the value observed for each case is not influenced by the value observed for any other case. For example, if we interview people, their individual responses to our questions must not be influenced by the responses given by other respondents. We can accomplish this by talking to each one alone. We have already detailed the procedures necessary for random sampling. So the second condition should be familiar. The third condition means that both the sample size (N) multiplied by the presumed population proportion (P) and the sample size multiplied by its counterpart Q (which equals $1 - P$) must be equal to or greater than 5. When these three conditions are met, the normal approximation to the exact binomial can be used with single-sample proportion problems. This is not a very restrictive set of conditions. A random sample of only 10 independent cases used to test an inference about a presumed population proportion of .50 would be sufficient.

Example 1

Suppose that we are working with a single-sample proportion problem that corresponds legitimately to a normal sampling distribution. Let us examine what this means. Imagine that we are interested in people's personal views toward life in a particular political district. Specifically, we want to know whether each person is optimistic about life or not. Although we may not know its value, there is one proportion describing how many of these people have an optimistic view about life. We can label this proportion as P. The proportion not having an optimistic view is labeled Q.

Suppose that we drew one random sample of a set size (say, 100) from this population and asked these people how they felt about life. Imagine that 55 of these individuals had an optimistic view and 45 did not. The sample proportion (p) having an optimistic view would be .55. Figure 9.1a illustrates this situation, showing some of the scores for both the population and sample.

FIGURE 9.1a Selecting One Random Sample of 100 Individuals from a Population for an Inference About a Proportion

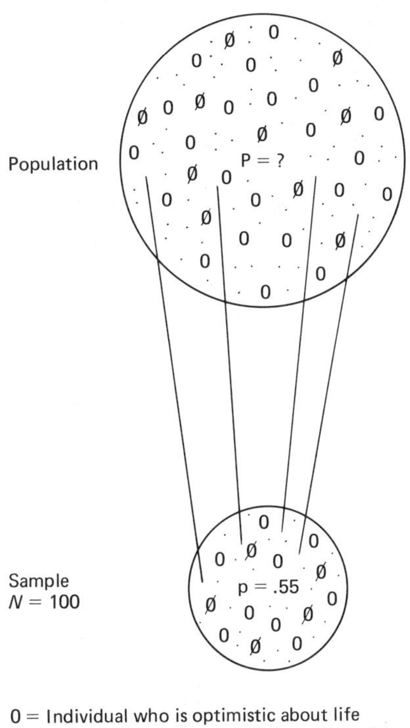

Now suppose that we selected another random sample of 100 individuals from the same population. This time the number who had an optimistic view might be only 51. Thus the second sample's proportion would be .51. The two sample proportions would differ because of sampling error. We can imagine the same process being repeated until every possible unique sample of 100 individuals had been selected from this one population. For each of these samples the proportion of individuals who had an optimistic view could be found. The distribution of all these sample proportions is highly predictable. If graphed, it would be shaped like a normal curve. Figure 9.1b illustrates this outcome.

Whereas an individual view toward life (optimistic or not) is based on a nominal measurement, a distribution of several sample proportions has ratio quality. The "scores" for the latter consist of a list of proportions: .55, .51, and so on. Because of this, a normally shaped distribution of sample proportions has a mean and standard deviation like any other distribution with interval or ratio measurement. The mean of our distribution is called the *mean of proportions* and symbolized as μ_p. It is mathematically equal to the population proportion, *P*. The standard deviation of a sampling distribution has a special name to distinguish it from other standard deviations. It is the *standard error*. The standard deviation of a sampling distribution of sample proportions is known as the *standard error of proportions*, σ_p. This standard error is mathematically equal to $\sqrt{PQ/N}$. The complete set of relevant distinctions and their symbols for a single-sample proportion test is summarized in Figure 9.2.

FIGURE 9.1b Sampling Distribution of All Possible Sample Proportions (p) from Samples with $N = 100$ Drawn Randomly from One Population

Sample	Sample Proportion (p)
1	.55
2	.51
3	.57
4	.52
⋮	⋮

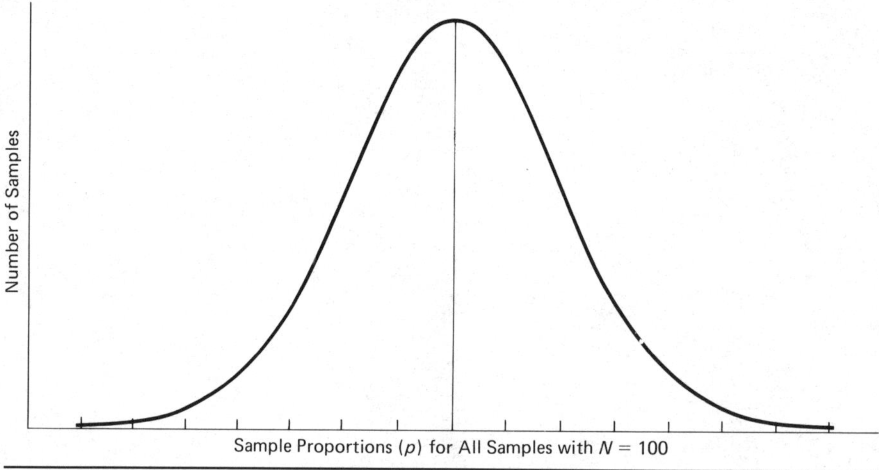

Let us now see how the standard normal distribution can be used to solve a single-sample proportion problem. Suppose a politician in our selected district has claimed that two-thirds ($P = .67$) of her constituents have an optimistic view toward life. However, an independent poll of the constituents found that .55 did. As before, imagine that the poll used a random sample of 100 adults to collect independent opinions. We want to test the politician's claim about constituent views toward life using the result from our sample. We will use the five-step process for making inferences outlined in Chapter 8.

FIGURE 9.2 Classification of Distribution Features for Single-Sample Proportion Test

Distribution	Features	
	Proportion	
One population	P	
One sample	p	
	Mean	Standard Deviation
Sampling distribution of proportions	μ_p	σ_p

Step 1. State the null and alternative hypotheses.
This is done as follows:

H_0: $P = .67$	The proportion of constituents in the population who have an optimistic view is .67. A random sample may differ from this because of sampling error.
H_1: $P \neq .67$	The proportion of constituents in the population who have an optimistic view is not .67. A random sample will differ from .67 beyond the expected influence of sampling error.

Step 2. Determine the criteria for rejecting H_0.

Suppose that we use the usual .05 level of significance. Our hypotheses represent a nondirectional, two-tailed test, so this alpha level will be split in half, with each part (.025) being applied to an extreme tail of the sampling distribution. But what sampling distribution is appropriate for our problem? The sample was randomly drawn and independent observations were gathered, so the first two conditions for normality are satisfied. Applying the formulas for the third condition, we find that $NP = 100\ (.67) = 67$ and $NQ = 100\ (1 - .67) = 100\ (.33) = 33$. Both are over 5, so we can use the standard normal as our sampling distribution. We can now consult Table D2 in Appendix D to establish our rejection regions.

As Figure 9.3a illustrates, our rejection regions are the extreme 2.5% of the area at both ends of a normal curve. This means that there is 47.5% of the area (50 − 2.5) between the mean and each rejection region's starting point. Table D2 tells us that a z score of exactly ± 1.96 defines the point at which the rejection regions begin. This z score is known as the *tabular statistic* or *critical value* against which our sample result can be compared. The rejection regions are often described in terms of the tabular statistic, rather than the level of significance itself. This practice would lead us to say that the right extreme rejection region is that area at and above a z score of $+1.96$. The left extreme rejection region is the area at and below -1.96. Together these would be described as the areas at and

FIGURE 9.3a Sampling Distribution of Single-Sample Proportions When Testing H_0: $\mu = .67$

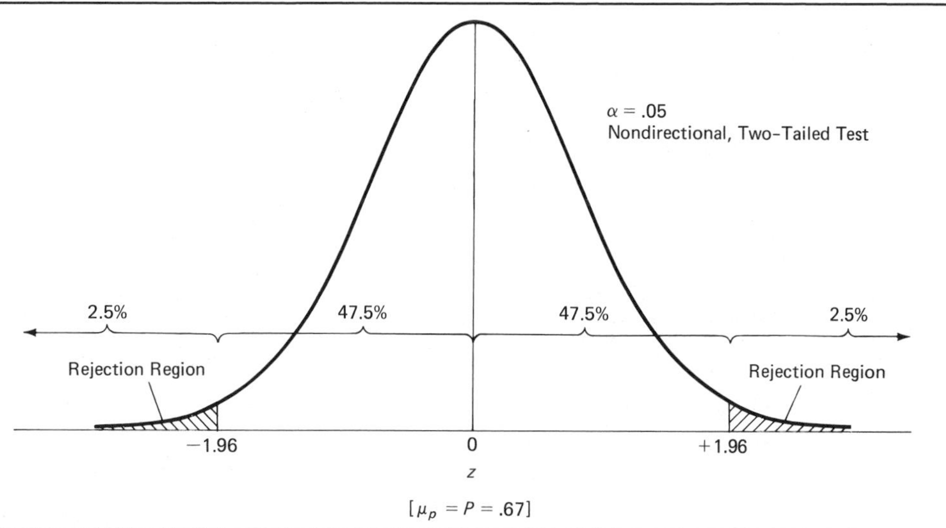

beyond both ± 1.96. If we find that our sample proportion is equivalent to a z score of ± 1.96 or beyond, we know that such a proportion has a probability of being the consequence of sampling error by no more than .05 assuming that H_0 is true. If this does happen, we will reject the null hypothesis. On the other hand, if we get a result equivalent to a z score between $+1.96$ and -1.96, we know that it would have a sampling error probability greater than .05. Such an outcome would lead us to fail to reject the null hypothesis.

Step 3. The test statistic.

The sample of 100 constituents consisted of 55 ($p = .55$) who had an optimistic life-view. However, the null hypothesis proposed that 67% ($P = .67$) actually had such a view. Thus our sample proportion differed from the proposed population proportion by $-.12$ ($.55 - .67$). We now need to know the probability that sampling error could cause us to miss the true proportion of support by at least .12 either above or below that true proportion. We must consider missing it in both directions because we are conducting a nondirectional test. Put another way, we want to know the probability that a result as extreme or more extreme than the one we have observed could occur even if the null hypothesis is correct. We are most interested in knowing whether that probability is equal to or less than our level of significance, .05. If it is, we will reject the null hypothesis. To make this determination, we must transform the sample and population proportions to a z score. The formula for this is

$$z = \frac{p-P}{\sigma_p}$$

z Transformation for a Single-Sample Proportion Test

where p = sample proportion
P = population proportion
σ_p = standard error of proportions

You may notice the similarity of this formula to the one we used in Chapter 7 to convert single case scores to z scores. In both, the numerator calculates a deviation (the difference) of an observed value (single score or sample proportion) from a mean score. Remember that $\mu_p = P$. The denominator is a standard deviation, S for single cases and σ_p for a sampling distribution of proportions.

We can apply the new z-score formula to our example once we have found the standard error using the following formula:

$$\sigma_p = \sqrt{\frac{PQ}{N}}$$

Standard Error of Proportions

where P = population proportion
$Q = 1 - P$
N = sample size

$$\sigma_p = \sqrt{\frac{(.67)(.33)}{100}} = \sqrt{\frac{.2211}{100}} = \sqrt{.002211} = .047$$

With this standard error now calculated the z transformation can be found:

$$z = \frac{p - P}{\sigma_p} = \frac{.55 - .67}{.047} = \frac{-.12}{.047} = -2.55$$

This calculated z score is based on the comparison of the sample statistic (p) to the hypothesized population parameter (P). It is known as the *test statistic*. This name is appropriate because we are using the sample data to test the null hypothesis.

Step 4. The decision about H_0.

This can be done in one of two equivalent ways, both of which you will find in the social science literature. In the first of these, illustrated in Figure 9.3b, we look in Table D2 to determine the probability that a test statistic like ours would occur because of sampling error and then compare it to the previously chosen level of significance. This means finding the area to the left of a z of -2.55 ($50 - 49.46 = .54$) and the area to the right of a z of $+2.55$ ($50 - 49.46 = .54$). Remember we must find both areas because we were using a nondirectional test. The two areas are then added: $.54 + .54 = 1.08\%$. The combined area is converted from its percentage form to a probability: $1.08\% = .0108$. We now compare this probability to the level of significance. Since the former (.0108) is less than the level of significance (.05), we know that sampling error could be considered to be the cause of our sample proportion differing from the hypothesized population proportion only about 1% of the time. This is such an unusual event that we decide to reject the null hypothesis. The decision rules for this method of making a decision were shown in Figure 8.1. Recall that we reject H_0 whenever the probability for a sample result is less than or equal to the level of significance.

The second method gives the same result but is shorter. It is illustrated in Figure 9.3c. We simply compare our test statistic (-2.55) to our tabular statistic (± 1.96). This can be done by locating our test statistic on a sketch of our sampling distribution showing the rejection regions. Our test statistic is in the rejection region to the left side in this example. Our conclusion is automatic: We reject the null hypothesis knowing the sample result would not occur with more than .05 probability assuming the null hypothesis to be true (i.e., $|\text{test } z| \geq |\text{tabular } z|$). The decision rules for this method are shown in Figure 9.4.

Either decision method leads us to reject the null hypothesis. We end our statistical analysis of the situation by concluding that it is improbable that the proportion of the politician's constituents who are optimistic is .67. We realize that we might be making a

FIGURE 9.3b Conclusion for H_0: $P = .67$ (First Method)

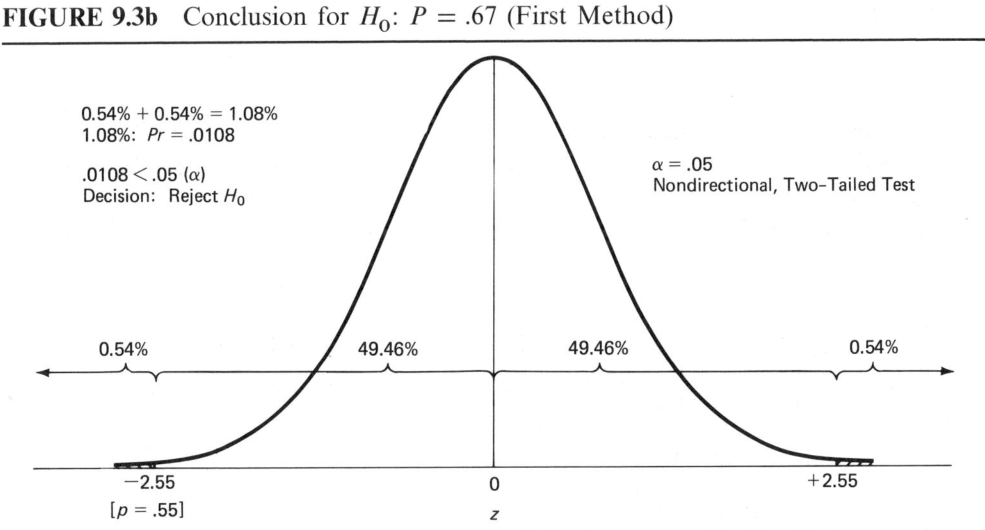

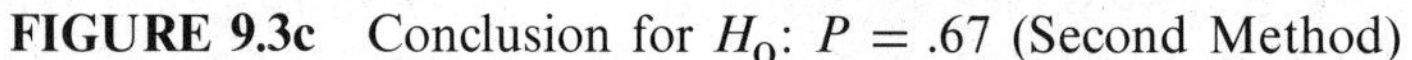

FIGURE 9.3c Conclusion for H_0: $P = .67$ (Second Method)

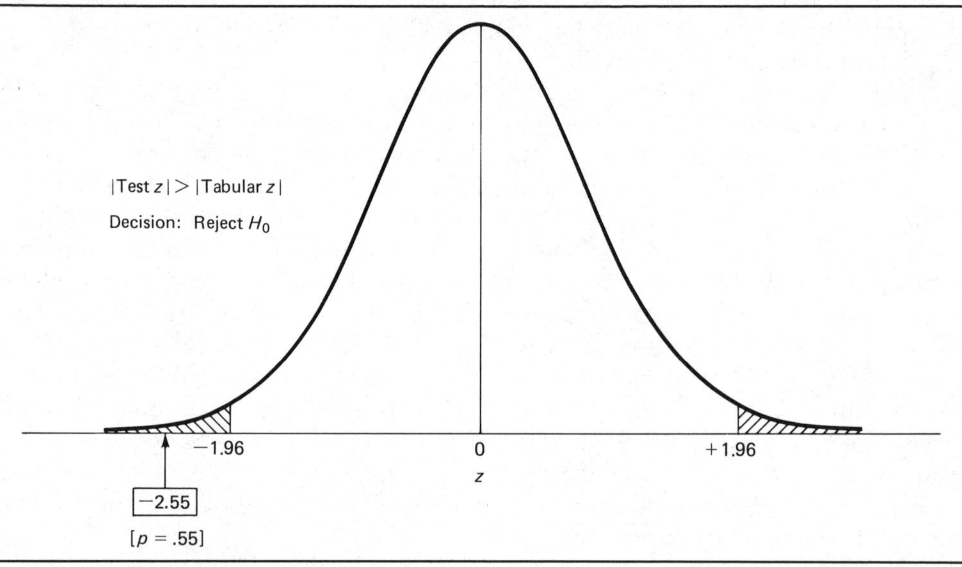

Type I error with this decision. Our sample may be one of those extreme results that occur less than about 1% of the time from a population with a proportion of .67. But we are doubtful about this interpretation. Instead, we decided that our sample proportion revealed a "statistically significant difference" from the hypothesized population proportion of .67. Researchers would commonly summarize all these details in the simple note "$z = -2.55$ ($p < .05$)." As you can see from this, the researcher reports the sampling distribution used by the "z" designation. The value found for the test statistic is reported (−2.55) and the level of significance is shown by noting that the test statistic had a probability less than the listed value (.05). Although we have been using Pr as the symbol for probability to distinguish it from other proportions (e.g., P and p), this is not often done in research reports. The lowercase p is the most commonly used symbol. You must determine from the context whether this is a probability or some other proportion.

Step 5. The conclusion.

Since our example did not have an elaborate discussion of a theoretical and real-world background, it is difficult to detail a substantive conclusion. But it would involve a consideration of the social-psychological theories of people's views toward life and of political theories of politician's perspectives on those views. Given the rejection of the null hypothesis, we would view the ideas behind our alternative hypothesis to be supported;

FIGURE 9.4 Making a Decision About a Null Hypothesis

If the |test statistic|	$\geq$	|tabular statistic| ,	then reject H_0
If the |test statistic|	$<$	|tabular statistic| ,	then fail to reject H_0

$P \neq .67$. This would include the practical implications of the research. Why do the people seem to have a different level of optimism than the politician claimed? Does the politician misapprehend public views or was she deliberately exaggerating them? A more direct answer to this last question would require a directional test.

Example 2

Let us look at another example of a single-sample proportion test. Suppose that a researcher is interested in studying race discrimination in the appointment of Spanish Americans to university faculty positions in the area of criminology. The researcher would need to determine the proportion of all university positions that have recently been filled with Spanish Americans. This kind of information would be available from several educational agencies. Suppose that this proportion was .20. The researcher expects less discrimination in criminology because of its association with the social sciences, which are assumed to be more sensitive than other disciplines to the issue of job discrimination.

The researcher can then draw a random sample of universities and determine the number of criminology positions that were filled with Spanish Americans. Suppose that the sample consists of 25 randomly chosen recent decisions in filling criminology positions. The researcher then looks for a significant difference in the two proportions. This would be only a crude test of apparent discrimination since it ignores other important factors, such as the availability and willingness of candidates to take jobs in criminology at the universities in the sample and the exact job qualifications for specific positions. But it will serve our present purposes of illustrating the statistical techniques.

Step 1. The hypotheses.

Since the researcher expects a sample proportion higher than the proposed population value (indicating less discrimination), it will create a directional test.

H_0: $P \leq .20$	The proportion of criminology positions filled by Spanish Americans is .20 or less. A random sample may differ from this because of sampling error.
H_1: $P > .20$	The proportion of criminology positions filled by Spanish Americans is greater than .20. A random sample will differ from .20 beyond the expected influence of sampling error.

In effect, this problem is asking whether criminology positions are filled with any less discrimination against Spanish Americans than are other university positions. The null hypothesis says that they are not. The alternative hypothesis says that they are.

Step 2. The criteria for rejecting H_0.

Suppose that the researcher is so confident of the less-discriminatory idea about hiring in criminology that a .01 level of significance can be used. This will be committed to one rejection region at the right extreme of the sampling distribution. If the conditions of independence and randomness are met, we can also look at the other condition for assuming a normal distribution. $NP = 25(.20) = 5$ and $NQ = 25(.80) = 20$. Thus the normal sampling distribution can be used. According to Table D2, .01 of the area remains to the left of a z score of approximately $+2.33$. This is our tabular statistic. The rejection region begins at $+2.33$ and extends to the right.

Step 3. The test statistic.

Now we evaluate the sample result. Suppose that the sample proportion was .30. First we find the standard error of proportions and then calculate the test statistic:

$$\sigma_p = \sqrt{\frac{(.20)(.80)}{25}} = \sqrt{\frac{.16}{25}} = \sqrt{.0064} = .08$$

$$z = \frac{.30 - .20}{.08} = \frac{.10}{.08} = 1.25$$

Step 4. The decision about H_0.

We can use the simpler of the two methods mentioned earlier to reach a conclusion about the null hypothesis. Figure 9.5 shows our test statistic (1.25) in comparison to the tabular statistic (2.33). We see that the test statistic does not fall into the rejection region that begins at a z of $+2.33$. This means that we cannot reject the null hypothesis. Our sample result was not one of those that would result from sampling error only 1% of the time or less. It had a larger probability than this. Therefore, we fail to reject the null hypothesis.

The statistical test could be summarized as $z = 1.25$ ($p > .01$, n.s.). A reader interprets this to mean that the test statistic of 1.25 was not significantly different from the value proposed by a null hypothesis using a .01 level of significance for a normal sampling distribution. This result is often referred to as being "not statistically significant," (n.s.). Some researchers may add the note that "the observed difference was in the predicted direction, but not significantly so." While such a statement is true, it does not lessen the conclusion that the null hypothesis was not rejected in this test. It does suggest that the conclusion might be changed if a less restrictive level of significance was used. Many researchers will do exactly that, although the practice is not strictly correct according to the traditional rules of Neyman-Pearson hypothesis testing. Regardless, the conclusion would remain as it is in this example at any usual level of alpha.

Step 5. The conclusion.

Now to the substance of the example. The outcome of the test means that the hiring practices in criminology are not significantly different from the general pattern believed to

FIGURE 9.5 Testing H_0: $P \leq .20$

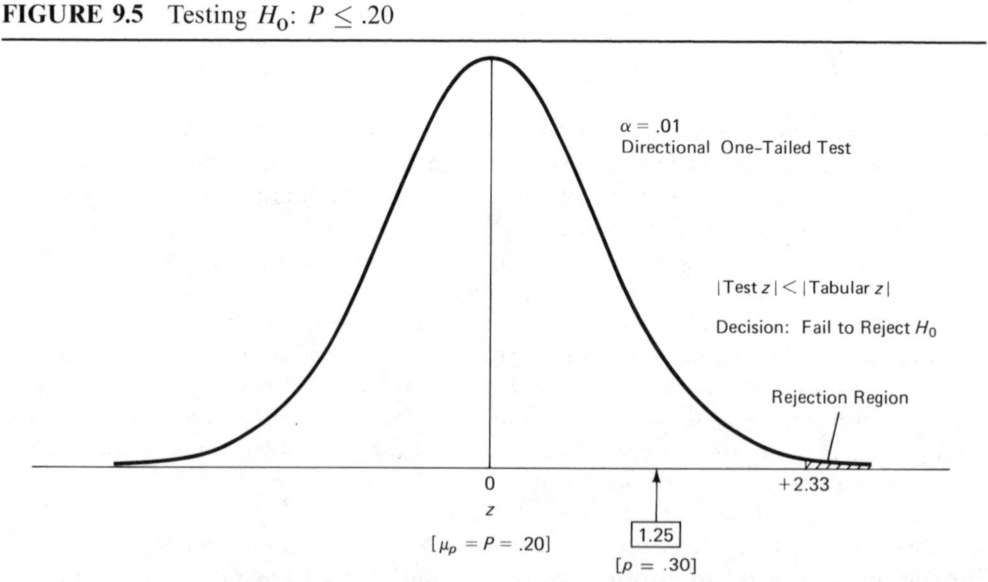

exist for all faculty positions. The .10 deviation (.30 − .20) was not sufficient with 25 random cases to support the contention that there is any such real difference in the larger population. We cannot reject the hypothesis that criminology positions are filled with just as much discrimination or nondiscrimination toward Spanish Americans as are other positions from the data we had to examine. It is possible that a larger sample size would alter the conclusion. A second test could pursue this possibility if the researcher wished.

Tests of a Single-Sample Mean

One of the more commonly used summary measures of a distribution is the mean. We looked at it extensively in Chaper 3 with other measures of central tendency as a descriptive statistic. Now we want to see how it is used to make inferences. There is considerable similarity between inferences concerning a mean score and those with a proportion. There is also more complexity because of the higher measurement level of a mean (interval versus nominal). We will look at the fundamental types of single-sample inference tests for means.

Sampling Distribution of Means

Like any inference test, a *single-sample mean test* is based on a sampling distribution. The sampling distribution of a single mean is the theoretical probability distribution of all sample means that could be found by drawing all possible samples of any particular size from that population. There are four important characteristics of such a sampling distribution.

1. If the population is normally distributed, the sampling distribution of single sample means will also be normally shaped regardless of the number of cases selected in each sample.
2. The mean of this sampling distribution, known as the *mean of the means* ($\mu_{\bar{X}}$), will be equal to the mean of the population (μ) from which the samples are drawn.
3. The standard error of this sampling distribution, known as the *standard error of the means* ($\sigma_{\bar{X}}$), becomes smaller as the sample size increases and is equal to $\sigma/\sqrt{N}$.
4. Regardless of the shape of the population's distribution, the sampling distribution of single sample means will approach a normal shape as the sample size becomes larger having a mean ($\mu_{\bar{X}}$) that approaches the population mean (μ) and a standard error ($\sigma_{\bar{X}}$) that is approximated by $\sigma/\sqrt{N}$.

Before exploring these characteristics of the sampling distribution of means, be sure that you are following the distinctions in names and symbols being used. These are shown in Figure 9.6. We are working with three distributions again: the population's, a single sample's, and the sampling distribution of means from all possible samples drawn from that population. The Greek symbols are used for the population, the Roman symbols are used for the sample, and the Greek symbols with Roman subscripts are used for the sampling distribution.

The first of the four characteristics of the sampling distribution of means should seem reasonable to you with a little reflection. If we think of drawing random samples

FIGURE 9.6 Classification of Distribution Features for Single-Sample Mean Test

Distribution	Mean	Standard Deviation
One population	μ	σ
One sample	$\overline{X}$	S
Sampling distribution of means	$\mu_{\overline{X}}$	$\sigma_{\overline{X}}$

from a population that is normally distributed, it seems natural that most of those samples would also be normally distributed since random samples are the most likely type to yield accurate representations of their populations. Therefore, we would expect the means from those samples to be normally distributed. The more notable part of this characteristic is the claim that the size of the samples would not alter the shape of the resulting sampling distribution. Whether we drew samples of 20, 200, or 2000, the sampling distribution of our sample means would be normally distributed when the population was also distributed normally.

The second characteristic refers to the expected value of the mean of the sampling distribution. It will equal the population mean because it involves the same individual scores. In the population these are added and then divided by their number to find μ. In the sampling distribution, smaller sets of these same individual scores are used to find separate sample means. These means are then added and divided by the number of samples. Although the second process is much more involved, it does use the same scores (although in multiple overlapping sets) and gives the same result.

The third characteristic considers the variability of the sampling distribution. It indicates that the standard error of the means is inversely related to the sample size. Larger samples result in sampling distributions that are more closely clustered. This should make intuitive sense to you. It means that larger samples will give us more accurate estimates (smaller standard errors) of the population mean.

The first three characteristics can be examined by a process of repeatedly drawing random samples from a normal population. This represents an empirical method of establishing the probability distribution. It is a simulation of the theoretical sampling distribution of all possible sample means from a population.

Suppose that we were researching features of chief executive officers for complex organizations, such as the salaries of bank presidents. Imagine that the entire population of such salaries was normally distributed with a mean of \$75,000 and a standard deviation of \$10,000. We might draw many independent and random samples from this population in which $N = 10$ and then graph the resulting distribution of sample mean salaries. It would be normally distributed with a mean of \$75,000 and a standard error of about \$3162 ($\sigma/\sqrt{N}$). The experiment could be repeated using a sample size of 25. Again the distribution of sample means would be normally shaped with a mean of \$75,000, but the standard error would now be about \$2000. A third demonstration of the process might be made using $N = 50$. This time the sampling distribution of sample means would be normally shaped with a mean of \$75,000 and a standard error of \$1414. As you see, the sampling distribution remains normal with a mean equal to the population's mean and the standard error decreases as the sample size increases. Figure 9.7 illustrates this series of demonstrations.

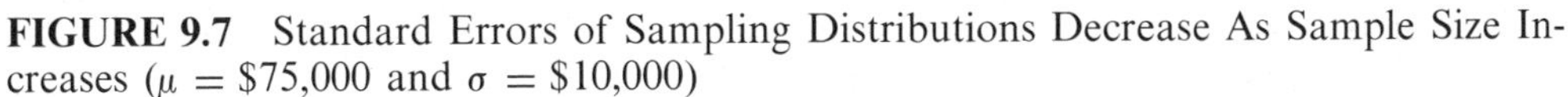
FIGURE 9.7 Standard Errors of Sampling Distributions Decrease As Sample Size Increases (μ = \$75,000 and σ = \$10,000)

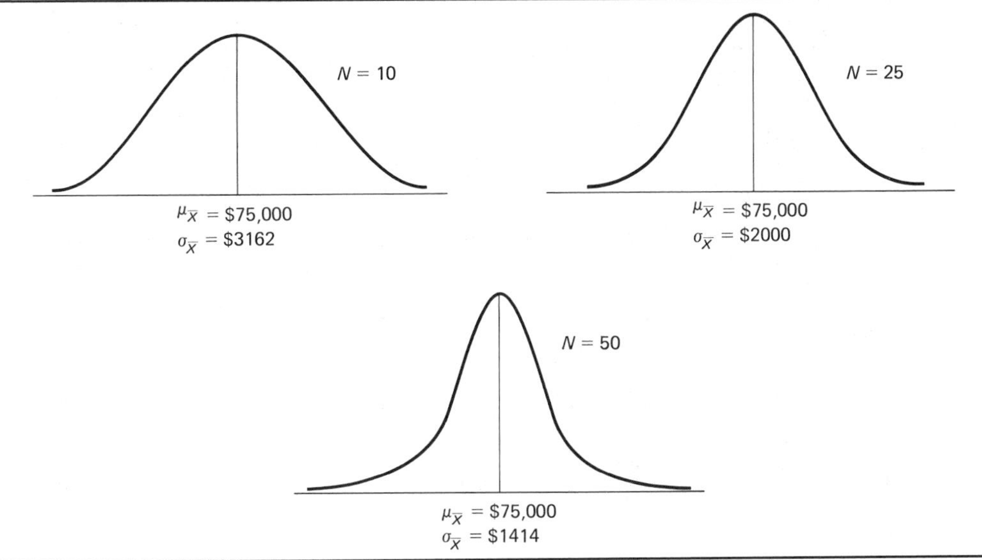

The fourth characteristic (known as the *Central Limit Theorem*) is the most amazing idea of the set. It says that sampling distributions may be nearly normal in shape even when the population being studied is not, provided that we have used a "large" sample size.

Think again about our hypothetical example of bank presidents and their salaries. Suppose that the distribution of these salaries still had a mean of \$75,000 and a standard deviation of \$10,000 but was not normally distributed. Instead, it was positively skewed, with a few presidents having extremely high salaries compared to the others. We might draw a random sample of 150 cases and find that the mean is \$60,000. A second random sample of 150 might have a mean of \$70,000. A third with $N = 150$ might have a mean of \$77,000. And so on. As always, we expect to see some differences between means for samples drawn from the same population because of sampling error. But the distribution of these sample means would approach a normal shape with a mean nearly equal to the true mean income of the population (\$75,000) and a standard deviation nearly equal to \$816 ($\$10{,}000/\sqrt{150}$). This sampling distribution would approach normality of shape because every sample mean would be equally as likely to include a few of the extremely high salaries. As a result of this equal-probability feature, the sample means would tend to be normally distributed even though the population of incomes was not. Figure 9.8 illustrates this.

The Central Limit Theorem also applies to several inference tests other than those with means. One version of it describes the single-sample proportion test we looked at before. It was not necessary to our earlier understanding of that test. But the theorem is basic to many other tests, including several used by social researchers.

One final point needs to be explored before we begin doing some single-sample mean tests. What exactly does "large" mean in the Central Limit Theorem? The answer depends on the distance that the population's distribution is from a normal shape. The

FIGURE 9.8 Sampling Distribution of Single-Sample Means ($N = 150$) When $\mu = \$75{,}000$ and $\sigma = \$10{,}000$ in a Non-normal Population

Sample	Sample Mean
1	$60,000
2	70,000
3	77,000
⋮	⋮

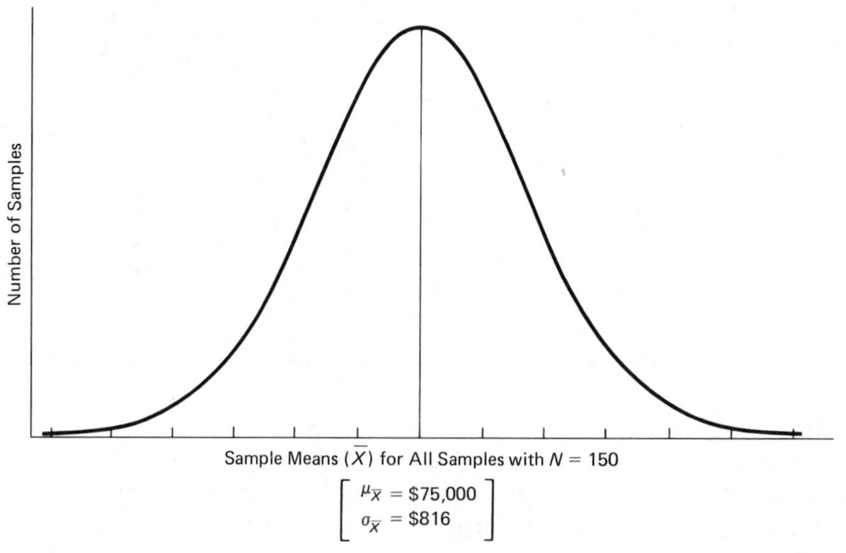

farther from normality it is, the larger the sample size must be in order for the sampling distribution to be nearly normal itself. Often the population's distribution is only vaguely known. This makes it difficult for a researcher to know when to make the assumption of a normal sampling distribution. Through experience a loose rule of thumb has been established to guide the researcher. If there is reason to believe that the population's distribution is not extremely different from normal and the sample size is at least 101, the Central Limit Theorem is often assumed to apply. But you should be prepared to find researchers assuming a normal sampling distribution for single-sample mean tests with samples as small as 31 cases. This may be justified if they have reason to be confident of the near-normality of the population's distribution. Unfortunately, those reasons are seldom revealed to the reader. It should make you cautious to accept the inference test and conclusions reached in these situations.

In summary, we can use the normal sampling distribution for single-sample mean tests whenever one of two sets of conditions are met. First, it can be used when the population is normally distributed. The sample size will not matter in this situation. Second, the normal sampling distribution can be used when the population's distribution is somewhat different from normal but the sample size is relatively large. For researchers this means that a normal sampling distribution can be assumed when σ is known and/or N is large. The population parameter σ may need to be known so that we can solve the formula for the standard error $\sigma_{\overline{X}}$ and perhaps to check the shape of the population's distribution as well. (More about this will appear in the section on t distributions.) When a normal sampling distribution can be assumed, the following z-score formula is used for single-sample mean tests:

$$z = \frac{\overline{X} - \mu}{\sigma_{\overline{X}}}$$ z Transformation for a Single-Sample Mean Test

where $\overline{X}$ = sample mean
μ = population mean
$\sigma_{\overline{X}}$ = standard error of means

and $$\sigma_{\overline{X}} = \frac{\sigma}{\sqrt{N}}$$ Standard Error of Means

where σ = population standard deviation
N = sample size

Normal Population with σ Known

Let's consider an example in which the normal sampling distribution can be used for a single-sample mean tests of an inference. Again suppose that we are working with a distribution of salaries for bank presidents which a banking association claims is normally shaped with $\mu = \$75{,}000$ and $\sigma = \$10{,}000$. A researcher has drawn a random sample of 67 salaries from this population in which $\overline{X} = \$80{,}000$ and $S = \$6{,}000$. Is the banking association's claim about the mean of bank presidents' salaries believable? We solve this inference problem using the same five steps that we have been following for previous problems.

Step 1. The hypotheses.

H_0: $\mu = \$75{,}000$	The population of bank presidents' salaries has a mean of \$75,000. Sampling error may cause a random sample from this population to have a different mean.
H_1: $\mu \neq \$75{,}000$	The population of bank presidents' salaries does not have a mean of \$75,000. Sampling error will not be the reason for a random sample to have a mean other than \$75,000.

These hypotheses describe a two-tailed test situation in which the alternative hypothesis includes two options opposed to the null hypothesis. Either the sample mean will be significantly greater than \$75,000 or less than it.

Step 2. The criteria for rejecting H_0.

Let us use the usual .05 level of significance. This is split equally between two rejection regions. The sampling distribution can be assumed to be normal given the information we have. Table D2 tells us that a z score of ± 1.96 is our tabular statistic. The rejection regions are those areas equal to and above $+1.96$ and equal to and below -1.96.

Step 3. The test statistic.

In order to use our z-transformation formula for a single-sample mean test, we must first calculate the standard error for our sampling distribution.

$$\sigma_{\overline{X}} = \frac{\sigma}{\sqrt{N}} = \frac{\$10{,}000}{\sqrt{67}} = \frac{\$10{,}000}{8.1854} = \$1221.687$$

Now we can calculate the test statistic:

$$z = \frac{\bar{X} - \mu}{\sigma_{\bar{X}}} = \frac{\$80{,}000 - \$75{,}000}{\$1221.687} = \frac{\$5000}{\$1221.687} = 4.09$$

Step 4. The decision about H_0.

The test statistic (4.09) exceeds the tabular statistic (± 1.96) and is in the right-hand rejection region of the sampling distribution. (See Figure 9.9.) This means that we decide to reject the null hypothesis. Our sample mean could be the result of sampling error less than 5 times in 100 similar tests. Results this unusual cause us to doubt the null hypothesis. Our sample mean is significantly different from the hypothesized population mean ($z = 4.09$, $p < .05$).

Step 5. The conclusion.

Since we rejected the null hypothesis, we draw the inference that the population does not really have a mean of \$75,000. We know that this decision includes some risk of a Type I error. Maybe the population truly does have a mean of \$75,000 and we had the misfortune to get a very unrepresentative sample from this population. But our test result tells us that we should be willing to risk this type of error given our criteria for rejecting the null hypothesis.

How could the difference between the sample and bank association be explained? It may mean that the information given by the banking association is out of date. It may mean that they were deliberately misstating the true mean salary of bank presidents. It may mean that our sample was not randomly drawn. Since the researcher's work is available to us, we would examine it carefully to be assured that the last of these explanations is not likely. We might investigate the possible explanations for the association's assertions, although this would require their cooperation. The final word for now is that the true

FIGURE 9.9 Testing H_0: $\mu = \$75{,}000$

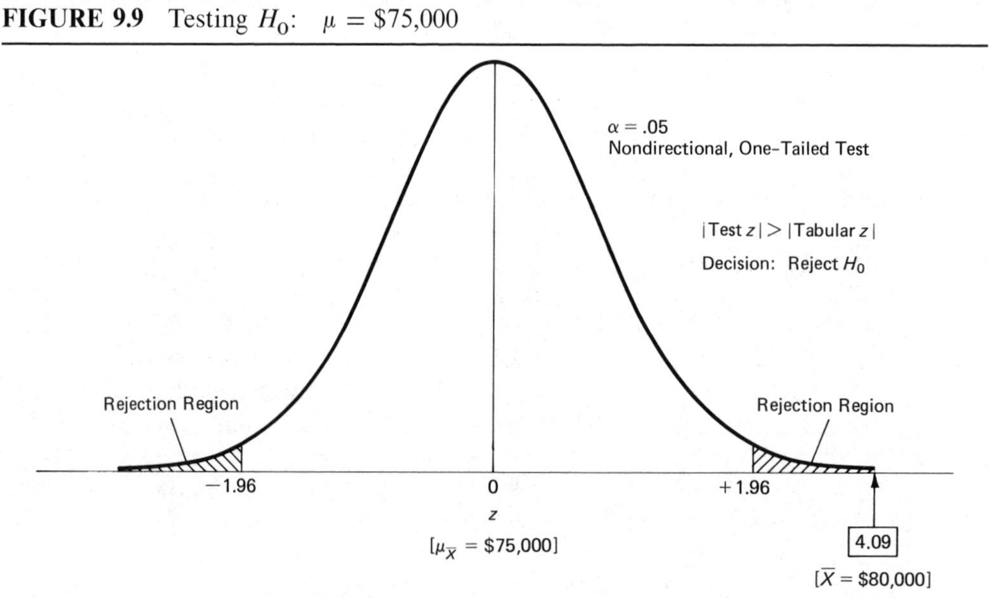

mean salary is not \$75,000 but the analysis has not told us the reason for the observed difference from the hypothesized value.

It is important for you to note that we cannot draw the conclusion that the true mean is *over* \$75,000. That would have required a directional, one-tailed test. Strictly speaking, we cannot change our hypotheses or criteria for rejecting the null hypothesis *after* we have seen the results of our sample. But the result does suggest that a directional test might be utilized in a future study.

Nonnormal Population with σ Not Known and Large Sample

Consider a new example. A researcher is interested in studying the ages of newspaper editors. The researcher assumes that the population of those ages is somewhat negatively skewed with relatively few editors being very young compared to the majority who reach such positions later in life. The researcher knows from census figures that the mean age of those in managerial positions of all kinds is 44 years. The researcher suspects that editors may need to be older than most managers because of a low turnover in such positions. A random sample of 200 editors is drawn in which the mean age is 45.5 with total variation Σx^2 equal to 44,775. Does this mean that newspaper editors differ in mean age from managers in general?

Step 1. The hypotheses.

H_0: $\mu \leq 44$ — The population of newspaper editors' ages has a mean of 44 years or less. A random sample may differ from this because of sampling error.

H_1: $\mu > 44$ — The population of newspaper editors' ages has a mean greater than 44 years. A random sample will differ from a mean of 44 beyond the expected influence of sampling error.

Note that this has created a one-tailed test because of the researcher's directional idea that newspaper editors tend to be older than managers of all kinds.

Step 2. The criteria for rejecting H_0.

Let us use the .05 level of significance. This will be committed to one rejection region at the extreme right of the sampling distribution. But what sampling distribution can be used? We do not have a normally distributed population of ages. But we do have a rather large sample ($N = 200$). The Central Limit Theorem says that conditions are right to assume that our sampling distribution of mean ages would be nearly normal. Using the normal approximation to our sampling distribution, we consult Table D2 to determine our test statistic. We find that a z score of $+1.64$ leaves 5.05% (50 − 44.95) of the area beyond it in the upper tail. A z score of $+1.65$ leaves 4.95% (50 − 45.05) of the area beyond it in the upper tail. These are equally close to our desired area of exactly 5%. Our best guess of the z score to use is to split the difference between these two contenders. We will use $+1.645$ as our test statistic. The rejection region is that area equal to and above $z = +1.645$.

Step 3. The test statistic.

As you saw in the preceding example, we need to find the standard error before we can calculate our test statistic. But we have a new problem. Our formula for $\sigma_{\bar{X}}$ requires us to know σ. In the present problem we do not know it. How can we proceed? Recall from Chapter 4 that s is an unbiased estimate of σ. It is found by dividing the total variation Σx^2 by the quantity $N - 1$ rather than dividing by N as we do for σ and S. Since the researcher has the raw data for the sample, s can be calculated and substituted for σ in this example.

$$s = \sqrt{\frac{\Sigma x^2}{N - 1}} = \sqrt{\frac{44{,}775}{199}} = 15$$

Our s equals 15. We can now use a new formula to estimate the standard error:

$$\hat{\sigma}_{\bar{X}} = \frac{s}{\sqrt{N}}$$ Standard Error of Means Estimated from s

where $\hat{\sigma}_{\bar{X}}$ = estimated standard error of means
s = estimated standard deviation of a population
N = sample size

In our example

$$\hat{\sigma}_{\bar{X}} = \frac{15}{\sqrt{200}} = \frac{15}{14.142} = 1.061$$

Now our test statistic can be found by a slightly modified version of the z-transformation formula above:

$$z = \frac{\bar{X} - \mu}{\hat{\sigma}_{\bar{X}}}$$ z Transformation for a Single-Sample Mean Test Using $\hat{\sigma}_{\bar{X}}$

$$= \frac{45.5 - 44}{1.061} = \frac{1.5}{1.061} = 1.41$$

Step 4. The decision about H_0.

The test statistic (1.41) does not exceed the tabular statistic (1.645) and is not in the rejection region. (See Figure 9.10.) Thus we fail to reject the null hypothesis. The probability that our sample mean differed from the proposed population value of 44 because of sampling error is greater than .05. This kind of sample result is one that causes us to be unable to reject the null hypothesis. Our sample mean does not differ significantly from the proposed population mean ($z = 1.41$, $p > .05$, n.s.).

Step 5. The conclusion.

Since we failed to reject the null hypothesis, we draw the inference that the true mean age of newspaper editors may really be 44 years, just like other people in managerial occupations. We know that this decision risks a Type II error but we must take that risk given the criteria we preset for testing the null hypothesis. There does not appear to be evidence in this test to support the idea that newspaper editors differ significantly in mean age from other managers. Perhaps there is just as much (or little) opportunity for people to move into editorships of newspapers as into other managerial jobs. Further empirical testing and theoretical reasoning will need to be given to the topic.

The Student's *t* Distributions

The two examples of bank presidents' incomes and newspaper editors' ages illustrate how we can test whether the mean for a single sample differs significantly from a proposed

FIGURE 9.10 Testing H_0: $\mu \leq 44$

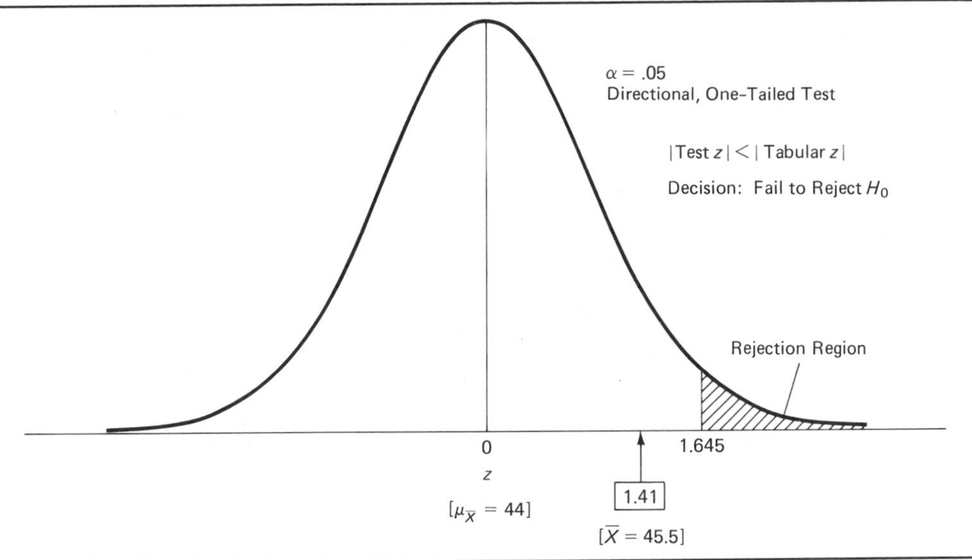

population mean when the sampling distribution can be assumed to be shaped like a normal curve. In the first example we knew that the population was normally distributed and we knew its standard deviation, σ. In the second example we did not have a normal population or know its standard deviation. But we did have a large sample and used its standard deviation s to estimate the standard error. That substitution of s for σ in the standard error formula is appropriate as long as the sample size is large, roughly $N > 100$.

But these two examples illustrate somewhat rare situations in social research. It is not often that we know such a parameter as σ and we may not have a large sample. We may have reason to believe that the population is distributed approximately like a normal curve, however. What then? When we must substitute s for σ from small samples representing nearly normal populations, the shapes of the appropriate sampling distributions are unimodal and symmetrical. However, they are slightly more variable compared to the standard normal curve. The smaller the sample is, the more variable the sampling distribution becomes. (See Figure 9.11.)*

*It now may seem that we are abandoning the first characteristic we listed earlier about the sampling distribution of the mean; namely, that it is normally distributed for samples of any size (even small ones) when the population is also normally distributed. That is not so. The sampling distributions with which we are working are actually the distributions of test statistics (e.g., $(\bar{X} - \mu)/\sigma_{\bar{X}}$), not of the sample statistics themselves (e.g., $\bar{X}$s). This distinction was not necessary before because the sampling distributions for both coincide in shape when σ is known. But when we do not know σ and must estimate it from s, the resulting estimate of $\sigma_{\bar{X}}$ ($\widehat{\sigma_{\bar{X}}}$) is itself a variable, not a constant as is $\sigma_{\bar{X}}$. The estimate ($\widehat{\sigma_{\bar{X}}}$) varies in value depending on the particular scores one has in the sample at hand. But there is only one value of the true $\sigma_{\bar{X}}$ for a sampling distribution. As a consequence of this, our test statistic is also an estimate. The sampling distribution of our estimated test statistic is more variable than the normal distribution (which does accurately describe *known* test statistics.) The actual sampling distribution for estimated test statistics concerning one- and two-sample means tests are shaped as t distributions. As sample size increases, the extent of differences between a normal distribution and a t distribution diminishes, until the two are identical at the theoretical point where $df = \infty$. This is the reason that some researchers will use the normal distribution as an approximate sampling distribution for tests in which σ is unknown. We have suggested that $N > 100$ is an appropriate guide for knowing when such an approximation can be used. But a t distribution remains the more accurate of the two possibilities whenever σ is unknown.

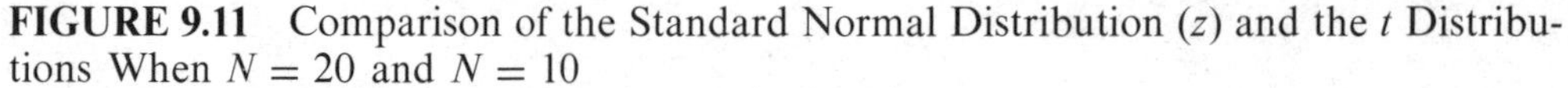

FIGURE 9.11 Comparison of the Standard Normal Distribution (z) and the t Distributions When $N = 20$ and $N = 10$

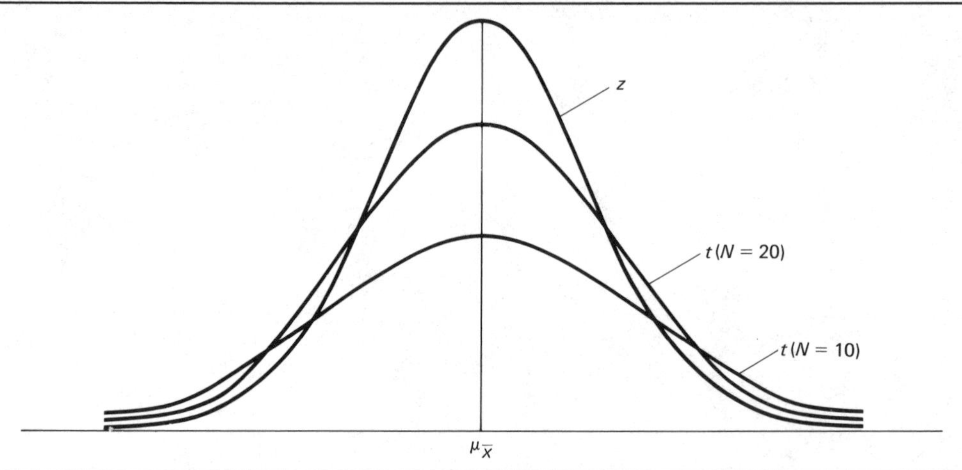

The tabular z scores from Table D2 will not accurately describe the appropriate rejection regions for these curves because they have more area in the tails than does the z curve. As a result of using those z scores we would decide to reject some null hypotheses when the sample data did not truly warrant such a decision (an increased probability of a Type I error).

Fortunately, the tabular values for these more variable sampling distributions are known and also recorded for our use. These distributions are known as *Student's t distributions.* "Student" was the pen name of William S. Gossett, who worked as a consultant to a brewery in Dublin, Ireland, around the turn of the twentieth century. Gossett calculated the tabular values for these sampling distributions with small samples and then published them under a pen name because of a contractual agreement he had with the brewery. The standard scores for these distributions are called *t scores.* They were calculated under the assumption that the population was normally distributed. Of course, the more closely a research situation conforms to this assumption, the more accurate the results of a test using t will be. As a rule, most researchers use the t statistic rather than the z whenever the sample size falls to 30 or less and the population is nearly normal but σ is not known. Because researchers are often unsure of the shape of the populations' distribution, they may use the t statistic with samples over 30 as well. Remember that the sampling distribution can be affected by both the normality of the population's distribution and the size of the sample. This is the reason for such apparent indecisiveness about a rule to follow for choosing z or t to test inferences. As a guide for you, we will use t whenever the population standard deviation (σ) is unknown, the sample size is 100 or less, *and* the population can be assumed to have an approximately normal shape.

As a consumer of social research you should anticipate seeing several uses of z and t tests. Some will meet the assumptions, some will not. Most will approximate them. This makes it difficult to evaluate the results. As it turns out, the t statistic will often result in correct decisions about null hypotheses even when its assumptions are not well met. Statisticians call such a test *robust.* Consequently, researchers will often use the t statistic even when it may seem that it is not appropriate. There are still limits, though. It should not be used when the population is known to depart markedly from normal, σ is not known, and the sample is very small (under 30). A nonparametric test is a much better alternative to use in such a situation. Some organization to this array of considerations is provided in Figure 9.12.

Tabular Values of Student's t

The tabular values of the t statistic are reported in Table D3 of Appendix D. To use this table you must know a quantity called the *degrees of freedom* (*df*) for each research problem. The degrees of freedom quantity is a mathematical idea that refers to the number of observations that are free to vary after we have placed some restriction on the data set. As an illustration, suppose that we had a set of five scores: 2, 4, 3, 5, and 1. These sum to 15 and have a mean of 3. Now let us require that the data set continue to have a mean of 3 but we do not insist that every observation keep its present value. How many of the observations can you change to a new value and still have their mean be 3? If you experiment with this, you will find that you can change any four of the observations to any other value you want, but the fifth one would not be free to become just any number. It would have to be one particular value so that the sum would still be 15 and the mean be 3.

For example, suppose that you changed the first four scores to: 3, 3, 4 and 4. You would have to make the fifth score a 1 or else the mean would not be 3. As this demonstrates, whenever we are working with the mean from a single data set we have $(N - 1)$ degrees of freedom. Table D3 is used for single-sample mean problems by calculating the degrees of freedom with the $(N - 1)$ formula. For example, a sample of 25 cases has 24 degrees of freedom for a single sample mean test.

Look now at Table D3. Note that it shows only some of the tabular t values for the many t distributions. Tabular values are shown for the levels of significance most often used with one- and two-tailed tests. A complete set of tabular values would require a separate table such as Table D2 for every degree of freedom (corresponding to every unique t sampling distribution). Instead, it is customary to rely on an abbreviated summary table like Table D3. For example, if we were testing a two-tailed null hypothesis at the .05 alpha level with a sample of 18 cases, we would look in the table opposite the row for 17 degrees of freedom and the column for a .05, two-tailed test. The tabular t value is ± 2.110. This is the value that the test statistic would need to equal or exceed in order for the null hypothesis to be rejected. That test statistic is found by a formula that was also used with our previous z test statistic:

FIGURE 9.12 Choosing an Inference Test for a Single-Sample Mean

Normally Distributed Population	**Known Population σ**	**Large Sample ($N > 100$)**	**Use**
Yes	Yes	Yes or No	z with $\sigma_{\bar{X}} = \sigma/\sqrt{N}$
No	Yes	Yes or No	z with $\sigma_{\bar{X}} = \sigma/\sqrt{N}$
Yes	No	Yes	z with $\hat{\sigma}_{\bar{X}} = s/\sqrt{N}$
No	No	Yes	z with $\hat{\sigma}_{\bar{X}} = s/\sqrt{N}$
Yes	No	No	t with $\hat{\sigma}_{\bar{X}} = s/\sqrt{N}$
No	No	No	No parametric test available

$$t = \frac{\overline{X} - \mu}{\widehat{\sigma_{\overline{X}}}}$$

t Transformation for a Single-Sample Mean Test

Let's now look at some examples of the situations in which researchers commonly use the *t* statistic to solve inference problems for single-sample means.

Normal Population with σ Not Known

Suppose that a test has been devised to measure aptitude toward social science. The test has been standardized to yield a normal distribution with a mean of 50. The test is given to a random sample of 20 college students, with the result being a mean of 58 and $s = 20$. Does this sample represent the population of student scores in which the mean is 50?

Step 1. The hypotheses.

H_0: $\mu = 50$ The population of scores has a mean of 50. A mean in a random sample may differ from this due to sampling error.

H_1: $\mu \neq 50$ The population of scores does not have a mean of 50. A random sample will have a different mean for reasons other than sampling error.

Step 2. The criteria for rejecting H_0.

We will use the .05 level of significance in this nondirectional, two-tailed test of H_0. Since we can assume that the population of test scores is normally distributed, but we do not know its standard deviation (σ) and we have a small sample ($N \leq 100$), we can use a *t* distribution as the approximation of the sampling distribution. In a one-sample mean test the exact *t* distribution to use is the one with $df = N - 1 = 20 - 1 = 19$. Table D3 indicates that the tabular *t* value for a .05, two-tailed test is ± 2.093. Our two rejection regions are the area equal to and above $+2.093$ and the area equal to and below -2.093. If the sample result falls into either of these areas, we will reject the null hypothesis.

Step 3. The test statistic.

To calculate the test statistic, we must first find the standard error. Since we do not know σ, we must substitute its estimate (s) in our formula.

$$\widehat{\sigma_{\overline{X}}} = \frac{s}{\sqrt{N}} = \frac{20}{\sqrt{20}} = \frac{20}{4.472} = 4.472$$

The test statistic can now be found:

$$t = \frac{\overline{X} - \mu}{\widehat{\sigma_{\overline{X}}}} = \frac{58 - 50}{4.472} = \frac{8}{4.472} = 1.79$$

Step 4. The decision about H_0.

The test statistic (1.79) does not exceed the tabular statistic (2.093); it is not in either rejection region. Therefore, we cannot reject the null hypothesis. (See Figure 9.13.) The probability is greater than .05 that sampling error could have caused us to draw a random sample from a normal population with a mean of 50, but that sample mean was 58 or more distant from 50. Our sample mean does not differ significantly from the proposed population mean; $t = 1.79$, $p > .05$, n.s.

FIGURE 9.13 Testing H_0: $\mu = 50$

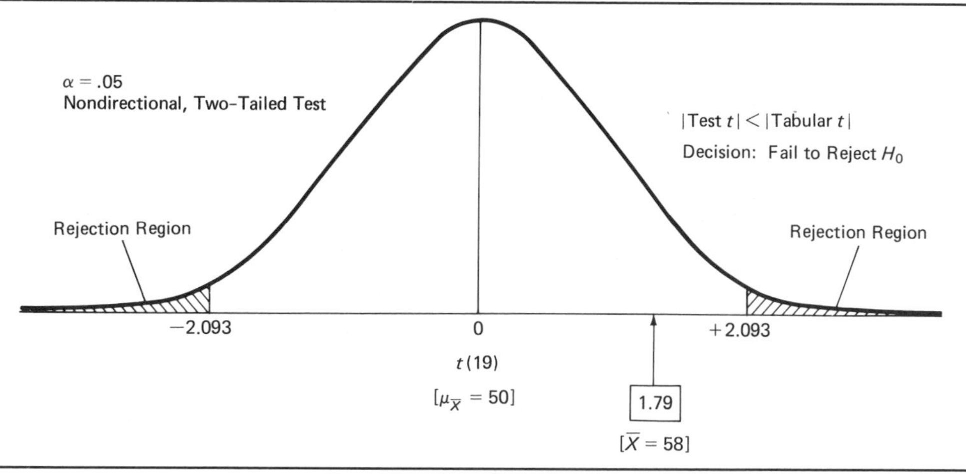

Step 5. The conclusion.

The standardized test does appear to have a rather stable mean of 50. The sample accurately represents this population within the limits of sampling error that are considered acceptable to our hypothesis test. We recognize the possibility that this conclusion includes some risk of a Type II error. The sample may not truly represent this population. But we are forced to take this risk given the conditions under which the test was conducted. The risk does not seem to be inordinate.

Let's consider another example of a single-sample test using the t statistic. A health maintenance official wants to know whether a new hospital's use of hospitalization for appendix removals is comparable to such use by other established hospitals in the region. The region's hospitalization stay for this type of surgery is normally distributed about a mean of 8 days. The new hospital is studied for a two-month period in which the mean stay for this procedure is 10 days with $s = 1.2$, calculated from a sample of 15 cases. Is there evidence that this new hospital differs significantly in this respect from the established hospitals in the region?

Step 1. The hypotheses.

H_0: $\mu = 8$ — The population of hospitalizations for appendix removals has a mean of 8 days. A random sample may differ from this because of sampling error.

H_1: $\mu \neq 8$ — The population of hospitalizations for appendix removals does not have a mean of 8 days. Sampling error is not the reason a random sample would differ from this.

Step 2. The criteria for rejecting H_0.

Let the level of significance be .01 for this two-tailed test. A t distribution with 14 degrees of freedom can be used as the sampling distribution. Table D3 shows that the tabular statistic for this situation is ± 2.977. We have two rejection regions. One begins at $+2.977$ and continues to the right. The other begins at -2.977 and goes to the left. A sample result in either rejection region will cause us to reject the null hypothesis.

Step 3. The test statistic.
The standard error can be estimated as follows:

$$\hat{\sigma}_{\bar{X}} = \frac{s}{\sqrt{N}} = \frac{1.2}{\sqrt{15}} = \frac{1.2}{3.8730} = .310$$

The test statistic is

$$t = \frac{10 - 8}{.310} = \frac{2}{.310} = 6.45$$

The test statistic (6.45) does exceed the tabular statistic (2.977); it is in the rejection region to the right. Thus we reject the null hypothesis. (See Figure 9.14.) Our sample result can be thought of as being the consequence of sampling error less than 5% of the time. This causes us to doubt the null hypothesis. Our sample mean does differ significantly from the proposed population mean ($t = 6.45$, $df = 14$, $p < .01$).

Step 5. The conclusion.

Since we rejected the null hypothesis, we draw the inference that the population of hospitalizations from which the sample could be randomly drawn is not one with a mean of 8 days. Our sample is significantly different from the pattern of hospitalizations for appendix removals at the other regional hospitals. It appears to represent a different population from the other hospitals. We could be making a Type I error with this decision, but the risk is not greater than a .01 probability. The health maintenance officer may want to investigate the new hospital further. Maybe the study came too early in the hospital's career. Perhaps it will conform to the regional pattern after it has been functioning for a while. But the results of the study also cause the official to be suspicious of the hospital and to watch for evidence that it truly has a different approach to treating patients undergoing appendix removals. Of obvious concern is whether this hospital is going to routinely use longer hospitalizations and therefore charge more for this operation than other hospitals. A follow-up study may be in order which uses a one-tailed test to check this idea. A larger sample (obtained by a longer study period) would be beneficial because of the smaller

FIGURE 9.14 Testing H_0: $\mu = 8$

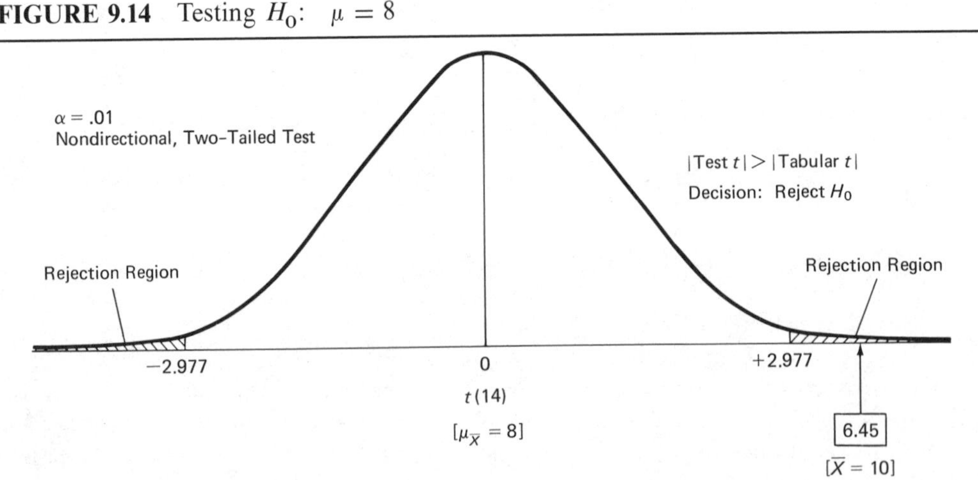

standard error that is likely to result. This would increase confidence in the sample mean without altering the other criteria for the test.

As the last two examples show, the process followed in a one-sample mean test using the t statistic is very similar to the process when the z statistic is used. Only the assumptions of the test and calculation of the standard error may differ. The same basic logic and steps are used for both tests. Remember that these t and z tests are parametric tests. They do require some assumptions about the underlying population's parameters. Later in this chapter we look at two nonparametric tests that can be used as alternatives to these parametric tests when the population assumptions are not reasonable for a research situation.

Estimating Parameters from Single Samples

It would not be unusual for a social researcher to have little idea of a population's true parameters. Indeed, the reason that a sample may be drawn is to estimate unknown parameters such as a mean or proportion. Researchers distinguish between two types of estimations: point estimation and confidence interval estimation.

A *point estimate* is a single estimated value of an unknown parameter based on a sample statistic. For example, we might wish to estimate the mean IQ for a school's students in the tenth grade. To do this we could draw a random sample of the school's tenth graders, administer a standardized IQ test, and calculate the sample's mean score, $\bar{X}$. Suppose that sample mean was 104. Our point estimate of the entire school's mean IQ for tenth graders would also be 104. A sample mean provides an unbiased point estimate of a population's mean, μ. We realize that the sampling error may cause our estimate to be incorrect, but we do not expect the point estimate to be necessarily above or below the true mean.

Suppose that we wanted to estimate the standard deviation for the IQs of those same students. Again, we could select a random sample of the students, measure their IQs, and calculate the descriptive standard deviation (S) for the sample. Suppose that this turned out to be 13. Would our best point estimate of the population's standard deviation (σ) be 13? No. Remember from our earlier discussions that S is a biased estimate of σ. That was the reason for the introduction of the inferential standard deviation, s. Their slight difference in computation (using N versus $N - 1$ in the denominator) was enough to make s an unbiased estimate of σ. Thus, if we wanted an unbiased estimate of the population's standard deviation in IQs (which we would want, of course), we would calculate s, not S. Suppose that s was 14.5. Our best point estimate of σ would be 14.5.

How accurate are these kinds of point estimates, even considering their unbiased character? You never really know unless you can determine the true value of the parameters being estimated. Researchers can enhance the credibility of estimates if they use confidence intervals instead of point estimates. A *confidence interval estimate* is a range of values centering on a sample statistic used to estimate an unknown parameter with a set degree of confidence. A confidence interval for our IQ problem above might be "92 to 116 with 95% confidence." Roughly, this means that we are quite certain that the true mean IQ of the school's tenth graders is between 92 and 116. We might widen the interval to "86 to 122 with 99% confidence." Now we are nearly certain that the true mean IQ is within these two limits. As this illustrates, we have increased confidence in intervals that are wider. That should seem reasonable.

Confidence intervals are calculated as rigorously as any other measure. For each particular parameter being estimated there is a separate formula, but each follows a general format:

$$\text{confidence interval} = \text{observed statistic} \pm \begin{pmatrix}\text{tabular} \\ \text{standard} \\ \text{score}\end{pmatrix}\begin{pmatrix}\text{standard} \\ \text{error}\end{pmatrix}$$ General Formula for a Confidence Interval

A tabular standard score (z or t) is multiplied by an appropriate standard error and then added to the sample statistic to find the upper limit of the confidence interval. The same product is also subtracted from the sample statistic to find the lower limit of the interval. The *level of confidence* for the interval is the probability (expressed in percentage form) that the interval is one of a set of such intervals that contain the parameter being estimated. It is found by the simple formula

$$\text{level of confidence} = (1 - \alpha)\,100$$ Level of Confidence

where α = level of significance

Thus a 95% level of confidence is used when the level of significance (α) is .05. A 99% level of confidence is used when the level of significance is .01, and so on. Let us see how confidence intervals are constructed for single-sample proportions and means.

Confidence Interval for a Proportion

Suppose a researcher has found that .58 of the 77 persons surveyed in a random sample of a community's households have read at least one book in the last year. What is the 95% confidence interval that the researcher can calculate to estimate the proportion of all persons in the community who have read a book?

A sampling distribution must be available for a confidence interval problem, just as it was for an inference test. We can use the same rules to decide on the appropriate sampling distribution for both types of procedures. Here we would want to calculate the NP and NQ products to see whether a normal distribution could be assumed. However, we do not know P (or Q); that is the reason for the interval estimate! But the sample proportion p is an unbiased estimate of the population's proportion P, so we can substitute the first for the second.

$$Np = 77(.58) = 44.66$$
$$Nq = 77(.42) = 32.34$$

Both of these are over 5, so a normal sampling distribution can be used for our problem. Next we determine the standard score for this distribution that goes with our level of confidence. A 95% confidence interval corresponds to a .05 level of significance. The z score that leaves .05 of the area under a normal curve in the tails is 1.96. (Note that confidence intervals use a two-tailed view of the sampling distribution.) Now we need the standard error. Again, we substitute the sample values p and q for the unknown P and Q values to estimate the standard error as shown:

$$\hat{\sigma}_p = \sqrt{\frac{pq}{N}} = \sqrt{\frac{(.58)(.42)}{77}} = \sqrt{\frac{.2436}{77}} = \sqrt{.00316} = .056$$

Now we can combine these pieces of information to find the sought-after confidence interval (CI):

$$CI = p \pm (z)(\sigma_p \text{ or } \hat{\sigma}_p)$$

Confidence Interval for a Single-Sample Proportion

$$\begin{aligned} 95\%\ CI &= .58 \pm (1.96)(.056) \\ &= .58 \pm .110 \\ &= .47 \text{ to } .69 \end{aligned}$$

Our 95% confidence interval estimate of the true proportion of the community's population which has read at least one book in the past year is .47 to .69, meaning between 47 and 69%.

Interpreting a Confidence Interval

What does our 95% confidence interval of .47 to .69 really mean? Does it mean that we have a 95% chance that the true proportion of book readers is between .47 and .69? Not exactly. Some researchers, and many nonresearchers, commonly misunderstand a confidence interval to be saying this. The probability that P is in this particular interval of .47 to .69 is either 1 or 0; either it is or it is not. Probability refers to outcomes that are variable. The true proportion of residents in the entire community who are book readers is a set value (for the one time period being studied); it does not fluctuate from one value to another. It is a *sample* proportion (p) that will vary due to sampling error.

To understand this you must consider a process of repeatedly drawing random samples from the population, finding each sample's proportion (p), and constructing a confidence interval around each of these sample values. If the process is repeated a very large number of times, we would approximate the sampling distribution of all possible confidence intervals that could be found from one population using samples of a set size while maintaining a certain level of confidence.

Such a hypothetical exercise is illustrated in Figure 9.15. Consider this entire set of 95% confidence intervals around sample proportions of book readers. Ninety-five percent of these intervals would contain the true population proportion of book readers and only 5% of the intervals would not; e.g., the one around p_{10}. Whether our one interval was one of the successful 95% or the 5% that were unsuccessful, we cannot know. Of course, it seems likely that our interval would be one of the successful ones since there are so many more of them, but it may not be.

How should the confidence interval constructed from a single sample be interpreted, then? There is the long answer that we have just gone through, and a short one. The long one can be summarized by a statement such as: If repeated random samples of 77 cases are drawn from this population of residents, 95% of the 95% confidence intervals constructed as ours was would contain the true proportion of book readers in the community.

The short interpretation often given by researchers is: We are 95% confident that *such an interval as* .47 to .69 contains the true proportion of book readers. The innocent-looking phrase "such an interval as" carries the more complex meaning of the long interpretation. If you remember the long meaning when you see the short statement, you will not be mislead. Without the long meaning in mind, it is very easy to form the impression that a single confidence interval has much more precision than is justified.

Confidence intervals around single-sample proportions are probably the most common type of single-sample inference. Most of the others are comparatively infrequent

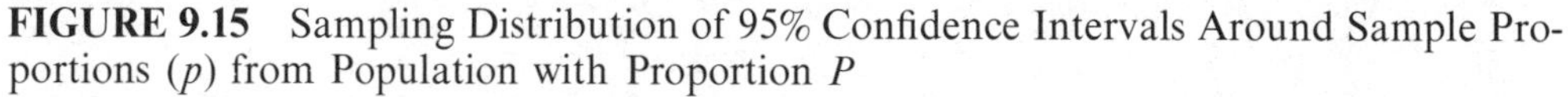

FIGURE 9.15 Sampling Distribution of 95% Confidence Intervals Around Sample Proportions (p) from Population with Proportion P

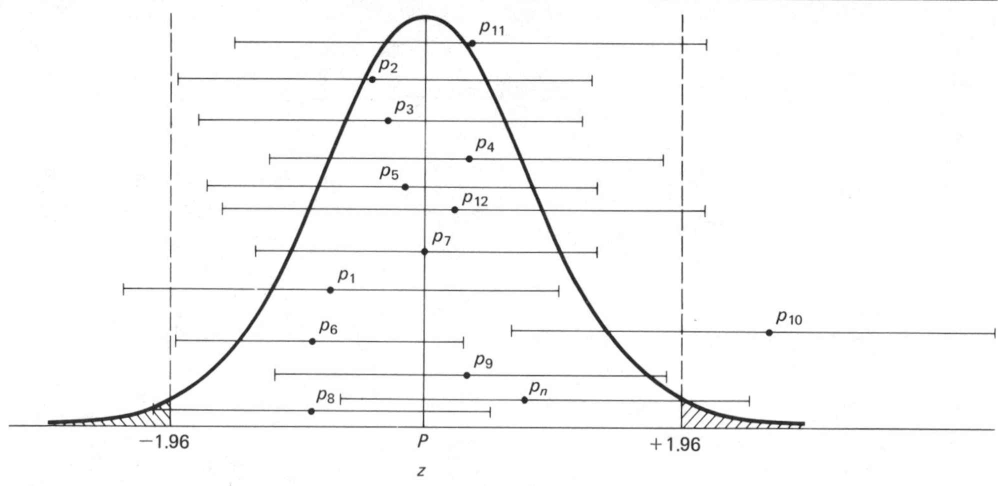

including those in this chapter. Research Examples 9.1 and 9.2 show the use of the single-sample proportion confidence interval statistic.

Research Example 9.1 illustrates the very familiar situation of a report from an opinion poll. It concerns the views by American adults about the state of black–white race relations in 1980 compared to earlier years. Whether viewed in the national or local contexts, both blacks and whites express the opinion that blacks were better off in 1980 than they had been before. The article points out that the percentages can be considered to be accurate within ±3 percentage points at the 95% confidence level. This is the same as saying that the 95% confidence interval around the results expressed as proportions would vary .03 on either side. That is, the tabular standard score ($z = 1.96$) multiplied by the standard error (σ_p) would be about .03. Notice that national public opinion polls typically must use stratified random samples of only about 1500 persons to achieve this degree of accuracy. As noted in Chapter 7, their validity has been remarkably high since sophisticated sampling techniques became standard.

RESEARCH EXAMPLE 9.1 Confidence Interval Estimate of a Single Proportion

From Everett Carll Ladd, "The State of Race Relations—1981," Public Opinion, *4 (2), April–May 1981:32ff. Reprinted by permission.*

When discussing a social problem where there is a sense that great progress has been made and yet, at the same time, where great difficulties remain, a commentator runs the risk of misplaced emphasis. If he stresses the former dimensions, he sounds like Pollyanna; if the latter, like some combination of Cassandra and Malthus. Race relations in the United States in the early 1980s, and specifically the way black and white Americans view recent developments in their interrelations, present precisely this challenge to the analyst. For the survey data, some of which are presented in the pages that follow, show emphatically that whites and blacks alike believe great progress has been made, and that the present climate is indeed one of greater tolerance than in the past. They also show that tensions remain.

Whites and blacks, alike, by large majorities now believe that the position of black Americans is better than a decade ago. Very few in either community think race relations in the area where they live have been deteriorating; many, especially among blacks, see the climate improving. . . .

Most of the responses shown in these surveys were gathered either by personal interviews (*Harris* and *Gallup* polls) or by telephone (*CBS/New York Times* and the *NBC/Associated Press* polls). Unless otherwise noted, the samples usually consist of approximately 1,500 voting age men and women, chosen to constitute a representative sample of the entire U.S. population. In the typical sample of 1,500 respondents, there is a 95 percent chance or better that the margin of error will not exceed ± 3 percent variation from the distribution which would appear if the nation's entire population were questioned. The possibilities for error are larger when numbers are displayed for subcategories of each sample. . . .

A DECADE OF PROGRESS IN THE NATION . . .

Question: Compared to ten years ago, do you think blacks in America are a lot better off, a little better off, about the same, a little worse off, or a lot worse off?

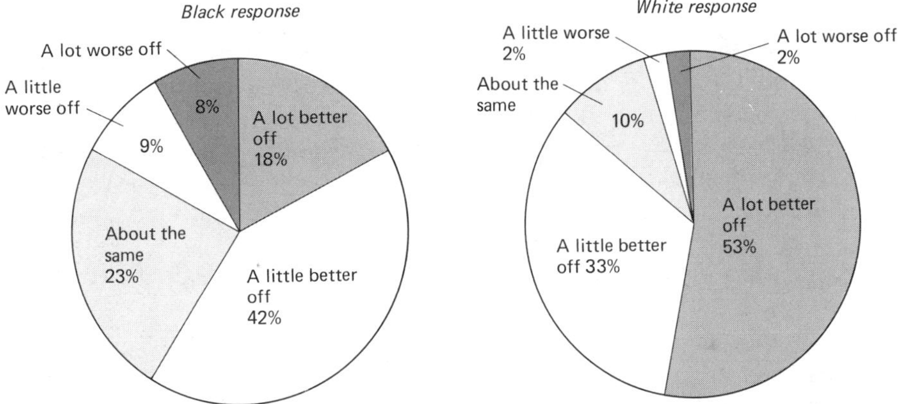

AND AT HOME

Question: Would you say that relations between whites and blacks in this area are better than they were a couple of years ago, worse than a couple of years ago, or just about the same as a couple of years ago?

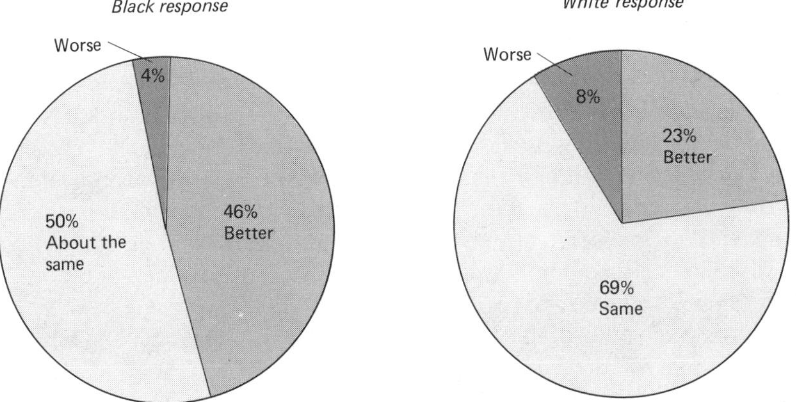

Source: top graphs: survey by NBC News/Associated Press, May 28–30, 1980; bottom graphs: survey by the Roper Organization (Roper Report 80-7), July 7–13, 1980.

Research Example 9.2 shows another confidence interval around a single-sample proportion. This one concerns the percentage of persons receiving two federal payments upon retirement: a civil service survivor annuity and a social security benefit. The study was done to help assess the possible effect of making it mandatory to participate in the largest social security program (Old Age, Survivors, and Disability Insurance). The results shown in the table represent a 10% random sample (16,893) from the possible population (168,930) of persons who were receiving dual benefits. The author points out that only 3.2% of the sample received a primary insurance amount (a type of social security benefit) of $300 or more. The standard error is said to be .01, and thus the 95% confidence interval around the 3.2 percentage was reported to be 3.0 to 3.4. Note, too, that the author gives the rather casual interpretation of the interval that you are not to interpret literally.

RESEARCH EXAMPLE 9.2 Confidence Interval Estimate of a Single Proportion

From Daniel N. Price, "Federal Civil Service Adult Survivor Annuitants and Social Security, December 1975," Social Security Bulletin, *44 (8), August 1981:3ff.*

The issue of universal coverage under the Old-Age, Survivors, and Disability Insurance (OASDI) program has become particularly prominent since the mid-1970s. One of the two largest groups of workers presently not covered under OASDI is the Federal civilian work force, although individual workers often obtain coverage through employment outside of their Federal careers. Examination of the experience of Federal civil service annuitants is one relevant way of informing policymakers and legislators on the universal coverage issue.

The main purpose of this article is to add to the picture already drawn of civil service retired worker annuitants by presenting a view of survivor annuitants—that is, those persons whose annuity derives from the death of a Federal worker or a Federal annuitant. By examining annuitant characteristics . . . the analysis helps define who would be affected if the Federal civil service retirement system and the Social Security program were coordinated.

The issue of gaps and overlays produced by the current lack of coordination between the two systems is also examined. This review shows, for example, the extent to which civil service survivor annuitants have employment and benefits under the OASDI program. Further, the data on wage-replacement rates throw light on unintended benefit subsidies that may occur because of joint receipt of civil service annuities and OASDI benefits.

This article reviews the recent experience of adult survivor annuitants—those persons who receive annuities as widows or widowers of Federal civil service employees or annuitants [except children]. The analysis is based on a 10-percent sample of the 310,310 widows and widowers on the civil service rolls as survivor annuitants on December 31, 1975. Information on earnings and benefits under the OASDI program was obtained from the files of the Social Security Administration. . . . Among the 267,800 civil service survivor annuitants . . . whose experience under the Social Security program was available, 168,930 or 63 percent, were receiving benefits under both programs. . . .

Distributions of dual beneficiaries by their PIA [Primary Insurance Amount] and family benefit amount under OASDI are given in [the table]. Very few (only 2.1 percent) dual beneficiaries had special age-72 or transitionally insured benefits (a

PIA of $69.60) under OASDI, or, at the other end of the range, a PIA of at least $300 (3.2 percent). The median PIA of all dual beneficiaries, $148, was much lower than that for the general population of Social Security beneficiaries, $221. . . .

Data in [the table] can be used to illustrate the minimal sampling error in the results presented in this article. The proportion of dual beneficiaries with a PIA of $300 or more was 3.2 percent of 16,893 sample cases. The standard error for 3.2 percent of the beneficiaries is 0.1 percent. Thus, with 95-percent confidence, the proportion of annuitants receiving a PIA under OASDI of $300 or more was 3.2 percent plus or minus 0.2 percent, or can be said to lie between 3.0 percent and 3.4 percent.

Primary Insurance Amount for Civil Service Survivor Annuitants, December 31, 1975[a]

Monthly Amount	Percent Receiving Primary Insurance Amount[b]
Total number	168,930
Total percent	100.0
Less than $101.40	2.1
$101.40	26.0
$101.50–$129.90	12.0
$130.00–$159.90	15.9
$160.00–$189.90	13.8
$190.00–$249.90	18.6
$250.00–$299.90	8.4
$300.00 or more	3.2
Median	$148

[a] Adapted from Table 5 in the original.
[b] Based on 10% random sample.

Confidence Interval for a Mean

Confidence intervals around a sample mean follow the same pattern as those for proportions. A first step is to determine the appropriate sampling distribution, either z or t. Then the appropriate tabular standard score from that distribution is found and multiplied by the standard error. This product is both added and subtracted from the observed mean in the sample to determine the limits of the confidence interval. Two examples can show this.

Suppose that in a random sample of 22 church members that each has been asked to reveal his or her highest level of education completed. The mean level was 10 years with an inferential standard deviation (s) of 2.1 years. What is the 99% confidence interval for this finding?

First we must decide whether a standard normal or one of the t curves is an appropriate sampling distribution. We do not know the population's standard deviation σ, so it will have to be estimated from s. We also have a small sample. These circumstances make it impossible to assume a standard normal sampling distribution. Can we assume that the population of church members has a normal distribution of levels of education? Although there may not seem to be an obvious answer to this, it is not unreasonable to believe that such a distribution would be nearly normal. Further, the t statistic is tolerant of some deviation from the normality assumption. So we might cautiously let our sampling distribution be a t distribution with 21 ($N - 1$) degrees of freedom.

We want a 99% confidence level, so we are working from the equivalent of a .01 level of significance. Table D3 indicates that the relevant tabular score at this level of significance and 21 df is 2.831. Our standard error is estimated as

$$\hat{\sigma}_{\overline{X}} = \frac{s}{\sqrt{N}} = \frac{2.1}{\sqrt{22}} = \frac{2.1}{4.6904} = .448$$

We can now calculate the confidence interval:

Confidence Interval for a Single-Sample Mean

$$CI = \overline{X} \pm (z \text{ or } t)(\sigma_{\overline{X}} \text{ or } \hat{\sigma}_{\overline{X}})$$

$$\begin{aligned} 99\%\ CI &= 10 \pm (2.831)(.448) \\ &= 10 \pm 1.268 \\ &= 8.73 \text{ to } 11.27 \end{aligned}$$

This means that we are 99% confident that an interval such as 8.73 to 11.27 will contain the mean level of education for the entire population of church members.

Take another example. To study the incidence of repetitions in petty law violations a researcher has randomly selected the overtime parking citations for 145 persons from the parking areas of a city's central business district. The present citations are cross-checked with past records over the previous five years. The mean number of citations for the 145 persons over this period is 4.6 with a s value of 1.3. What is the 90% confidence interval estimate for the population of persons receiving citations for the same area and time period?

We have no information about the population of citations. Thus we do not know the shape of that distribution or its mean and standard deviation. But we do have a rather large sample. The Central Limit Theorem tells us that we can use the standard normal sampling distribution for this problem.

The 90% level of confidence corresponds to a .10 level of significance. Table D2 tells us that the z score which leaves two extreme tails of 5% each (10% total) is 1.645. We will again have to estimate the standard error using s as a substitute for σ:

$$\hat{\sigma}_{\overline{X}} = \frac{s}{\sqrt{N}} = \frac{1.3}{\sqrt{145}} = \frac{1.3}{12.0416} = .108$$

Now we can find the confidence interval:

$$\begin{aligned} 90\%\ CI &= 4.6 \pm (1.645)(.108) \\ &= 4.6 \pm .178 \\ &= 4.42 \text{ to } 4.78 \end{aligned}$$

If repeated samples of 145 cases were randomly drawn from this population of citations and confidenceeintervals were constructed for each of them as this one was, we expect that 90% of these intervals would contain the true mean number of overtime parking citations received by the relevant population.

ON THE SIDE 9.1 William S. Gossett ("Student")

William S. Gossett (1876–1937) was born in Canterbury, England, and educated at Winchester and New College, Oxford, where he studied chemistry and mathematics. His career was centered on his work for the brewery of Arthur Guinness Son and Co. Ltd. in Dublin, Ireland, where he began work in 1899. The brewery had decided to employ a number of scientists to improve the quality of its product. The problems presented by the brewery concerned the variable quality of the raw materials (barley and hops) used, the temperature fluctuations in the facility, and the necessity for conducting short experiments to study the production process. Gossett was the most trained in mathematics of the new scientists and so became the focus of many questions. Early on, he realized the limitations of large-sample theory when applied to the circumstances in the brewery and set about to discover a correct approach for small samples.

Gossett was not highly trained as a statistician but he solved the small-sample problems mostly on his own. He did correspond and meet with such reknown figures as Karl Pearson (who was slow to show interest in the issues) and R. A. Fisher (who was much more sympathetic at the start). Gossett produced a report on sampling errors in small samples for Guinness in 1904 and the ideas have been used ever since in the brewery. He was given a year's leave to study at the Biometric Laboratory with Pearson at University College in 1906. In 1908, in *Biometrika*, he published his most famous work, entitled "The Probable Error of a Mean" under the pen name "Student," to avoid any contractual conflict with Guinness. In it he showed the distribution of a sampling experiment based on small samples ($N = 4$ to 10) from a normal population in which the mean and standard deviation of the samples were independent. He introduced the formula for what we now call the t test and made the distinction between sample and population standard deviations, which previously had not been appreciated.

A year later (1909) Gossett published a paper on the probability of the correlation coefficient, showing how it could be evaluated in small samples, as he had before with the mean. He published other papers on such subjects as nonrandom samples, the Poisson distribution, an extension of his tables for small sample means with $N = 2$ to 30, and issues concerning experimental design in agricultural contexts.

Although Gossett's reputation was made in statistics, his life was not concerned mainly with this field. He was more involved in the routine work of the Guinness brewery, including inspections of the barley fields and work in plant genetics. He was considered to be an unassuming man who modestly asked for the opinions of others but who had a way of "getting there first," ahead of those more highly trained than himself. He was a hard worker who preferred to do his own calculations, even to the point of deriving theoretical quantities himself rather than locating them in a published table. His students, clients, and colleagues found him to be willing to give them as much time as they desired and always to be in a kindly and tolerant mood. He left Dublin in 1935 to take charge of a new Guinness brewery in London but died suddenly two years later at the early age of 61. It is clear from the published evidence that he was a most loved and respected person, with a wide circle of friends and acquaintances.

Based on E. S. Pearson, "6. W. S. Gossett ('Student')," 348–351; L. McMullen, "(1) 'Student' as a Man," 355–360; and E. S. Pearson, "(2) 'Student' as a Statistician," 360–403, all in E. S. Pearson and M. G. Kendall, *Studies in the History of Statistics and Probability*. London: Charles Griffin & Company Ltd., 1970.

Nonparametric Tests for Single Samples

The z and t tests for single-sample means are parametric tests. They require certain assumptions about population parameters; either its distribution must be normal in shape or its standard deviation σ must be known. What if these assumptions cannot be met? The z test for single-sample proportions is a nonparametric test, as would be the binomial solution for small-sample, dichotomous proportions questions (as in Chapter 7) if they were put into a hypothesis-testing framework for the purpose of drawing an inference. Neither of these requires the researcher to assume that the population distribution has a particular shape or standard deviation. However, not all problems involving nominal-level data (and, hence, statistics of frequencies and proportions) are dichotomous. Nor may we want to reduce the categories of measurement to only two possibilities so that those earlier techniques could be applied. Instead, we may want to conduct an inference test for multiple-category measurement at the nominal (or, even, ordinal) level. Two nonparametric tests are introduced here to meet these two types of inference needs. The chi-square single-sample test (χ^2-I) is applied to nominal data in the form of frequencies and can be used in the place of our proportions tests. The Kolmogorov-Smirnov single-sample test (KS-I) is applied to ordinal data in the form of cumulative proportions and can be used in place of our means tests or our proportions tests if we can satisfy the higher level of measurement requirement.

Chi-Square Single-Sample Test (χ^2-I)

In Chapter 5 we looked at the chi-square statistic as a method to determine whether there was any association between two nominal variables. A special version of this same statistic is often used to determine whether a single sample's distribution on one nominal variable is significantly different from the distribution that was theoretically expected to occur. In this form it is called a *chi-square goodness-of-fit test.* We can use this as a nonparametric inference test for data at the nominal level provided that we have a random sample of independent observations. It can be used in place of a z test for single-sample proportions when we are working with a dichotomous variable. We can also use it when we have a nominal distribution across several categories of measurement. Of course, more complex measurements can be reduced to one of these if it creates a meaningful situation to analyze. Let's look at an example.

Suppose that we are examining public perception of the relative importance of several national political issues. We have drawn a random sample of 60 adults from a small town and asked them to choose one of four issues as the most important to them. The distribution of these first choices is shown in Table 9.1.

TABLE 9.1 Distribution of Most Important National Issue

Issue	f
Inflation	23
Taxation	15
Energy	12
Social welfare	10
Total	60

Step 1. The hypotheses.

The null hypothesis represents the idea that for the entire town's population of adults, the importance choices would be equally distributed across the four national issues. The frequencies expected (f_e) for the issues are all the same. If the observed frequencies (f_o) match this, the calculated chi-square value will be zero. Only sampling error should cause any fluctuation from this outcome. The alternative hypothesis represents the idea that the population's importance choices would be unequally distributed across the four national issues. Some issues would be more important than others to a greater extent than could be attributed to sampling error.

H_0: $\chi^2 = 0$ The population of importance choices is equally distributed across the issues. Sampling error may cause observed frequencies in a random sample to have an unequal distribution.

H_1: $\chi^2 \neq 0$ The population of importance choices is unequally distributed across the issues. A random sample will have an unequal distribution beyond the expected influence of sampling error.

Step 2. The criteria for rejecting H_0.

Let's use the .05 level of significance. The sampling distribution is one of a family of chi-square distributions. These are positively skewed with the skew diminishing as the degrees of freedom increases. (See Figure 9.16.)

FIGURE 9.16 Sampling Distributions of Chi-Square for $df = 1, 3, 5,$ and 10

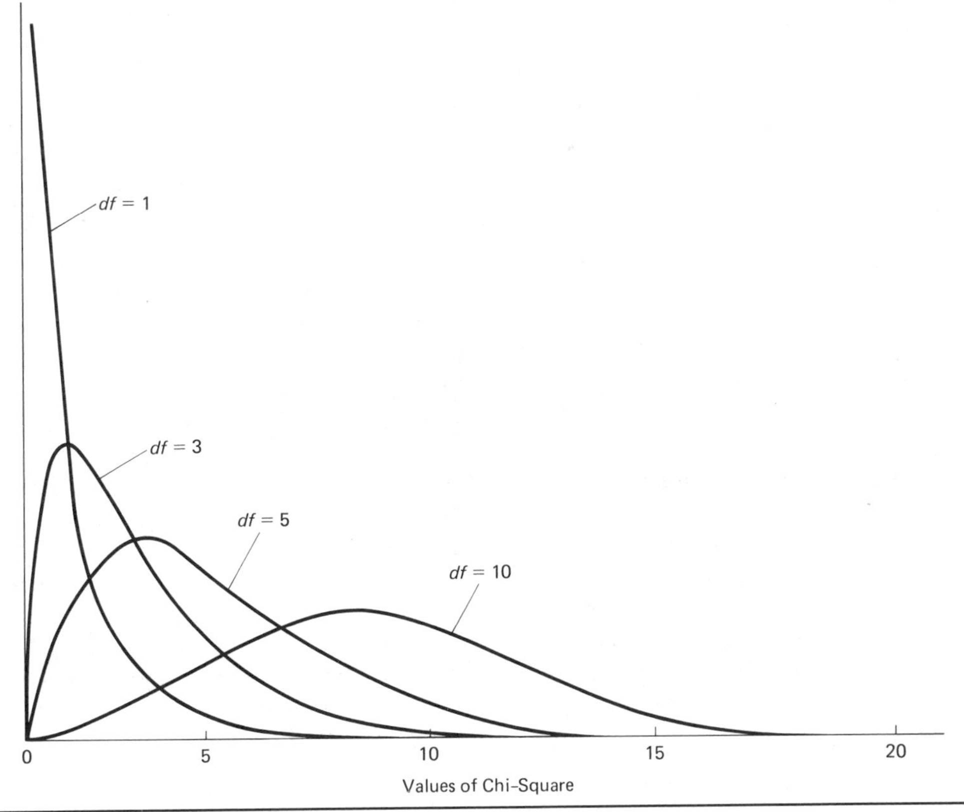

Although there are no parametric assumptions to meet, the chi-square test does have some simple nonparametric requirements. We must have independent observations from a random sample. The data need be only nominal, but the number of cases must not be too small or large. Historically, researchers have insisted that the minimum was 5 for any expected (not observed) frequency when there were only two measurement categories. In situations that failed to meet this standard researchers have commonly used a measure called Yates' corrected chi-square. However, the practice has been severely criticized by some. A more universally accepted alternative to the regular chi-square test for these situations is the Fisher's exact test. Both of these are discussed further in the section *Related Statistics* at the end of the chapter. If there are more than two measurement categories, many researchers have required that no more than 20% of those categories can have expected frequencies smaller than 5. They will combine or drop categories to achieve this. Further, an upper limit of 500–1000 is often cited as a maximum N since the calculated chi-square tends to be statistically significant even for modest patterns of expected-observed differences in very large samples. These assumptions concerning chi-square are new to our consideration of it as an inference test. They were not relevant to its use as a descriptive measure. The small-sample restriction is somewhat controversial and bears additional attention.

One difficulty with small samples (and thus low expected frequencies) relates to the continuity of the chi-square sampling distributions. It is possible for a sampling distribution to have a smooth continuous shape only when there are a very large number of plotting points. This occurs with the larger sample sizes because so many unique combinations of sample outcomes are created when we consider the complete set that is possible. With very small samples, there are only a limited number of possible outcomes, and thus the sampling distribution is described by a series of discrete (separated) plotting points. The binomial sampling distribution with small sample sizes is an illustration of this kind of discrete (not continuous) distribution. The tabled values for the chi-square statistic are calculated under the assumption that the relevant sampling distributions are continuous. If they are not, it is feared that the table may mislead us about the true probabilities for our observed results.

Tests of this thesis and the historically recommended Yates' corrected chi-square tend to show that both are not well supported (Camilli and Hopkins, 1978, 1979; Bradley et al., 1979). In usual research situations the regular chi-square test performs very adequately. Further, the Yates' measure is so extremely conservative that it renders most hypothesis tests meaningless. These issues are important because the continuity concern has become part of the culture of social statistics. You will find the Yates' measure in common use in spite of the evidence against it. So as we take our first look at chi-square as an inference test, you should be aware of the controversy about this analysis practice that is common to social research. As a guide for the proper use of the chi-square test, it now seems that it can be used with samples as small as 8 when there are only two measurement categories and 20 when there are more categories, provided that the sample's distribution is not highly skewed (overwhelmingly concentrated in only a few of the available categories.) Distributions from large samples (over 500) should be examined visually to verify the statistical analysis based on a chi-square test.

The tabular value of chi-square which serves as a tabular statistic defining a rejection region comes from Table D4 in Appendix D. Like the t table (Table D3), we must know the relevant degrees of freedom to use this table. For a chi-square goodness-of-fit test the following formula is used:

$$df = k - 1$$
where k = number of categories

Degrees of Freedom for χ^2-*I* Test

As Table 9.1 shows, we have four categories in our example. Thus

$$df = 4 - 1 = 3$$

Table D4 indicates the tabular chi-square value for various levels of significance and degrees of freedom. That table shows that the tabular statistic when $\alpha = .05$ is 7.815. The rejection region is that area beginning at 7.815 and extending above it to the right. (See Figure 9.17a.) Notice that a chi-square distribution uses only one right-side tail to define a rejection region, even for a nondirectional alternative hypothesis (the only common type for a chi-square test).

Step 3. The test statistic.

We calculate chi-square as we did in Chapter 5 by comparing the observed frequencies to the expected frequencies. The expected frequencies are now those presumed by the null hypothesis. For an equal distribution of expected frequencies, we simply divide N by the number of categories, $60/4 = 15$ in this example. We use the same formula as before:

$$\chi^2 = \Sigma \frac{(f_o - f_e)^2}{f_e}$$

Test Statistic Value of χ^2-*I*

The actual calculations are shown in Table 9.2. Our test statistic is found to be 6.53. Notice that a separate computation of a standard error is not necessary.

Step 4. The decision about H_0.

The test statistic (6.53) does not exceed the tabular statistic (7.815); it is not in the rejection region. Thus we must decide to fail to reject the null hypothesis. (See Figure

FIGURE 9.17a Criteria for Testing H_0: $\chi^2 = 0$

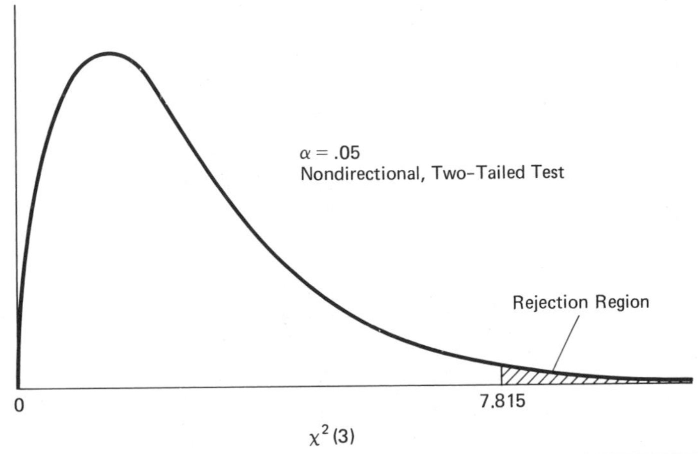

TABLE 9.2 Worksheet for Chi-Square Single-Sample Test

Issue	f_o	f_e
Inflation	23	15
Taxation	15	15
Energy	12	15
Social welfare	10	15

$$\chi^2 = \sum \frac{(f_o - f_e)^2}{f_e}$$
$$= \frac{(23 - 15)^2}{15} + \frac{(15 - 15)^2}{15} = \frac{(12 - 15)^2}{15} + \frac{(10 - 15)^2}{15}$$
$$= \frac{(8)^2}{15} + \frac{(0)^2}{15} + \frac{(-3)^2}{15} + \frac{(-5)^2}{15}$$
$$= \frac{64}{15} + \frac{0}{15} + \frac{9}{15} + \frac{25}{15}$$
$$= 4.2\overline{6} + 0 + 0.60 + 1.6\overline{6}$$
$$= 6.5\overline{3}$$

9.17b.) Although the sample data clearly show that different frequencies of people chose different issues as the most important, those observed differences were within the limits that can be attributed to sampling error at the .05 level. The differences between the sample's distribution and the hypothesized one are not statistically significant; $\chi^2 = 6.53$, $df = 3$, $p > .05$, n.s.

Step 5. The conclusion.

Since we failed to reject the null hypothesis, we draw the inference that the population from which this sample was drawn may well have an equal distribution of importance choices among the four national issues. How could this happen given our sample results? They are so apparently different from an equal distribution. The reason may be a conse-

FIGURE 9.17b Conclusion for H_0: $\chi^2 = 0$

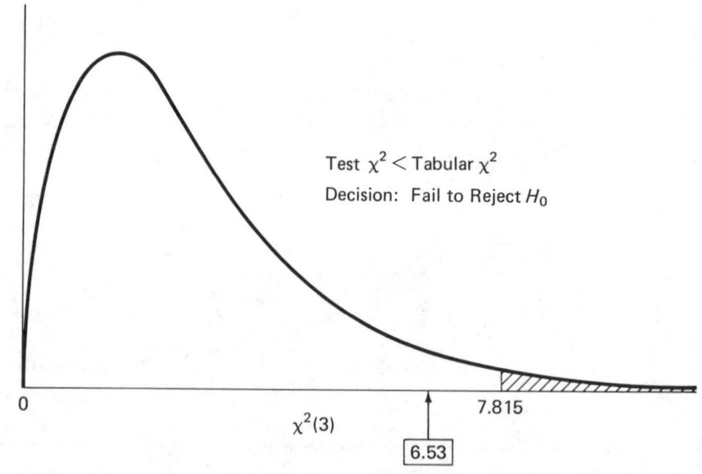

quence of the sample size. The smaller the sample is, the greater the observed differences must be in order for a statistically significant difference to be inferred. A larger sample in a replication study might find such a difference to be significant. But for now we must accept the results of this sample. We have insufficient evidence to doubt the null hypothesis. Apparently, there is no one or more of the issues that is of greater or lesser importance to the entire population of adults in this small town.

Research Example 9.3 presents an illustration of a chi-square goodness-of-fit test. The example concerns a study of climatic and sociocultural affects on seasonal variations in conceptions and births. Two rural populations were studied, one in an area of Uganda and the other in highland Mexico. The table in the example shows the total number of births and conceptions recorded over several years for the two areas. Chi-square tests were applied separately to each location's distribution across the 12 months of a year. The expected frequencies for each month were the mean number for all months combined. Thus 184.5 was the expected frequency against which the actual frequencies were compared for Uganda. In Mexico the expected frequency for each month was 827.5, its monthly mean. In both places the test determined that there was a significant difference between the observed and expected frequencies. Notice that the extent of significance for these findings was extremely high (.0001). This is, in part, a reflection of the very large sample sizes. The test chi-square can take on an inflated value when N is very large. However, visual inspection of the two distributions suggests that there truly were wide fluctuations in the frequencies for the 12 months.

RESEARCH EXAMPLE 9.3 Chi-Square Goodness-of-Fit Test

From Richard W. Thompson and Michael C. Robbins, "Seasonal Variation in Conception in Rural Uganda and Mexico," American Anthropologist, *75 (3), June 1973:676ff. Reprinted by permission.*

. . . [We] wish to assess the combined impact of certain sociocultural and climatic variables on the seasonal variation in conception in two geographically and cultural-historically separate rural populations. . . . Data for the present study were collected from rural peasant populations in the Buganda region of Uganda by Robbins, and by Thompson in highland Mexico. These populations are quite similar in their degree of technological sophistication, seasonal variability of climate, and the absence of strictly observed seasonal periods of sexual abstinence and birth control. Individuals in both regions are, for the most part, sedentary agriculturalists lacking easy access to modern material items such as contraceptives, television, automobiles, and air conditioners. The major climatic factors in both regions are alternating wet and dry seasons. Temperature variations in Uganda are slight, but somewhat more marked in highland Mexico. The population from Uganda, selected for this analysis, consists primarily of Baganda, a well-known East African people for whom several excellent accounts of their sociocultural system are available. . . . The population of Mexico is comprised largely of Otomi Indians and Mestizos for whom there are also several ethnographic accounts. . . .

The data used to estimate the frequency of conceptions and births were derived from birth records. Monthly birth records covering the eight-year period 1963-70 were obtained from the Municipio of Amealco, Queretero, in Mexico. Monthly birth records of Uganda were obtained from a sub-county headquarters in Buddu county,

Masaka District, and cover a nine-year span 1957-66. [The table] contains the aggregated total number of monthly births and conceptions for these time spans in each region. . . .

Total Monthly Births and Conceptions[a]

Month of Birth	Uganda (Sub-County)	Mexico (Municipio)	Month of Conception
January	254	807	May
February	145	693	June
March	188	809	July
April	203	999	August
May	174	928	September
June	193	898	October
July	188	881	November
August	157	772	December
September	177	747	January
October	133	805	February
November	199	803	March
December	205	788	April
Monthly mean	184.5	827.5	
Chi-square	59.9[b]	94.9[b]	

[a] Table I in the original.
[b] Indicates differences significant beyond .0001 level.

Tests for differences in frequencies of conception among the months were calculated for both populations. One-by-twelve chi-square tests of significance were employed using the monthly means of 184.5 and 827.5 as expected values for Uganda and Mexico respectively. Differences among the months were statistically significant beyond the .0001 level for both populations (see the table). . . .

Kolmogorov–Smirnov Single-Sample Test (*KS-I*)

The *Kolmogorov–Smirnov single-sample test* is another nonparametric test we often see used. It examines data at the ordinal level in the form of cumulative proportions. Like the χ^2-*I* test, it determines whether an observed distribution is significantly different from a theoretical distribution that could have occurred. The theoretical distribution is most often one in which the proportions are equally divided among the categories of measurement. To incorporate the ordered quality of the data into the test, the proportions are accumulated from the lowest to the highest categories in both the observed and theoretical distributions. If the observed distribution matches the theoretical, their cumulative proportions for all categories will be alike. The differences between these two sets of cumulative proportions are described by a set of sampling distributions calculated for exact sample sizes. Tabular values, known as *D,* are identified for selected levels of significance and used as the tabular statistics in the test. Beyond the ordinal level of measurement requirement, this test needs independent observations from a random sample as did the chi-square-I test.

TABLE 9.3 Distribution of Overtime Parking Citations by Day

Day	f
Monday	9
Tuesday	8
Wednesday	4
Thursday	2
Friday	2
Total	25

Let's consider a slight variation in our earlier problem concerning the incidence of citations for overtime parking. This time we are interested in knowing whether the citations were issued to any different extent according to the number of days that have passed during the workweek (Monday through Friday). Suppose that our sample contains 25 cases and the citations are distributed as shown in Table 9.3.

Step 1. The hypotheses.

H_0: $cP_o = cP_e$ The cumulative proportions of citations for the population are distributed equally through the workweek. A random sample may differ from this because of sampling error.

H_1: $cP_o \neq cP_e$ The cumulative proportions of citations for the population are not distributed equally through the workweek. A random sample will have such an unequal distribution which is beyond the expected influence of sampling error.

Step 2. The criteria for rejecting H_0.

We can use the .05 significance level and a nondirectional test for the new sampling distribution of Kolmogorov–Smirnov D values. Tabular values of D are differences between observed and expected cumulative proportions that must be equaled or exceeded in order for H_0 to be rejected. The more often used tabular D values are presented in Table D5 of Appendix D. That table indicates that a tabular D of .27 is the appropriate value for an alpha level of .05 when the sample size is 25.

Step 3. The test statistic.

The *KS-I* test requires a computational procedure for the test statistic that is similar in some ways to that for chi-square. It is shown in Table 9.4. There you see both the

TABLE 9.4 Worksheet for Kolmogorov–Smirnov Single-Sample Test

Day	f_o	$<cf_o$	$<cp_o$	f_e	$<cf_e$	$<cp_e$	$\lvert <cp_o - <cp_e \rvert$
Monday	9	9	.36	5	5	.20	.16
Tuesday	8	17	.68	5	10	.40	.28
Wednesday	4	21	.84	5	15	.60	.24
Thursday	2	23	.92	5	20	.80	.12
Friday	2	25	1.00	5	25	1.00	.00

distributions of observed (from the sample) and theoretical (expected by H_0) citations: f_o and f_e. A less-than cumulative frequency distribution is determined from each of these: $<cf_o$ and $<cf_e$. These are then converted to distributions of less-than cumulative proportions: $<cp_o$ and $<cp_e$. The absolute value of the difference between these latter two values for each category (day) is found. The *KS-I* test identifies the largest of these differences as *LD*. In this problem *LD* is .28. The largest difference between observed and expected cumulative proportions was for Tuesday.

$$LD = |<cp_o - <cp_e|$$ Kolmogorov-Smirnov-I Test Statistic *LD*

where LD = largest difference between observed and expected cumulative proportions

Step 4. The decision about H_0.

Since the test statistic (.28) does exceed the tabular statistic (.27), we decide that the null hypothesis should be rejected. We would get an observed distribution like ours fewer than 5 times in 100 from a population with an equal distribution because of sampling error. Thus sampling error is not a sufficient explanation for our sample's result. The observed distribution of cumulative proportions does differ significantly from the theoretically expected equal distribution ($LD = .28$, $p < .05$).

Step 5. The conclusion.

Since we rejected the null hypothesis, we can draw the inference that the cumulative proportions of citations in the population are not distributed equally across the five days of the workweek. This may mean that some further information is needed to explain why this happens. Is it because of different volumes of vehicles in the area on different days? If so, we should redo the problem using rates, not raw frequencies. If not, we might ask whether the police are working from a quota which they try to fill relatively early in the week so that they can "coast" during the later part. We could not simply assert this explanation. We would have to find some supporting evidence of it other than the statistical coincidence. Still, it is an important possibility in studying policing behavior. Alternatively, maybe it is the public, not the police, who behave differently on different days. Perhaps they are more preoccupied earlier in the week and forget to watch their parking meters. The possibilities can go on from here to many interesting ideas about a rather ordinary problem in social behavior. It requires a partnership between theory and research to begin to understand it.

Keys to Understanding

There are three main issues to consider as you interpret and evaluate inferences from single-samples. These are (1) the meaning of the distinction between hypothesis testing and estimation, (2) the contrast between statistical significance and theoretical or practical importance, and (3) understanding the model behind each inference technique.

Key 9.1. Distinguish between hypothesis testing and estimation. Although this chapter has shown specific technical differences between hypothesis testing and estimation, you may not yet realize how important the distinction is for interpreting social research. Hypothesis testing forces us to make a decision about an exact description of a population proposed in a null hypothesis. Estimation does not use any prior notion of a population. Which is more realistic for understanding the social world?

If a proposed description of a population is an important possibility to be tested, there is great potential benefit in using hypothesis-testing procedures. It may be truly useful to know whether a suggested value of a population parameter is likely to be accurate. Is the claim that .67 of the constituents have an optimistic life-view true? Are bank presidents really receiving a mean salary of $75,000?

But other times the proposed population value against which a sample value is compared is only arbitrarily selected. The researcher may not have a very clear idea about the value of a parameter, and to pick one for the purpose of a hypothesis test can be merely an empty exercise. In this situation it would be more realistic to use the interval estimation technique instead. A random sample can be selected and the results used to develop a probable view of a parameter. Although a hypothesis test provides a more definitive conclusion than does an interval estimate, there is little merit in having such a conclusion if it is about an arbitrary inference. Why try to be definitive concerning a hypothesis that has unknown importance?

This raises the second key to understanding. *Key 9.2.* Distinguish between a result that is statistically significant and one that is important. A statistically significant result is one that differs from a hypothesized value with a minimal degree of sampling error. That degree of error is usually determined in a conventional manner that has only a vague attachment to the specific research situation. But whether this result is important theoretically or practically is a different issue. Results can be statistically significant but not important to either a theory of social science or the practical world of the people being studied. The reverse can be true as well. An important and real theoretical difference might be found to be not statistically significant in a sample because of a small sample size or some other reason. Step 5 in our hypothesis testing procedure sensitizes you to these possibilities. Do not overlook this step as you read social research. It is often the one that makes the other parts of the effort worthwhile. (For further discussion see Selvin, 1957 and Atkins and Jarrett, 1979:87–109.)

Key 9.3. Examine the *model* behind each inference technique. The model consists of the assumptions and methods used to arrive at the inference. The models for each of the six single-sample statistics used in this chapter are summarized for you in Inference Models 9.1 to 9.6. You can use these both for a review of each technique and for a guide in interpreting their use. A major part of each model to keep uppermost in your thoughts concerns the assumptions that determine whether the technique is parametric or nonparametric. Whenever the model specifies some requirement about a population's distribution or parameters, it is a parametric test. Otherwise, it is nonparametric. Remember from Chapter 8 that the distinction is important because of their comparative power. A parametric test is more likely to reject false null hypotheses than is a nonparametric test. But the use of a parametric technique when there are large violations of its assumptions gives ambiguous results. We cannot be sure of its conclusions. Nonparametric techniques are less likely to discover a false null hypothesis, but we can be more confident that the conclusion to reject H_0 is not contaminated by failure to meet the assumptions of the test. Choosing between them is an example of the proverbial "balancing act" with alternatives.

Inference Model 9.1 Single-Sample Proportion Test

Purpose:

To determine whether a sample proportion differs significantly from a proposed population proportion.

Sampling Distribution:

Standard normal (z).

Assumptions of the Test:

1. Independent and random observations collected in one sample.
2. Nominal level of measurement.
3. $NP \geq 5$ and $NQ \geq 5$.

Typical Hypotheses:

Nondirectional Test	Directional Test
H_0: $P =$ (some proposed value)	H_0: $P \geq$ or $\leq$ (some proposed value)
H_1: $P \neq$ (some proposed value)	H_1: $P <$ or $>$ (some proposed value)

Tabular Statistic:

z at a predetermined level of significance from Table D2.

Standard Error:

$$\sigma_p = \sqrt{\frac{PQ}{N}} \quad \text{or} \quad \hat{\sigma}_p = \sqrt{\frac{pq}{N}}$$

Test Statistic:

$$z = \frac{p - P}{\sigma_p} \quad \text{or} \quad \frac{p - P}{\hat{\sigma}_p}$$

Inference Model 9.2 Single-Sample Mean Test

Purpose:

To determine whether a sample mean differs significantly from a proposed population mean.

Sampling Distribution:

A. Standard normal (z), or

B. Student's t.

Assumptions of the Test:

1. Independent and random observations collected in one sample.
2. Interval level of measurement.

3A. σ is known or N is large (> 100).

3B. Normally distributed population when σ is unknown.

Typical Hypotheses:

Nondirectional Test	Directional Test
H_0: $\mu =$ (some proposed value)	H_0: $\mu \geq$ or $\leq$ (some proposed value)
H_1: $\mu \neq$ (some proposed value)	H_1: $\mu <$ or $>$ (some proposed value)

Tabular Statistic:

A. z at a predetermined level of significance from Table D2.

B. t at $df = N - 1$ and a predetermined level of significance from Table D3.

Standard Error:

$$\sigma_{\bar{X}} = \frac{\sigma}{\sqrt{N}} \text{ (if } \sigma \text{ is known)} \quad \text{or} \quad \hat{\sigma}_{\bar{X}} = \frac{s}{\sqrt{N}} \text{ (if } \sigma \text{ is unknown)}$$

Test Statistic:

A. $z = \dfrac{\bar{X} - \mu}{\sigma_{\bar{X}}}$ or $\dfrac{\bar{X} - \mu}{\hat{\sigma}_{\bar{X}}}$

B. $t = \dfrac{\bar{X} - \mu}{\hat{\sigma}_{\bar{X}}}$

Inference Model 9.3 Single-Sample Frequencies—Chi-Square Goodness-of-Fit Test (χ^2-*I*)

Purpose:

To determine whether a sample's observed frequency distribution differs significantly from that which would be expected based on an assumed population distribution.

Sampling Distribution:

Chi-square.

Assumptions of the Test:

1. Independent and random observations collected in one sample.
2. Nominal level of measurement.
3. N must not be extremely small.

Typical Hypotheses:

Nondirectional Test	Directional Test
H_0: $\chi^2 = 0$	(Not ordinarily used)
H_1: $\chi^2 \neq 0$	

Tabular Statistic:

Chi-square at $df = k - 1$ and a predetermined level of significance from Table D4.

Standard Error:

Included in test statistic formula; separate calculation not required.

Test Statistic:

$$\chi^2 = \Sigma \frac{(f_o - f_e)^2}{f_e}$$

Inference Model 9.4 Single-Sample Cumulative Proportions—Kolmogorov-Smirnov Single-Sample Test (*KS-I*)

Purpose:

To determine whether a sample's distribution of cumulative proportions differs significantly from a proposed population's distribution of cumulative proportions.

Sampling Distribution:

Kolmogorov-Smirnov.

Assumptions of the Test:

1. Independent and random observations collected in one sample.
2. Ordinal level of measurement.

Typical Hypotheses:

Nondirectional Test	Directional Test
H_0: $cP_o = cP_e$	(Not ordinarily used)
H_1: $cP_o \neq cP_e$	

Tabular Statistic:

Kolmogorov-Smirnov D at N and a predetermined level of significance from Table D5.

Standard Error:

Included in test statistic formula; separate calculation not required.

Test Statistic:

$$LD = \text{largest } |{<}cp_o - {<}cp_e|$$

Inference Model 9.5 Confidence Interval Estimate of a Single Proportion

Purpose:

To determine an interval estimate of a population's proportion from a sample's proportion with a known level of confidence.

Sampling Distribution:

Standard normal (z).

Assumptions of the Estimate:

Same as single-sample proportion test.

Construction of the Estimate:

1. Choose level of confidence and determine tabular z.
2. Calculate standard error: σ_p or $\widehat{\sigma}_p$.
3. Calculate the estimate:

$$CI = p \pm (z)(\sigma_p \text{ or } \widehat{\sigma}_p)$$

Inference Model 9.6 Confidence Interval Estimate of a Single Mean

Purpose:

To determine an interval estimate of a population's mean from a sample's mean with a known level of confidence.

Sampling Distribution:

A. Standard normal (z), or

B. Student's t.

Assumptions of the Estimate:

Same as single-sample mean test.

Construction of the Estimate:

1. Choose level of confidence and determine tabular z or t.
2. Calculate standard error: $\sigma_{\overline{X}}$ or $\widehat{\sigma}_{\overline{X}}$.
3. Calculate the estimate:

A. $CI = \overline{X} \pm (z)(\sigma_{\overline{X}} \text{ or } \widehat{\sigma}_{\overline{X}})$

B. $CI = \overline{X} \pm (t)(\widehat{\sigma}_{\overline{X}})$

KEY TERMS

Standard Error

Population Proportion (P)

Sample Proportion (p)

Mean of Proportions (μ_p)

Standard Error of Proportions (σ_p)

Population Mean (μ)

Sample Mean ($\overline{X}$)

Mean of Means ($\mu_{\overline{X}}$)

Standard Error of Means ($\sigma_{\overline{X}}$)

Population Standard Deviation (σ)

Sample Standard Deviation (S)

Inferential Standard Deviation (s)

Tabular Statistic

Test Statistic

Sampling Distribution of Proportions

Sampling Distribution of Means

Central Limit Theorem

Student's t Distributions

Robust

Degrees of Freedom (df)

Point Estimate

Confidence Interval Estimate

Level of Confidence

Single-Sample Proportions Test

Single-Sample Means Test

Chi-Square Goodness-of-Fit Test (χ^2-I)

Kolmogorov–Smirnov Single-Sample Test (KS-I)

EXERCISES

1. A private agency claims that half ($P = .50$) of its work force is composed of females. A random sample of 90 workers finds that .35 are females. Determine whether there is a significant difference between the sample result and the agency's claim at the .05 level of significance using a nondirectional test. (a) Write the hypotheses symbolically and verbally. (b) Find NP and NQ, determine the sampling distribution, the tabular statistic, and rejection regions(s). (c) Find the standard error and test statistic. (d) Make a decision about H_0. (e) State a conclusion.
2. An advertising company says its historical pattern has been to have 10% minority employment ($P = .10$). But in the past year 25% of its 22 new employees have been minority members ($p = .25$). Determine if there has been a significant increase in hiring minorities during the past year in comparison to the company's history at the .01 level of significance using a directional test. (a) Write the hypotheses symbolically and verbally. (b) Find NP and NQ, determine the sampling distribution, the tabular statistic, and rejection region(s). (c) Find the standard error and test statistic. (d) Make a decision about H_0. (e) State a conclusion.
3. A recruiter for a small college claims that its students have consistently had a mean of 600 on a standardized entrance exam. A random sample of the records for 1000 current students reveals they have a mean of 540 on the exam. This translates into a test statistic of $z = -2.50$. Assume that the null hypothesis was $\mu = 600$ and that a nondirectional test was performed. (a) If alpha = .05, what is the tabular statistic? (b) If alpha = .05, should H_0 be rejected? (c) If alpha = .01, what is the tabular statistic? (d) If alpha = .01, should H_0 be rejected? (e) If H_0 were rejected, does it make you doubt the accuracy of the recruiter's claim? (f) If H_0 were rejected, does it mean that the sample had a lower mean than the entire population of students? (g) If H_0 were not rejected, does it make you suspect the accuracy of the recruiter's claim? (h) If H_0 were not rejected, does it mean that the sample mean differed significantly from the proposed population mean of 600?
4. A radio company claims that its products have a mean life of 10 years. A random sample of 250 products finds that they had a mean life of 8.5 years. This translates into a test statistic of $z = -1.86$. Assume the null hypothesis was that $\mu = 10$ and a nondirectional test was performed. (a) If alpha = .05, what is the tabular statistic? (b) If alpha = .05, should H_0 be rejected? (c) If alpha = .10, what is the tabular statistic? (d) If alpha = .10, should H_0 be rejected? (e) If H_0 were rejected, would it make you suspect the accuracy of the company's claim? (f) If H_0 were rejected, does it mean that the sample had a significantly lower mean than the claimed population of all company products? (g) If H_0 were not rejected, does it make you suspect the company's claim? (h) If H_0 were not rejected, does it mean that the sample mean does not differ significantly from the claimed mean for all company products?
5. In selling a house in a "quiet neighborhood," a real estate agent assures the buyer that no more than 50 cars a day pass the house (as a mean). A random sample of 10 days finds that a mean of 150 cars a day passed the house. This translates into a test statistic of $t = 5.21$. Assume the null hypothesis was that $\mu \leq 50$ and a directional test was performed. (a) If alpha = .05, what is the tabular statistic? (b) If alpha = .05, should H_0 be rejected? (c) If alpha = .01, what is the tabular statistic? (d) If alpha = .01, should H_0 be rejected? (e) If H_0 were rejected, would it make you suspect the real estate agent's claim? (f) If H_0 were rejected, does it mean that the sample mean was significantly greater than the mean proposed by the agent? (g) If H_0 were not rejected, does it make you suspect the agent's claim? (h) If H_0 were not rejected, does it mean that the proposed population mean is significantly lower than the sample mean?
6. A random sample of 200 is selected for a single-sample mean test. (a) Find *df.* (b) What sampling distribution would be used? (c) If alpha = .05, what is the tabular statistic for a nondirectional test? (d) If alpha = .05, what is the tabular statistic for a directional test?
7. A random sample of 15 is selected for a single-sample mean test. (a) Find *df.* (b) What sampling distribution would be used? (c) If alpha = .05, what is the tabular statistic for a nondirectional test? (d) If alpha = .01, what is the tabular statistic for a directional test?
8. A random sample of 18 is selected for a single-sample mean test. (a) Find *df.* (b) What sampling distribution would be used? (c) If alpha = .10, what is the tabular statistic for a directional test? (d) If alpha = .01, what is the tabular statistic for a nondirectional test?

9. A random sample of 500 is selected for a single-sample mean test. (a) Find *df.* (b) What sampling distribution would be used? (c) If alpha = .10, what is the tabular statistic for a nondirectional test? (d) If alpha = .01, what is the tabular statistic for a directional test?
10. The results from a random sample of 100 persons in a community indicate that 80% ($p = .80$) are in favor of public housing for the elderly. (a) Could a normal sampling distribution be assumed? Explain. (b) Find the estimated standard error. (c) Find the 95% confidence interval from this sample result. (d) Find the 99% confidence interval from this sample result.
11. A random sample of 200 persons in an urban neighborhood finds that 70% ($p = .70$) believe they have inadequate city services. (a) Could a normal sampling distribution be assumed? Explain. (b) Find the estimated standard error. (c) Find the 95% confidence interval from this sample result. (d) Find the 90% confidence interval from this sample result.
12. The 95% confidence interval constructed around a sample proportion of .3 was .2 to .4. This was based on a large random sample of children who were asked if they thought summer school should be mandatory. Which of the following would be known to be true from this? (a) 95% of the children questioned said "yes" to the question. (b) There is a 95% probability that the sample proportion is between .2 and .4. (c) The proportion of the population that thinks summer school should be mandatory is .3. (d) There is 95% confidence that such an interval as .2 to .4 contains the population's proportion that thinks summer school should be mandatory. (e) If repeated random samples of the same size were drawn from this same population, 95% of the 95% confidence intervals that could be found would contain the population's proportion that thinks summer school should be mandatory.
13. Suppose that the 99% confidence interval constructed around a sample mean of 12 was 10.7 to 13.3. This was based on asking a random sample of adults how many years of education they had completed. Which of the following would be known to be true from this? (a) 99% of the people questioned had between 10.7 and 13.3 years of education. (b) There is a 99% probability that the mean of the sample is between 10.7 and 13.3 years of education. (c) The mean educational level in the population being sampled is 12 years. (d) There is 99% confidence that such an interval as 10.7 to 13.3 contains the mean educational level of the population being sampled. (e) If repeated random samples of the same size were drawn from this population, 99% of the 99% confidence intervals that could be found would contain the population's mean.
14. A random sample of 196 Chicanos have a mean income of $12,000 with $s = \$7000$. (a) What sampling distribution could be assumed in this situation? (b) Find the estimated standard error. (c) Find the 95% confidence interval from this sample result. (d) Find the 80% confidence interval from this sample result.
15. A random sample of 60 male health club members have a mean weight of 170 pounds with $s = 30$. (a) Find the estimated standard error. (b) If a normal sampling distribution is used, what is the appropriate tabular statistic for a 95% confidence interval? (c) Find the 95% confidence interval using your answers to parts (a) and (b). (d) If a normal sampling distribution is not used, what is the appropriate tabular statistic for a 95% confidence interval? (e) Find the 95% confidence interval using your answers to parts (a) and (d).
16. A chi-square goodness-of-fit test for a random sample of people has found a test statistic of 11.5. The data concerned a comparison across five categories of religious affiliation. H_0 was that $\chi^2 = 0$. (a) Find *df.* (b) If alpha = .05, find the tabular statistic. (c) If alpha = .05, should H_0 be rejected? (d) If alpha = .01, find the tabular statistic. (e) If alpha = .01, should H_0 be rejected? (f) If H_0 were rejected, would you conclude that the observed distribution of religious affiliation was the same as the theoretically expected distribution? (g) If H_0 were rejected, would you conclude that there was a significant difference between the observed and expected distributions? (h) If H_0 were not rejected, would you conclude that there was not a significant difference between the observed and expected distributions?
17. A chi-square goodness-of-fit test for a random sample of high schools has found a test statistic of 4.5. The data concerned an enrollment comparison across three areas of programs emphasis. H_0 was $\chi^2 = 0$. (a) Find *df.* (b) If alpha = .05, find the tabular statistic. (c) If alpha = .05, should H_0 be rejected? (d) If alpha = .10, find the tabular statistic. (e) If alpha = .10, should H_0 be rejected? (f) If H_0 were rejected, would you conclude that the observed distribution of

program enrollments differed significantly from the expected distribution? (g) If H_0 were not rejected, would you conclude that the observed and theoretical distributions were the same except for the effects of sampling error? (h) If H_0 were rejected, would you conclude that one or more program enrollments occurred significantly more or less often than the others?

18. The following distribution of political party membership was found for a random sample of 60 voters.

Party	f
Democratic	22
Republican	14
Other	5
None	19
Total	60

Determine if this observed distribution differs significantly from a hypothesized equal distribution of the 60 party affiliations. Use the chi-square goodness-of-fit test for H_0: $\chi^2 = 0$. (a) Find df and the tabular statistic if alpha = .05. (b) What would be the expected frequencies according to the null hypothesis? (c) Find the test value of chi-square. (d) What is your decision about H_0? (e) What is df and the tabular statistic if alpha = .10? (f) What is your decision about H_0 if alpha = .10?

19. The following distribution of miles run by a random sample of joggers was found for the seven days of one randomly chosen week.

Day	f **(miles)**
Monday	9
Tuesday	7
Wednesday	5
Thursday	8
Friday	12
Saturday	18
Sunday	18
Total	77

Use a Kolmogorov–Smirnov-I test to determine if the observed distribution differs significantly from an equal distribution. H_0 is that there is no such difference. (a) Find the tabular statistic (D) if alpha = .05. (b) Find the test statistic (LD). (c) What is your decision about H_0? (d) Find the tabular statistic (D) if alpha = .10. (e) What is your decision about H_0 if alpha = .10?

20. The observed distribution for the incidence of accidental food poisoning from equal-sized random groups of families representing four social classes was as follows:

Social Class	f
Upper	0
White-collar middle	2
Blue-collar middle	8
Lower	20
Total	30

Determine if this frequency distribution differs significantly from an equal distribution of the 30 incidents across the four social classes. Use the chi-square goodness-of-fit test for H_0: $\chi^2 = 0$. (a) Find df and the tabular statistic if alpha = .05. (b) What would be the expected frequencies according to the null hypothesis? (c) Find the test value of chi-square. (d) What is your decision about H_0? (e) What is df and the tabular statistic if alpha = .01? (f) If alpha = .01, what is your decision about H_0?

21. Using the same data as in Exercise 20, do a Kolmogorov–Smirnov-I test. (a) Find the tabular statistic (D) if alpha = .05. (b) Find the test statistic (LD). (c) What is your decision about a null hypothesis of no difference? (d) Find the tabular statistic (D) if alpha = .01. (e) What is your decision about a null hypothesis of no difference if alpha = .01?

22. Compare your answers to Exercises 20 and 21. (a) Do both tests reject a null hypothesis of no difference between observed and expected distributions at the .05 level? (b) Do both reject that hypothesis at the .01 level? (c) Would the results of χ^2-I and KS-I tests always compare as they did here for every distribution of observed scores and an expected distribution of equal scores? (d) Explain the difference between the two tests for the problem in Exercise 20.

23. Locate a published example of an opinion poll based on a single sample in which a parameter estimate is reported. Determine whether an indication of its accuracy is included (e.g., "results are within ±3 percentage points of the true population value"). Attempt to assess the accuracy of the estimate by calculating a standard error and a 95% confidence interval.

RELATED STATISTICS

Yates' Corrected Chi-Square

When $df = 1$ and any expected (not observed) frequency is less than 5, the chi-square distribution tends not to be truly continuous. As a result the test chi-square may exceed the tabular chi-square when it should not. Therefore, a null hypothesis is falsely rejected. The Yates' corrected chi-square ("Yates' correction for continuity") reduces the value of the test statistic so that this should not happen. It uses the following formula:

$$\chi^2 = \Sigma \frac{(|f_o - f_e| - .5)^2}{f_e}$$

Yates' corrected chi-square has been in use for many years and it is often used by researchers for small-sample problems that produce small expected frequencies. However, the practice is criticized by many as being unreliable since it "corrects" by the same amount (−.5) regardless of the size of the deviation of an expected frequency from the minimum of 5 and for all cells of a table even if only one has small frequencies. Further, chi-square has been shown to yield reliable test results with sample sizes as low as 8, which usually create many small expected frequencies. For these reasons, Yates' corrected chi-square is not recommended in this text.

See: Blalock (1979:290–292) on its use and Camilli and Hopkins (1978, 1979) on its critique.

Fisher's Exact Test

The Fisher's exact test is another alternative to chi-square when $df = 1$ and the expected frequencies are very small. Unlike Yates' corrected chi-square, the Fisher test provides an exact statistic for each specific situation. Its computation is rather cumbersome to do by hand and therefore is nearly always done by a computer program.

See: Blalock (1979:292–297).

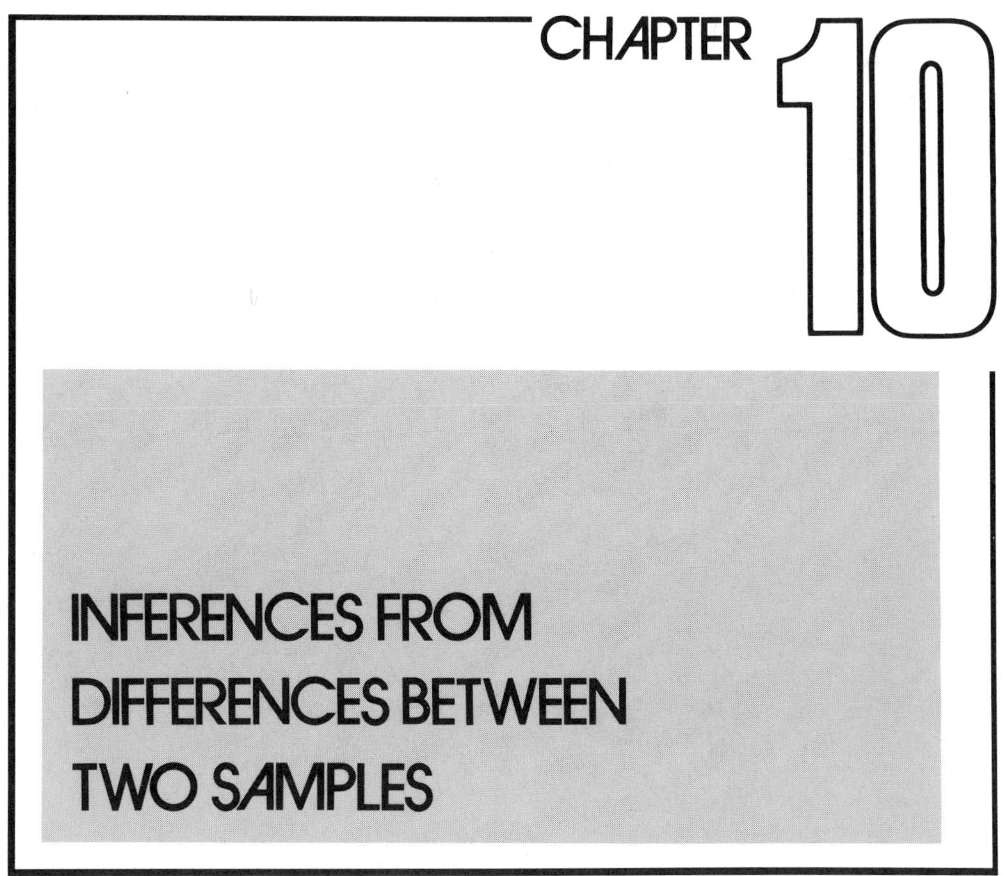

CHAPTER 10

INFERENCES FROM DIFFERENCES BETWEEN TWO SAMPLES

We will now expand our look at inferences from those based on a single sample to those based on differences between two samples. Again, we will consider some parametric and nonparametric tests. Inferences based on the tests of differences between two samples represent a more common statistical situation in social research than do inference tests from a single sample. However, the basic logic for both types of inferences is the same.

A Test of Difference Between Two Proportions

Suppose a researcher has found that in one country (Avalon) the proportion of people who have never married is .22 for a random sample of 30 adults. A random sample of 30 adults in another country (Brandan) has a comparable proportion of .29. Do the sample results mean that the entire countries differ significantly in their proportions of never-marrieds? This is the kind of problem that is ideal for a difference-in-two-proportions test. Although neither country's overall proportion was known, the researcher expected them to differ. The null hypothesis that would be formally tested is the opposite idea that the two countries do not differ in such proportions.

This example is diagrammed in Figure 10.1 from the point of view of the null hypothesis. It shows representative individual scores in both populations and samples.

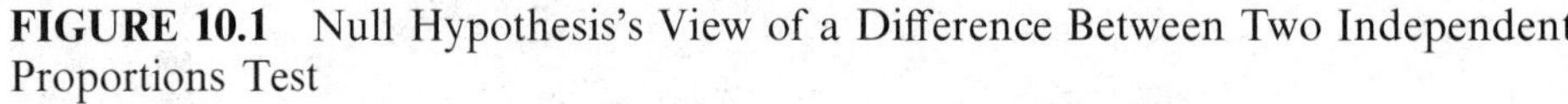

FIGURE 10.1 Null Hypothesis's View of a Difference Between Two Independent Proportions Test

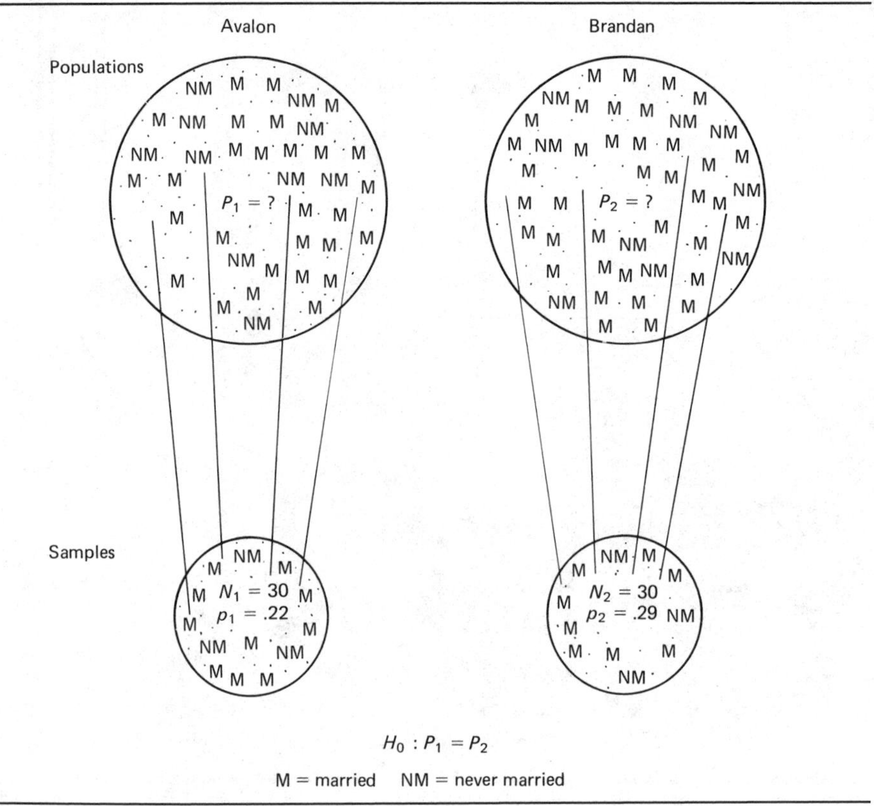

There you see that the first country's proportion of never-marrieds is designated as P_1 and the second's as P_2. The sample proportions are labeled p_1 and p_2, respectively.

Sampling Distribution of Differences Between Two Proportions

We can continue with our example of the proportions of never-marrieds in Avalon and Brandan. The observed difference in the two sample proportions is .22 − .29 = −.07. If two other random samples of the same size were drawn independently from the same two populations, their observed difference in proportions might be .25 − .24 = .01. A third pair of random and independent samples might differ by .27 − .25 = .02. And so on for the differences between all possible pairs of sample proportions that could be found for these two populations. The distribution of these differences is illustrated in Figure 10.2.

Like all other sampling distributions, the *sampling distribution of differences in proportions* has a mean (the *mean of the differences in proportion,* designated $\mu_{p_1-p_2}$) and a standard deviation (the *standard error of the differences in proportions,* designated $\sigma_{p_1-p_2}$). The mean of this sampling distribution is theoretically equal to the difference in the two populations' proportions, $\mu_{p_1-p_2} = P_1 - P_2$. In our example this would be zero since the null hypothesis considers P_1 and P_2 to be equal. The distinctions in features and symbols that are used for a difference in proportions test are summarized in Figure 10.3. Give it careful study; it is your translation guide.

The sampling distribution is normally shaped if four familiar assumptions can be met.

1. The variable on which the two groups are to be compared need be measured only at the nominal level. Our example does this.
2. The researcher must have independent and random observations. We have discussed the meaning of independent and random observations before and our example does conform to this requirement.
3. These observations must be contained in two independent samples. Two samples are considered to be independent of each other whenever the selection of cases for one sample is not influenced by the selection of cases for the other. This can be done in one of two ways. Two entirely separate samples can be drawn as our example describes. Alternatively, one sample can be drawn and then split into two independent groups or subsamples. For instance, our researcher could have

FIGURE 10.2 Sampling Distribution of Differences Between Two Independent Proportions

Proportions for Samples from Avalon p_1	Proportions for Samples from Brandan p_2	Difference $(p_1 - p_2)$
.22	.29	−.07
.25	.24	.01
.27	.25	.02
⋮	⋮	⋮

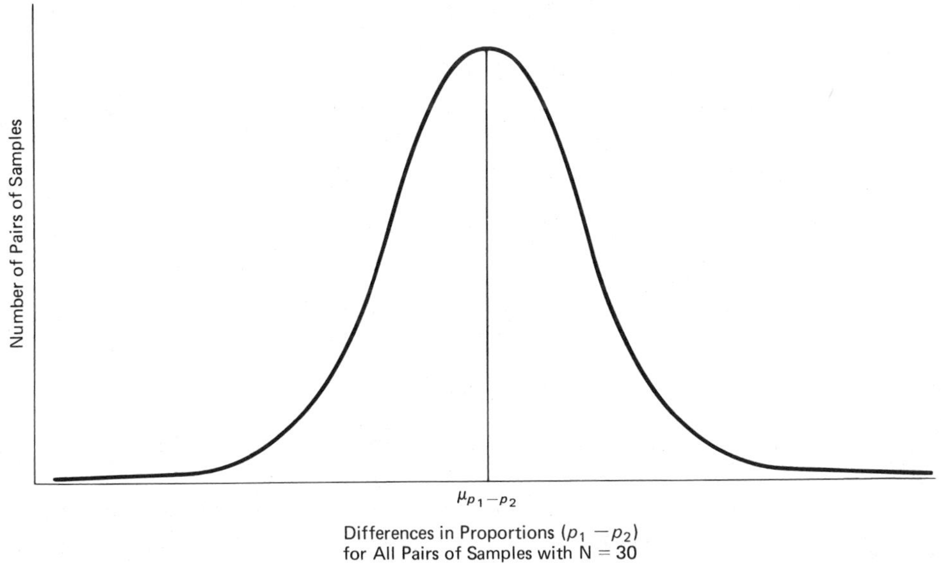

FIGURE 10.3 Classification of Distribution Features for Two-Sample Proportions Test

Distribution	Features	
	Proportion	
First population	P_1	
Second population	P_2	
First sample	p_1	
Second sample	p_2	
	Mean	Standard deviation
Sampling distribution of differences between two proportions	$\mu_{p_1-p_2}$	$\sigma_{p_1-p_2}$

selected one random sample from the two countries combined and then divided it into two subsamples by placing all the cases from Avalon in one group and all those from Brandan in the other.

4. The quantities *NP* and *NQ* are at least 5. For a difference-in-two proportions test, *P* and *Q* represent the weighted arithmetic means of P_1 and P_2 and of their counterparts, Q_1 and Q_2, respectively. As our example illustrates, these are seldom known. In their place the researcher uses weighted means based on the sample data, known as p and q.

$$\hat{P} = p = \frac{N_1 p_1 + N_2 p_2}{N_1 + N_2} \qquad \hat{Q} = q = 1 - p$$

Estimation of *P* and *Q* from Two Samples

where p_1 = proportion in first sample
p_2 = proportion in second sample
N_1 = number of cases in first sample
N_2 = number of cases in second sample

In our example:

$$p = \frac{(30)(.22) + (30)(.29)}{30 + 30} = \frac{6.6 + 8.7}{60} = \frac{15.3}{60} = .255$$

$$q = 1 - .255 = .745$$

We can now find the $N\hat{P}$ and $N\hat{Q}$ values*:

$$N\hat{P} = (30)(.255) = 7.65 \qquad N\hat{Q} = (30)(.745) = 22.35$$

Since both of these exceed 5, the fourth and final assumption has been met in our example; a normal sampling distribution is appropriate for our inference test.

*If N_1 and N_2 had differed, their mean would have been used as N for these computations.

The standard error of this sampling distribution can be calculated directly from the population proportions (P_1 and P_2), if they are known. The method uses the same combined values of P and Q as mentioned above. When the null hypothesis states that $P_1 = P_2$, either one could be used as the value of P and $(1 - P)$ would be used as the value of Q:

$$\sigma_{p_1-p_2} = \sqrt{PQ\left(\frac{1}{N_1} + \frac{1}{N_2}\right)}$$

Standard Error of the Differences in Proportions

If P_1 and P_2 are not known (as in our example), the standard error is estimated by the combined sample proportions using the values p and q that we saw above. This method of estimating the standard error is known as using the "pooled estimate" since information from two samples is put together.

$$\widehat{\sigma}_{p_1-p_2} = \sqrt{pq\left(\frac{1}{N_1} + \frac{1}{N_2}\right)}$$

Estimated Standard Error of the Differences in Proportions

The test statistic for evaluating the difference between two sample proportions compares their observed difference ($p_1 - p_2$) to the expected difference between the two population proportions ($P_1 - P_2$). In the present example (and most real situations, as well) that expected difference is 0 since P_1 and P_2 are hypothesized (by H_0) to be equal. This simplifies the general form of the test statistic formula:

$$z = \frac{(p_1 - p_2) - (P_1 - P_2)}{\sigma_{p_1-p_2}}$$

z Transformation for a Difference in Proportions Test

when $P_1 = P_2$:

$$z = \frac{(p_1 - p_2) - 0}{\sigma_{p_1-p_2}} = \frac{p_1 - p_2}{\sigma_{p_1-p_2}}$$

where $\sigma_{p_1-p_2}$ can be estimated by $\widehat{\sigma}_{p_1-p_2}$

We can now apply these formulas to our example in a formal test of difference between the two proportions of never-marrieds.

Step 1. The hypotheses.

H_0: $P_1 = P_2$ — The proportion of never-marrieds in Avalon is equal to the proportion of never-marrieds in Brandan. Sampling error may cause the proportions for random samples from these two populations to differ.

H_1: $P_1 \neq P_2$ — The proportion of never-marrieds in Avalon is not equal to the proportion of never-marrieds in Brandan. Proportions for random samples from these two populations will reflect this difference beyond the expected influence of sampling error.

Step 2. The criteria for rejecting H_0.

The alternative hypothesis suggests two ways in which the null hypothesis may be incorrect. Either P_1 is greater than P_2 or P_1 is less than P_2. These two possibilities create a nondirectional, two-tailed test of H_0. If the researcher uses the .05 level of significance with this nondirectional test, the tabular statistic will be a z of ± 1.96. This creates the usual two rejection regions beginning at these two tabular values and continuing to the extremes of the sampling distribution.

Step 3. The test statistic.

To find the test statistic, the standard error must first be calculated. It must be estimated in this example. We already have the values of p and q from above.

$$\begin{aligned}\hat{\sigma}_{p_1-p_2} &= \sqrt{pq\left(\frac{1}{N_1} + \frac{1}{N_2}\right)} \\ &= \sqrt{(.255)(.745)\left(\frac{1}{30} + \frac{1}{30}\right)} = \sqrt{(.189975)\left(\frac{2}{30}\right)} \\ &= \sqrt{(.189975)(.06\overline{6})} = \sqrt{.012665} = .1125\end{aligned}$$

The test statistic is found to be

$$\begin{aligned}z &= \frac{p_1 - p_2}{\hat{\sigma}_{p_1-p_2}} \\ &= \frac{.22 - .29}{.1125} = \frac{-.07}{.1125} = -.62\end{aligned}$$

Step 4. The decision about H_0.

The absolute value of the test statistic ($-.62$) does not exceed the absolute value of the tabular statistic (1.96); it does not fall into either rejection region. (See Figure 10.4.) Thus the researcher fails to reject the null hypothesis. This means that the sample proportions could be the result of sampling error more than 5% of the time when drawing pairs of

FIGURE 10.4 Testing $H_0: P_1 = P_2$

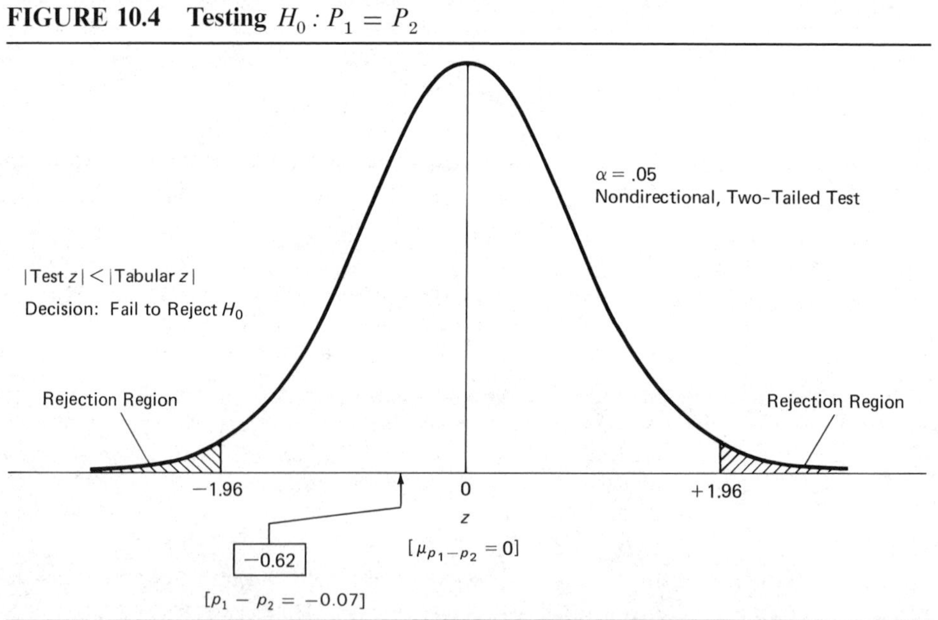

random samples from two populations with identical proportions. The observed difference in sample proportions is not statistically significant; $z = -.62$, $p > .05$, n.s.

Step 5. The conclusion.

Since the researcher fails to reject the null hypothesis, he or she concludes that the two populations do not differ in proportions. Avalon and Brandan may well have the same proportions of never-marrieds. Whatever reason led the researcher to believe otherwise has not been supported by these data. Of course, the researcher may be making a Type II error in drawing this conclusion. The null hypothesis may actually be false and it therefore should be rejected. The two countries may truly differ in such proportions but the two samples were unrepresentative of this difference. However, this test suggests otherwise.

Tests of Difference Between Two Means

In Chapter 9 we saw how there are several specific kinds of significance tests with single sample means depending on which population parameters are known and the size of the sample. As we now look at tests of difference between two means you will discover that there are again many versions of one basic procedure. We will look at four types of tests for the difference between two means: (1) a z test of difference between two independent means, (2) a t test of difference between two independent means for populations with homogeneous variances, (3) a t test of difference between two independent means for populations with heterogeneous variances, and (4) a t test for differences between two dependent means.

Sampling Distribution of Differences Between Two Independent Means

The *sampling distribution of differences between means* for two independent samples is much like the one for differences between two proportions. Imagine a researcher who believes that females and males differ in feelings of autonomy. A standardized scale can be used to measure such feelings at the interval level. The researcher would test a null hypothesis which stated that separate populations of females and males do not differ in mean autonomy scores ($\mu_1 = \mu_2$). The researcher would draw a random sample of females, measure their individual feelings of autonomy, and calculate the sample mean ($\bar{X}_1$). The process would be repeated independently for a sample of males to find $\bar{X}_2$. Suppose that these samples each had 100 cases and the means were 9 and 12, respectively. This situation is illustrated in Figure 10.5 from the point of view of the null hypothesis. It shows representative individual scores for both populations and samples.

The difference between the two sample means ($9 - 12 = -3$) is just one of many possible differences that the researcher could have observed. We can imagine finding all possible pairs of female and male samples each based on 100 cases, calculating their means, and finding the differences between those means. The distribution of all these differences is our sampling distribution. Figure 10.6 shows this.

This sampling distribution has a mean called the *mean of the differences in means.* It uses $\mu_{\bar{X}_1-\bar{X}_2}$ as its symbol and equals in value the difference between the two population means, $\mu_1 - \mu_2$. The standard deviation of this distribution is the *standard error of the differences in means* and uses the symbol, $\sigma_{\bar{X}_1-\bar{X}_2}$. Its value is found by one of several possible formulas depending on the circumstances of the situation, as you will see later. These distinctions in means and standard deviations among the populations, samples, and sampling distribution are shown in Figure 10.7.

FIGURE 10.5 Null Hypothesis's View of a Difference Between Two Independent Means Test (Values Shown Represent Autonomy Scores)

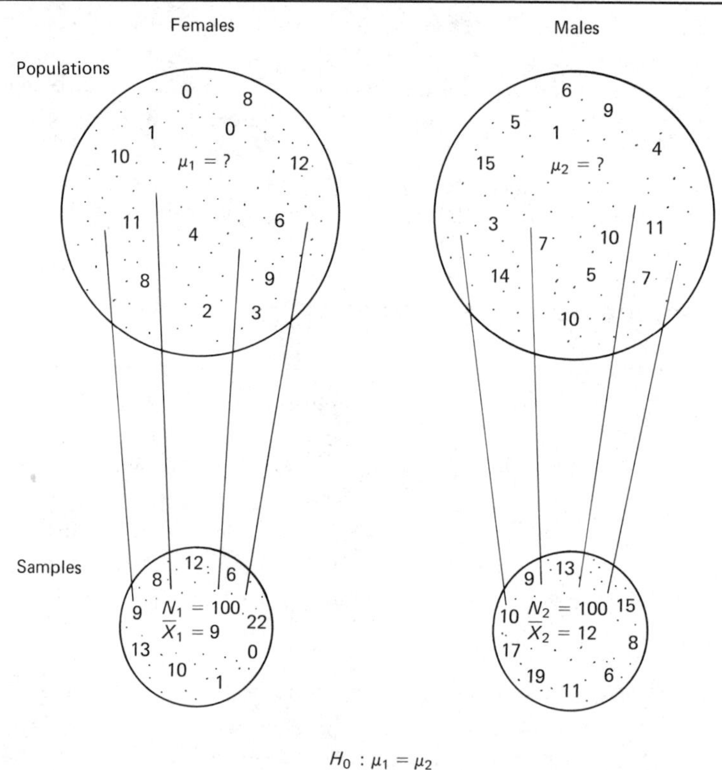

The sampling distribution of differences between means is normally distributed if four assumptions are met: (1) independent and random observations, (2) two independent samples (or subsamples), (3) interval level of measurement, and either (4a) known standard deviations in the two populations (σ_1 and σ_2) or (4b) large samples ($N_1 + N_2 > 100$). If neither of the fourth assumptions are satisfied but the two populations are normally distributed, evaluation of the sampling distribution is well approximated by using a Student's t curve at $df = N_1 + N_2 - 2$. Notice that this formula for the degrees of freedom combines the df value for two single-sample means tests ($N_1 - 1$ and $N_2 - 1$). Tabular values of t are again found in Table D3 and those for z are in Table D2 of Appendix D.

The test statistic for a difference in means test compares the observed difference between the two sample means ($\bar{X}_1 - \bar{X}_2$) to the expected difference between the two population means ($\mu_1 - \mu_2$) and divides this by the standard error of the sampling distribution:

$$z = \frac{(\bar{X}_1 - \bar{X}_2) - (\mu_1 - \mu_2)}{\sigma_{\bar{X}_1 - \bar{X}_2}}$$

z Transformation for a Difference in Means Test

When the population means are hypothesized (by H_0) to be equal the formula simplifies to

$$z = \frac{(\bar{X}_1 - \bar{X}_2) - 0}{\sigma_{\bar{X}_1 - \bar{X}_2}} = \frac{\bar{X}_1 - \bar{X}_2}{\sigma_{\bar{X}_1 - \bar{X}_2}}$$

Calculating the Standard Error of the Differences in Means

The standard error of the differences between two independent means can be found exactly when the standard deviations of the two populations are known:

$$\sigma_{\bar{X}_1 - \bar{X}_2} = \sqrt{\frac{\sigma_1^2}{N_1} + \frac{\sigma_2^2}{N_2}}$$ Standard Error of the Differences in Means

where σ_1 = standard deviation of first population
σ_2 = standard deviation of second population
N_1 = number of cases in first sample
N_2 = number of cases in second sample

FIGURE 10.6 Sampling Distribution of Differences Between Two Independent Means

Means for Samples of Females $\bar{X}_1$	Means for Samples of Males $\bar{X}_2$	Difference $(\bar{X}_1 - \bar{X}_2)$
9	12	−3
10	11	−1
10	9	1
7	10	−3
11	9	2
⋮	⋮	⋮

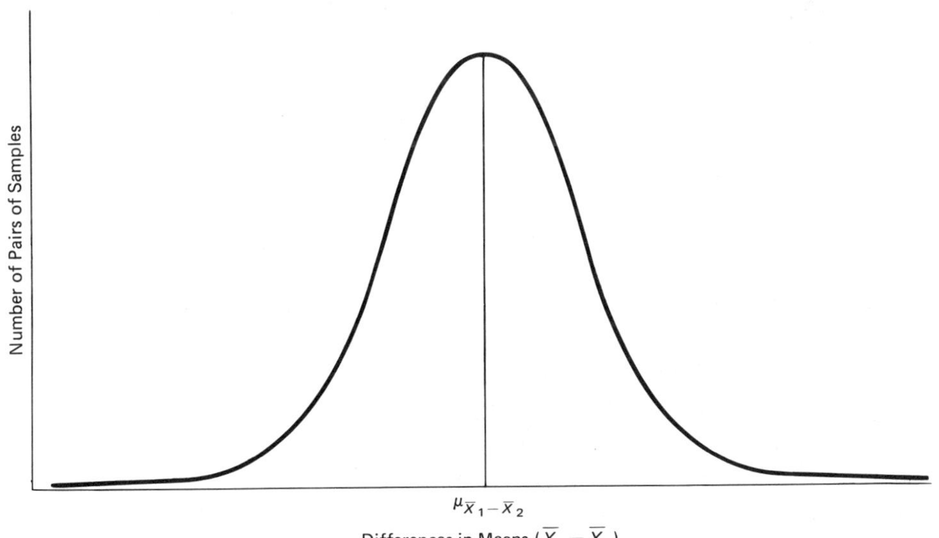

FIGURE 10.7 Classification of Distribution Features for Two-Sample Means Test

Distribution	Mean	Standard Deviation
First population	μ_1	σ_1
Second population	μ_2	σ_2
First sample	$\overline{X}_1$	s_1
Second sample	$\overline{X}_2$	s_2
Sampling distribution of differences between two means	$\mu_{\overline{X}_1-\overline{X}_2}$	$\sigma_{\overline{X}_1-\overline{X}_2}$

Notice that this formula *adds* the two values representing the populations even though we are working with a *difference* in means test. This apparent paradox is actually reasonable. Each population is contributing to the possible variation in differences between pairs of sample means. Some differences represent a very high sample mean drawn from the first population and a very low sample mean drawn from the second population. These kinds of outcomes increase the size of sampling error for the entire sampling distribution. Thus the variation in the differences between sample means will be larger than the variation in either population alone. The formula for the standard error of the differences combines the variances of the two populations, each as a ratio with the size of the sample drawn from that population.

Often the population standard deviations are unknown. The standard error of the differences in means can be estimated from sample data, but the exact procedure depends on the two variance parameters. Researchers must consider whether the variances for the two populations are *homogeneous* (alike) or *heterogeneous* (unlike). To determine whether the two variances are homogeneous requires a separate statistical test that is described in the section *Related Statistics* at the end of this chapter.

If the population variances are homogeneous, a pooled estimate (s_p^2) of their shared value is calculated, much as we did before with two proportions. This estimate combines the two sample variances to give the researcher a more reliable estimate of the value of the population variances than would either sample variance alone. These are weighted by their respective degrees of freedom ($N - 1$ for each sample):

$$s_p^2 = \frac{(N_1 - 1)s_1^2 + (N_2 - 1)s_2^2}{N_1 + N_2 - 2} \qquad \text{Pooled Estimate of Population Variance}$$

where s_1^2 = inferential variance of first sample
s_2^2 = inferential variance of second sample
N_1 = number of cases in first sample
N_2 = number of cases in second sample

This pooled estimate is then used in the formula for estimating the standard error:

$$\hat{\sigma}_{\bar{X}_1-\bar{X}_2} = \sqrt{s_p^2\left(\frac{1}{N_1} + \frac{1}{N_2}\right)}$$

Estimated Standard Error of Differences in Means When $\sigma_1^2 = \sigma_2^2$

If the population variances are very different from each other (heterogeneous), a pooled estimate like s_p^2 cannot be calculated. Instead, the variances from the two samples are kept separate in finding the estimated standard error:

$$\hat{\sigma}_{\bar{X}_1-\bar{X}_2} = \sqrt{\frac{s_1^2}{N_1} + \frac{s_2^2}{N_2}}$$

Estimated Standard Error of Differences in Means When $\sigma_1^2 \neq \sigma_2^2$

The estimate of the standard error for this situation will be less reliable than with homogeneous variances. Further, the resulting inference test is approximated (with only fair accuracy) by a t distribution whose degrees of freedom are estimated by the weighted means of the separate degrees of freedom from each sample:

$$df = \frac{N_1 df_1 + N_2 df_2}{N_1 + N_2}$$

Estimated Degrees of Freedom for a Difference in Means Test When $\sigma_1^2 \neq \sigma_2^2$

where df_1 = degrees of freedom from first sample ($N_1 - 1$)
df_2 = degrees of freedom from second sample ($N_2 - 1$)

These difficulties in describing the appropriate sampling distribution make this situation undesirable. Researchers attempt to avoid it by using large samples of equal size. This tends to control the effect of heterogeneity in population variances on a test where homogeneity is assumed. That test is then often used, rather than one assuming heterogeneous variances.

A z Test of Difference Between Two Independent Means

An inference test for the difference between two means from independent samples can assume a normal sampling distribution if the standard deviations from the two populations being sampled are known or if we have such large samples that the Central Limit Theorem applies. It would be very unusual for a researcher to know the population standard deviations but not know the means that were the object of a potential inference test. It is most likely that either both or neither would be known. However, let us imagine a situation in which someone is providing us with information about the population standard deviations but not the population means.

Suppose that an industrial sociologist wanted to examine possible sex discrimination against female employees in hourly wages paid by a very large manufacturing company. The industrial sociologist knows someone in the company who is able to find out that the standard deviation for hourly wages paid to female employees is \$2.40 and it is \$2.30 for male employees in comparable assembly line jobs. With the same insider's help, a random sample of the company's assembly line employees is selected, their hourly wages are determined, and the cases are split into two equal groups of 200 based on sex. The mean hourly wage of the female sample is \$5.50 and it is \$6.90 for the males. Is the female mean significantly lower than the male mean?

Step 1. The hypotheses.

H_0: $\mu_1 \geq \mu_2$ The mean hourly wage of all female assembly line employees of the company is either equal to or greater than the mean hourly wage of the company's male assembly line employees. Sampling error may cause the means for random samples of female and male employees to differ.

H_1: $\mu_1 < \mu_2$ The mean hourly wage of all female assembly line employees of the company is less than the mean hourly wage of the company's male assembly line employees. Mean hourly wages for random samples of female and male employees will reflect this directional difference beyond the expected influence of sampling error.

Step 2. The criteria for rejecting H_0.

The assumptions of independent and random observations in two independent samples and interval level of measurement are met in this example. In addition, the standard deviations of the two populations are known and the sample sizes are large ($N_1 + N_2 > 100$). For all these reasons combined the sampling distribution can be considered to be normally shaped. If the researcher uses the traditional .05 level of significance, the tabular statistic is -1.645 for this directional, one-tailed test. The rejection region begins at this value and extends to the extreme left of the distribution.

Step 3. The test statistic.

The standard error is found by formula as follows:

$$\sigma_{\bar{X}_1-\bar{X}_2} = \sqrt{\frac{\sigma_1^2}{N_1} + \frac{\sigma_2^2}{N_2}}$$

$$= \sqrt{\frac{(2.40)^2}{200} + \frac{(2.30)^2}{200}} = \sqrt{\frac{5.76}{200} + \frac{5.29}{200}} = \sqrt{.02880 + .02645}$$

$$= \sqrt{.05525} = .2351$$

The test statistic can now be found:

$$z = \frac{\bar{X}_1 - \bar{X}_2}{\sigma_{\bar{X}_1-\bar{X}_2}}$$

$$= \frac{5.50 - 6.90}{.2351} = \frac{-1.40}{.2351} = -5.95$$

Step 4. The decision about H_0.

As Figure 10.8 shows, the test statistic (-5.95) falls into the rejection region since its absolute value is greater than that of the tabular statistic (-1.645). This means that the sample means would differ as they did less than 5% of the time because of sampling error assuming the null hypothesis were true. A result this unusual leads the industrial sociologist to reject the null hypothesis. The female mean is significantly lower than the male mean; $z = -5.95$, $p < .05$.

Step 5. The conclusion.

Since the null hypothesis has been rejected, the alternative hypothesis is considered to be a reasonable description of the situation. The sample data lend support to the idea

FIGURE 10.8 Testing H_0: $\mu_1 \geq \mu_2$

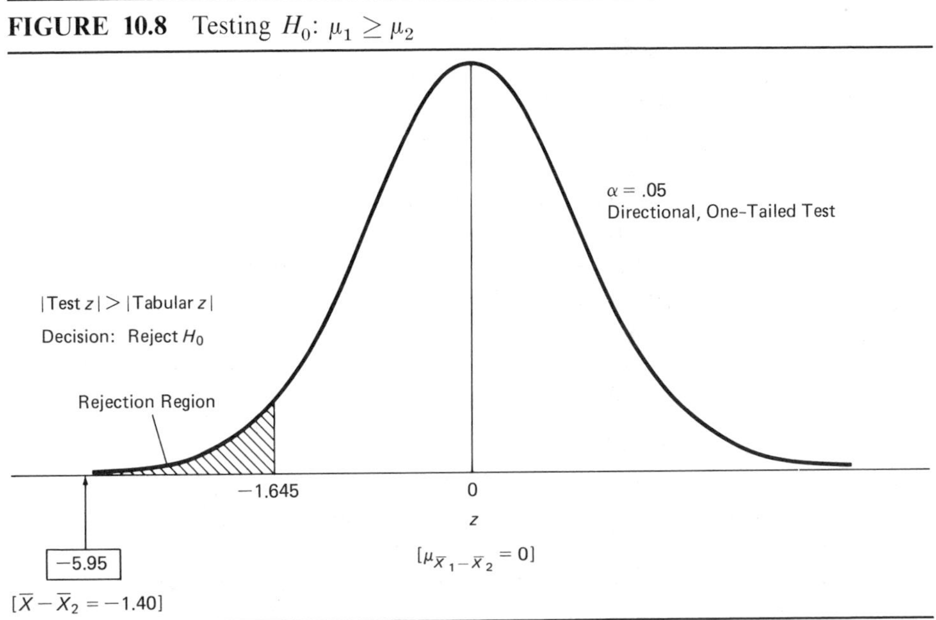

that the female assembly line workers are paid less than their male counterparts, at least as measured by mean wage. The researcher could be wrong, however. The conclusion may represent a Type I error. The female population mean may not be less than the male population mean. The observed difference in sample means may represent those very unusual outcomes that can occur even though H_0 is true. But the probability of this is low (less than .05) and the researcher was willing to risk making such a mistake, so the decision to reject H_0 stands.

This conclusion means that sex discrimination looms as a possible explanation of the findings. But more study needs to be given to the issue. Other sources of differences, such as experience, education, and age, must also be considered before a firm conclusion of sex discrimination can be reached.

A *t* Test of Difference Between Two Independent Means with Homogeneous Population Variances

You will more often find a researcher using a t rather than a z test for drawing an inference about the difference between two independent means. The population standard deviations are seldom known and the sample sizes may be rather small. The t test does require the researcher to assume that the populations have approximately normal distributions for the interval variable on which they are being compared. Experienced researchers are able to make a sound judgment about this assumption for a situation. Remember, too, that the t test is robust; it tolerates modest shortcomings in meeting this assumption. If the population distributions depart substantially from normality, a nonparametric test should be used instead.

Recall from Chapter 8 our illustration of a comparison between urban and rural people in their feelings of powerlessness. The researcher believed that there would be a difference in the mean powerlessness scores for the two populations. A random sample of urban residents had a mean score of 15 and a random sample of rural residents had a mean of 10. Suppose that $N_1 = 16$ and $N_2 = 9$, and $s_1 = 3.1$ and $s_2 = 2.8$. Is this difference in sample means sufficient to warrant the inference that the populations of urban and rural residents differ in mean powerlessness at the .01 level?

Step 1. The hypotheses.

H_0: $\mu_1 = \mu_2$ — The populations of urban and rural residents have the same mean powerlessness scores. The means for random samples of urban and rural residents may differ within the limits of sampling error.

H_1: $\mu_1 \neq \mu_2$ — The population of urban and rural residents do not have the same mean powerlessness scores. The means for random samples of urban and rural residents will reflect this difference beyond the expected influence of sampling error.

Step 2. The criteria for rejecting H_0.

The assumptions for a t distribution are met. The exact t distribution is the one at $df = N_1 + N_2 - 2 = 16 + 9 - 2 = 23$. At the .01 level of significance for a nondirectional, two-tailed test and $df = 23$, the tabular statistic is ± 2.807. The two rejection regions begin at these two t scores and extend to the extremes of the distribution.

Step 3. The test statistic.

Let's assume that the two population variances are nearly equal (homogeneous). To estimate the standard error, the researcher must first find the pooled variance estimate, s_p^2.

$$s_p^2 = \frac{(N_1 - 1)s_1^2 + (N_2 - 1)s_2^2}{N_1 + N_2 - 2}$$

$$= \frac{(16 - 1)(3.1)^2 + (9 - 1)(2.8)^2}{16 + 9 - 2} = \frac{(15)(9.61) + (8)(7.84)}{23}$$

$$= \frac{144.15 + 62.72}{23} = \frac{206.87}{23} = 8.994$$

$$\hat{\sigma}_{\bar{X}_1 - \bar{X}_2} = \sqrt{s_p^2\left(\frac{1}{N_1} + \frac{1}{N_2}\right)}$$

$$= \sqrt{8.994\left(\frac{1}{16} + \frac{1}{9}\right)} = \sqrt{8.994(.0625 + .11\bar{1})}$$

$$= \sqrt{8.994(.1736\bar{1})} = \sqrt{1.561458} = 1.250$$

The test statistic is now found to be

$$t = \frac{\bar{X}_1 - \bar{X}_2}{\hat{\sigma}_{\bar{X}_1 - \bar{X}_2}} = \frac{15 - 10}{1.250} = \frac{5}{1.250} = 4.00$$

Step 4. The decision about H_0.

The test statistic (4.00) exceeds the tabular statistic (2.807) and is therefore in the right-side rejection region. The researcher rejects the null hypothesis, knowing that the probability that this conclusion is wrong is less than .01. (See Figure 10.9.) The observed difference in means is statistically significant; $t = 4.00$, $df = 23$, $p < .01$.

FIGURE 10.9 Testing H_0: $\mu_1 = \mu_2$

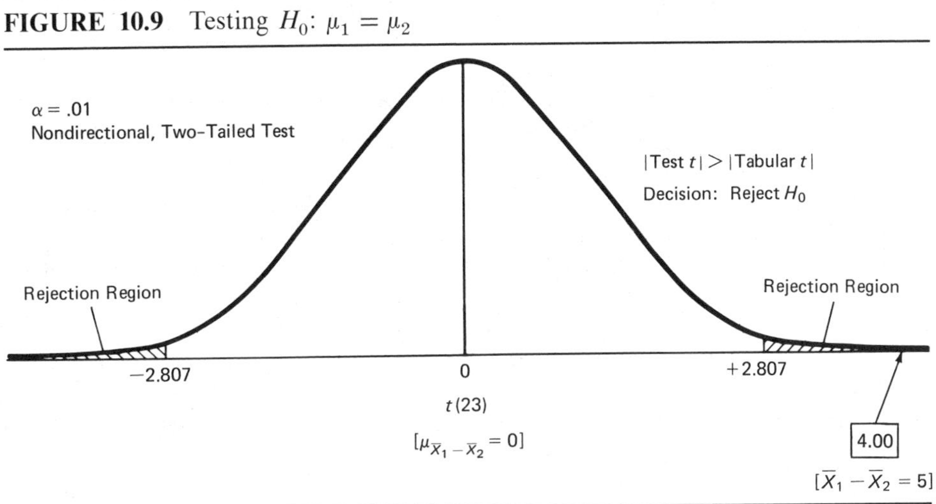

Step 5. The conclusion.

The researcher's expectation that urban and rural residents differ significantly in mean powerlessness is supported by the test. Replications of this study would add confidence to this conclusion if they had the same outcome. The present result may please the researcher, but it does not necessarily explain the finding. Theoretical reasons for the urban–rural difference must be considered. What do theories of social alienation say about the comparison? Do particular living conditions cause the powerlessness difference? Do cultural conditions cause the difference? The researcher must now work to devise tests to distinguish between various competing explanations. Finding a statistically significant difference is often only one part of a long task in reaching an understanding of social phenomena.

*A *t* Test of Difference Between Two Independent Means with Heterogeneous Population Variances

A researcher may find that he or she is unable to assume that the variances for the two populations being sampled are homogeneous (equal). The same researcher may be working with small samples but is able to assume that the two populations are normally distributed. A t test can be used as an appropriate significance test in this situation. The procedure is similar to the one in the last example except for the identification of the sampling distribution and the computation of the standard error. Consider this example.

An archaeologist is interested in the relative extent of discovery of a certain kind of stone artifact from two excavation sites. Each site was examined using a random selection of 1-square-meter areas for study. At the first site the mean number of artifacts recovered from 13 areas was 36 with an s value of 9. At the second site the mean number of the artifacts recovered from 9 areas was 31.5 with an s value of 6. The researcher had expected the first site to have a higher mean than the second, based on archaeological theory. Do the findings indicate that there was a significant difference of the type the researcher expected at the .05 level?

Step 1. The hypotheses.

H_0: $\mu_1 \leq \mu_2$ The population of artifacts at the first site has a mean that is either equal to or less than the mean for the population of artifacts at the second site. Random samples from the two sites may differ in means due to sampling error.

H_1: $\mu_1 > \mu_2$ The population of artifacts at the first site has a mean that is greater than the mean for the population of artifacts at the second site. Means for random samples of these two sites will reflect this difference beyond the expected influence of sampling error.

Step 2. The criteria for rejecting H_0.

If the population distributions are normal in shape, a t distribution can be used here. Let's assume that the population variances are heterogeneous. The degrees of freedom for the appropriate t curve are estimated as follows:

$$df = \frac{N_1 df_1 + N_2 df_2}{N_1 + N_2}$$

$$= \frac{13(12) + 9(8)}{13 + 9} = \frac{156 + 72}{22} = \frac{228}{22} = 10.4$$

For a .05, directional test, the closest tabular t ($df = 10$) in Table D3 is 1.812. Our sampling distribution has one rejection region at the right side that begins at $t = +1.812$.

Step 3. The test statistic.

Given that the population variances are heterogeneous, the standard error is estimated as follows:

$$\hat{\sigma}_{\bar{X}_1 - \bar{X}_2} = \sqrt{\frac{s_1^2}{N_1} + \frac{s_2^2}{N_2}}$$

$$= \sqrt{\frac{(9)^2}{13} + \frac{(6)^2}{9}} = \sqrt{\frac{81}{13} + \frac{36}{9}} = \sqrt{6.2307692 + 4}$$

$$= \sqrt{10.2307692} = 3.199$$

The test statistic is

$$t = \frac{\bar{X}_1 - \bar{X}_2}{\hat{\sigma}_{\bar{X}_1 - \bar{X}_2}}$$

$$= \frac{36 - 31.5}{3.199} = \frac{4.5}{3.199} = 1.41$$

Step 4. The decision about H_0.

The test statistic (1.41) does not exceed the tabular statistic (1.812); it does not fall into the rejection region. Thus the archaeologist fails to reject the null hypothesis. (See Figure 10.10.) The difference in the two sample means is not statistically significant; $t = 1.41$, $df = 10$, $p > .05$, n.s.

FIGURE 10.10 Testing H_0: $\mu_1 \leq \mu_2$

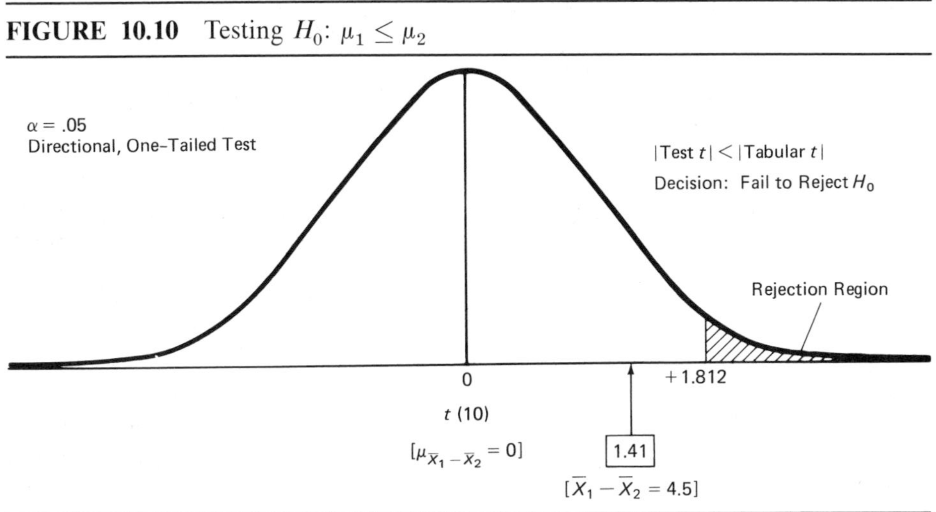

Step 5. The conclusion.

The archaeologist has not found support for the expectation that the first site would have a higher mean number of the stone artifacts than the second. Although the first *sample* had a higher mean than the second, the inference that the first *population* has a higher mean than the second population is not warranted from the data. Whatever theoretical reasons made the archaeologist believe the first site had a higher mean should now be opened to reexamination.

Research Examples of Testing Difference Between Two Independent Means

Research Examples 10.1 and 10.2 present illustrations of tests of significant difference between two independent means. In R.E. 10.1 the researchers were investigating the relationship between marital happiness and a concept they called "integrative complexity." This concept means the ability of the individual to be adaptive and flexible in dealing with his or her spouse. The techniques of measurement are carefully explained in the example. The researcher's overall hypothesis was a directional idea that couples who had high integrative complexity would be happier than couples who had low integrative complexity. The hypothesis is reported to have been supported in each of three specific types of tests. Notice that the authors do not specifically tell the reader that the tests are between independent means nor that the variances are assumed to be homogeneous or heterogeneous. It is rather simple to determine that the means are independent from reading the description of the sample and methods of data collection. However, the assumption made about the variances cannot be determined. As a reader, you are left to trust that the appropriate assumption has been made and incorporated with the computations. This is not atypical of many published examples of significance tests concerning means.

What is the population for which inferences from these results can be drawn? The answer is not obvious in this example. The sample was "a group of Princeton University faculty and their wives" who "were asked to participate in the study." This does not appear to be a random sample. It is possible that the sample may represent the population of Princeton University faculty and wives (spouses) or even a larger population of upper middle-class couples. However, further research using samples with better known representativeness is needed to establish the proper population. As noted before, the appropriate population is often described only vaguely in social research. This limits the interpretation and comparability of many studies.

RESEARCH EXAMPLE 10.1 Difference Between Two Independent Means Test

From Bryant Crouse, Marvin Karlins, and Harold Schroder, "Conceptual Complexity and Marital Happiness," Journal of Marriage and the Family, *30 (4), November 1968:643ff. Reprinted by permission.*

. . . Many investigators have postulated a relationship between marital happiness and the ability of the individual to be adaptive and flexible in dealings with his spouse. . . . One personality theory that deals with behavioral adaptivity is the conceptual systems approach of Schroder and his associates. In this theory, information-processing ability varies among individuals and is measured in terms of its *integrative complexity.* Low levels of integrative complexity are associated with intolerance of ambiguity, dogmatism, rigidity, and closed mindedness. . . . An integratively complex individual, on the other hand, is flexible in his dealings with the environment. He is a flexible explorer of this world: he does not close fast under uncertainty and is attuned, adaptive, and sensitive to environmental change.

[I]t would be expected that integratively complex individuals, in comparison to persons lower in integrative complexity, would be more likely to be happily married—due to their superior ability to adapt to the flexible demands of interpersonal dynamics within the marital relationship. . . . [I]t would seem that when husband and wife are both *high* in integrative complexity (HiHi couple), the possibility of their being happy would be maximized; while the chances of happiness for a husband and wife both low in integrative complexity (LoLo couple) would be minimized. . . . It is predicted, therefore, that couples in which both husband and wife are high in integrative complexity (as measured by the Paragraph Completion Inventory) will be happier (as assessed by a standard measure of marital happiness) than couples in which husband and wife are both low in integrative complexity. . . .

A group of Princeton University faculty members and their wives ($N = 42$ couples) were asked to participate in this study. . . . Data collection was accomplished in two phases. In phase one a measure of marital happiness was administered to 128 couples: the General Satisfaction of Self and Concept of Mate's Satisfaction in Marriage Test devised by Burgess and Wallin. . . . [A]t no time was a participating subject informed of his spouse's responses on any testing instrument. Twenty of the 128 participating couples in phase one were eliminated from the remainder of the investigation because they were of foreign nationality. Forty-two of the remaining 108 couples agreed to serve as subjects in phase two of the experiment. During the phase-two testing session, the measure of integrative complexity (Paragraph Completion Inventory or PCI) was administered. . . .

It was hypothesized that couples in which both partners were high in integrative complexity (HiHi) would be happier than couples in which husband and wife were both low in integrative complexity (LoLo). Subjects were dichotomized into high and low complexity groups by the median split procedure. Since measures of integrative complexity were available for both interpersonal and marriage stimuli, three independent analyses were run on the basis of (1) standard integrative complexity scores, (2) marital intergrative complexity scores, and (3) the sum of the integrative complexity and marital integrative complexity scores. . . . *T* Tests were used in each analysis (one-tailed). The results indicate that (1) HiHi couples (determined by the standard measure of integrative complexity) were significantly happier ($t = 2.18$, $p < .025$) than LoLo couples, marital happiness being determined by scores on the Burgess and Wallin marital happiness test; (2) HiHi couples (determined on the marital stimuli test of integrative complexity) were significantly happier ($t = 2.07$, $p < .025$) than LoLo couples as assessed by their responses on the marital happiness test; (3) when complexity scores (standard and marital) are combined, HiHi couples are still significantly happier ($t = 2.13$, $p < .05$) than LoLo couples on the Burgess and Wallin measure. . . . [T]he findings of this study support the hypothesized relationship between marital happiness and an individual's ability to be adaptive (flexible) in the marriage.

Research Example 10.2 concerns a study of rape victims. It compares those who did report a rape to those who did not on the basis of their perceptions of how police handle rape cases. The victims were asked to rate police behavior toward rape victims on five semantic continua. Mean scores for the two groups of rape victims were calculated and compared using the t test for independent samples. All five sets of mean differences were found to be significant using .05 as the minimum alpha level.

The authors point out two important circumstances in the study for our interpretation. First, these authors tell us that the sample of victims was not randomly selected. It was based on the availability of cases. This was a necessity given the low rate of rape. The selection of the specific 17-month time period and the Colorado location were not explained. Having made no attempt to have a random sample, the authors realized that inferences could not legitimately be drawn. Thus they note the second point of interest to us. The t tests shown in the table are used to illustrate the inferences that would have been possible *if* the sample had been randomly drawn from some population. The text of the article does not contain this same cautionary tone but proceeds to describe the mean scores to be statistically significant. Further, comparisons in the apparent degree of significance are made on the basis of those tests. The article is more candid than most in its acknowledgment of the use of a nonrandom method to select its sample. But it does attempt to benefit from the implied importance that a significance test gives to research. Seemingly, there is a trend toward this practice becoming more widespread.

RESEARCH EXAMPLE 10.2 Difference Between Two Independent Means Test

From Richard L. Dukes and Christine L. Mattley, "Predicting Rape Victim Reportage," Sociology and Social Research, *62 (1), October 1977:63ff. Reprinted by permission.*

. . . Even in light of recent research, knowledge about victims of rape remains limited. One of the principal reasons for this state of affairs is that many victims do not report the crime to police or to other professionals. . . . Clearly, if the overall understanding of victims is to be increased, the understanding of the victims' decision-making processes about whether or not to report the crime to police is a first step. . . .

The sample for this investigation consisted of 45 victims who were raped in the Colorado Springs, Colorado region between January 1, 1973 and June 1, 1975. Of the 45 respondents, 25 had reported the crime to police and 20 had not reported. Victims of assaults to commit forcible rape are not included in the sample. Victims of statutory rape (without force) are not included in the sample. The age-range of victims included in the sample was mid-teens to mid-fifties.

"Theoretical sampling" of both reporting and non-reporting victims is necessary for comparative purposes. . . . In this type of sampling (sometimes referred to as "availability" sampling) the investigator attempts to obtain as many cases as possible within categories of theoretical interest. No attempt is made to ensure that the sample is representative of some larger population. . . . In order to obtain respondents for the investigation, victims were invited to volunteer to be interviewed. The invitation was extended to victims first on radio and television through public service announcements. Later, paid advertising was used. . . .

The interviewers for this investigation were all volunteers from the Colorado Springs Rape Crisis Service who were trained to use the interview schedule. Their counselling experience and understanding of the problems confronting victims proved to be an extremely important factor in their success. Most interviews were conducted over the telephone. . . . The interview schedule . . . contained open-ended questions as well as pre-scaled fixed-alternative-response items. . . .

The preconceptions held by a victim concerning the manner in which police handle rape cases should influence whether or not she reports the rape to them. In order to assess the influence of these preconceptions, victims were asked to score them along five continuua: sympathetic-unsympathetic, concerned-unconcerned, believing-doubting, efficient-inefficient, and considerate-humiliating.

Each of these continuua yielded differences in mean scores of close to one scale point or more, and each difference was statistically significant beyond the .05 level [see the table]. Obtained t's were highest for the continuua of concerned-unconcerned and considerate-humiliating. Obtained values for t were lowest for the continuua efficient-inefficient. It appears from the data, that perceptions of the *way victims are treated* when they report are more important than perceptions of how efficiently cases are processed. . . .

Mean Scores and Levels of Significance by Reporting and Non-reporting Victims for Semantic Differential Items for the Question: "Before the rape, what was your impression of police handling of rape cases?"[a]

Item	Mean Scores for Reporting Victims	Mean Scores for Non-reporting Victims	t	Level of Significance[b]
Sympathetic/unsympathetic	2.83	1.79	2.65	$p \leq .01$
Concerned/unconcerned	3.37	1.89	3.94	$p \leq .0005$
Believing/doubting	2.74	1.89	2.26	$p \leq .025$
Efficient/inefficient	3.46	2.65	2.01	$p \leq .05$
Considerate/humiliating	2.86	1.72	3.08	$p \leq .005$

[a] Table 1 in the original.

[b] Tests of significance are presented in this research for illustrative purposes. They indicate the confidence in conclusions which would be justified given a random sample of the same size.

A *t* Test of Difference Between Two Dependent Means

In each of the previous tests of difference between two mean scores those means represented samples (or subsamples) that were independent of each other. There was no connection between the cases selected for each sample. However, social researchers often work with samples that are dependent or related to each other. Typically, this occurs in one of two kinds of experimental research designs: matched samples or before–after samples.

First, the subjects in one sample may be purposely matched with those in a second sample on such characteristics as sex, age, or verbal ability. This is done to control for the possible influence of these characteristics on the scores for the variable on which the two samples are being compared. For example, one sample may be the "experimental group" which receives a certain stimulus, while the other sample is the "control group" which does not receive that stimulus. The researcher wants to compare the two sets of scores. Because the cases in the two groups have been matched, their scores are not independent of each other. By deliberately matching the cases on some characteristic, their scores on the variable under study are linked. They are likely to be similar because they come from cases with similar backgrounds. They do not correspond to a random pairing of scores from cases with varied characteristics.

Second, in another type of experimental design one group of subjects is measured at one point in time, a stimulus is administered, and the same subjects are measured again to determine whether a change has occurred. This is called a before–after study. The cases in the first group are the same persons as those in the second group. Obviously, the scores from the two groups are not independent of each other. An individual will tend to score more alike at each data collection point than would two different individuals.

The general outline before of the procedure for conducting a difference-between-two means test still applies in the situation of dependent samples. However, an adjustment must be made in the determination of the sampling distribution and the computation of its standard error. The test continues to require that the scores are at the interval level of measurement, random (within each group), and independent of each other (within each group). But the two sets of scores are considered to be dependent.

Sampling Distribution of Differences Between Two Dependent Means

When the two samples are dependent they are treated as belonging to one group (not two) for which the researcher has two observations per case (not one per case). This is illustrated in Figure 10.11 from the point of view of the null hypothesis. The sampling distribution is conceptualized as being the set of all possible differences between related group means that could be found from one population of paired scores. That is, the researcher imagines selecting two observations at a time for each case in every possible sample of a set size from the population. The mean is found for the set of first observations in each sample and a separate mean is found for the set of second observations in each sample. Then the difference is calculated between each mean based on a set of first observations and each mean based on a set of second observations. The distribution of these differences will again approach a normal shape if the standard deviations of a subpopulation of scores representing all first scores and the subpopulation of all second scores are known or if the sample size is large (> 100). When neither of these conditions is satisfied, the sampling distribution is assumed to be a t distribution at $df = N - 1$, where N means the number of pairs of observations. As always, the population of all pairs of observations must be normally distributed for the researcher to use the t distribution. These are the

FIGURE 10.11 Null Hypothesis's View of a Difference Between Two Dependent Means

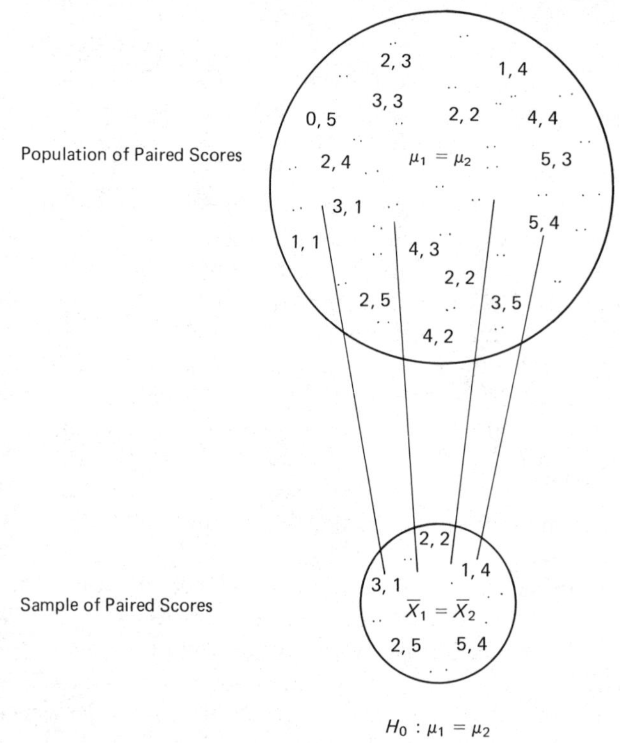

usual conditions we have discussed before, although they may appear strange because we are applying them to dependent samples. (See Figure 10.12.)

The sampling distribution will have a mean that is again labeled $\mu_{\bar{X}_1-\bar{X}_2}$ and is equal in value to the true difference between the mean of all first observations and the mean of all second observations (μ_1 and μ_2, respectively). If the researcher was using the null hypothesis that the two sets of scores have the same mean, the mean of this sampling distribution is zero. The standard error of this distribution is again labeled $\sigma_{\bar{X}_1-\bar{X}_2}$. The definitional formula for this quantity is

$$\sigma_{\bar{X}_1-\bar{X}_2} = \sqrt{\sigma^2_{\bar{X}_1} + \sigma^2_{\bar{X}_2} - 2\rho\, \sigma_{\bar{X}_1}\sigma_{\bar{X}_2}}$$

Definitional Formula for Standard Error of the Differences Between Two Dependent Means

This formula instructs the researcher to find the standard error for each subpopulation of first and second observations separately ($\sigma_{\bar{X}_1}$ and $\sigma_{\bar{X}_2}$), square and add these, and then subtract a quantity representing the correlation between the two sets of observations. This later quantity is zero when the sets are independent, and therefore it was not shown to you before for the tests involving independent samples. But now we need to include it because our samples are dependent.

Fortunately, there is a much simpler way to find the standard error for dependent means than indicated by this formula. The following uses the "direct-difference method" to find it exactly from population data or to estimate it from sample data:

$$\sigma_{\bar{X}_1-\bar{X}_2} = \sqrt{\left[\frac{\Sigma D^2}{N} - \left(\frac{\Sigma D}{N}\right)^2\right]\frac{1}{N}}$$

Computational Formula for Standard Error of the Differences Between Two Dependent Means When Population Data Are Used

or

$$\hat{\sigma}_{\bar{X}_1-\bar{X}_2} = \sqrt{\frac{\Sigma D^2 - \frac{(\Sigma D)^2}{N}}{N(N-1)}}$$

Computational Formula for Standard Error of the Differences Between Two Dependent Means When Sample Data Are Used

where D = difference within each pair of observations
N = number of cases (pairs of observations)

FIGURE 10.12 Sampling Distribution of Differences Between Two Dependent Means

Means for Samples of First Observations $\bar{X}_1$	Means for Samples of Second Observations $\bar{X}_2$	Difference $(\bar{X}_1 - \bar{X}_2)$
3	4	−1
2	2	0
4	1	3
4	3	1
⋮	⋮	⋮

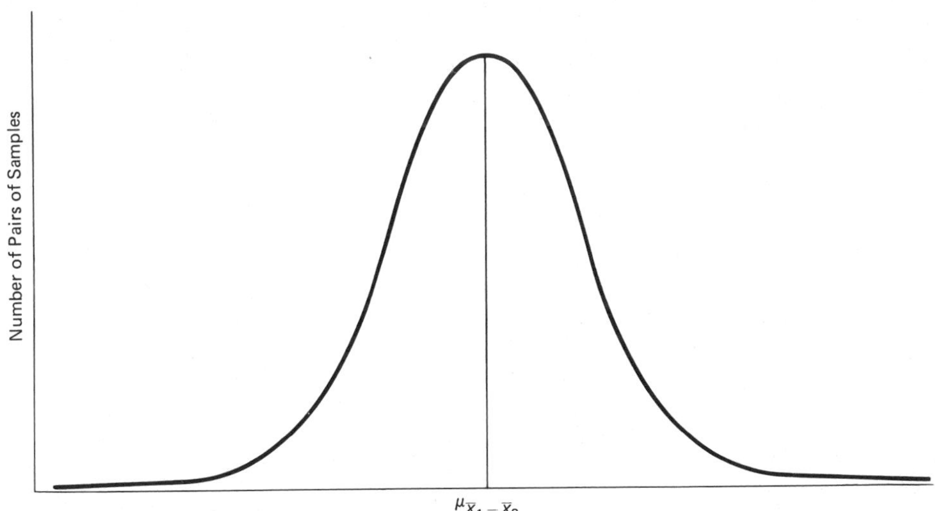

Since a researcher most often works from sample data, the second of these formulas is the one commonly used.

Let's look at an example to see how these adaptations are made to a test of difference between two dependent means. Suppose that a social psychologist has devised an experiment in which 10 subjects are given a test measuring their feelings of helpfulness, then required to do a meaningless task as a group, and finally remeasured on the helpfulness test. The researcher suspects that the meaningless task will diminish the subjects' feelings of helpfulness from the first to the second measurement. The mean helpfulness score before the task is 44, whereas the mean after the task is 32. Is the difference statistically significant at the .05 level?

Step 1. The hypotheses.

H_0: $\mu_1 \leq \mu_2$ The mean of the subpopulation of helpfulness scores before the meaningless task is either equal to or less than the mean of the subpopulation of helpfulness scores after the task. Sampling error may cause means from a random sample to differ from this.

H_1: $\mu_1 > \mu_2$ The mean of the subpopulation of helpfulness scores before the meaningless task is greater than the mean of the subpopulation of helpfulness scores after the task. Means from a random sample will differ in this direction beyond the influence of sampling error.

Step 2. The criteria for rejecting H_0.

The standard deviations of the two subpopulations (before and after scores) are unknown and the sample size is small. Thus a normal sampling distribution cannot be assumed. But it is reasonable to believe that the subpopulations would be normally distributed, especially if the helpfulness test has been standardized. At the .05 level of significance for this directional, one-tailed test and with $df = N - 1 = 10 - 1 = 9$, the tabular t score is $+1.833$. The sampling distribution has one rejection region at its extreme right side.

Step 3. The test statistic.

We find the standard error using the computational formula for sample data. Those data and the interim calculations are shown in Table 10.1. Notice how the difference is found between each subject's before and after score and then each difference is squared. The total of these is the quantity ΣD^2 (2348). The total of the differences without squaring is the quantity ΣD (120).

$$\hat{\sigma}_{\bar{X}_1 - \bar{X}_2} = \sqrt{\frac{\Sigma D^2 - \dfrac{(\Sigma D)^2}{N}}{N(N-1)}}$$

$$= \sqrt{\frac{2348 - \dfrac{(120)^2}{10}}{10(9)}} = \sqrt{\frac{2348 - 1440}{90}} = \sqrt{\frac{908}{90}}$$

$$= \sqrt{10.08\overline{8}} = 3.176$$

Now the researcher finds the test statistic:

$$t = \frac{\bar{X}_1 - \bar{X}_2}{\hat{\sigma}_{\bar{X}_1 - \bar{X}_2}} = \frac{44 - 32}{3.176} = \frac{12}{3.176} = 3.78$$

TABLE 10.1 Helpfulness Scores Before and After a Meaningless Task ($N = 10$)

Subject	Before X_1	After X_2	Difference D	D^2
1	40	25	15	225
2	29	36	−7	49
3	38	38	0	0
4	41	33	8	64
5	55	34	21	441
6	52	40	12	144
7	48	34	14	196
8	44	33	11	121
9	43	15	28	784
10	50	32	18	324
	440	320	$\Sigma D = 120$	$\Sigma D^2 = 2348$
	$\bar{X}_1 = 44$	$\bar{X}_2 = 32$		

Step 4. The decision about H_0.

The test statistic (3.78) does exceed the tabular statistic (1.833), falling into the rejection region. Therefore, the null hypothesis is rejected. (See Figure 10.13.) Such a decision runs the risk of being a Type I error with no more than a probability of .05. The before mean is significantly greater than the after mean; $t = 3.78$, $df = 9$, $p < .05$.

Step 5. The conclusion.

Having rejected the null hypothesis the researcher is able to conclude that the alternative hypothesis has been given support by the test. The researcher believed that the meaningless task would reduce the subjects' feelings of helpfulness and apparently it did. This may confirm some theoretical ideas and everyday experiences of the researcher. But there are other possible explanations still to be investigated. Did the subjects represent a cross section of all people or were they concentrated within a few personal and social

FIGURE 10.13 Testing H_0: $\mu_1 \leq \mu_2$

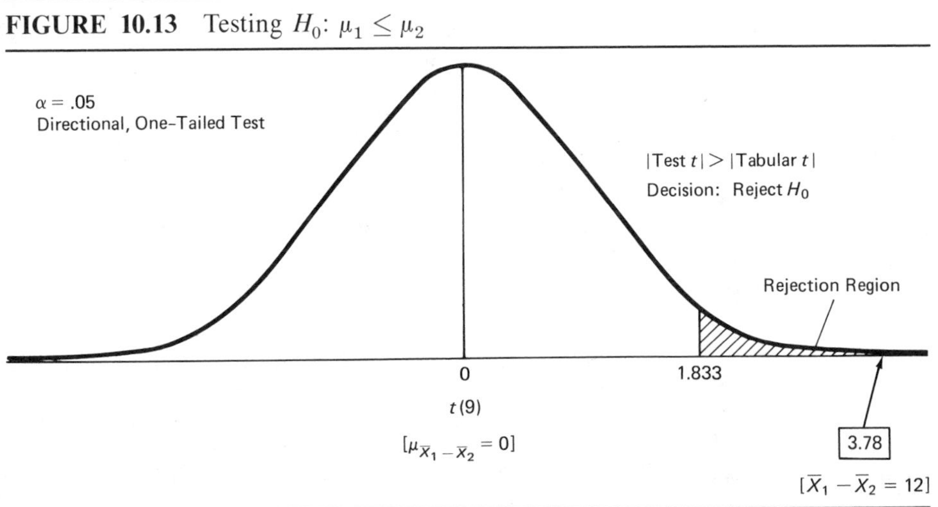

characteristics? With a sample of only 10, it would be impossible to represent a very diverse population. The study should be repeated with a large sample and more careful controls on potentially important features such as age, sex, race, and so forth. Without this the results cannot be fairly generalized to much extent.

Research Example 10.3 presents an illustration of a test of significant difference between two dependent means. The study concerned the affects of forced relocation on a group of 39 elderly persons. These people were asked to name other persons with whom they felt "close" both before and after they moved from one home for the aged to another in Toronto, Canada. The move occurred because the first home was closed for safety reasons. Comparisons were made between the before and after numbers of close residents, close staff, close family and friends, and a total. The table in the example shows that a t test was used for "paired samples" and it found significant differences in the mean number of close residents ($p < .001$), mean number of close staff ($p < .05$), and the mean total number ($p < .001$). All of these differences were toward fewer primary relationships after the move. The difference in mean number of close family and friends was not significant.

The example demonstrates how this type of significance test is often used to suggest inferences. Notice that the sample size diminished from the before to the after measurement. Further, the sample was not randomly selected but used the judgment technique instead. These two factors create some ambiguity about the appropriate population to which the results can be generalized.

RESEARCH EXAMPLE 10.3 Difference Between Two Dependent Means Test

From Lilian Wells and Grant Macdonald, "Interpersonal Networks and Post-Relocation Adjustment of the Institutionalized Elderly," The Gerontologist, *21, (2), April 1981:177ff. Reprinted by permission.*

Involuntary relocation creates major disruptions in the lives of elderly people. For many in this particulary vulnerable group, relocation constitutes a threatening event which may manifest itself in undesirable physical, emotional and social consequences. Much of the literature on relocation reflects the seriousness of the problem. A number of researchers have reported that extensive environmental change can lead to behavioral, psychological and physical deterioration in elderly people. . . . On the other hand, there is evidence to suggest that a stimulating new environment may increase life satisfaction and functioning for those people who are able to cope with the change. . . . [This study's] objective is to explore the extent of disruption in close interpersonal networks created by inter-institutional relocation and to determine if there is a link between close relationships prior to the move and successful physical and psychological adjustment following it.

The relocation which forms the basis of this study was brought about by the closure of Hilltop Acres which, having been constructed in 1902, no longer met safety standards. The building was one of eight Homes for the Aged, owned and operated by the Municipality of Metropolitan Toronto's Dept. of Social Services. . . .

We provided some control to assure uniformity of the health and environmental variables. The study sample was restricted to those residents requiring "extended care" and those who chose to move to other homes within the system. There were 74 residents who met these criteria, but 3 refused to participate, 8 were unable to complete the structured interview, 2 did not speak English and 5 moved before we could interview them. . . .

Network information.—Residents were asked to identify other residents, staff, family and friends outside the home with whom they felt "close." It has been assumed for this study that the notion of "closeness" is an acceptable translation of the sociological concept of a primary relationship, or a personal tie involving support and affectional concern . . . Using this method, the total range of primary relationships was obtained for the three network categories: (1) close residents; (2) close staff, (3) close family and friends. . . .

Sample attrition.—At the time of follow-up, 5 subjects (9% of the sample) had died, 3 refused to be interviewed and 9 were disoriented and unable to participate. Complete follow-up data were obtained for . . . 39 (69%) for the network information. . . .

The results indicate that relocation substantially disrupts the primary relationship networks of many of the residents. The 39 subjects, for which there was network information available before and after the move, showed a significant loss in terms of their range of close ties. As a group, these 39 subjects identified, prior to the move, a total of 165 persons whom they described as close. Following relocation, these same subjects named only 114 persons as close, a reduction in the total number of primary relationships of 31%. The mean number of relationships for this group prior to the move was 4.2 compared with 2.9 after, a loss which was statistically significant (t-value $= 3.45$, d.f. $= 38$, $p < .001$). This reduction in range was largely attributable to the loss of close relationships with residents and staff from the former home which were not replaced after relocation. The number of family and friends outside the home remained constant over the period of the move [see the table]. One might speculate that the 31% reduction in the total range of primary relationships will eventually disappear as residents find new friends amongst the residents and staff of their new homes. However, the figure clearly reflects the extent of the disruption in their close social interactions created by relocation. . . .

Comparison of the Mean Number of Primary Relationships in Different Network Categories Before and After Relocation ($N = 39$)[a]

Network Categories	Before Relocation Mean	Before Relocation S.D.	After Relocation Mean	After Relocation S.D.	t-Value[b]	Significance
Number of close residents	1.00	1.17	0.31	0.47	3.69	$p < 0.001$
Number of close staff	0.90	1.47	0.26	0.71	2.48	$p < 0.05$
Number of close family and friends	2.33	1.78	2.36	1.58	−0.14	N.S.S.[c]
Total number of close relationships	4.23	3.07	2.92	1.98	3.45	$p < 0.001$

[a] Table 2 in the original.
[b] t-test between variables for paired samples.
[c] Not statistically significant.

Nonparametric Tests for Differences Between Two Samples

It is not always possible to meet the parametric assumptions required for the z and t tests we have been using with means. A researcher may not know the standard deviations or shapes of population distributions. Further, a difference-between-two-proportions test may not be appropriate because the variable under study is not dichotomous. Nonparametric tests are available to conduct tests of difference in these situations. We will look at our old friends, the chi-square test and the Kolmogorov–Smirnov test, in modified forms that suit the two sample circumstance. In between these will be a new face, the median test.

Chi-Square Test of Independence for Two Samples (χ^2-*II*)

We have seen how the chi-square test was used in Chapter 5 to determine whether there was any association between two nominal variables in a descriptive sense. In Chapter 9 we saw how it can be used to determine whether a single sample's frequency distribution on one nominal variable is significantly different from a distribution that is theoretically expected to occur. Now we will see how it can be used to determine whether two independent groups are different from each other in their relative frequency distributions. As you will discover, this use is only minimally different from those before.

The *chi-square test of independence* is a nonparametric significance test that compares the frequency distributions for two (or more) independent samples on a nominal variable. The test assumes that the observations are independent of each other and randomly chosen. This can be accomplished by actually drawing two (or more) random samples from different populations or by drawing one random sample and splitting it into independent groups. Like all chi-square procedures, the total number of cases must not be too small or large, although the exact limitations are debated.

The tabular chi-square value against which a test value is compared is found by a new formula for the appropriate degrees of freedom. It is based on the size of the table containing the frequency distributions:

$$df = (r - 1)(c - 1)$$

Degrees of Freedom for Chi-Square-II Test

where r = number of rows
c = number of columns

The test statistic uses the same formula and definitions of terms as that in Chapter 5:

$$\chi^2 = \Sigma \frac{(f_o - f_e)^2}{f_e}$$

Chi-Square-II Test Statistic

The expected frequency (f_e) for each cell in a table is found by multiplying its two marginal totals (row by column) and dividing the product by the overall total. An example should refresh your memory about this process.

Suppose that a legislative aide has asked a random sample of adults whether they believe that possession of small amounts of marijuana should be decriminalized. The researcher splits the sample by age into two groups of younger and older adults. This produces the distribution shown in Table 10.2. Do the two groups differ significantly at the .05 level?

TABLE 10.2 Distribution of Marijuana Possession Decriminalization Attitude by Two Adult Groups

Attitude Toward Decriminalization of Marijuana Possession	Younger Adults	Older Adults
Agree	55	19
Not certain	31	22
Disagree	14	39

Step 1. The hypotheses.

The null hypothesis expects the true (population) chi-square to be zero since the observed frequencies would match the expected frequencies. The alternative hypothesis expects these sets of frequencies to differ causing chi-square not to be zero.

H_0: $\chi^2 = 0$ The population of younger adults will not differ from the population of older adults in their frequency distributions of attitudes toward the decriminalization of possessing small amounts of marijuana. Sampling error may cause distributional differences for random samples from these two populations.

H_1: $\chi^2 \neq 0$ The population of younger adults will differ from the population of older adults in their frequency distributions of attitudes toward the decriminalization of possessing small amounts of marijuana. Distributions for random samples from these two populations will reflect this difference beyond the expected influence of sampling error.

Step 2. The criteria for rejecting H_0.

Our problem meets all the requirements for a chi-square distribution; independent and random observations for two independent subsamples on nominal frequencies. Our table has three rows and two columns, so the degrees of freedom are

$$\begin{aligned} df &= (3 - 1)(2 - 1) \\ &= (2)(1) \\ &= 2 \end{aligned}$$

At the .05 level of significance and 2 *df*, the tabular chi-square is 5.991. A single rejection region to the right side is used with this type of chi-square problem. The test is nondirectional; the alternative hypothesis does not predict either set of frequencies to exceed the other. Thus it would be satisfied with a difference either way. However, the chi-square computation produces only positive values (through the squaring of differences) other than zero. The test chi-square will be zero when the two frequency distributions (compared in relative terms) are identical. When they are different it will be some positive value, regardless of which set of frequencies exceeds the other. As a result, a single rejection region describes those results so different from zero as to be considered beyond the expected influence of sampling error. Ours begins at 5.991 and extends to the right.

TABLE 10.3 Worksheet for Chi-Square-II Test

Attitude	Younger Adults	Older Adults	Total
Agree	55 (41.1)	19 (32.9)	74
Not certain	31 (29.4)	22 (23.6)	53
Disagree	14 (29.4)	39 (23.6)	53
Total	100	80	180

$$\chi^2 = \sum \frac{(f_o - f_e)^2}{f_e}$$

$$= \frac{(55-41.1)^2}{41.1} + \frac{(19-32.9)^2}{32.9} + \frac{(31-29.4)^2}{29.4} + \frac{(22-23.6)^2}{23.6} + \frac{(14-29.4)^2}{29.4} + \frac{(39-23.6)^2}{23.6}$$

$$= \frac{(13.9)^2}{41.1} + \frac{(-13.9)^2}{32.9} + \frac{(1.6)^2}{29.4} + \frac{(-1.6)^2}{23.6} + \frac{(-15.4)^2}{29.4} + \frac{(15.4)^2}{23.6}$$

$$= \frac{193.21}{41.1} + \frac{193.21}{32.9} + \frac{2.56}{29.4} + \frac{2.56}{23.6} + \frac{237.16}{29.4} + \frac{237.16}{23.6}$$

$$= 4.7010 + 5.8726 + .0871 + .1085 + 8.0667 + 10.0492$$

$$= 28.88$$

Step 3. The test statistic.

The test chi-square is calculated as we did in Chapter 5. The expected and obtained frequencies are compared using the formula as shown in Table 10.3. In our example, the test chi-square is 28.88.

Step 4. The decision about H_0.

The test statistic (28.88) does exceed the tabular statistic (5.991); it does fall into the rejection region. Thus the legislative aide rejects the null hypothesis. (See Figure 10.14.) The differences between the observed and expected frequencies are beyond those that would occur because of sampling error 5% of the time. The two subsamples do differ significantly; $\chi^2 = 28.88$, $df = 2$, $p < .05$.

Step 5. The conclusion.

Having rejected the null hypothesis, the legislative aide decides that younger and older adults have different attitudes toward the decriminalization of possessing small amounts of marijuana. This makes it difficult to formulate a single stand on the issue that would satisfy both groups. The politician for whom the aide works will have to take a position with the inferred knowledge that either the younger or older adults will be offended.

Median Test for Two Samples (*Mdn-II*)

The median test for two samples is a nonparametric relative of the chi-square test of independence for two samples. The median-II test compares the frequency distributions for two samples on a variable that has been measured at the ordinal level or higher. The scores on that variable for both samples are combined and the median found for the unified distribution. The scores are then split into two categories for each sample using the median as the dividing point. All scores above the median are put in one category and all those at or below the median are put in a second category. This creates a 2 × 2 table to which the usual chi-square-II test is applied. The result is a test of the null hypothesis that the two samples represent two populations with the same medians. If that hypothesis is true, there will be equal proportions of scores above (and not above) the common median in each sample.

Imagine that a metropolitan arts council decides to see whether there is a difference in the number of cultural events attended during a year by those residents from the inner city and those from the suburban areas. The council expects that there is a difference. A random sample of 20 inner-city residents and a random sample of 30 suburban residents are asked how many times they have visited or attended a city concert, play, museum, or art gallery during the past 12 months. The individual results are shown in Table 10.4. Do they differ significantly at the .05 level?

In considering this problem the person doing the research needs to decide whether a parametric or a nonparametric test should be performed. The number of events attended is a ratio-level variable, the samples are randomly selected, and means could be compared between them. The logical parametric choice would be a *t* test. But are the populations of scores normally distributed? That is, is the distribution of the number of events attended for all inner-city residents shaped like a normal curve, and similarly for all suburbanites? Not likely. We should expect these population distributions to be very positively skewed with most people seldom attending and a few people attending much more often than do the majority. If this is true, the *t* test is not appropriate. Let's try a median test instead. It makes no requirements about the distributions of scores in the populations.

FIGURE 10.14 Testing H_0: $\chi^2 = 0$

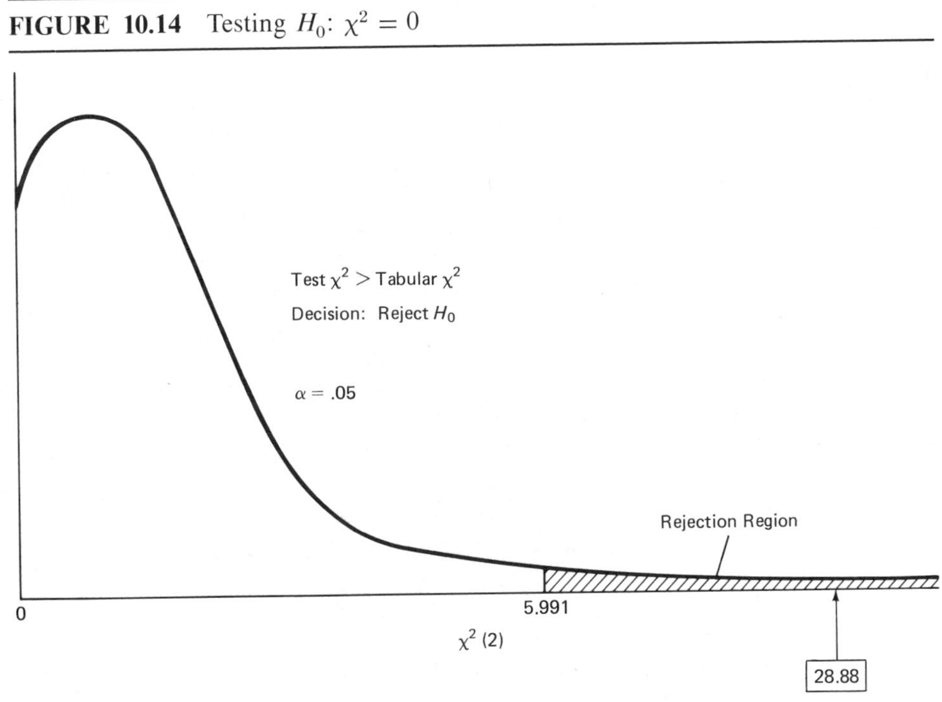

TABLE 10.4 Distribution of Number of Cultural Events Attended in Past 12 Months by Two Groups of Residents

Inner-City Residents ($N = 20$)		Suburban Residents ($N = 30$)		
1	0	1	3	10
0	0	1	1	2
2	6	4	3	3
3	3	0	4	4
0	2	4	0	3
0	1	1	5	0
4	5	2	6	6
1	0	7	1	4
1	2	5	4	4
0	3	2	8	5

Step 1. The hypotheses.

H_0: $Mdn_1 = Mdn_2$ The median number of events attended by the population of inner-city residents is equal to the median number of events attended by the population of suburban residents. Random samples from these populations may differ in medians due to sampling error.

H_1: $Mdn_1 \neq Mdn_2$ The median number of events attended by the population of inner-city residents is not equal to the median number of events attended by the population of suburban residents. Random samples from these populations will differ in medians beyond the expected influence of sampling error.

Step 2. The criteria for rejecting H_0.

The median test requirements of independent and random observations on a variable at the ordinal level (or higher) in two independent samples are met. Because we use the chi-square distribution to perform the test, we must also meet its assumptions about case size. This, too, is satisfied. The degrees of freedom are determined in the same manner as a chi-square-II test. Here $df = (2 - 1)(2 - 1) = 1$ since our table will be 2×2. At $df = 1$ and a level of significance of .05 for this nondirectional test, the tabular value of chi-square is 3.841.

Step 3. The test statistic.

The computations for the test value of chi-square are shown in Table 10.5. First the scores are put into an ordered array to find the overall median. It is 2.5. Then a tally is made of the number of inner-city scores that are above 2.5 and not above it. These frequencies are 6 and 14, respectively. A separate tally is made of the number of suburban scores that are above and not above 2.5. These frequencies are 19 and 11, respectively. The test value of chi-square is calculated for the 2×2 table containing these frequencies. It is 5.33.

Step 4. The decision about H_0.

The test statistic (5.33) does exceed the tabular statistic (3.841), so the null hypothesis is rejected. (See Figure 10.15.) The proportions of scores above and not above the common median do differ between the two groups beyond what was expected from sampling error

TABLE 10.5 Worksheet for a Median Two-Sample Test

1. Find the common median for the two sets of scores combined:

   0 0 0 0 0 0 0 0 0 0 1 1 1 1 1 1 1 1 1 2 2 2 2 2 2 | 3 3 3 3 3 3 3 4 4 4 4 4 4 4 4 5 5 5 5 6 6 6 7 8 10

   Position of median $= \frac{N+1}{2} = \frac{50+1}{2} = 25.5\text{th score}$

   Median = 2.5

2. Construct a 2 × 2 table for median test:

Attendance	Residents Inner-City	Residents Suburban
Above 2.5	6	19
Not Above 2.5	14	11

3. Find chi-square for the table:

Attendance	Residents Inner-City	Residents Suburban	
Above 2.5	6 (10)	19 (15)	25
Not Above 2.5	14 (10)	11 (15)	25
	20	30	50

$$\chi^2 = \frac{(6-10)^2}{10} + \frac{(19-15)^2}{15} + \frac{(14-10)^2}{10} + \frac{(11-15)^2}{15}$$

$$= \frac{(-4)^2}{10} + \frac{(4)^2}{15} + \frac{(4)^2}{10} + \frac{(-4)^2}{15}$$

$$= \frac{16}{10} + \frac{16}{15} + \frac{16}{10} + \frac{16}{15}$$

$$= 1.6 + 1.0\overline{6} + 1.6 + 1.0\overline{6}$$

$$= 5.3\overline{3}$$

at the .05 level. The two sets of frequencies are significantly different; $\chi^2 = 5.33$, $df = 1$, $p < .05$.

Step 5. The conclusion.

Having rejected the null hypothesis, the research worker concludes that the two populations of residents have different median attendance at cultural events. The arts council seems to be correct in assuming that there was a difference between inner-city and suburban residents in their attendance. If this is not a desirable situation to the council, it must begin to determine why this is happening and to devise methods to change it.

Research Example 10.4 presents an application of the median test. The example shows an attempt to evaluate the utility of one particular kind of research program sponsored by the National Science Foundation. The program was designed to promote research that would have direct applications for meeting national needs. The author sought

FIGURE 10.15 Testing H_0: $Mdn_1 = Mdn_2$

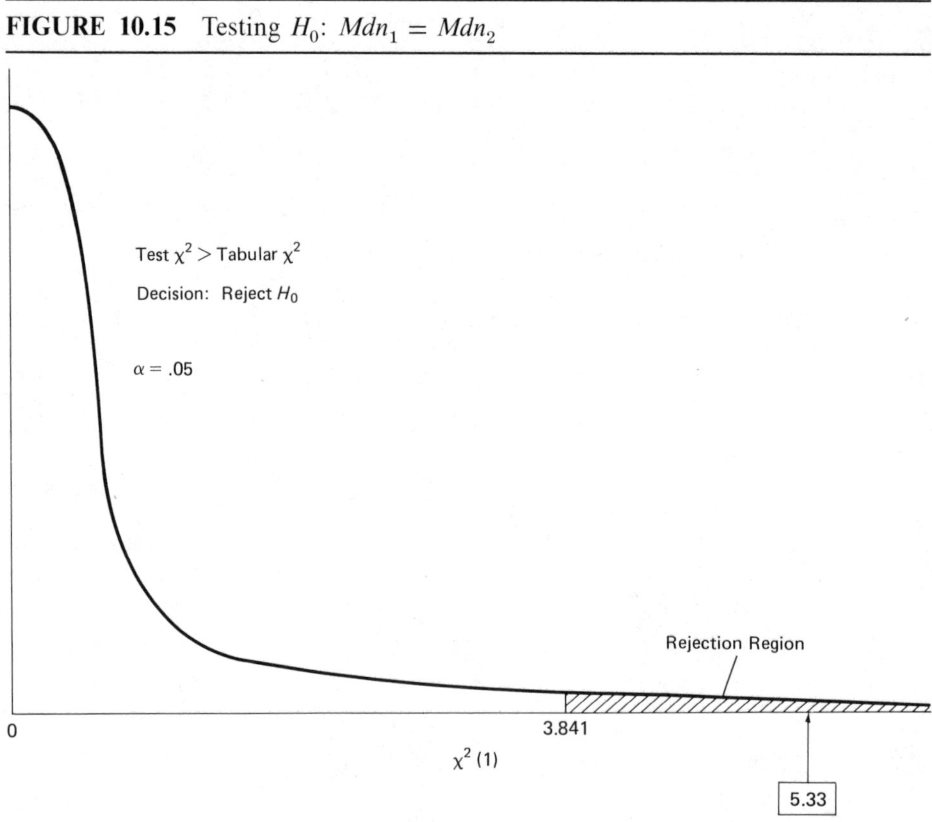

to evaluate the success in reaching this goal for 100 randomly selected projects. This was done by comparing the perceived utility of the projects to their evaluations by peer reviewers when they had been first proposed for funding. The 30 projects that were judged to have the most utility were compared to the 30 judged to have the least utility. These are described as the "top 30" and "bottom 30," respectively. The other 40 projects were ignored in this analysis. The overall median peer evaluation score was determined for the 60 projects and used to split each group of 30 into two categories (above and below the overall median evaluation score of 26). A chi-square test was then applied. It revealed that there was no significant difference between the top 30 and bottom 30 projects in median peer evaluation. The author concludes that this means that the peer evaluations do not correspond to the subsequent perceived usefulness of the projects. This is an important conclusion since the projects were intended to produce applied research that would have utility. Although the measurement in the example may not be easily understood at first reading, the use of the median test is clearly demonstrated.

Again, the sample is a point of interest in our interpretations of the example. Although the projects were randomly selected, the peer evaluations and utilization ratings of those projects were not. It is these evaluations and ratings which constitute the observations that are the basis of the test. It may have been impractical to use randomly selected observations because the populations are small or vaguely known. Research that relies on "expert" judgments often finds it difficult to use a random sample. Sometimes this causes the inferential results to be ambiguous.

Notice that the author refers to the test statistic as "chi-squared." Although this practice is unusual, it does make sense since quantities are being squared. Also being shown is the exact probability that the test statistic would occur because of sampling error (.78) rather than the alpha level (e.g., .05). You will find many researchers doing this. It allows you to alter the researcher's level of significance to one of your own choosing. Then you can determine whether the null hypothesis would have been rejected by the new criterion. There are times when this is helpful in understanding the research and comparing it to other studies.

RESEARCH EXAMPLE 10.4 Median Test

From Samuel J. Raff, "RANN Research at NSF, Some Results of an Evaluation," Evaluation Quarterly, *3 (3), August 1979:497ff. Reprinted by permission.*

The Research Applications Directorate of the National Science Foundation (NSF) was initiated in 1972. Its program was called Research Applied to National Needs (RANN) and in recent years its annual budget has been between 60 and 90 million dollars. . . . The program focused on selected problems of national importance for the purpose of contributing to their timely practical solution. . . .

Clearly, RANN research must ultimately be evaluated, qualitatively at least, by its impact on national needs. One hopes ultimately to find important products, processes, legislation, and improvements in our quality of life, to which RANN research has made demonstrably significant contributions. Process evaluation procedures, however, are required for more rapid feedback to management even though these may be coarse and imprecisely related to impact on national needs. Accordingly, evaluation experimentation started along two tracks: reviewer-based evaluation and utilization-based evaluation.

The former . . . is based on the opinions of qualified evaluators. . . . [Utilization-based evaluation] must start by identifying the uses made of the results of research projects. . . .

The survey (Lindsey and Lessler, 1976) was conducted by the Research Triangle Institute of North Carolina in 1975. It is based on a stratified random sample of 100 of the 524 new projects funded by RANN in 1972 and 1973. . . . The principal investigator (NSF parlance for the leader of the project) of each of these 100 projects was asked to identify up to ten people who, he thought, might be using the results of his research. Many could not identify ten. The average was about seven. The approximately 700 people so identified were then questioned about their use of the results of the research.

The questionnaire used was eight pages in length and contained 29 questions of the "check a box" type . . . and 14 dealt with the utilization of the research by the respondent's organization and its effect on that organization. Some of the key questions used were, "Are you/is your organization currently using these research results?" and "Do you/does your organization plan to use these research results at some point in the future?" . . . By telephone follow-up, a response rate of approximately 80% was achieved. . . . The 100 projects were ordered according to score [based on the users' responses], and statistically significant differences in a number of project characteristics were sought between the top 30 and the bottom 30 projects. . . .

As a first step in seeking significant differences between the 30 most utilized and the 30 least utilized projects, we examined the peer review scores received by the proposals which led to these projects. The proposal peer review process normally used by RANN involves soliciting comments and an overall adjective rating (Poor, Fair, Good, Very Good or Excellent) by mail from four or more knowledgeable people. Some are researchers in the particular field or proximate fields, some are potential users of the research, and some are managers of related research projects. The adjective ratings were quantified on the basis of Poor = 0, Fair = 10, . . . Excellent = 40, and for each of the 60 projects an average score was compiled from the peer review records. The median of all 60 average scores was 26 and a 2 × 2 contingency table was prepared for a test of significant differences between the top 30 and bottom 30 projects with respect to average proposal review scores.

The table and chi-squared results are presented in [the table]. The chi-squared value is only .077 and a value that large or larger could occur randomly with 78% probability. Therefore, the data furnishes no basis to believe that proposals receiving higher average peer review scores will result in more utilized research. . . .

Top 30 Versus Bottom 30 by Peer Review Rating[a]

	Utilization Rating of Projects[c]	
Peer Review Rating[b]	**Top 30**	**Bottom 30**
Above 26	13	13
Below 26	14	12

Chi-squared = .077, probability = .78.

[a] Adapted from Table 2 in the original.
[b] Median of all average ratings = 26
[c] 2 of Top 30 and 5 of Bottom 30 omitted because peer review method differed from the others.

Kolmogorov-Smirnov Two-Sample Test (*KS-II*)

The Kolmogorov-Smirnov two-sample test is a nonparametric significance test that compares the distributions of cumulative proportions for two independent samples (or subsamples) on an ordinal variable. It is not limited by a sample size restriction, such as a median test is, since it does not use the chi-square distribution. For this reason some researchers prefer the *KS-II* test to the median test.

The *KS-II* test makes the usual assumptions. The observations are independent and random from each other. There are two independent samples (or subsamples). The variable on which the samples are being compared is measured at the ordinal level.

The tabular value (D) against which a test value is compared is found by one of several formulas from Table D6 in Appendix D as determined by the level of significance. The basic format of these formulas is:

$$D = (\text{constant}) \sqrt{\frac{N_1 + N_2}{N_1 N_2}}$$ Tabular Value of D for *KS-II*

where N_1 and N_2 = two sample sizes
constant = value from Table D6

TABLE 10.6 Distribution of Social Class by Two Groups

Social Class	Native Americans f_1	Recent Refugees f_2
Lower	25	30
Middle	13	15
Upper	2	5
Total	40	50

The test statistic must equal or exceed this value of D in order for the null hypothesis to be rejected. The test statistic (LD) is the largest difference found between cumulative proportions for the two samples. An example will illustrate the procedure.

A sociologist is interested in the comparative social class standings (measured as an ordinal variable) for Native Americans and recent refugees to the country. A random sample of each population is selected. Information about income, education, and occupation are used to categorize each case in one of three social classes: lower, middle, and upper. The two distributions are shown in Table 10.6. Do they differ significantly at the .10 level?

Step 1. The hypotheses.

H_0: $cP_1 = cP_2$ — The population of Native Americans will not differ from the population of recent refugees in their cumulative-proportion distributions of social class. Sampling error may cause random samples from these populations to differ in such distributions.

H_1: $cP_1 \neq cP_2$ — The population of Native Americans will differ from the population of recent refugees in their cumulative-proportion distributions of social class. Distributions for random samples from these populations will reflect this difference beyond the expected influence of sampling error.

Step 2. The criteria for rejecting H_0.

The .10 level of significance has been established for this nondirectional test. Table D6 indicates that the tabular statistic, D, is found by the following formula:

$$D = 1.22\sqrt{\frac{N_1 + N_2}{N_1 N_2}}$$

$$= 1.22\sqrt{\frac{40 + 50}{(40)(50)}} = 1.22\sqrt{\frac{90}{2000}} = 1.22\sqrt{.045}$$

$$= 1.22(.212) = .259$$

This is the value that must be equaled or exceeded by the test statistic if the null hypothesis is to be rejected.

Step 3. The test statistic.

TABLE 10.7 Worksheet for Kolmogorov-Smirnov II Test

	Native Americans			Recent Refugees			
Social Class	f_1	$<cf_1$	$<cp_1$	f_2	$<cf_2$	$<cp_2$	$\lvert <cp_1 - <cp_2 \rvert$
Lower	25	25	.625	30	30	.600	.025
Middle	13	38	.950	15	45	.900	.050
Upper	2	40	1.000	5	50	1.000	.000

$LD = .050$

The computation of the test statistic (LD) is shown in Table 10.7. There you see that the less-than cumulative frequencies and cumulative proportions are found for each group. The absolute difference between the cumulative proportions is calculated for each category of the social class variable. In this example the largest of these differences (LD) is .050, which occurs for the middle-class proportions.

Step 4. The decision about H_0.

The test statistic (.050) does not exceed the tabular statistic (.259). The sociologist fails to reject the null hypothesis. The slight differences observed in cumulative proportions between the two samples are the kind that can be explained as the consequence of sampling error when the null hypothesis is true. The two samples do not differ significantly in social class ($LD = .050$; $p > .10$, n.s.).

Step 5. The conclusion.

Having failed to reject the null hypothesis, the researcher must now begin to explain the lack of a difference between the groups. Consideration of theories about minority groups in American society should offer some helpful ideas about this finding. Both Native Americans and recent refugees are similar in having minority status and therefore are likely to have relatively low social class standing. There are many characteristics that separate the two groups, but these test data suggest that social class is not one of them.

Keys to Understanding

As you read research about inferences from differences between two samples, keep in mind the following four keys to understanding.

Key 10.1. Remember that all the techniques discussed in this chapter have been inferences of the hypothesis-testing type. No examples of interval estimation were included. (However, some are listed in the section *Related Statistics*.) This was done because hypothesis testing for differences between two samples is the more common kind of inference. This is the opposite pattern of differences based on one sample that we saw in Chapter 9. It is a reflection of the common interest in social research to compare groups. Much of social research is based on an examination of the differences between people. A simple method to make the examination is to compare groups of people on a characteristic of curiosity. This leads rather naturally to an alternative hypothesis that often expects the groups to differ and a null hypothesis that says they do not. Hence a hypothesis test is used in the study.

It certainly is possible to examine differences between two samples using interval estimation. Some even consider it preferable because it avoids the necessity of having a specific hypothesis to test and because it can allow some indication of the strength of an association between the characteristic distinguishing the samples and the variable on which they are being compared. However, the technique is not often utilized in social research.

Key 10.2. Like any hypothesis test, those concerning differences between two samples should be based on substantive reasoning. You must be careful to distinguish between statistical significance on the one hand, and importance in the social world of thought and actions on the other. Knowing that two samples differ significantly tells you that they differ beyond the allowable limits of sampling error. Knowing that there is an important difference between the samples is determined by the theory and policy connected to the research. Following an inference test, the researcher must still interpret the results. What do they mean for theory? What do they suggest about real-world concerns? Ask yourself these questions as you read reports of research.

Key 10.3. Compare each two-sample hypothesis test to its model of ideal situations and procedures. To understand the merits of any of the tests you must understand its model. The models for the tests discussed in this chapter are summarized for you in Inference Models 10.1 to 10.6. Consult them as often as you need to clarify the uses and techniques of the tests. As before, be especially conscious of the assumptions of each test.

Inference Model 10.1 Difference Between Two Independent Proportions Test

Purpose:

To determine whether the proportions for two independent samples differ significantly from the proposed difference between two population proportions.

Sampling Distribution:

Standard normal (z).

Assumptions of the Test:

1. Independent and random observations.
2. Two independent samples.
3. Nominal level of measurement.
4. $NP \geq 5$ and $NQ \geq 5$.

Typical Hypotheses:

Nondirectional Test	Directional Test
H_0: $P_1 = P_2$	H_0: $P_1 \geq$ or $\leq P_2$
H_1: $P_1 \neq P_2$	H_1: $P_1 <$ or $> P_2$

Tabular Statistic:

z at predetermined level of significance from Table D2.

Standard Error:

$$\sigma_{p_1-p_2} = \sqrt{PQ\left(\frac{1}{N_1} + \frac{1}{N_2}\right)} \qquad \text{(if } P \text{ and } Q \text{ are known)}$$

$$\hat{\sigma}_{p_1-p_2} = \sqrt{pq\left(\frac{1}{N_1} + \frac{1}{N_2}\right)} \qquad \text{(if } P \text{ and } Q \text{ are unknown)}$$

$$\text{with } p = \frac{N_1 p_1 + N_2 p_2}{N_1 + N_2}$$

Test Statistic:

$$z = \frac{p_1 - p_2}{\sigma_{p_1-p_2}} \quad \text{or} \quad \frac{p_1 - p_2}{\hat{\sigma}_{p_1-p_2}}$$

Inference Model 10.2 Difference Between Two Independent Means Test

Purpose:

To determine whether the means for two independent samples differ significantly from the proposed difference between two population means.

Sampling Distribution:

A. Standard normal (z), or

B. Student's t.

Assumptions of the Test:

1. Independent and random observations.
2. Two independent samples.
3. Interval level of measurement in dependent variable.

4A. σ_1 and σ_2 are known or $N_1 + N_2$ is large (> 100).

4B. Normally distributed populations.

Typical Hypotheses:

Nondirectional Test	Directional Test
H_0: $\mu_1 = \mu_2$	H_0: $\mu_1 \geq$ or $\leq \mu_2$
H_1: $\mu_1 \neq \mu_2$	H_1: $\mu_1 <$ or $> \mu_2$

Tabular Statistic:

A. z at a predetermined level of significance from Table D2.

B. t at a predetermined level of significance from Table D3 with $df = N_1 + N_2 - 2$ for homogeneous variances or

$$df = \frac{N_1 df_1 + N_2 df_2}{N_1 + N_2} \qquad \text{for heterogeneous variances}$$

Standard Error:

$$\sigma_{\bar{X}_1 - \bar{X}_2} = \sqrt{\frac{\sigma_1^2}{N_1} + \frac{\sigma_2^2}{N_2}} \qquad \text{(if } \sigma_1 \text{ and } \sigma_2 \text{ are known)}$$

$$\hat{\sigma}_{\bar{X}_1 - \bar{X}_2} = \sqrt{s_p^2\left(\frac{1}{N_1} + \frac{1}{N_2}\right)} \qquad \text{(if } \sigma_1 \text{ and } \sigma_2 \text{ are unknown and } \sigma_1^2 \text{ and } \sigma_2^2 \text{ are homogeneous)}$$

$$\text{with } s_p^2 = \frac{(N_1 - 1)s_1^2 + (N_2 - 1)s_2^2}{N_1 + N_2 - 2}$$

$$\hat{\sigma}_{\bar{X}_1 - \bar{X}_2} = \sqrt{\frac{s_1^2}{N_1} + \frac{s_2^2}{N_2}} \qquad \text{(if } \sigma_1 \text{ and } \sigma_2 \text{ are unknown but } \sigma_1^2 \text{ and } \sigma_2^2 \text{ are heterogeneous)}$$

Test Statistic:

$$\text{A. } z = \frac{\bar{X}_1 - \bar{X}_2}{\sigma_{\bar{X}_1 - \bar{X}_2}} \quad \text{or} \quad \frac{\bar{X}_1 - \bar{X}_2}{\hat{\sigma}_{\bar{X}_1 - \bar{X}_2}}$$

$$\text{B. } t = \frac{\bar{X}_1 - \bar{X}_2}{\hat{\sigma}_{\bar{X}_1 - \bar{X}_2}}$$

Inference Model 10.3. Difference Between Two Dependent Means Test

Purpose:

To determine whether the means for two dependent samples differ significantly from the proposed difference between two population means.

Sampling Distribution:

A. Standard normal (z), or

B. Student's t.

Assumptions of the Test:

1. Independent and random observations within each sample.
2. Two dependent samples.
3. Interval level of measurement in dependent variable.

4A. σ_1 and σ_2 are known or N is large (> 100).

4B. Normally distributed populations.

Typical Hypotheses:

Nondirectional Test	Directional Test
H_0: $\mu_1 = \mu_2$	H_0: $\mu_1 \geq$ or $\leq \mu_2$
H_1: $\mu_1 \neq \mu_2$	H_1: $\mu_1 <$ or $> \mu_2$

Tabular Statistic:

A. z at a predetermined level of significance from Table D2.

B. t at $df = N - 1$ and a predetermined level of significance from Table D3.

Standard Error:

$$\sigma_{\bar{X}_1-\bar{X}_2} = \sqrt{\left[\frac{\Sigma D^2}{N} - \left(\frac{\Sigma D}{N}\right)^2\right]\frac{1}{N}} \quad \text{(if population data are used)}$$

$$\hat{\sigma}_{\bar{X}_1-\bar{X}_2} = \sqrt{\frac{\Sigma D^2 - \dfrac{(\Sigma D)^2}{N}}{N(N-1)}} \quad \text{(if sample data are used)}$$

Test Statistic:

$$\text{A. } z = \frac{\bar{X}_1 - \bar{X}_2}{\sigma_{\bar{X}_1-\bar{X}_2}} \quad \text{or} \quad \frac{\bar{X}_1 - \bar{X}_2}{\hat{\sigma}_{\bar{X}_1-\bar{X}_2}}$$

$$\text{B. } t = \frac{\bar{X}_1 - \bar{X}_2}{\hat{\sigma}_{\bar{X}_1-\bar{X}_2}}$$

Inference Model 10.4 Difference Between Two Independent Distributions of Frequencies—Chi-Square Test of Independence for Two Samples (χ^2-*II*)

Purpose:

To determine whether the observed frequency distributions for two independent samples differ significantly from those which would be expected based on two assumed population distributions.

Sampling Distribution:

Chi-square distributions.

Assumptions of the Test:

1. Independent and random observations.
2. Two independent samples.
3. Nominal level of measurement.
4. N must not be extremely small.

Typical Hypotheses:

Nondirectional Test	Directional Test
H_0: $\chi^2 = 0$	(Not ordinarily used)
H_1: $\chi^2 \neq 0$	

Tabular Statistic:

Chi-square at $df = (r - 1)(c - 1)$ and a predetermined level of significance from Table D4.

Standard Error:

Included in test statistic formula: separate calculation not required.

Test Statistic:

$$\chi^2 = \Sigma \frac{(f_o - f_e)^2}{f_e}$$

Inference Model 10.5 Difference Between Two Independent Medians—Median Two-Sample Test (*Mdn-II*)

Purpose:

To determine whether the medians for two independent samples differ significantly from those proposed for two populations.

Sampling Distribution:

Chi-square distributions.

Assumptions of the Test:

1. Independent and random observations.
2. Two independent samples.
3. Ordinal level of measurement in dependent variable.
4. N must not be extremely small.

Typical Hypotheses:

Nondirectional Test	Directional Test
H_0: $Mdn_1 = Mdn_2$	(Not ordinarily used)
H_1: $Mdn_1 \neq Mdn_2$	

Tabular Statistic:

Chi-square at $df = (r - 1)(c - 1)$ and a predetermined level of significance from Table D4.

Standard Error:

Included in test statistic formula: separate calculation not required.

Test Statistic:

$$\chi^2 = \Sigma \frac{(f_o - f_e)^2}{f_e}$$

Inference Model 10.6 Difference Between Two Independent Distributions of Cumulative Proportions—Kolmogorov-Smirnov Two-Sample Test (*KS-II*)

Purpose:
To determine whether the cumulative proportions for two independent samples differ significantly from those proposed for two populations.

Sampling Distribution:
Kolmogorov–Smirnov.

Assumptions of the Test:
1. Independent and random observations.
2. Two independent samples.
3. Ordinal level of measurement in dependent variable.

Typical Hypotheses:

Nondirectional Test	Directional Test
H_0: $cP_1 = cP_2$	(Not ordinarily used)
H_1: $cP_1 \neq cP_2$	

Tabular Statistic:
Kolmogorov–Smirnov D at a predetermined level of significance and a formula from Table D6.

Standard Error:
Included in test statistic; separate calculation not required.

Test Statistic:

$$LD = \text{largest } |{<}cp_1 - {<}cp_2|$$

Notice how the reviews of the various research examples in this chapter gave attention to the assumptions of the test being used. As it happened, many of the examples had difficulty meeting the basic requirement of a random sample. So do a lot of other studies in social science. This seriously limits their ability to make defensible generalizations to larger populations.

Key 10.4. Consider the appropriateness of a selected test to the research situation and character of the data. Researchers face this problem all the time. As you learn about more tests, it becomes a salient task for you as well. There are dozens of inference tests for two-sample differences in widespread use by social researchers. Although it may seem that we looked at more than enough of them, we only began. How do you learn when to use each one? How do you decide to choose one over another?

There are three main issues for you to consider in making this decision.

1. Parametric versus nonparametric. Does the research situation meet the assumptions of a parametric test? If so, ordinarily it should be used. If not, a nonparametric test should be used.
2. Independent versus dependent samples. As you have seen with the t test, the techniques differ when the samples are independent or dependent. The same is true for nonparametric tests. Although we did not discuss them, there are nonparametric tests for dependent samples (See the section *Related Statistics*.)
3. Level of measurement. Each test assumes a certain minimal level of measurement for the variable on which the samples are being compared. A researcher can

FIGURE 10.16 Common Two-Sample Tests of Difference in Social Research

Minimum Level of Measurement	Parametric Test		Nonparametric Test	
	Independent Samples	**Dependent Samples**	**Independent Samples**	**Dependent Samples**
Interval	z and t tests for independent means F test for independent variances[a]	z and t tests for dependent means	Randomization test for independent samples[a]	Randomization test for dependent samples[a] Wilcoxon test[a]
Ordinal			Median test Kolmogorov–Smirnov-II test Wald–Wolfowitz runs test[a] Mann–Whitney U test[a]	Sign test[a]
Nominal			z test for independent proportions Chi-square-II test Yates' corrected chi-square[a] Fisher's exact test[a]	McNemar test[a]

[a] Listed in the section *Related Statistics.*

reduce the level in order to use a desired test, but generally he or she cannot raise it. Remember that there is continuing controversy about the level of measurement for rating scales, such as those that use arbitrary whole numbers. Many of the examples, hypothetical and actual, have shown how these measurements are often deemed to have interval quality. As a result the researcher may use a parametric test such as t. But others may criticize the research for the practice.

The tests discussed in this chapter and listed in the section *Related Statistics* are classified by these three issues in Figure 10.16. This should help bring some order to the array of tests commonly used to compare two samples in social research.

KEY TERMS

Sampling Distribution of Differences Between Two Proportions

Mean of the Differences in Proportions ($\mu_{p_1-p_2}$)

Standard Error of the Differences in Proportions ($\sigma_{p_1-p_2}$)

Population Proportions (P_1, P_2)

Sample Proportions (p_1, p_2)

Estimated Standard Error of the Differences in Proportions ($\widehat{\sigma}_{p_1-p_2}$)
Sampling Distribution of Differences Between Two Independent Means
Mean of the Differences in Means ($\mu_{\bar{X}_1-\bar{X}_2}$)
Standard Error of the Differences in Means ($\sigma_{\bar{X}_1-\bar{X}_2}$)
Population Means (μ_1, μ_2)
Population Standard Deviations (σ_1, σ_2)
Sample Means ($\bar{X}_1$, $\bar{X}_2$)
Sample Standard Deviations (s_1, s_2)
Estimated Standard Error of the Differences in Means ($\widehat{\sigma}_{\bar{X}_1-\bar{X}_2}$)
Pooled Estimate of Population Variances (s_p^2)
Homogeneous Variances
Heterogenous Variances
Sampling Distribution of Differences Between Two Dependent Means
Chi-Square Test of Independence for Two Samples (χ^2-*II*)
Median Test for Two Samples (*Mdn-II*)
Kolmogorov–Smirnov Two-Sample Test (*KS-II*)

EXERCISES

For Exercises 1 to 6, indicate whether each concerns a difference between (a) proportions or means, (b) independent or dependent samples, and (c) whether a z or t sampling distribution should be assumed.

1. A random sample of 150 houses in Apple Canyon had a mean sales price of \$100,000 while another random sample of 200 houses in Maple Valley had a mean sales price of \$90,000. The researcher compared the findings.
2. Fifty percent of the 80 randomly selected people in one company like their work while only 40% of the 65 randomly selected people in another company like their work. The industrial sociologist compared the two groups.
3. A group of 14 child abusers participate in an intervention program. A standardized self-concept scale is administered to them before and after the program and the mean scores are compared.
4. The students in one section of a course in basic statistics ($N = 25$) are taught using a lecture–discussion method. Those in another section of the same course ($N = 20$) are taught using a computer–tutorial program. The mean semester grades are compared.
5. Eight feminists and 12 nonfeminists are shown a picture for four seconds of a female federal judge sitting in court. In describing the picture, two of the feminists and five of the nonfeminists say the judge was a man. The results are compared.
6. Twelve male students of known mathematical ability are compared to 12 female students of equal mathematical ability on their mean scores on a screening test to gain entrance to an accelerated class.
7. An opinion poll is taken in a Western state and another is taken independently in an Eastern state. In the first poll ($N_1 = 100$), 40% ($p_1 = .40$) of the adults randomly questioned said they wished pollution standards were better enforced. In the second poll ($N_2 = 100$), 52% ($p_2 = .52$) expressed the same sentiment. Determine whether there is a significant difference between these sample results and a null hypothesis of no population difference ($P_1 = P_2$). Use a nondirectional test with a .05 level of significance. (a) State the hypotheses symbolically and verbally. (b) Find p and q and then Np and Nq. (c) Determine the sampling distribution, tabular statistic, and rejection region(s). (d) Find the standard error and test statistic. (e) Make a decision about H_0. (f) State a conclusion.
8. A random sample of 100 people living in a home for the elderly finds that 65% ($p_1 = .65$) of the 40 males and 60% ($p_2 = .60$) of the 60 females have hearing losses. Determine whether there is a significant difference between the proportions of males and females with hearing losses in the home compared to a null hypothesis of no such population differences ($P_1 = P_2$). Use a nondirectional test with a .05 level of significance. Answer the same question parts (a)-(f) as in Exercise 7.
9. (a) Using the data in Exercise 7, redo parts (c), (e), and (f) for a .01 level of significance with the same nondirectional test.

(b) Using the data in Exercise 7, redo parts (c), (e), and (f) for a .05 level of significance with a directional test of H_0: $P_1 \geq P_2$.

10. (a) Using the data in Exercise 8, redo parts (c), (e), and (f) for a .01 level of significance with the same nondirectional test.

(b) Using the data in Exercise 8, redo parts (c), (e), and (f) for a .05 level of significance with a directional test of H_0: $P_1 \leq P_2$.

11. Two national polls have asked independent random samples of 1000 adults whether they believe that nuclear energy production should be encouraged. In the first poll, 30% ($p_1 = .30$) believe it should be. In the second poll, 20% ($p_2 = .20$) believe it should be. Determine whether there is a significant difference between these two poll results and a null hypothesis of no such population difference ($P_1 = P_2$). Let the level of significance be .001 and use a nondirectional test. Answer the same question parts (a)–(f) as in Exercise 7.

12. (a) Using the data in Exercise 11, redo parts (c), (e), and (f) for a level of significance of .01 with the same nondirectional test.

(b) Using the data in Exercise 11, redo parts (c), (e), and (f) for a level of significance of .05 with the same nondirectional test.

13. Two independent, large, and random samples find that .45 and .60, respectively, of those questioned believe the cost of a college education is too low. This translated into a test statistic of $z = -2.40$. Assume the null hypothesis was that $P_1 = P_2$ and a nondirectional test was performed. (a) If the level of significance is .05, what is the tabular statistic? (b) If the level of significance is .05, should H_0 be rejected? (c) If the level of significance is .01, what is the tabular statistic? (d) If the level of significance is .01, should H_0 be rejected? (e) If H_0 were rejected, would it make you doubt that the two populations had identical proportions? (f) If H_0 were rejected, does it mean the first sample's proportion is significantly lower than the second sample's? (g) If H_0 were not rejected, does it make you doubt that the two populations had identical proportions? (h) If H_0 were not rejected, does it mean that the two samples' proportions differed significantly from each other?

14. Use the same information as Exercise 13 except let the null hypothesis be that $P_1 \leq P_2$. Suppose that the first sample represented people having college degrees and the second sample represented people not having such degrees. Assume that the test statistic was still $z = -2.40$ but a directional test was performed. Answer the same question parts (a)–(h) as Exercise 13.

15. A random sample of 51 minority children attending integrated public schools has a mean academic achievement score of 81 with $s_1 = 3$. An independent random sample of 101 minority children attending a segregated public school has a mean academic achievement score of 79.9 with $s_2 = 3$. Assume that the two population variances are homogeneous. Determine whether there is a significant difference in mean academic achievement for these samples compared to a null hypothesis of no such population differences ($\mu_1 = \mu_2$). Let the level of significance be .05 and use a nondirectional test. (a) State the hypotheses symbolically and verbally. (b) Find *df* and determine the sampling distribution, tabular statistic, and rejection region(s). (c) Find the standard error and test statistic. (d) Make a decision about H_0. (e) State a conclusion.

16. Using the same data as Exercise 15, redo parts (a), (b), (d), and (e) with a .01 level of significance for a directional test of H_0: $\mu_1 \leq \mu_2$.

17. A random sample of elected, local government officials discovers that the mean amount of stock ownership is $5252 with $s_1 = \$210$ for those 30 officials who were over 70 inches tall. The mean was $5276 with $s_2 = \$213$ for the 20 officials who were not over 70 inches tall. Assume that the two population variances are homogeneous. Determine whether there is a significant difference in mean stock ownership for these subsamples compared to a null hypothesis of no such population difference in means. Let the level of significance be .05 and use a nondirectional test. Follow the same five steps in Exercise 15.

18. Using the same data in Exercise 17, redo parts (a), (b), (d), and (e) for a .10 level of significance with the same nondirectional test.

***19.** Twenty-two unmatched students were randomly assigned to either an experimental group or control group. The 11 in the first group (experimental) were told they had been carefully selected to participate in a test of a new teaching exercise. The 11 in the second group (control) were told nothing special. The experimental subjects gave the exercise a mean rating of 9 (highly favorable) with $s_1 = 1$. The control subjects gave the exercise a mean rating of 7 (moderately favorable) with $s_2 = 1.5$. Assume that the population variances are heterogenous. Determine whether there is a significant difference between the samples compared to a null hypothesis of no population difference in means. (a) Find the standard error. (b) Find the test statistic. (c) What are the estimated degrees of freedom? (d) If the level of significance is .05, what is the tabular statistic for a nondirectional test? (e) Given part (c), should H_0 be rejected? (f) If the level of significance is .01, what is the tabular statistic for a nondirectional test? (g) Given part (e), should H_0 be rejected? (h) Suppose that H_0 is $\mu_1 \leq \mu_2$ and the level of significance is .005. What is the tabular statistic for a directional test? (i) Given part (g), should H_0 be rejected?

20. A random sample of 11 single parents is given a standardized happiness test and its mean happiness score is 4. The parents complete a series of five weekly seminars on coping with single parenthood. They retake the happiness test and have a mean score of 6. The data for individuals are as follows:

Parent	X_1	X_2
A	5	7
B	5	5
C	6	9
D	3	6
E	3	5
F	4	3
G	7	10
H	5	8
I	1	4
J	2	5
K	3	4

Determine whether there is a significant difference between these two sample means compared to a null hypothesis that the first mean would not be lower than the second mean ($\mu_1 \geq \mu_2$). (a) Find the standard error. (b) Find the test statistic. (c) If the level of significance is .05, what is the tabular statistic for a directional test? (d) Given part (c), should H_0 be rejected? (e) If the level of significance is .01, what is the tabular statistic for a directional test? (f) Given part (e), should H_0 be rejected?

21. A random sample of five heterosexual males has a mean life expectancy of 25 more years. A random sample of five homosexual males (which has been individually matched on age, race, occupation, and social class to the sample of heterosexuals) has a mean life expectancy of 21 more years. The individual scores are as follows:

Matched Pair	Heterosexuals X_1	Homosexuals X_2
AA	10	9
BB	30	32
CC	25	21
DD	33	23
EE	27	20

Determine whether there is a significant difference between these two sample means compared to a null hypothesis of no such population difference. (a) Find the standard error. (b) Find the test statistic. (c) If the level of significance is .05, what is the tabular statistic for a nondirectional test? (d) Given part (c), should H_0 be rejected? (e) If the level of significance is .10, what is the tabular statistic? (f) Given part (e), should H_0 be rejected?

22. A union organizer wondered whether support for organizing a union differed by job tenure. A random sample of a company's employees were asked their feelings about a union and this was compared in two subsamples of job tenure. Apply a chi-square-II test to the results shown below as a bivariate frequency distribution:

Union Support	**Job Tenure (Years)** 0–4	5+
Favorable	31	24
Not Favorable	14	26

H_0 was that there is no significant difference between the job tenure subsamples in union support. (a) If the level of significance is .05, what is the tabular value of chi-square? (b) If the level of significance is .01, what is the tabular value of chi-square? (c) Calculate the test value of chi-square. (d) Should H_0 be rejected at the .05 level? (e) Should H_0 be rejected at the .01 level? (f) State a conclusion assuming that H_0 was rejected.

23. A cultural anthropologist is interested in possible differences in place of residence for two subsamples of native peoples classified by type of family organization. A random sample of native people from an Asian country is interviewed, providing the following table of frequencies:

Family Organization	Ancestral Home	**Place of Residence** Other Isolated Place	Urban Place
Father—Dominant	24	18	18
Mother—Dominant	6	12	17

H_0 is that there is no significant difference in place of residence by family organization. Apply a chi-square-II test to these data. Answer the same question parts (a)–(e) of Exercise 22 (f) State a conclusion assuming that H_0 was *not* rejected.

24. A candidate for governor is concerned that support for election may not be uniform in different areas of the state. Opinion poll results are examined by area with the following results in bivariate frequencies:

Support	**Area** North	South
Strong	10	18
Moderate	21	22
Weak	14	10

H_0 is that support does not differ significantly by area. Apply a chi-square-II test to these data. (a) If the level of significance is .05, what is the tabular statistic? (b) If the level of significance is .01, what is the tabular statistic? (c) Calculate the test value of chi-square. (d) Should H_0 be rejected at the .05 level? (e) Should H_0 be rejected at the .01 level? (f) State a conclusion assuming that H_0 was rejected.

25. A family sociologist is interested in the common idea that Catholics have more children than Protestants. The number of children for a random sample of 30 families is found as displayed below by religious affiliation:

Number of Children

Catholic Families		Protestant Families	
0	2	4	1
2	10	2	2
1	7	0	2
5		2	3
4		1	4
4		1	3
2		2	5
6		6	3
5		0	3

Use a median test to determine whether there is a significant difference between the two religious groups in median number of children compared to a null hypothesis of no such population difference. (a) Calculate the overall median. (b) Calculate the test statistic (chi-square). (c) If the level of significance is .05, should H_0 be rejected? (d) If the level of significance is .10, should H_0 be rejected? (e) State a conclusion assuming that H_0 was rejected.

26. Twenty randomly selected 10-year-old children are given a quiz in a situation which allows them to cheat by looking at the quiz answers posted at the back of the room. A hidden observer records the number of times each child glances at the posted answers during a monitor's intentional absence from the room. The individual results were as follows for the boys and girls in the experiment:

Number of Glances

Boys		Girls	
0	3	1	4
4	8	2	6
1	0	0	1
5	3	3	3
2	1	10	5

Use a median test to determine whether there is a significant difference between the two gender groups in median number of cheating glances compared to a null hypothesis of no such population difference. Answer the same question parts (a)–(e) of Exercise 25.

27. Refer to the data in Exercise 24. Use a Kolmogorov–Smirnov-II test to determine whether there is a significant difference in cumulative proportions of support between the North and South groups of individuals compared to a null hypothesis of no such population difference. (a) Find the tabular statistic (D) for a .05 level of significance. (b) Find D for a .01 level of significance. (c) Calculate the test statistic (LD). (d) At the .05 level, should H_0 be rejected? (e) At the .01 level, should H_0 be rejected? (f) State a conclusion assuming that H_0 was rejected.

28. A random sample of students receiving academic scholarships is compared to one of students receiving athletic scholarships on mean grade point. The results are shown below in frequencies. Answer the same question parts (a)–(f) as in Exercise 27. Use a *KS-II* test of the null hypothesis of no significant difference in cumulative proportion of mean grade points between the two populations.

	Type of Scholarship	
Mean Grade	Academic	Athletic
D	0	1
C	2	12
B	14	15
A	20	4

29. Locate an example of one parametric and one nonparametric test discussed in this chapter. Discuss their use in relationship to the goals of the research being conducted and the *Keys to Understanding* in this chapter.

RELATED STATISTICS

Randomization Test for Independent Samples

The randomization test for independent samples is a nonparametric test for comparing such samples on an interval variable. It is used only for very small samples: $N_1 + N_2 \leq 12$. The test statistic's exact probability is calculated directly in each application to determine a conclusion about the null hypothesis.

See: Champion (1981:176–181).

Randomization Test for Dependent Samples

The randomization test for dependent samples is a nonparametric test for comparing such samples on an interval variable. Again, it is used only for very small samples: $N_1 + N_2 \leq 12$.

See: Champion (1981:182–187).

Wilcoxon Test

The Wilcoxon test is a nonparametric test for comparing two dependent samples on an interval variable. The samples can be of any size. It uses its own tabular values up to $N = 50$. (See Table D12.) After that a z approximation is used.

See: Runyon and Haber (1980:339–340) and Champion (1981:187–190).

Wald-Wolfowitz Runs Test

The Wald–Wolfowitz runs test is a nonparametric test for comparing two independent samples on an ordinal variable. It is not a very powerful test but it is easy to apply. It is adversely affected by tied scores. It uses its own tabular values up to $N = 20$. (See Table D13.) After that it is approximated by a z distribution.

See: Champion (1981:262–266).

Mann-Whitney *U* Test

The Mann–Whitney *U* Test is a nonparametric test for comparing two independent samples on an ordinal variable. It is more powerful than the *KS-II* and Wald–Wolfowitz tests but is adversely affected by tied scores. It uses its own set of tabular values up to $N = 20$. (See Tables D14.1 to 14.4.) After that it is approximated by a z distribution.

See: Runyon and Haber (1980:333–337) and Champion (1981:271–276).

Sign Test

The sign test is a nonparametric test for comparing two dependent samples on an ordinal variable. It is not a very powerful test but it is easy to apply. It uses its own set of tabular values up to $N = 25$. (See Table D15.) After that it is approximated by a z distribution. See: Champion (1981:276–280) and Runyon and Haber (1980:337–338).

Yates' Corrected Chi-Square

The Yates' correction is often applied to the chi-square-II test for comparing two independent samples on a nominal variable that is dichotomous. The correction is identical to that described in the section *Related Statistics* of Chapter 9.

Fisher's Exact Test

Fisher's exact test can be used in place of the chi-square-II test for small independent samples in a 2×2 table comparison. Its use is identical to that described in the section *Related Statistics* of Chapter 9.

McNemar Test

The McNemar test is a nonparametric test based on the chi-square distribution. It is used for the before–after type of dependent samples comparison on a dichotomous nominal variable. Its use is not appropriate for the matched type of dependent samples test. See: Champion (1981:240–243).

***F* Test for Homogeneous Variances**

The z and t tests for independent means requires that the population variances be known to be either homogeneous or hetereogenous. A parametric test of homogeneity between two variances can be conducted using the F statistic, which is discussed extensively in Chapter 11. The test is based on the F distribution found in Table D7. The formula for the test statistic is

$$F = \frac{s_1^2}{s_2^2}$$

where s_1^2 = larger sample variance
s_2^2 = smaller sample variance

See: Runyon and Haber (1980:242–243).

Confidence Interval Estimate of a Difference Between Two Proportions

A confidence interval estimate of the difference between two population proportions can be found from the observed difference between two sample proportions by the formula

$$CI = (p_1 - p_2) \pm (z)(\sigma_{p_1-p_2} \text{ or } \widehat{\sigma}_{p_1-p_2})$$

See: Mattson (1981:147–148).

Confidence Interval Estimate of a Difference Between Two Means

A confidence interval estimate of the difference between two population means can be found from the observed difference between two sample means by the formula

$$CI = (\overline{X}_1 - \overline{X}_2) \pm (z \text{ or } t)(\sigma_{\overline{X}_1-\overline{X}_2} \text{ or } \widehat{\sigma}_{\overline{X}_1-\overline{X}_2})$$

See: Spatz and Johnston (1981:208–210) and Mattson (1981:148–149).

Computational Formula for Chi-Square in 2 × 2 Tables

The test value of the chi-square statistic can be found by a special computational formula when the data are in a 2×2 table. The formula works directly from the four observed frequencies and four marginal totals without having to calculate the expected frequencies. Except for rounding errors, the formula shown above in the chapter for chi-square-II yields the same answer as that for this computational formula.

$$\chi^2 = \frac{N(AD - BC)^2}{(A + B)(C + D)(A + C)(B + D)}$$

where A = observed frequency in upper-left cell
B = observed frequency in upper-right cell
C = observed frequency in lower-left cell
D = observed frequency in lower-right cell
N = total number of cases

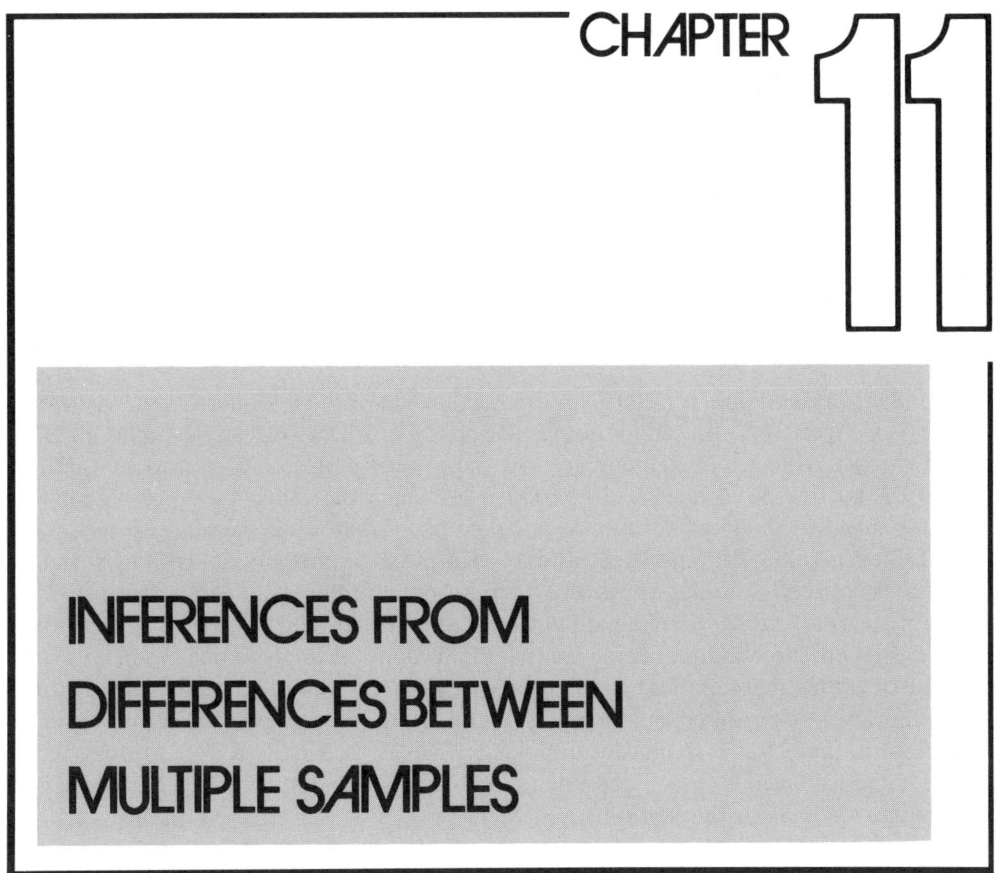

CHAPTER 11

INFERENCES FROM DIFFERENCES BETWEEN MULTIPLE SAMPLES

We saw in Chapters 9 and 10 that social researchers use one set of specialized techniques to draw inferences from single samples and a different set for inferences from differences between two samples. You might reluctantly imagine that this pattern continues, with different tests being used for comparisons between three samples, others for four samples, and so on. Not so. Statistical methods have been developed that allow inferences to be drawn from any number of samples. Some of these even overlap with the two-sample comparisons we have already examined. These are called multiple-sample or k-sample tests. The "k" may stand for any number of samples equal to or greater than 2, although it usually stands for 3 or more. We will look at the more common parametric and nonparametric k-sample procedures in social research.

Parametric Multiple-Sample Means Tests

Analysis of Variance

Analysis of variance (ANOVA) is a very versatile and sometimes complex statistical technique. You will see it used in diverse ways and in alternative forms. Many of these are not covered here but are highlighted in the section *Related Statistics* at the end of the

TABLE 11.1 ANOVA Summary Table

Source of Variation	Sum of Squares	*df*	Mean Squares	*F*
Between groups	31.6	2	15.80	13.17*
Within groups	14.4	12	1.20	
Total	46.0	14		

$^{*}p < .05$.

chapter. Our present look at ANOVA is as an extension of the difference of means test (t) we saw in Chapter 10. It is used in those situations in which two or more samples differing in one characteristic are being compared by their mean scores on a dependent variable. This is the basic type of analysis of variance on which the other types are based. It is known as *one-way analysis of variance* or *single-factor analysis of variance.*

Table 11.1 shows the finished product of a possible one-way analysis of variance. You may have seen a similar-appearing table in some published research. It is called an ANOVA summary table. What can a knowledgeable reader determine from such a table? First, there were three samples being compared. Second, the analysis was based on a total of 15 cases. If the three samples had identical case sizes, each contained five cases. Third, if all 15 cases were considered as belonging to one group, that group's variation would be 46.0. Fourth, there was a significant difference between at least two of the three sample means at the .05 level. How is it possible to know all this from the table? That is what we are going to look at in this section.

The Basics of ANOVA

Analysis of variance is a test of differences between independent *means,* but it is based on a comparison of *variances* (the squared version of a standard deviation). This is a result of mathematical convenience and necessity. The usual null hypothesis is that the means from the separate populations are identical. For example, with three groups: H_0: $\mu_1 = \mu_2 = \mu_3$. The alternative hypothesis is that this is not true; H_1: not $[\mu_1 = \mu_2 = \mu_3]$. It is proper to represent the alternative hypothesis this way to indicate that at least two of the means are different from each other. The idea that all three differ would be shown as $\mu_1 \neq \mu_2 \neq \mu_3$. But that is not necessarily what the researcher has in mind, nor is it the logical opposite to the null hypothesis of no difference. For example, maybe $\mu_1 = \mu_2$ and $\mu_1 = \mu_3$, but $\mu_2 \neq \mu_3$.

As you reflect about testing such a null hypothesis as the one above, it may occur to you that a series of t tests for each pair of means might be used rather than going to a new technique. We might think of testing whether $\mu_1 = \mu_2$, $\mu_1 = \mu_3$ and $\mu_2 = \mu_3$ with three t tests. Compared to analysis of variance, this would be both inefficient and potentially misleading.

ANOVA can test for any difference between several means in one unified procedure. Further, the t tests for pairs of means would not be completely independent of each other. After testing $\mu_1 = \mu_2$ and $\mu_1 = \mu_3$, the test of $\mu_2 = \mu_3$ is mathematically dependent of the first two tests. This increases the probability that a Type I error will occur beyond any preset level of significance. So although a researcher might use a .05 level of significance, the actual probability of making a Type I error in the three t tests would actually be greater than .05. This is an unacceptable circumstance in research. We must know the true probability associated with our inferences or we cannot fairly place confidence in them. ANOVA has no such problem.

Analysis of variance is based on a comparison of two independent estimates of the variance in the populations. One of these examines the differences *between** the sample means. If the null hypothesis of no difference between means is true, the sample means will be alike or differ only within the expected limits of sampling error. If the null hypothesis is false, the sample means will differ by more than is customarily expected when population means are equal. The second variance estimate examines the score differences *within* the samples and combines them into a pooled estimate of the common variance in the populations. This estimate will be a reliable indicator of the population variances regardless of whether the sample means are the same or different. A ratio can then be made of the first variance estimate to the second. If the null hypothesis is true, the two estimates will be approximately equal and their ratio will be nearly 1.0. If the null hypothesis is false, the first estimate will be larger than the second and their ratio will be greater than 1.0.

This should be familiar to you. The two-sample t test described a similar situation. With a null hypothesis of no difference between population means, its formula was

$$t = \frac{\overline{X}_1 - \overline{X}_2}{\sigma_{\overline{X}_1 - \overline{X}_2}}$$

The numerator of this formula is a measure of difference *between* means for the two groups being compared. The denominator is a measure of differences *within* the two groups. Their ratio is a t score. So t is a special version of the general type of test we use to compare means between multiple groups. In fact, t^2 is equivalent to the analysis of variance test statistic (F) for the two-sample situation. A major difference in their computations reflect t's specific focus on means and standard deviations while the more general method uses variances.

Variance ratios are called *F ratios* which are distributed according to a family of sampling distributions known as F distributions. The "F" refers to Ronald A. Fisher, their developer and a very influential figure in the history of modern statistics. The F distributions are based on the same assumptions as those for the t distributions. The observations must be random and independent of each other. The samples (or subsamples) must be independent. Populations are assumed to be normally distributed and their variances to be homogeneous.

These F distributions are shaped approximately like chi-square distributions, to which they are related. They begin at 0 and are positively skewed. However, the F distributions are described by two types of degrees of freedom, not one. The first is known as the *degrees of freedom between groups* (*dfb*) and the second as the *degrees of freedom within groups* (*dfw*).

*Grammatically, it would be more correct to use the phrase "among groups" rather than "between groups" since more than two groups may be involved. But the historical pattern has been to use the latter, so you should expect to see it instead.

Their formulas are quite simple:

	Degrees of Freedom
$dfb = k - 1$	Between Groups and
$dfw = N_T - k$	Within Groups

where k = number of samples (groups)
N_T = total number of cases in all samples combined

The between-groups degrees of freedom illustrate how one degree of freedom is lost when there is a single restriction on the scores. We can use the sample means to calculate a quantity known as the *grand* or *total mean* ($\bar{X}_T$). The total mean is the mean for all cases combined from the separate groups. If we require the total mean to be a constant (which it is), all but one of the group means is free to vary in describing the same total mean. But that last one is completely determined by the values of the other group means and the total mean. For the within-groups degrees of freedom one degree of freedom is lost in each group as its individual scores are used to define their own group mean. Thus there are as many degrees of freedom lost in combination as there are group means. Overall, *dfw* equals the number of individual observations minus the number of groups, since each group has one mean. There is a third type of degrees of freedom, *df total* (*dft*). This is the *df* value for all scores combined used to define the total mean. Because there is only one total mean, $dft = N_T - 1$.

Analysis of variance proceeds by finding quantities needed to calculate the two variance estimates. These include measures of variation (known as *sum of squares*) and the degrees of freedom between and within groups. The observed ratio of variances (test F) is compared to a tabular ratio (tabular F) which is determined by a level of significance and the same two *df* values using Table D7 in Appendix D. As before, the null hypothesis is rejected only when the test statistic equals or exceeds the tabular statistic, since it would then represent a predefined, unusual set of sample outcomes. Otherwise, the null hypothesis is retained.

An Illustration of ANOVA

Let's examine in more detail the procedures of analysis of variance using our earlier example summarized in Table 11.1. The data for this example are shown in Table 11.2a. They concern a comparison in mean annual weeks of paid vacation for samples of office workers, laborers, and managers which each contain five cases.

TABLE 11.2a Number of Weeks of Paid Vacation in Random Samples of Workers from Three Occupations

Office Workers	Laborers	Managers
1	4	4
2	4	6
2	5	7
1	6	5
4	4	5

Step 1. The hypotheses.

H_0: $\mu_1 = \mu_2 = \mu_3$ The three populations of office workers, laborers, and managers have the same mean number of weeks of annual paid vacation. Random samples from these populations may differ in such means within the expected limits of sampling error.

H_1: not $[\mu_1 = \mu_2 = \mu_3]$ The three populations of office workers, laborers, and managers do not have the same mean number of weeks of annual paid vacation. Random samples from these populations will differ in such means beyond the expected influence of sampling error.

Step 2. The criteria for rejecting H_0.

The assumptions for using an F distribution and test in this problem are all met if we believe that the distributions of vacation weeks are nearly normal for the three populations. This would not be an unreasonable assumption. Our two degrees of freedom are found according to the earlier formulas. We have three samples ($k = 3$) and there are five cases in each ($N_T = 15$):

$$\begin{aligned} dfb &= k - 1 \\ &= 3 - 1 \\ &= 2 \\ dfw &= N_T - k \\ &= 15 - 3 \\ &= 12 \end{aligned}$$

Using these two degrees of freedom, we can find a tabular value of F for the appropriate sampling distribution from Table D7. We also need a level of significance, as always. Ours is .05, as the footnote to Table 11.1 suggests ("$p < .05$"). Looking in Table D7, we find that our tabular F is 3.88. A common method of labeling this in social research is: $F(2, 12, .05) = 3.88$. The two degrees of freedom and level of significance are given in the parentheses.

Like all previous tabular statistics (z, t, *and* χ^2), our tabular F represents a point in a sampling distribution which defines the beginning of a rejection region. Our use of the F distribution will be confined to those situations in which a nondirectional test is being conducted of H_0. As noted above, F distributions begin at 0 and they have only one rejection region at the right side (for our purposes.) Our distribution, is sketched in Figure 11.1a. It shows the rejection region beginning at 3.88 and continuing to positive infinity.

Step 3. The test statistic.

From the raw data in Table 11.2a you can see the individual weeks of annual paid vacation for the 15 workers. We begin the quest for our test value of F by finding three measures of variation known as the *sum of squares* values. Before calculating these, it may be helpful if you think in terms of some ideas basic to both regression-correlation and analysis of variance.

We examined the regression meaning of correlation in Chapter 6. There we saw that it involved an analysis of a quantity known as the total variation in a dependent variable. Total variation was the sum of all individually squared deviations from the mean for that variable. In a formula, it is $\Sigma(X - \bar{X})^2$. This quantity was split into two components: explained and unexplained variation. The explained variation was that part of the total

FIGURE 11.1a Criteria for Testing H_0: $\mu_1 = \mu_2 = \mu_3$

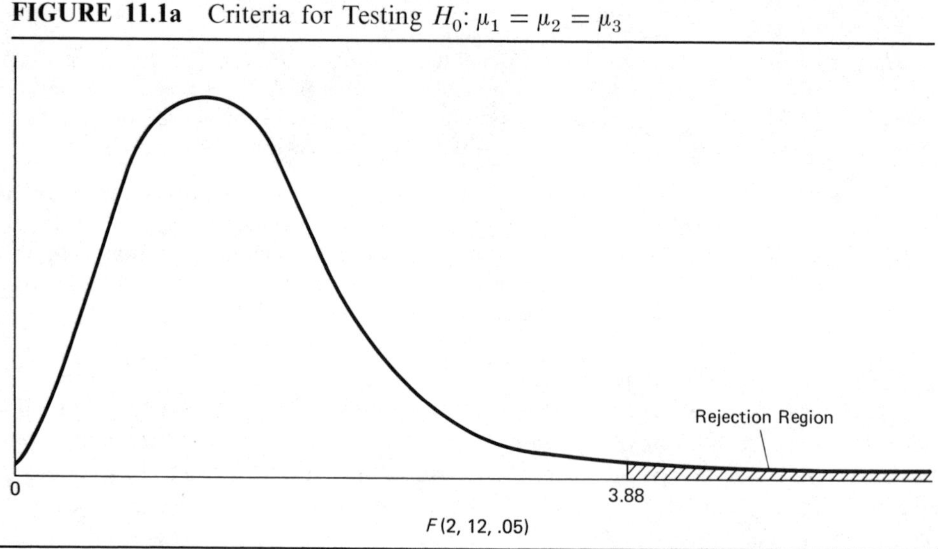

that could be attributed to some independent variable. The unexplained was that part of the total variation that could not be attributed to that independent variable. The ratio of explained to total variation was the square of the correlation coefficient: r^2 = (explained variation/total variation). From this we could determine how closely associated an independent and dependent variable were.

There is a strong parallel between this and analysis of variance. ANOVA also works with the total variation in a dependent variable that is split into two components. This time the total is called *sum of squares total* (*SST*). *SST* is still defined as total variation; $\Sigma(X - \overline{X})^2$. But in analysis of variance each sample has its own mean. Which of these is the $\overline{X}$ in the total variation formula? None of them! In ANOVA such a mean refers to the mean for all cases combined, the total mean. For this reason it is clearer to write the formula for *SST* as

$$SST = \Sigma(X - \overline{X}_T)^2$$ Definitional Formula for Sum of Squares Total

where X = each individual score

$\overline{X}_T$ = mean of all scores combined (total mean)

We can demonstrate this formula with the data for our example. The preliminary computations are shown in Table 11.2b. By summing the subtotals from the three samples we see that all 15 scores total 60 weeks of paid vacation. Thus the total mean is 4.0. This total mean is subtracted from each individual's score ($X - \overline{X}_T$) in columns labeled in deviation form, "x_T," and then squared in columns labeled "x_T^2." The subtotals from the three samples are summed to determine the value of *SST*:

$$SST = 26 + 5 + 15 = 46$$

The sum of squares total is divided into one part called *sum of squares between* (*SSB*). In experimental research, this component is often called *sum of squares treatment,*

TABLE 11.2b Preliminary Calculations for ANOVA from Table 11.2a Using Definitional Formulas

Office Workers					Laborers					Managers				
X_1	x_T	x_T^2	x_1	x_1^2	X_2	x_T	x_T^2	x_2	x_2^2	X_3	x_T	x_T^2	x_3	x_3^2
1	−3	9	−1	1	4	0	0	−0.6	.36	4	0	0	−1.4	1.96
2	−2	4	0	0	4	0	0	−0.6	.36	6	2	4	0.6	.36
2	−2	4	0	0	5	1	1	0.4	.16	7	3	9	1.6	2.56
1	−3	9	−1	1	6	2	4	1.4	1.96	5	1	1	−0.4	.16
4	0	0	2	4	4	0	0	−0.6	.36	5	1	1	−0.4	.16
10	—	26	0	6	23	—	5	0	3.20	27	—	15	0	5.20

$\bar{X}_1 = 10/5 = 2$ $\quad\quad$ $\bar{X}_2 = 23/5 = 4.6$ $\quad\quad$ $\bar{X}_3 = 27/5 = 5.4$

$$\bar{X}_T = \frac{10 + 23 + 27}{15} = \frac{60}{15} = 4$$

which refers to a measure of the influence of some experimental variable (the "treatment") on the subjects. It is equivalent to explained variation in regression. The second part of SST is called *sum of squares within* (SSW) or, alternatively, *sum of squares error* (SSE). It is equivalent to unexplained variation in regression.

SSB is a measure of variation between the groups being compared. It does this by quantifying the extent of difference between the group (sample) means compared to the total mean:

$$SSB = \Sigma[N_i(\bar{X}_i - \bar{X}_T)^2]$$ Definitional Formula for Sum of Squares Between

where $\bar{X}_i$ = mean of each sample
$\bar{X}_T$ = mean of all scores combined (total mean)
N_i = number of cases in each sample

For each sample, the deviation of its own mean ($\bar{X}_i$) from the total mean ($\bar{X}_T$) is squared and multiplied by its own sample size (N_i). These products are summed from all samples to find SSB. If the samples were from populations with identical means, each would have nearly equivalent means and SSB would be zero. Our three means were 2.0, 4.6, and 5.4, as shown in Table 11.2b. SSB can then be found as just described:

$$\begin{aligned} SSB &= 5(2 - 4)^2 + 5(4.6 - 4)^2 + 5(5.4 - 4)^2 \\ &= 5(-2)^2 + 5(0.6)^2 + 5(1.4)^2 \\ &= 5(4) + 5(.36) + 5(1.96) \\ &= 20 + 1.8 + 9.8 \\ &= 31.6 \end{aligned}$$

The sum of squares within represents a measure of variation within the groups (samples) being compared. It is defined as a process in which the usual variation measure is found for each sample separately and these are then summed:

$$SSW = \Sigma x_1^2 + \Sigma x_2^2 + \Sigma x_3^2 + \cdots + \Sigma x_k^2$$

Definitional Formula for Sum of Squares Within

where $\Sigma x_1^2 = \Sigma (X - \overline{X}_1)^2$ for first sample

$\Sigma x_2^2 = \Sigma (X - \overline{X}_2)^2$ for second sample

$\Sigma x_3^2 = \Sigma (X - \overline{X}_3)^2$ for third sample

$\cdot \quad \cdot\cdot$

$\cdot \quad \cdot\cdot$

$\Sigma x_k^2 = \Sigma(X - \overline{X}_k)^2$ for last sample

The within-group deviations, their squares, and their subtotals are shown in Table 11.2b. From these SSW is as follows:

$$SSW = 6 + 3.2 + 5.2 = 14.4$$

Several alternative formulas have been developed to ease the computation of these three sum of squares values. These formulas are entirely equivalent mathematically to the definitional formulas shown above. However, no deviation scores are required for these new formulas. The only difference in answers, if any, will be due to rounding errors. One commonly used set of computational formulas is as follows:

$$SST = \Sigma X_T^2 - \frac{(\Sigma X_T)^2}{N_T}$$

Computational Formulas for Sum of Squares Total

$$SSB = \left[\Sigma \frac{(\Sigma X_i)^2}{N_i}\right] - \frac{(\Sigma X_T)^2}{N_T}$$

Sum of Squares Between

$$SSW = \Sigma\left[\Sigma X_i^2 - \frac{(\Sigma X_i)^2}{N_i}\right]$$

Sum of Squares Within

where ΣX_T^2 = sum of all squared scores

$(\Sigma X_T)^2$ = square of all scores summed

ΣX_i^2 = sum of squared scores in each sample

$(\Sigma X_i)^2$ = square of scores summed in each sample

N_T = number of scores in all samples combined

N_i = number of scores in each sample

We can demonstrate these computational formulas for our example. Some of these preliminary calculations are shown in Table 11.2c.

$$\begin{aligned} SST &= \Sigma X_T^2 - \frac{(\Sigma X_T)^2}{N_T} \\ &= (26 + 109 + 151) - \frac{(10 + 23 + 27)^2}{15} \\ &= 286 - \frac{(60)^2}{15} = 286 - \frac{3600}{15} = 286 - 240 \\ &= 46 \end{aligned}$$

TABLE 11.2c Preliminary Calculations for ANOVA from Table 11.2a Using Computational Formulas

Office Workers		Laborers		Managers	
X_1	X_1^2	X_2	X_2^2	X_3	X_3^2
1	1	4	16	4	16
2	4	4	16	6	36
2	4	5	25	7	49
1	1	6	36	5	25
4	16	4	16	5	25
10	26	23	109	27	151

$$SSB = \left[\sum \frac{(\Sigma X_i)^2}{N_i}\right] - \frac{(\Sigma X_T)^2}{N_T}$$

$$= \left[\frac{(10)^2}{5} + \frac{(23)^2}{5} + \frac{(27)^2}{5}\right] - \frac{(10 + 23 + 27)^2}{15}$$

$$= \left[\frac{100}{5} + \frac{529}{5} + \frac{729}{5}\right] - 240$$

$$= (20 + 105.8 + 145.8) - 240$$

$$= 271.6 - 240$$

$$= 31.6$$

$$SSW = \Sigma\left[\Sigma X_i^2 - \frac{(\Sigma X_i)^2}{N_i}\right]$$

$$= (26 - \frac{100}{5}) + (109 - \frac{529}{5}) + (151 - \frac{729}{5})$$

$$= (26 - 20) + (109 - 105.8) + (151 - 145.8)$$

$$= 6 + 3.2 + 5.2$$

$$= 14.4$$

One further note about the computations of the sum of squares values is useful. You can check the computations by realizing that $SST = SSB + SSW$. The calculated values for SSB and SSW can be added to see if they equal the calculated value of SST. If not, find the mistakes before continuing. Check:

$$SST = SSB + SSW$$
$$46 = 31.6 + 14.4$$
$$= 46$$

Analysis of variance is based on a comparison of the *variance* between the groups (samples) to the *variance* within the groups (samples). So the measures of *variation* in the form of sum of squares must be converted to *variances.* This is done by dividing each by its own degrees of freedom. The results are called *mean sum of squares* or *mean squares,* for short.

$$MSB = \frac{SSB}{dfb} \qquad \text{Mean Square (Variance) Between}$$

$$MSW = \frac{SSW}{dfw} \qquad \text{Mean Square (Variance) Within}$$

The between and within versions of sum of squares in our example can now be converted to variances or mean squares:

$$MSB = \frac{31.6}{2} = 15.80$$

$$MSW = \frac{14.4}{12} = 1.20$$

The F test statistic used in ANOVA is found by the simple ratio of these two mean squares:

$$F = \frac{MSB}{MSW} \qquad \text{Test Statistic } F$$

Our test value of F is

$$F = \frac{15.80}{1.20} = 13.17$$

This means that the ratio of the two variances is 13.17 to 1. The between-groups variance is more than 13 times as large as the within-groups variance. This is the F value shown in the ANOVA summary table (Table 11.1).

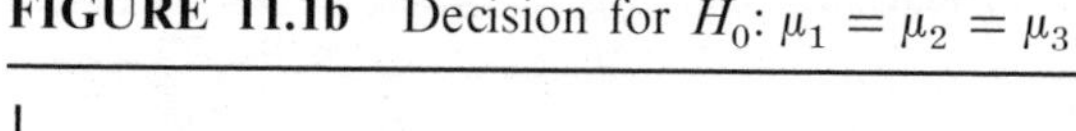

FIGURE 11.1b Decision for H_0: $\mu_1 = \mu_2 = \mu_3$

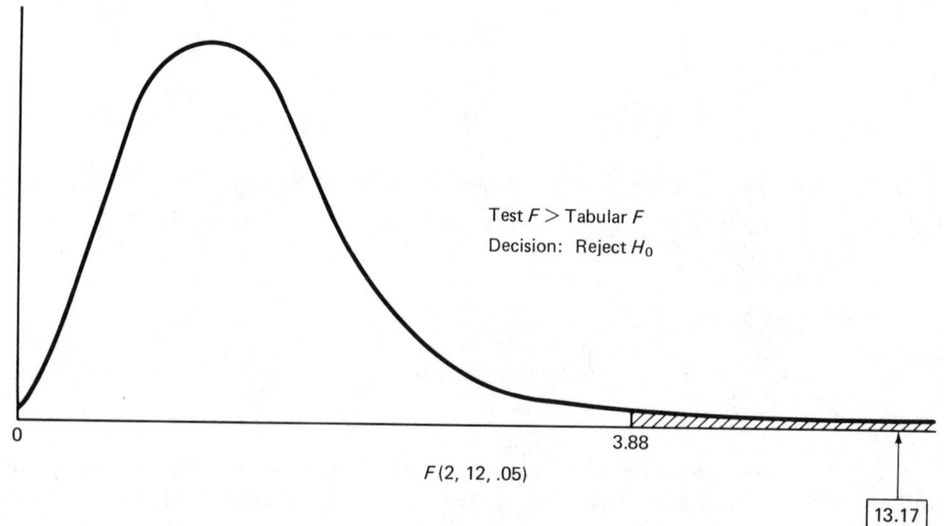

Step 4. The decision about H_0.

The test statistic (13.17) does exceed the tabular statistic (3.88). The null hypothesis is, therefore, rejected. (See Figure 11.1b.) There is a significant difference between at least two of the three sample means at the .05 level; $F = 13.17$, $df = 2$ and 12, $p < .05$.

Step 5. The conclusion.

Our researcher has found evidence that at least two of the three means in number of weeks of annual paid vacation differ. This lends credence to whatever theoretical reasons lead the researcher to propose such an alternative hypothesis. It suggests that the amount of vacation is different among these three types of occupations. Whether all three differ from each other, or only some of them differ is not known from this *F* test. That determination requires another test which will be discussed later (Tukey).

A Second Illustration of ANOVA

Let's try another example to ensure that you understand how the *F* test is conducted without all the intermediate explanations. Suppose that an academic counselor has formed the informal impression that students seem to have different levels of test anxiety at different stages in their college careers. To test this idea she draws a stratified random sample of 24 college students from a university population such that there are six students in each of the four classifications, freshmen through seniors. Each student is given a standardized interview measuring text anxiety. The sample results are shown in Table 11.3a. Are there significant differences in mean test anxiety between any of the four groups at the .01 level?

Step 1. The hypotheses.

H_0: $\mu_1 = \mu_2 = \mu_3 = \mu_4$ — The four subpopulations of freshmen, sophomores, juniors, and seniors have the same mean test anxiety. Random subsamples may differ in such means within the expected limits of sampling error.

H_1: not $[\mu_1 = \mu_2 = \mu_3 = \mu_4]$ — The four subpopulations of freshmen, sophomores, juniors, and seniors do not have the same mean test anxiety. Random subsamples will differ in such means beyond the expected influence of sampling error.

TABLE 11.3a Test Anxiety Scores for College Students by Classification

Freshmen	Sophomores	Juniors	Seniors
5	4	2	2
4	4	1	3
3	5	3	4
5	3	1	3
5	4	2	1
4	3	2	3

Step 2. The criteria for rejecting H_0.

The four subsamples of students are independent groups representing four subpopulations. The individual scores are random and independent of each other. An F distribution can be used as the appropriate sampling distribution since the standardized interview would be expected to yield normal distributions of anxiety scores in the subpopulations. The two relevant degrees of freedom are

$$dfb = k - 1 = 4 - 1 = 3$$
$$dfw = N_T - k = 24 - 4 = 20$$

The tabular F at the .01 level is therefore 4.94. (See Table D7.)

Step 3. The test statistic.

First, we work from the preliminary calculations shown in Table 11.3b to find the three sums of squares using the computational formulas.

$$SST = \Sigma X_T^2 - \frac{(\Sigma X_T)^2}{N_T}$$

$$= (116 + 91 + 23 + 48) - \frac{(26 + 23 + 11 + 16)^2}{24}$$

$$= 278 - \frac{(76)^2}{24} = 278 - \frac{5776}{24} = 278 - 240.6\overline{6}$$

$$= 37.33$$

$$SSB = \left[\sum \frac{(\Sigma X_i)^2}{N_i}\right] - \frac{(\Sigma X_T)^2}{N_T}$$

$$= \left[\frac{(26)^2}{6} + \frac{(23)^2}{6} + \frac{(11)^2}{6} + \frac{(16)^2}{6}\right] - \frac{(26 + 23 + 11 + 16)^2}{24}$$

$$= \left[\frac{676}{6} + \frac{529}{6} + \frac{121}{6} + \frac{256}{6}\right] - 240.6\overline{6}$$

$$= (112.6\overline{6} + 88.1\overline{6} + 20.1\overline{6} + 42.6\overline{6}) - 240.6\overline{6}$$

$$= 263.6\overline{6} - 240.6\overline{6}$$

$$= 23.00$$

$$SSW = \Sigma\left[\Sigma X_i^2 - \frac{(\Sigma X_i)^2}{N_i}\right]$$

$$= (116 - 112.6\overline{6}) + (91 - 88.1\overline{6}) + (23 - 20.1\overline{6}) + (48 - 42.6\overline{6})$$

$$= 3.3\overline{3} + 2.8\overline{3} + 2.8\overline{3} + 5.3\overline{3}$$

$$= 14.33$$

Check:

$$SST = SSB + SSW$$
$$37.33 = 23.00 + 14.33$$
$$= 37.33$$

TABLE 11.3b Preliminary Calculations for ANOVA from Table 11.3a

Freshmen		Sophomores		Juniors		Seniors	
X_1	X_1^2	X_2	X_2^2	X_3	X_3^2	X_4	X_4^2
5	25	4	16	2	4	2	4
4	16	4	16	1	1	3	9
3	9	5	25	3	9	4	16
5	25	3	9	1	1	3	9
5	25	4	16	2	4	1	1
4	16	3	9	2	4	3	9
26	116	23	91	11	23	16	48

Next we find the mean squares (the two variances).

$$MSB = \frac{SSB}{dfb} = \frac{23}{3} = 7.67$$

$$MSW = \frac{SSW}{dfw} = \frac{14.33}{20} = .717$$

Now the test statistic itself is found.

$$F = \frac{MSB}{MSW} = \frac{7.67}{.717} = 10.70$$

Step 4. The decision about H_0.

The test statistic (10.70) exceeds the tabular statistic (4.94); it falls into the rejection region. We therefore reject the null hypothesis at the .01 level of significance: $F = 10.70$, $df = 3$ and 20, $p < .01$. Figure 11.2 shows our decision in the context of the appropriate

FIGURE 11.2 Testing H_0: $\mu_1 = \mu_2 = \mu_3 = \mu_4$

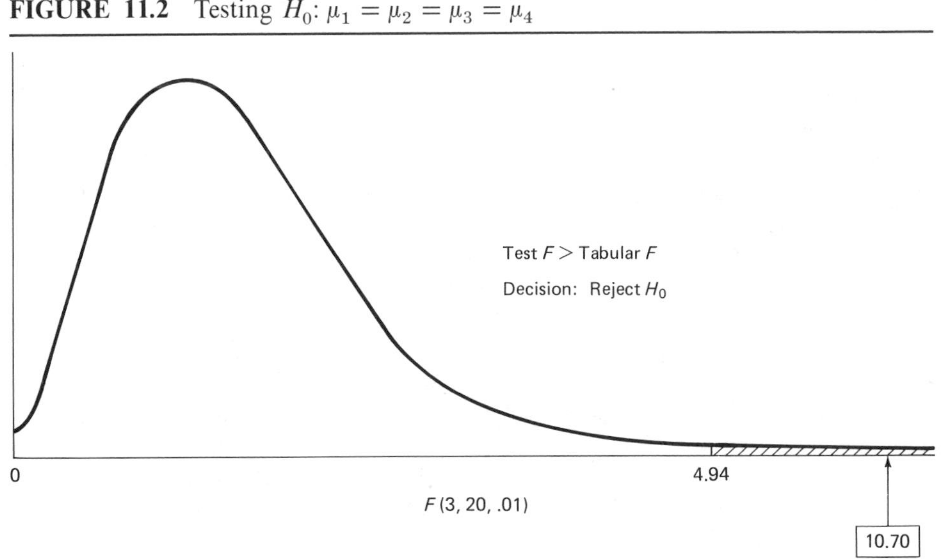

TABLE 11.4 ANOVA Summary Table for Test Anxiety Among College Students

Source of Variation	Sum of Squares	*df*	Mean Squares	*F*
Between groups	23.00	3	7.67	10.70*
Within groups	14.33	20	.717	
Total	37.33	23		

*$p < .01$.

sampling distribution and criteria for the test. The summary table for this analysis is shown in Table 11.4.

Step 5. The conclusion.

The null hypothesis of no difference between any of the four means has been rejected. We conclude that at least two of the four class means in test anxiety do differ. The counselor's idea has been supported by these data. Test anxiety is not the same for the four categories of student classification. We must be cautious, however, not to interpret this to mean that test anxiety changes for *individuals* from some years to others. We examined four sets of independent students. Individual change can only be shown from panel studies of the same individuals through time.

Two illustrations of one-way analysis of variance with actual data are shown in R.E. 11.1 and 11.2. In R.E. 11.1 the researchers wanted to compare attitudes toward community growth by county commissioners in Florida and Georgia. After some preliminary analysis

RESEARCH EXAMPLE 11.1 One-Way Analysis of Variance *F* Test

From Vincent L. Marando and Robert D. Thomas, "County Commissioners' Attitudes Toward Growth: A Two-State Comparison," Social Science Quarterly, *58 (1), June 1977:129ff. Reprinted by permission.*

Ingrained into American society is the tenet that growth means community vitality and therefore is "good." Recently, however, policy debates have ensued as to whether community growth should be encouraged, stabilized, or discouraged. . . . Growth advocates are certain that community health and prosperity is dependent upon continued expansion. . . . Those calling for moderate expansion or a rollback view growth as detrimental to the quality of life. They are certain that too much growth drains resources. . . . County commissioners are at the center of this controversy at the local level. They have key authorities to encourage or stabilize growth through decisions on land use planning, subdivision regulations, zoning, and the issuance and enforcement of building codes. . . . This article is about Florida and Georgia commissioner attitudes toward growth. Our data are derived from questionnaires received from 253 Florida and Georgia county commissioners and from U.S. census information. The 253 commissioners represent 27 percent of the 938 county commissioners in Florida and Georgia. While our data set does not constitute a random sample, it was found to be representative of the universe of county commissioners when compared with the available biographical information kept on record by the Associations of County Commissioners of Florida and Georgia. . . .

Commissioner growth attitudes were measured on a scale which ranged from an attitude favoring the promotion of county growth to an attitude favoring the stabilization of county growth. Commissioners were asked if they agreed, tended to agree, tended to disagree, or disagreed with seven growth questions . . . Utilizing [factor analysis], four items which were well associated with each other . . . were constructed into a scale. . . .

The commissioner's attitude toward growth is conditioned by the environment of his governing jurisdiction. Therefore, we selected four comprehensive indicators which encompass the setting within which commissioners face growth issues: urbanization, community wealth, governmental complexity, and the state in which the county is located. . . .

The state in which a commissioner resides was found to be a stronger indicator of growth attitudes than urbanization, wealth, and governmental complexity . . . How then do growth attitudes differ between Florida and Georgia commissioners? We found Florida commissioners to be more oriented toward stabilizing growth than Georgia commissioners. As [the table] shows, the mean score for Florida commissioners was 7.2 on the 13 interval attitude scale, while the mean score for Georgia commissioners was 4.3. Moreover, this variation in growth attitudes is accounted for by the difference between Florida and Georgia rather than by the differences within each state. . . .

Analysis of Growth Attitude Variance of County Commissioners Between and Within Florida and Georgia ($N = 249$)[a]

	Degrees of Freedom	Sum of Squares	Mean Squares	*F* Ratio	*F* Probability
Between states	1	502.8	502.8	78.5	0
Within states	247	1582.1	6.4		
Total	248				

State	*N*	Mean[b]	Standard Deviation	95 Percent Confidence Interval for Mean
Florida	114	7.2	3.0	6.6 to 7.8
Georgia	135	4.3	2.0	4.0 to 4.7
Total	249	5.6	2.9	5.3 to 6.0

[a]Table 2 in the original.
[b]The higher the mean, the more stabilization-oriented is the attitude.

The differences between Florida and Georgia commissioners' growth attitudes suggest that we should examine state level policy differences between and among states. Recent growth policies in Florida appear to have permeated commissioner attitudes even among those commissioners least expected to be influenced. We found Florida commissioners from counties which lost population to be more stablization-oriented than Georgia commissioners whose counties had gained more than 25 percent in population.

Florida and Georgia have had different growth experiences, resulting in different state policies. Throughout its contemporary history, Florida has been concerned with growth. Until recently, the major objective was to encourage immigration by making the state inhabitable for development. The state attempted to accomplish this objective through homestead exemption and drainage of wetlands program. In the past five years, problems have arisen which necessitated that changes be made in state policy: a severe drought, destruction of the Everglades, and rapid population increases. . . . Georgia has not experienced the same historical evolution of growth nor have recent growth events in Georgia been nearly as intense. . . .

These policy differences between Florida and Georgia suggest that state policies may affect local activities as profoundly as demographic and socioeconomic conditions. . . .

it was determined that the state of residence was a more important determinant of differences in this attitude than some other conditions in the social environment. The one-way analysis of variance F test indicated that the difference in growth attitude between the two states was much greater than the difference in growth attitude within the states. The test value of F (78.5) was clearly significant. The sampling error probability of this F value was so small that it was reported to be zero. Of course, there would be some nonzero (although very small) probability attached to this F value. It was merely rounded to zero in the research report.

There are two other items for you to notice in this example. First, the data are described as not constituting a random sample, but being representative, nonetheless. What the authors mean by this is that they did not select the sample using a random technique. However, when they compared available information for all area commissioners to those in their sample, there was a close match. This is a common method used in surveys to support the use of inferential statistics for nonrandom samples. It has become an acceptable practice. It can still be mistaken, though. In this example, the sample may represent the larger population in certain demographic characteristics (age, sex, education, and so forth) but be unrepresentative in the variable under study (growth attitude). It is not possible to be sure whether this problem exists for any particular example because the required information is unknown.

Second, the measurement of the dependent variable in this example is carefully described so that the reader can have confidence in its quality. Here the authors explain that the variable is a composite of responses to four separate questions. The four questions were selected from an original set of seven following the use of factor analysis. Earlier it was indicated that factor analysis is one of the best and most widely used techniques for the creation of a variable with truly interval quality. It is important that the dependent variable have interval quality because the F test is based on an examination of interval differences between and within the sample groups. Here those two groups were commissioners in the states of Florida and Georgia.

Research Example 11.2 also shows the F test in use. This example concerns separate analyses of four attitudes by prisoners from three categories of prison confinement. The attitudes were those toward prison culture (prisonization), lawyers, law and the judicial system, and police. The three comparison groups were prisoners in maximum-, medium-, and minimum-security facilities in the state of Washington. The analysis found that there were no significant differences between the three groups in attitudes toward prisonization and lawyers (Tables 1 and 2 in the example). There were significant differences in attitudes toward law and the judicial system and toward the police (Tables 3 and 4 in the example).

RESEARCH EXAMPLE 11.2 One-Way Analysis of Variance *F* Test

Geoffrey P. Alpert, "Prisons as Formal Organizations: Compliance Theory in Action," Sociology and Social Research, *63 (1), October 1978:112ff. Reprinted by permission.*

Compliance theory, as applied to prisons as complex organizations of the total institutional type, can be empirically tested only if due attention is paid to significant features of institutional variation. Two relevant strands that serve as necessary background for the use of institutional variation as a major independent variable in the study reported here are the concept of total institutions as developed by Goffman (1961) and compliance theory as elaborated by Etzioni (1961, 1970, 1975) on the basis of fruitful suggestions set forth by Simmel toward the end of the 19th century. . . .

Prisons exemplify the kind of coercive organization which requires compliance but at the same time embitters and alienates their inmates. This results from the fact that the goals of the prison, namely custody and confinement, are diametrically opposed to the primary goal of inmates: freedom. Prison subculture is indeed in conflict with the institutional goals. . . .

Other studies provide concurrence with the view that when the perceived level of coercion is reduced, alienation of prisoners declines, staff-prisoner interaction increases and chances for rehabilitation should increase. . . . If this line of reasoning is correct, it is reasonable to anticipate that as the degree of coercion is altered, corresponding variation in attitudes should follow. Historically, prisons have been distinguished from each other on the basis of custody classification (maximum, medium, and minimum). Correspondingly, one would expect that prisoners incarcerated in a maximum security prison for several months would develop more negative attitudes than those incarcerated in medium or minimum security prisons. In other words, prisoners' attitudes should reflect variation in institutional coercion. It is the purpose of this paper to investigate the effects of this independent variable, institutional diversity, on the attitudes of a cohort of prisoners over a six-month period.

Data were collected from prisoners at the time of their admission to the Washington State prison system between June and September, 1974 and again, after they had spent six months in the institution to which they were transferred. . . . While our cohort was relatively heterogeneous at our first contact, its members were transferred to separate units on the basis of age, criminal sophistication and history of confinement. . . . In other words, by T_2 [time two], each subpopulation was relatively homogeneous.

We administered four psychometric attitudinal scales to determine the way prisoners felt toward the system by which they were incarcerated and toward their actual incarceration. Specifically, we questioned prisoners on their attitudes toward police, law and the judicial system, lawyers and prison culture (prisonization). The scales include items representing various aspects of each attitudinal domain to which respondents could answer strongly agree (5) to strongly disagree (1). Scale scores were derived by unit weighting and summing the item scores. Thus, the higher a scale score, the more positive the attitude. . . . Total scale scores were used to determine statistical independence. . . .

In order to ascertain the effect, if any, of institutional diversity on the several attitudinal measures, two analyses were used. First, we subtracted the score at T_2

from the T_1 score creating a change score. A single one way analysis of variance was performed on the change in each attitudinal domain. [The second analysis is not reprinted here.] . . .

TABLE 1. Analysis of Variance of Prisonization by Institution[a]

Source	*df*	*SS*	Mean *S*	*F*	Significance of *F*
Between-group change	2	124.192	31.048	1.554	.187
Within-group change	185	3855.873	19.979		

[a]Adapted from Table 1 in the original.

[Table 1 shows] the results of applying the above procedures to ascertain the effect of variations in the patterns of coercion on prisonization. . . . Significant differences were not indicated by the *F*-statistical value. . . . This statistic tests the hypothesis that all changes are the same . . . When one uses the generally accepted .05 significance level, one is unable to *accept* [emphasis added] . . . [this hypothesis.]

TABLE 2. Analysis of Variance of Attitudes Toward Lawyers by Institution[a]

Source	*df*	*SS*	Mean *S*	*F*	Significance of *F*
Between-group change	2	33.692	8.423	.524	.722
Within-group change	185	3105.061	16.088		

[a]Adapted from Table 3 in the original.

TABLE 3. Analysis of Variance of Attitudes Toward the Law and Judicial System by Institution[a]

Source	*df*	*SS*	Mean *S*	*F*	Significance of *F*
Between-group change	2	12.820	30.705	2.756	.029
Within-group change	185	2150.433	11.142		

[a]Adapted from Table 5 in the original.

Variations in the patterns of coercion have no effect on attitudes toward lawyers as indicated by the data in [Table 2]. . . . [Table 3 reveals] the effects of institutional diversity on attitudes toward law and the judicial system. In [the table], the *F*-ratio indicates that at the .05 level, one cannot accept the hypothesis that average change is the same for all institutions. . . . Data on institutional diversity and attitudes toward police are presented in [Table 4.] The table reveals highly significant changes in attitudes toward police. . . .

TABLE 4. Analysis of Variance Toward Police by Institution[a]

Source	*df*	*SS*	Mean *S*	*F*	Significance of *F*
Between-group change	2	330.924	82.731	3.860	.005
Within-group change	185	4136.938	21.435		

[a]Adapted from Table 7 in the original.

Our data suggest that institutional variations in patterns of coercion have very little effect on either the degree of prisonization experienced by inmates, or on their attitudes toward lawyers. It does affect attitudes toward law and the judicial system and attitudes toward police. . . .

You can verify these decisions about null hypotheses of no difference by comparing the reported "significance level of *F*" to the .05 alpha level. You will note that the first two tables show probabilities over .05, while the last two are lower than .05. Notice, also, that the author mistakenly states that the null hypothesis for the first table is not *accepted* when actually it is not *rejected*. The error is apparently isolated to the one sentence and not in an interpretation of the findings. You cannot verify the calculations of the mean squares because special analysis techniques were being used.

This example also used a composite scoring system to measure each of the four dependent variables. The exact measurement of these variables as reported may seem confusing to you at first. Notice that the scores represent the difference between a score at the initial interview and the score at a second interview following confinement in a specific institution. In this way, the before–after sets of scores are transformed to a single set of change scores. The same one-way analysis of variance we have been discussing can then be used. A special version of ANOVA for dependent scores is not needed. This technique is similar to that of using the "direct difference method" with the *t* test for the difference between two dependent means as discussed in Chapter 10.

ON THE SIDE 11.1 Ronald A. Fisher

Ronald A. Fisher (1890–1962) is credited with originating and developing many techniques that are standard to modern statistics. He was English-born and, like Carl F. Gauss, considered to be a genius who had an early interest in astronomy and made mathematics his field of primary study.

His first publication (in 1912 at age 22) on fitting frequency curves began a series of correspondence with W. S. Gossett ("Student") that lasted until Gossett's death in 1937. A 1915 publication presented the distribution of the coefficient of correlation and brought him to the attention of most statisticians, including Karl Pearson, the reigning giant. Pearson did not accept the work and set out an elaborate numerical investigation which ultimately failed to discredit Fisher's work.

After college and some early jobs, Fisher was offered two positions at once: one with Pearson at University College and one at the Rothamsted Experimental Station. He chose the latter and it began what many

consider the most remarkable career of any statistician. The decision also marked Fisher's great interest in both applied and theoretical statistics and hardened a lifelong feud with Pearson over several statistical issues beyond the original one in 1915.

Fisher's work at Rothamsted put him in contact with several agricultural and genetics researchers. He gladly gave attention to the practical problems their areas presented. This created the occasions for him to develop his ideas on designing experiments and evaluating their results for which he is best known. Over the next 20 years he produced more than 120 papers, three major books, and a set of tables describing the distributions of various statistical measures. In 1925 he published the first edition of the historic *Statistical Methods for Research Workers.* It became a classic and has gone through 14 editions. It also became part of his conflict with Pearson. Pearson had refused Fisher permission to reproduce a table of chi-square values in the book. Pearson had his (financial) reasons, but Fisher considered it an affront he never forgot.

Like some other works by Fisher, the book was criticized by theoretical statisticians for its lack of mathematical proofs and the like. Fisher did not waver; he had written it to assist practicing researchers, not mathematicians. Actually, Fisher's mathematical ability was excellent. He could imagine and prove sophisticated ideas using geometrical solutions in a remarkably complex and elegant manner.

The book led to Fisher's development of a set of techniques for experiments. He clarified numerous issues about control and experimental groups with proper controls for environmental and background factors. Illustrative examples concerned plant yields considering the influences of rainfall, various fertilizers, plant varieties, and nonuniformity of soil fertility in the experimental plots. These issues were pursued for everything from tea tasting and the weight of marigold roots to barley yields and the sugar content of sugar beets. Another whole area of Fisher's interest concerned human genetics, which led him to a commitment to eugenics.

During the period 1920 to 1940, Fisher's preeminence in statistics was so great that he was considered the primary source for most topics in the field. This included not only correlation and experiments but also regression, multiple and partial correlations, and the analysis of variance. The latter showed the integration between an entire series of problems, only vaguely seen as similar previously. The probability distributions for variance ratios carry Fisher's initial (F) in his honor, as they should since he was their chief developer.

Later in life Fisher produced an especially disappointing work on smoking and cancer. Fisher was correctly concerned with the problems of concluding cause and effect from nonexperimental data. But his analysis failed to clarify what type of data would be required to settle on a causal conclusion to the issue.

On retirement Fisher moved to Australia, where he died in 1962. He is remembered as a man who could be personally difficult and rude but also affable to his friends. A more contemporary statistician of very high note, William Cochran, illustrated Fisher's mix of self-assurance and self-disregard with an anecdote. Cochran was walking with Fisher in London when they wanted to cross a busy street. Cochran feared for Fisher's safety because of Fisher's poor eyesight. But the impatient Fisher suddenly stepped into a minor traffic gap calling back: "Oh, come on, Cochran. A spot of natural selection won't hurt us."

Based on Jerzy Neyman, "R. A. Fisher (1890–1962): An Appreciation" and "Footnote by William G. Cochran," *Science,* 156 (3781), June 16, 1967:1456–1462; M. G. Kendall, "Ronald Aylmer Fisher, 1890–1962," *Biometrika,* 50 (1 and 2), June 1963:1–15; S. E. Fienberg and D. V. Hinkley (eds.), *R. A. Fisher: An Appreciation.* New York: Springer-Verlag, 1980; Joan Fisher Box, *R. A. Fisher, The Life of a Scientist.* New York: John Wiley & Sons, Inc., 1978.

Tukey *HSD* Test

The analysis of variance F test tells the researcher whether any two means are significantly different from among those being compared. If the test value of F is not significant, none of the means differ statistically. But if that F value is significant, then at least two of them do differ. However, the F value does not indicate which means differ nor how many of them differ. There are a few techniques used by researchers to determine which means do differ. We will look at one of these which is commonly used for equal-sized samples following the finding that the test value of F is significant.

The Tukey *HSD* test determines which pairs of means differ significantly. It uses a special statistic referred to as the *HSD* value, an "honestly significant difference." The name refers to the problem in repeated applications of the t test to pairs of means in which a difference between two means may appear to be significant but actually is not because of the overlapping probabilities of Type I errors. This was mentioned earlier. Tukey's test finds a number representing the true (honest) minimum by which two means must differ to be a statistically significant difference, that is, warrant the rejection of a null hypothesis of no difference. The test requires three steps.

Step 1. Identify all differences between pairs of sample means.

Let's use the data from our first illustration above of an F test. A simple method to identify all of the differences between pairs of means is to put them in a matrix as shown in Table 11.5. This matrix shows the absolute value of the differences for the pairs only in its upper right triangle. The remainder of the matrix is ignored since it is either not relevant (the diagonal) or is redundant (the lower left triangle.) In the matrix you see that there are three sets of differences between pairs of the three sample means.

Step 2. Find the *HSD* value.

HSD if found by the following formula:

$$HSD = q_{\alpha}\sqrt{\frac{MSW}{N_i}} \qquad \text{Tukey } HSD \text{ Statistic}$$

where q_{α} = tabular value (q) at a level of significance (α) for k (number of samples) and dfw (degrees of freedom within)

MSW = mean square within from ANOVA

N_i = number of scores in each sample

The values of q_{α} appear in Table D8 of Appendix D under the name of Studentized range values. In our example we enter the table at $dfw = 12$ and $k = 3$. We use the .05

TABLE 11.5 Absolute Differences Between Pairs of Mean Vacation Weeks for Workers in Three Occupations

	$\bar{X}_1$	$\bar{X}_2$	$\bar{X}_3$
$\bar{X}_1 = 2.0$	—	2.6*	3.4*
$\bar{X}_2 = 4.6$	—	—	.8
$\bar{X}_3 = 5.4$	—	—	—

$^*p < .05$.

TABLE 11.6 Absolute Differences Between Pairs of Mean Test Anxiety for Student Classifications

	$\bar{X}_1$	$\bar{X}_2$	$\bar{X}_3$	$\bar{X}_4$
$\bar{X}_1 = 4.33$	—	.50	2.50*	1.66
$\bar{X}_2 = 3.83$	—	—	2.00*	1.16
$\bar{X}_3 = 1.83$	—	—	—	.83
$\bar{X}_4 = 2.67$	—	—	—	—

*$p < .01$.

level of significance since that was used in the prior *F* test. The tabular value of q_α at those levels is 3.77. We now substitute this in our *HSD* formula.

$$HSD = 3.77\sqrt{\frac{1.2}{5}} = 3.77\sqrt{.24}$$
$$= 3.77(.49) = 1.85$$

What does this number mean? It is the minimum difference between a pair of means in our example which could be considered to be statistically significant.

Step 3. Identify those pairs of means that honestly differ.

From the matrix we see that $\bar{X}_1$ versus $\bar{X}_2$ and $\bar{X}_1$ versus $\bar{X}_3$ both exceed the *HSD* value of 1.85, so they are significantly different. The $\bar{X}_2$ versus $\bar{X}_3$ difference does not exceed 1.85. There is a statistically significant difference in the mean number of annual paid vacation weeks between office workers and laborers and between office workers and managers. But the means for laborers and managers do not differ significantly.

Let us try it again using the data from our second example above. This time the *F* test told us that there was a significant difference in test anxiety between at least two student classifications.

Step 1. Identify all differences between pairs of means.

The pairs of mean differences are shown in Table 11.6. There are six such pairs for the four sample means.

Step 2. Find *HSD*.

The level of significance is .01. The degrees of freedom within is 20 and *k* is 4. The value of q_α from Table D8 is, therefore, 5.02. Our value of *HSD* is now found:

$$HSD = 5.02\sqrt{\frac{.717}{6}} = 5.02\sqrt{.1195}$$
$$= 5.02(.346) = 1.74$$

Step 3. Identify pairs that honestly differ.

The means in this example must differ by 1.74 or more to be significantly different. Only two of the six pairs of means differs by at least 1.74, $\bar{X}_1$ versus $\bar{X}_3$ and $\bar{X}_2$ versus $\bar{X}_3$. We conclude that there is a significant difference in mean test anxiety between freshmen and juniors and between sophomores and juniors. All the other pairs of means differ within the limits of sampling error at the .01 level of significance. This throws some additional light on the example. The rather general conclusion we drew for the *F* test suggested that there might be a difference in mean test anxiety across the entire four-year span from freshmen through seniors. In fact, there is evidence only for a difference between those in their first and third years of school and those in their second and third years. The counselor's idea is supported by these data only in this limited way.

Nonparametric Tests with Multiple Samples

There are four main nonparametric tests used for multiple-sample comparisons in the social sciences. All are based on the chi-square sampling distributions and two of them are expanded versions of nonparametric tests we saw in Chapter 10 for comparisons between two samples. The four tests are: chi-square for k samples, median test for k samples, Kruskal–Wallis H test, and Friedman two-way analysis of variance test.

Chi-Square Test of Independence for *k* Samples (χ^2-k)

The chi-square-II test used to compare two independent samples on a nominal dependent variable is easily expanded to situations in which there are more than two (k) samples. The extension is so straightforward that the technique is not usually identified as a separate test. For organizational purposes here, we will call it the chi-square-k test (χ^2-k). A short example will illustrate it.

Suppose that a researcher is interested in the early socialization of children. He investigates the topic by comparing three samples of preschool children on their ability to recognize a dollar sign ($), a common symbol in adult culture. The samples represent populations from three social classes: lower, middle, and upper. The frequency distributions showing the number of children in each sample who do and do not recognize the dollar sign are presented in Table 11.7. Do these distributions differ at the .05 level of significance?

Step 1. The hypotheses.

H_0: $\chi^2 = 0$ — The populations of preschool children from lower, middle, and upper social classes do not differ in their frequency distributions of ability to recognize the dollar sign ($). Sampling error may cause distributional differences for random samples from these three populations.

TABLE 11.7 Recognition of the Dollar Sign ($) by Preschool Children from Three Social Class Backgrounds

Recognition of Dollar Sign ($)	Social Class Backgrounds for Samples of Preschool Children			
	Lower Class	Middle Class	Upper Class	
Yes	13	15	12	40
No	7	10	3	20
	20	25	15	60

H_1: $\chi^2 \neq 0$ The populations of preschool children from lower, middle, and upper social classes do differ in their frequency distributions of ability to recognize the dollar sign ($). Distributions for random samples from these three populations will reflect this difference beyond the expected influence of sampling error.

Step 2. The criteria for rejecting H_0.

Like all other inference uses of chi-square, there is a set of conditions to be met in order for the researcher to use the chi-square distribution and test. The data must be collected randomly and independently from independent samples (or subsamples). We need only nominal level of measurement in the dependent variable. We cannot have too few cases. All these conditions are met in this example.

The appropriate tabular value of chi-square is found by knowing the degrees of freedom and the level of significance. As with chi-square-II, $df = (r - 1)(c - 1)$. Here this is $(2 - 1)(3 - 1) = 2$. The alpha level is .05. The tabular chi-square from Table D4 at these values is 5.991. As before, a single rejection region at the right side of the sampling distribution is used.

Step 3. The test statistic.

The test value of chi-square is found in the same manner as a chi-square-II test. The expected frequency for each of the cells in the raw data table is found and compared to the actual (obtained) frequencies according to the chi-square formula. This is shown in Table 11.8. There you see that the test chi-square is 1.741.

TABLE 11.8 Worksheet for Chi-Square *k*-Sample Test

Recognition of Dollar Sign ($)	**Social Class Backgrounds for Samples of Preschool Children**			
	Lower Class	Middle Class	Upper Class	
Yes	13 (13.3)	15 (16.7)	12 (10)	40
No	7 (6.7)	10 (8.3)	3 (5)	20
	20	25	15	60

$$\chi^2 = \Sigma \frac{(f_o - f_e)^2}{f_e}$$

$$= \frac{(13 - 13.3)^2}{13.3} + \frac{(15 - 16.7)^2}{16.7} + \frac{(12 - 10)^2}{10} + \frac{(7 - 6.7)^2}{6.7} + \frac{(10 - 8.3)^2}{8.3} + \frac{(3 - 5)^2}{5}$$

$$= \frac{(-.3)^2}{13.3} + \frac{(-1.7)^2}{16.7} + \frac{(2)^2}{10} + \frac{(.3)^2}{6.7} + \frac{(1.7)^2}{8.3} + \frac{(-2)^2}{5}$$

$$= \frac{.09}{13.3} + \frac{2.89}{16.7} + \frac{4}{10} + \frac{.09}{6.7} + \frac{2.89}{8.3} + \frac{4}{5}$$

$$= .007 + .173 + .400 + .013 + .348 + .800$$

$$= 1.741$$

Step 4. The decision about H_0.

The test value of chi-square (1.741) does not exceed the tabular statistic (5.991); it does not fall into the rejection region. Therefore, the null hypothesis is not rejected. There is no significant difference between the three groups; $\chi^2 = 1.741$, $df = 2$, $p > .05$, n.s.

Step 5. The conclusion.

Having failed to reject the null hypothesis, the researcher concludes that the populations of lower-, middle- and upper-class preschool children do not differ in their abilities to recognize the dollar sign. That ability appears to have been equally a part of the socialization processes for all three groups. It may mean that such socialization is part of a mass cultural system that is broader than any one social class. Social class has no apparent influence as examined by this test. Of course, it may have had an even earlier influence or its influence may be masked by some other variables that are not being considered here. But by this limited test, social class seems not to be relevant to the recognition ability in question.

Median *k*-Sample Test (*Mdn-k*)

The median test for two independent samples can also be expanded for comparisons between any number of independent samples. Its procedures are identical to the two-sample situation except for the calculation of the test statistic. At that point the chi-square-*k* test value is found rather than the chi-square-II test value. As you have already seen, this is only a very modest change. Another short example will illustrate this.

Suppose that our researcher studying early socialization is still persuaded that there are some social class differences between preschoolers in ability to recognize common symbols of adult culture, such as the dollar sign. This time the researcher selects 15 symbols and determines how many are recognized by the children in each of the three samples representing the social class categories of lower, middle, and upper. The frequency distributions are shown in Table 11.9. Do they differ significantly in median ability?

The dependent variable is the number of symbols recognized by the children. As such, it has ratio measurement quality. An analysis of variance using the *F* statistic could be used for a parametric test. But here we are thinking in terms of using a nonparametric test. Why would a researcher do this? Maybe the parametric assumptions for the *F* test cannot be met. For example, the populations of scores (number of symbols recognized) may not be normally distributed. Alternatively, the researcher simply prefers a nonparametric test because of uncertainty about some aspects of the measurement in the research. Regardless, let's see how the median-*k* test would be applied here.

TABLE 11.9 Number of Common Symbols Recognized from List of 15 by Samples of Preschoolers from Three Social Class Backgrounds

Lower Class ($N_1 = 10$)		Middle Class ($N_2 = 15$)		Upper Class ($N_3 = 12$)	
5	12	9	9	15	11
4	2	8	4	6	10
6		14	15	12	11
3		10	12	13	12
5		7	9	7	
2		10	5	10	
7		13	11	14	
1		11		9	

Step 1. The hypotheses.

H_0: $Mdn_1 = Mdn_2 = Mdn_3$

The median number of symbols recognized by populations of lower-, middle-, and upper-class preschool children are the same. Random samples from these populations may differ in medians within the expected limits of sampling error.

H_1: not $[Mdn_1 = Mdn_2 = Mdn_3]$

The median number of symbols recognized by populations of lower-, middle-, and upper-class preschool children are not all the same. Random samples from these populations will differ in medians beyond the expected influence of sampling error.

Step 2. The criteria for rejecting H_0.

The median *k*-sample test requirements are the same as those for the chi-square-*k* test except that the dependent variable must have at least ordinal level of measurement. These requirements are met in this example.

The degrees of freedom are again $(r - 1)(c - 1)$. Here $df = (2 - 1)(3 - 1) = 2$. (See Step 3 in Table 11.10.) Let's use the .10 level of significance since our earlier experience with this general research problem has shaken the confidence of the researcher in the idea that social class is relevant to the ability to recognize common adult symbols. The tabular statistic from Table D4 at 2 *df* and $\alpha = .10$ is 4.605. The test statistic must equal or exceed this in order for H_0 to be rejected.

Step 3. The test statistic.

The raw data from Table 11.9 are put into a unified distribution and its overall median is found. This work is shown in Table 11.10. There you see that the overall median is 9. Now the number of scores above and not above 9 are tallied for each sample and entered into a bivariate table. Then a test value of chi-square is calculated. It is 9.635.

Step 4. The decision about H_0.

The test statistic (9.635) does exceed the tabular statistic (4.605), so the null hypothesis is rejected. The proportions of scores above and not above the common median of 9 do differ among the three samples beyond what could be attributed to sampling error at the .10 level of significance. The three populations are not all alike; $\chi^2 = 9.635$, $df = 2$, $p < .10$.

Step 5. The conclusion.

Having rejected the null hypothesis, the researcher now concludes that the three populations of preschoolers do not have the same median number of symbols recognized. What does this mean, especially in light of the earlier finding that the same three populations did not differ in ability to recognize the one symbol of a dollar sign? A researcher might answer this way.

Perhaps recognition of the dollar sign is atypical of recognition for the universe of all common symbols used by adults. Although there may be no social class differences in recognition of that one symbol, there are differences in the number of all such symbols that preschoolers are able to recognize.

However, this interpretation is not completely persuasive. It contains some "sleight of hand" reasoning. It is not really fair to compare the findings for one statistical test using a .05 level of significance to another using a .10 level. When the significance level changes,

TABLE 11.10 Worksheet for a Median k-Sample Test Using Data from Table 11.9

1. Unified distribution:

																		Mdn ↓				
1	2	2	3	4	4	5	5	5	6	6	7	7	7	8	9	9	9	9	10	10	10	10
L	L	L	L	L	M	L	L	M	L	U	L	M	U	M	M	M	M	U	M	M	U	U

11	11	11	11	12	12	12	12	13	13	14	14	15	15
M	M	U	U	L	M	U	U	M	U	M	U	M	U

2. Find median:

$$\text{Position of } Mdn = \frac{N+1}{2} = \frac{37+1}{2} = \frac{38}{2} = 19\text{th score}$$

$$Mdn = 9$$

3. Construct a bivariate table:

	Lower Class	Middle Class	Upper Class	
Above 9	1 (4.9)	8 (7.3)	9 (5.8)	18
9 and Below	9 (5.1)	7 (7.7)	3 (6.2)	19
	10	15	12	37

4. Calculate chi-square:

$$\chi^2 = \frac{(1-4.9)^2}{4.9} + \frac{(8-7.3)^2}{7.3} + \frac{(9-5.8)^2}{5.8} + \frac{(9-5.1)^2}{5.1} + \frac{(7-7.7)^2}{7.7} + \frac{(3-6.2)^2}{6.2}$$

$$= \frac{(-3.9)^2}{4.9} + \frac{(0.7)^2}{7.3} + \frac{(3.2)^2}{5.8} + \frac{(3.9)^2}{5.1} + \frac{(-0.7)^2}{7.7} + \frac{(-3.2)^2}{6.2}$$

$$= \frac{15.21}{4.9} + \frac{0.49}{7.3} + \frac{10.24}{5.8} + \frac{15.21}{5.1} + \frac{0.49}{7.7} + \frac{10.24}{6.2}$$

$$= 3.104 + 0.067 + 1.766 + 2.982 + 0.064 + 1.652$$

$$= 9.635$$

so do the criteria for rejecting a null hypothesis. We can understand how a researcher might decide to use a less restrictive level for a new significance test following an earlier failure to reject a null hypothesis. But it is a mistake if the findings are to be compared. Our researcher should have stayed with the .05 level if the two tests are to be viewed together. In this example, we see from Table D4 that the test statistic would have been 5.991 at the .05 level. Thus, the null hypothesis would yet be rejected, leading to the conclusion already drawn. This "fortunate" outcome does not always occur in research.

Kruskal–Wallis *H* Test

A much more powerful nonparametric test than either the chi-square-k or median-k tests is often used in social research to compare multiple samples. It is the Kruskal–Wallis H test. It is the most powerful nonparametric test that can be used as an alternative to the parametric analysis of variance F test. The Kruskal–Wallis H test compares three or more independent samples on a variable that has been measured at the ordinal level or higher in the form of ranks. It is based on a new statistic known as H. Sampling distributions of H

are closely approximated by chi-square distributions at $df = k - 1$, where k = the number of independent samples or subsamples. The tabular value of chi-square is used as the tabular statistic for the H test.

The test statistic uses a new formula and procedure. The raw scores are converted to one set of ranks across all samples and substituted in the following formula:

$$H = \frac{12}{N(N+1)}\left[\sum \frac{(\Sigma R_i)^2}{N_i}\right] - [3(N+1)]$$

Test Statistic for Kruskal–Wallis H

where ΣR_i = sum of ranks in each sample
N = number of scores in all samples combined
N_i = number of scores in each sample

The constants 12, 3, and 1 occur in the formula for every application; they are not related to either the number of samples or scores in any particular problem. The calculated value of the test statistic is compared to the tabular value of H as estimated by the tabular chi-square to reach a decision about the null hypothesis being tested. Ordinarily, that null hypothesis reflects the idea that the samples do not differ in ranks (calculated as a mean) on some dependent variable.

Let us consider an example. Suppose that a gerontologist is examining the extent of bereavement by widows. The gerontologist has selected a random sample of 24 women who became widows at least two years prior to the study. Each was asked about the length (in days) of her feelings of profound bereavement following her husband's death. This information is shown in Table 11.11 for the widows, divided into subsamples by their ages. The researcher wants to know whether there is a significant difference between the three subsamples in length of bereavement at the .05 level of significance.

Step 1. The hypotheses.

H_0: $H = 0$ The subpopulations of younger, middle-aged, and older widows have the same mean ranked lengths of bereavement. Random subsamples may differ in such ranks within the expected limits of sampling error.

TABLE 11.11 Length of Bereavement (Days) for Widows of Three Age Categories

Younger Widows $N_1 = 6$	Middle-Aged Widows $N_2 = 10$	Older Widows $N_3 = 8$
31	26	32
15	74	38
60	62	45
25	75	20
37	54	89
45	41	60
	67	45
	90	42
	57	
	51	

TABLE 11.12 Preliminary Calculations for a Kruskal-Wallis *H* Test Using Data from Table 11.11

Younger Widows		Middle-Aged Widows		Older Widows	
X_1	R_1	X_2	R_2	X_3	R_3
31	20	26	21	32	19
15	24	74	4	38	17
60	7.5	62	6	45	13
25	22	75	3	20	23
37	18	54	10	89	2
45	13	41	16	60	7.5
	$\Sigma R_1 = 104.5$	67	5	45	13
		90	1	42	15
		57	9		$\Sigma R_3 = 109.5$
		51	11		
			$\Sigma R_2 = 86$		

H_1: $H \neq 0$ The subpopulations of younger, middle-aged, and older widows have different mean ranked lengths of bereavement. Random subsamples will differ in such ranks beyond the expected influence of sampling error.

Step 2. The criteria for rejecting H_0.

The assumptions for a Kruskal–Wallis *H* test are met in this example. The degrees of freedom are $(k - 1) = (3 - 1) = 2$. The tabular value of *H* is estimated from the chi-square distribution with 2 degrees of freedom at the .05 level of significance. This value from Table D4 is 5.991. H_0 will be rejected if the test value of *H* is 5.991 or greater.

Step 3. The test statistic.

The initial step in calculating the test value of *H* is to convert the raw scores (days) from Table 11.11 into ranks. This is shown in Table 11.12. The ranks are assigned by ignoring which sample contains each score. The highest score from all the samples is given rank "1." The second highest overall is given rank "2," and so on. When there are ties, they are given the mean of the ranks that they occupy in the list, as we did in Chapter 5 with the Spearman's coefficient of correlation. For example, there are two raw scores of 60, one in the sample of younger widows and the second in the sample of older widows. These occupy the seventh and eighth positions among all scores. They are each given the mean of 7 and 8; $(7 + 8)/2 = 7.5$. The next raw score is given rank 9 since it is the ninth score in descending order. Notice that there is a three-way tie at the raw score of 45. These occupy the 12th, 13th, and 14th positions; $(12 + 13 + 14)/3 = 13$.

After ranking the scores across the samples, totals are found for the ranks within each sample separately. The six ranks for the younger widows sum to 104.5, for instance. The totals are substituted into the formula for the test statistic:

$$H = \frac{12}{N(N+1)}\left[\sum \frac{(\Sigma R_i)^2}{N_i}\right] - [3(N+1)]$$

$$= \frac{12}{24(24+1)}\left[\frac{(104.5)^2}{6} + \frac{(86)^2}{10} + \frac{(109.5)^2}{8}\right] - [3(24+1)]$$

$$= \frac{12}{24(25)}\left[\frac{10920.25}{6} + \frac{7396.00}{10} + \frac{11990.25}{8}\right] - [3(25)]$$

$$= \frac{12}{600}[1820.042 + 739.600 + 1498.781] - [75]$$

$$= (.020)(4058.423) - (75)$$

$$= 81.168 - 75$$

$$= 6.168$$

This formula is very sensitive to the effects of rounding error. Keep your calculations accurate to one-thousandths, at a minimum, as shown here.

Step 4. The decision about H_0.

The test statistic (6.168) exceeds the tabular statistic (5.991). The null hypothesis is rejected. There is a significant difference among the three groups; $H = 6.168$, $df = 2$, $p < .05$.

Step 5. The conclusion.

Rejecting the null hypothesis allows the gerontologist to make the inference that the three age groupings of widows differ in length of bereavement. There are two technical points to note about this so that you are not misled. First, we have tested for a difference in length of bereavement using ranks as our measure of time. It is not means, frequencies, or other forms of time that are being compared. Whenever the form of the data is changed, we cannot be sure that the same conclusion would be reached for all forms possible. For example, an analysis of variance *F* test using means for these same data might reach a different conclusion. Second, concluding that there is a significant *H* value does not tell us whether all possible pairs of samples differ in length of bereavement or only some of them. As we saw with the earlier *k*-sample tests, we only know that at least two of the samples differ significantly. Maybe more pairs of samples differ, maybe not. The answer would require a separate test appropriate for two-sample comparisons such as those we discussed in Chapter 10. There are no nonparametric tests for this task parallel to Tukey's *HSD* for the *F* test.

Friedman Two-Way Analysis of Variance

The final nonparametric test we are going to examine closely here is the Friedman two-way analysis of variance test. It is designed for comparing three or more related (dependent) samples on an ordinal variable. The related scores can be from matched samples or from one sample measured three or more times. The test is a reasonably powerful nonparametric alternative to the parametric analysis of variance *F* test for dependent samples, which is listed in the section *Related Statistics* at the end of this chapter.

Like the Kruskal–Wallis *H* test, this one is approximated by the chi-square distribution. The appropriate degrees of freedom are found by the formula of $(k - 1)$, where k is the number of conditions under which the measurements are taken. This test also converts raw scores to ranks, although the exact procedure differs from the Kruskal–Wallis method. Here the ranks are assigned across the rows (for each case), not across all scores (for the entire set of cases at once).

The test uses a new statistic labeled χ_r^2 and is known as "chi-*r*-square." The "*r*" stands for the use of ranks in the test. The test statistic is found by the following formula:

$$\chi_r^2 = \frac{12}{Nk(k+1)}[\Sigma(\Sigma R_i)^2] - [3N(k+1)]$$

Test Statistic for Friedman Two-Way ANOVA

where ΣR_i = sum of ranks in each measurement condition
k = number of conditions
N = total number of cases (not scores)

The calculated value of the test statistic is compared to a tabular value of χ_r^2 which is estimated by the tabular chi-square. This determines the conclusion for the test of H_0. The usual null hypothesis tested is that the scores do not differ under the various conditions.

An example will illustrate the use of this test. A social psychologist is studying the effects of law school education on feelings of self-esteem by students. The researcher is aware that graduate schools, such as a law school, are notorious (sometimes undeservingly) for creating highly stressful conditions. Assignments and exams are often characterized as being impossibly difficult. Many new students experience suddenly deflated self-esteem with slow recovery as they learn to adapt to the requirements. The researcher is not willing to endorse this precise pattern of increasing self-esteem over time. But she does believe that there may be a significant difference in self-esteem scores between the first, second, and third years of law school for those who receive their degrees.

A random sample of 10 law school students is selected and each is given a standardized measure of self-esteem at the midyear point during each of the three years they are in school. The results are shown in Table 11.13. The researcher would use the following procedure to test for a significant difference.

Step 1. The hypotheses.

H_0: $\chi_r^2 = 0$ — The population of law school students have self-esteem scores that are equal in rank for each of the three years of their studies. A random sample from this population may differ in such ranks within the expected limits of sampling error.

H_1: $\chi_r^2 \neq 0$ — The population of law school students have self-esteem scores that are not equal in rank for each of the three years of their studies. A random sample from this population will differ in such ranks beyond the expected influence of sampling error.

Step 2. The criteria for rejecting H_0.

The Friedman test makes the same assumptions as the Kruskal–Wallis test except that the scores represent dependent, not independent, measurements. These assumptions

TABLE 11.13 Self-Esteem Scores for Students in Law School

Student	First Year	Second Year	Third Year
Z	10	15	25
Y	5	12	10
X	14	18	22
W	20	15	21
V	7	14	24
U	17	10	8
T	9	13	25
S	4	19	17
R	6	14	26
Q	8	16	19

TABLE 11.14 Preliminary Calculations for Friedman Two-Way ANOVA Test

	First Year		Second Year		Third Year	
Student	X_1	R_1	X_2	R_2	X_3	R_3
Z	10	3	15	2	25	1
Y	5	3	12	1	10	2
X	14	3	18	2	22	1
W	20	2	15	3	21	1
V	7	3	14	2	24	1
U	17	1	10	2	8	3
T	9	3	13	2	25	1
S	4	3	19	1	17	2
R	6	3	14	2	26	1
Q	8	3	16	2	19	1
		$\Sigma R_1 = 27$		$\Sigma R_2 = 19$		$\Sigma R_3 = 14$

are met in this example. The tabular value of χ_r^2 is estimated from the chi-square distribution with $(k - 1)$ degrees of freedom. Here that *df* value is $(3 - 1) = 2$. If the researcher uses the .05 level of significance, that tabular statistic from Table D4 is approximately 5.991. The test statistic needs to equal or exceed this to cause rejection of H_0.

Step 3. The test statistic.

To find the test statistic, the raw scores from Table 11.13 must be transformed into ranks across the rows. This preliminary work is shown in Table 11.14. Notice that the ranks "1," "2," and "3" are assigned for each individual's set of three self-esteem scores. The highest rank ("1") is given to the highest score for each individual. Ranks are not assigned across the entire table as was done with the Kruskal–Wallis test. Our example did not have any ties. If it had, they would have been given mean ranks for the tied positions as it is done for other tests.

The ranks are summed down the columns for each condition (year). If the null hypothesis were confirmed exactly, these sums would be the same. We can see that they are not the same here, but that does not mean the null hypothesis is wrong. Remember, sampling error will cause some differences even though H_0 is correct.

The sums of the ranks and other information are entered in the test statistic formula as follows:

$$\chi_r^2 = \frac{12}{Nk(k+1)}[\Sigma(\Sigma R_i)^2] - [3N(k+1)]$$

$$= \frac{12}{10(3)(3+1)}[(27)^2 + (19)^2 + (14)^2] - [3(10)(3+1)]$$

$$= \frac{12}{10(3)(4)}[729 + 361 + 196] - [3(10)(4)]$$

$$= \frac{12}{120}(1286) - (120)$$

$$= .100(1286) - 120$$

$$= 128.600 - 120$$

$$= 8.6$$

Like the Kruskal–Wallis H test statistic formula, this one is also sensitive to rounding errors. Again, accuracy to one-thousandths at the minimum should be maintained.

Step 4. The decision about H_0.

The test statistic (8.6) exceeds the tabular statistic (5.991). The null hypothesis is rejected. There is a significant difference among the three sets of measurements; $\chi_r^2 = 8.6$, $df = 2$, $p < .05$.

Step 5. The conclusion.

Rejection of the null hypothesis means that the researcher can conclude that there is some significant difference in ranks of self-esteem scores between at least two of the years of law school. To determine how many pairs of these time periods differ significantly would require a two-sample test for dependent scores. But for now, the social psychologist can draw the inference that there is a significant difference in self-esteem felt by students in some of the three years of law school. In this general way, the researcher's idea has been supported.

Keys to Understanding

There are three general keys to understanding differences based on k-sample tests of difference. The first two of these are identical to two of the keys for two-sample tests. The third one is new.

Key 11.1. You must know the model behind each test in order to understand its use. The model tells you what assumptions are being made, what sampling distribution is being used, what hypothesis is being tested, and how the test statistic is to be calculated and interpreted. The models for the six major tests discussed in this chapter are presented for you in Inference Models 11.1 to 11.6. They make a handy summary for your study and use of the tests. Note that all the tests assume random samples containing independent observations. This remains a difficult assumption for social researchers to actualize. The subject matter often occurs in diverse forms and is studied in natural settings that are not subject to control. Researchers are forced to use sampling techniques that combine elements of availability with completely random selection. Continue to look for explanations of the samples used and shape your interpretations of the findings to match them.

(Text continues on page 427.)

Inference Model 11.1 Difference Between Two or More Independent Means—Analysis of Variance F Test

Purpose:

To determine whether the means from two or more (k) independent samples differ significantly from the proposed difference between two or more (k) population means.

Sampling Distribution:

F distributions.

Assumptions of the Test:

1. Independent and random observations.
2. Two or more independent samples.
3. Interval level of measurement in dependent variable.
4. Normally distributed populations.
5. Homogeneous population variances.

Typical Hypotheses:

Nondirectional Test

H_0: $\mu_1 = \mu_2 = \mu_3 = \cdots = \mu_k$

H_1: not $[\mu_1 = \mu_2 = \mu_3 = \cdots = \mu_k]$

Directional Test

(Not ordinarily used)

Tabular Statistic:

F at $dfb = k - 1$, $dfw = N_T - k$, and a predetermined level of significance from Table D7 (k = number of samples; N_T = total number of cases).

Test Statistic:

1. Sums of Squares (Computational Formulas):

$$SST = \Sigma X_T^2 - \frac{(\Sigma X_T)^2}{N_T}$$

$$SSB = \left[\Sigma \frac{(\Sigma X_i)^2}{N_i}\right] - \frac{(\Sigma X_T)^2}{N_T}$$

$$SSW = \Sigma\left[\Sigma X_i^2 - \frac{(\Sigma X_i)^2}{N_i}\right]$$

2. Mean Squares (Variances):

$$MSB = \frac{SSB}{dfb} \qquad MSW = \frac{SSW}{dfw}$$

3. Test Value of F:

$$F = \frac{MSB}{MSW}$$

Inference Model 11.2 Identification of Differing Pairs of Means from Sets of Three or More Independent Means—Tukey *HSD* Test

Purpose:

To identify those pairs of means that differ significantly among sets of three or more independent means following the F test.

Sampling Distribution:

HSD statistic distributions (also known as the Studentized range distributions).

Assumptions of the Test:

Same as ANOVA F test and equal-sized samples.

Typical Hypotheses:

Nondirectional Test

H_0: $\mu_1 = \mu_2 = \mu_3 = \cdots = \mu_k$

H_1: one or more pairs of means differ (e.g., $\mu_1 = \mu_2$ and/or $\mu_1 = \mu_3$ and/or etc.)

Directional Test

(Not ordinarily used)

Tabular Statistic:

HSD based on q_α at dfw, k, and a predetermined level of significance from Table D8 ($dfw = N_T - k$; k = number of samples).

Test Statistic:

$$HSD = q_\alpha \sqrt{\frac{MSW}{N_i}}$$

Inference Model 11.3 Difference Between Three or More Independent Distributions of Frequencies—Chi-Square Test of Independence for *k* Samples (χ^2-*k*)

Purpose:

To determine whether the frequency distributions for three or more independent samples differ significantly from those proposed for three or more populations.

Sampling Distribution:

Chi-square distributions.

Assumptions of the Test:

1. Independent and random observations.
2. Three or more independent samples.
3. Nominal level of measurement in dependent variable.
4. N must not be extremely small.

Typical Hypotheses:

Nondirectional Test	Directional Test
H_0: $\chi^2 = 0$	(Not ordinarily used)
H_1: $\chi^2 \neq 0$	

Tabular Statistic:

Chi-square at $df = (r - 1)(c - 1)$ and a predetermined level of significance from Table D4.

Test Statistic:

$$\chi^2 = \Sigma \frac{(f_o - f_e)^2}{f_e}$$

Inference Model 11.4 Difference Between Three or More Independent Medians—Median *k*-Sample Test (*Mdn-k*)

Purpose:

To determine whether the medians for three or more independent samples differ significantly from those proposed for three or more populations.

Sampling Distribution:

Chi-square distributions.

Assumptions of the Test:

1. Independent and random observations.
2. Three or more independent samples.
3. Ordinal level of measurement in dependent variable.
4. N must not be extremely small.

Typical Hypotheses:

Nondirectional Test	Directional Test
H_0: $Mdn_1 = Mdn_2 = Mdn_3 = \cdots = Mdn_k$	(Not ordinarily used)
H_1: not $[Mdn_1 = Mdn_2 = Mdn_3 = \cdots = Mdn_k]$	

Tabular Statistic:

Chi-square at $df = (r - 1)(c - 1)$ and a predetermined level of significance from Table D4.

Test Statistic:

$$\chi^2 = \Sigma \frac{(f_o - f_e)^2}{f_e}$$

Inference Model 11.5 Difference Between Three or More Independent Distributions of Ranks—Kruskal-Wallis *H* Test

Purpose:

To determine whether the distributions in ranks for three or more independent samples differ significantly from those proposed for three or more populations.

Sampling Distribution:

H statistic distributions as estimated by chi-square.

Assumptions of the Test:

1. Independent and random observations.
2. Three or more independent samples.
3. Ordinal level of measurement (expressed as ranks) in dependent variable.
4. N must not be extremely small.

Typical Hypotheses:

Nondirectional Test	Directional Test
H_0: $H = 0$	(Not ordinarily used)
H_1: $H \neq 0$	

Tabular Statistic:

H statistic as estimated by chi-square with $df = k - 1$ from Table D4.

Test Statistic:

$$H = \frac{12}{N(N+1)}\left[\sum \frac{(\Sigma R_i)^2}{N_i}\right] - [3(N+1)]$$

Inference Model 11.6 Difference Between Three or More Dependent Distributions of Ranks—Friedman Two-Way Analysis of Variance

Purpose:

To determine whether the distributions in ranks for three or more dependent samples differ significantly from those proposed for three or more populations.

Sampling Distribution:

χ_r^2 statistic distributions as estimated by chi-square.

Assumptions of the Test:

1. Independent and random observations within each sample.
2. Three or more dependent samples.
3. Ordinal level of measurement (expressed as ranks) in dependent variable.
4. N must not be extremely small.

Typical Hypotheses:

Nondirectional Test	Directional Test
H_0: $\chi_r^2 = 0$	(Not ordinarily used)
H_1: $\chi_r^2 \neq 0$	

Tabular Statistic:

χ_r^2 statistic as estimated by chi-square with $df = k - 1$ from Table D4.

Test Statistic:

$$\chi_r^2 = \frac{12}{Nk(k+1)}[\Sigma(\Sigma R_i)^2] - [3N(k+1)]$$

FIGURE 11.3 Common *k*-Sample Tests of Difference in Social Research

Minimum Level of Measurement in the Dependent Variable	Parametric Test: Independent Samples	Parametric Test: Dependent Samples	Nonparametric Test: Independent Samples	Nonparametric Test: Dependent Samples
Interval	ANOVA *F* Test for Independent Means Tukey *HSD* Test Newman–Keuls Test[a] Scheffé Test[a]	ANOVA *F* Test for Repeated Measures[a] ANOVA *F* Test for Matched Groups[a]		
Ordinal			Median-*k* Test Kruskal–Wallis *H* Test	Friedman Two-Way ANOVA
Nominal			Chi-Square-*k* Test	Cochran *Q* Test[a]

[a]Listed in the section *Related Statistics*.

Key 11.2. Determine that an appropriate inference test has been selected. Each inference test is used for a different situation. Sometimes the difference is major and of obvious importance. Sometimes the difference is minor and so technical that many readers cannot appreciate it. We have limited our discussion to major differences. A guide to the correct selection of a *k*-sample test is shown in Figure 11.3 for the more common tests used in social research. Like the guide shown in Chapter 10, this one is organized around three criteria: parametric versus nonparametric tests, independent versus dependent samples, and the level of measurement in the dependent variable.

Key 11.3. Remember that except for the post-ANOVA *F* procedures (e.g., Tukey *HSD*), all the *k*-sample tests presented are general tests of difference. They either indicate that no pairs of groups differ or that at least one pair differs. If any do differ, the tests do not tell you how many pairs differ or which pairs it is that differ. That determination requires the use of a subsequent test, usually one for comparing only two samples at a time. This influences the conclusion that legitimately can be drawn from the tests. A researcher who compares four groups with a *k*-sample test and finds significant differences is not justified in claiming that the four groups differ. Maybe only two of them do. Knowing this distinction will help you to understand better the real meaning of the results from a *k*-sample test.

KEY TERMS

Analysis of Variance
One-Way ANOVA
F Ratio
Degrees of Freedom Between (*dfb*)
Degrees of Freedom Within (*dfw*)
Degrees of Freedom Total (*dft*)
Sum of Squares
Sum of Squares Total (*SST*)
Sum of Squares Between (*SSB*)
Sum of Squares Within (*SSW*)

Total Mean ($\bar{X}_T$)
Mean Squares
Mean Square Between (MSB)
Mean Square Within (MSW)
Tukey *HSD* Test
Chi-Square-*k* Test (χ^2-k)
Median-*k* Test (Mdn-k)
Kruskal–Wallis *H* Test
Friedman Two-Way Analysis of Variance (χ_r^2)

EXERCISES

1. Four random samples, each of six equal-sized neighborhoods, record the following number of psychiatric hospital admissions during the four seasons of one year.

Spring	Summer	Fall	Winter
2	1	2	3
1	2	3	5
3	0	5	6
4	2	2	4
2	0	3	5
0	1	3	7

Use an analysis of variance *F* test to determine whether there is a significant seasonal difference in the mean number of admissions at the .05 level. (a) State the hypotheses. (b) Establish the criteria for rejecting H_0. (c) Calculate the test statistic. (d) Make a decision about H_0. (e) State a conclusion.

2. Apply a Tukey *HSD* test to the data in Exercise 1 at the .05 level of significance. (a) Create a matrix of the pairs of mean differences. (b) Calculate *HSD*. (c) Identify the pairs of significant differences.

3. Apply a median *k*-sample test to the data in Exercise 1 at the .05 level of significance. Follow the usual five steps as listed in Exercise 1, (a)–(e).

4. Apply a Kruskal–Wallis *H* test to the data in Exercise 1 at the .05 level of significance. Follow the usual five steps as listed in Exercise 1, (a)–(e).

5. Compare your findings for the analyses from Exercises 1 to 4.

6. A foreign country has three regions: west, central, and east. The following shows the number of marriages in one year in three random samples, each composed of eight native villages selected from a region.

West	Central	East
0	5	2
1	6	4
4	10	5
2	3	4
1	7	3
2	12	8
1	11	3
0	4	2

Use an analysis of variance *F* test to determine whether there is a significant difference in the mean number of regional marriages at the .05 level. Follow the usual five steps as listed in Exercise 1, (a)–(e).

7. Apply a Tukey *HSD* test to the data in Exercise 6 at the .05 level. Follow the three steps listed in Exercise 2, (a)–(c).
8. Apply a median *k*-sample test to the data in Exercise 6 at the .05 level. Follow the usual five steps listed in Exercise 1, (a)–(e).
9. Apply a Kruskal–Wallis *H* test to the data in Exercise 6 at the .05 level. Follow the usual five steps as listed in Exercise 1, (a)–(e).
10. Compare your findings for the analyses from Exercises 6 to 9.
11. Suppose that an analysis of variance *F* test has compared three groups of elementary school children in reading ability. Each group contained 15 children. The three sums of squares were calculated to be 100 (total), 24 (between) and 76 (within). Construct a complete ANOVA summary table based on this information.
12. Using the information in Exercise 11, what is the tabular value of *F* at the (a) .05 level of significance? (b) .01 level of significance? Under which of these two conditions would the test value of *F* from Exercise 11 be considered significant?
13. The following ANOVA summary table is reported for a comparison of cities in mean unemployment levels.

Source	*SS*	*df*	*MS*	*F*
Between	60	3	20	4.37
Within	440	100	4.58	
Total	500	103		

(a) How many groups were being compared? (b) How many cases were being compared? (c) What is the variation for all cases considered together? (d) Is there a significant difference in means at the .05 level? (e) Is there a significant difference in means at the .01 level?

14. Suppose that samples of whites, blacks, Chicanos, and others have been compared in their support for a presidential candidate with the following frequency results:

	Sample			
Support	Whites	Blacks	Chicanos	Others
Yes	6	4	20	5
Uncertain	14	25	6	0
No	20	1	4	0

Apply a chi-square-*k* test to these data. Follow the usual five steps as listed in Exercise 1, (a)–(e). Use the .05 level of significance.

15. Samples of employees are compared in job satisfaction according to the type of authority the employers hold over them with the following frequency results:

	Basis of Authority		
Job Satisfaction	Remunerative	Normative	Coercive
High	44	33	11
Middle	21	35	14
Low	15	32	25

Apply a chi-square-*k* test to these data. Follow the usual five steps as listed in Exercise 1, (a)–(e). Use the .05 level of significance.

16. Student opinions of public financing for colleges are collected before, during, and after graduation with the following results. (Higher scores indicate a more favorable opinion.)

Student	Before	During	After
2349	4	8	2
1067	5	10	4
4002	6	8	1
2139	3	9	6
0991	6	7	5
2920	7	10	2
3487	5	7	1
1119	3	6	4
2255	2	5	3
3066	5	10	8

Apply a Friedman two-way ANOVA test to these data. Follow the usual five steps as listed in Exercise 1, (a)–(e). Use the .05 level of significance.

17. In an experiment nine subjects were asked to read a series of four comparable vignettes (each describing a person in trouble) under four conditions. Condition A had soothing music playing in the background, condition B had a heated argument occurring in the next room, condition C had a series of interruptions by a "lost subject," and condition D had no special stimuli (control). The subject was asked how the person in each vignette should be treated by an authority figure. These responses were scored on a 1 (low) to 5 (high) scale of helpfulness with the following results:

	Condition			
Subject	A	B	C	D
1	5	1	2	3
2	4	1	4	3
3	4	1	3	3
4	5	1	2	4
5	3	1	4	2
6	5	2	4	3
7	4	1	2	4
8	4	1	2	2
9	5	1	3	3

Apply a Friedman two-way ANOVA test to these data. Follow the usual five steps as listed in Exercise 1, (a)–(e). Use the .01 level of significance.

18. Suppose that a one-way ANOVA F test has rejected H_0: $\mu_1 = \mu_2 = \mu_3$ for samples of Republicans, Democrats, and Independents in mean education at the .05 level. (a) Does this mean that Republicans and Democrats differed significantly in mean education? (b) Does this mean that Democrats and Independents differed significantly in mean education? (c) Does this mean that Republicans and Independents differed significantly in mean education? (d) Does this mean that at least one of the comparisons listed in parts (a), (b), and (c) must be a significant difference?

19. Suppose that a Kruskal–Wallis H test has failed to reject H_0: $H = 0$ for the same three samples named in Exercise 18 at the .05 level. (a) Does this mean that the three samples probably

represent identical populations in education levels? (b) Does this mean that any of the three groups is significantly different from the others? (c) Does this mean that H_0 would have been rejected at the .01 level?

20. Search the social science literature for an example of a one-way analysis of variance. Summarize the researcher's goals, describe how analysis of variance was applied, and identify and defend a nonparametric alternative that could have been applied in the same situation.

RELATED STATISTICS

Newman–Keuls Procedure

The Newman–Keuls procedure is a slightly more complex application of the Tukey *HSD* test for determining significant differences between pairs of means following an ANOVA *F* test. It requires the means to be arranged from smallest to largest in a matrix. The differences between means are compared to a different critical (tabular) value of *HSD* for each row in that matrix depending on the number of means represented.

See: Champion (1981:201–203).

Scheffé Test

The Scheffé test is another common alternative to the Tukey *HSD* test. It uses its own tabular and test statistic based on the prior *F* test. It is a more conservative test than either the Tukey *HSD* or the Newman–Keuls procedure, yielding fewer pairs of significant differences. However, it is more versatile, being suitable to unequal sample sizes and tolerant of violations to the assumptions of the *F* test.

See: Champion (1981:203–206) and Hays (1973:606–609).

Cochran *Q* test

The Cochran *Q* test is a nonparametric test for comparing two or more dependent samples on a nominal variable. The nominal variable must be a dichotomy, but the samples can be either matched or the same subjects measured repeatedly. It uses the chi-square sampling distribution.

See: Champion (1981:246–251).

Elaborations on the One-Way Analysis of Variance *F* Test

Random versus fixed factors: The independent variable in ANOVA is called a "factor." When it is a discrete variable (gender, religious affiliation, political party), it is called a "fixed factor." When it is a continuous variable (age, education, degree of opinion), it is called a "random factor."

Levels of factors: The level of a factor refers to the number of categories in the independent variable. A "four-level, one-way analysis of variance" means that the groups being compared are based on one independent variable which has been divided into four categories, each identifying a group.

Dependent samples: Special versions of the ANOVA *F* test are required for comparing dependent (not independent) samples. One version is used with matched groups and another with repeated observations of the same subjects.

See: Hays (1973:564–574).

Analysis of covariance: Analysis of covariance is a technique used to "correct" for any pretreatment variations on the dependent variable. This ensures that the posttreatment differences observed in the research are not due to any differences that predated the study, and therefore could not be due to the independent variable (treatment).

See: Blalock (1979:510ff) and Hays (1973:654–659).

Higher-order analyses of variance: Often researchers compare groups that differ on more than one independent variable (factor), each with at least two levels. If two independent variables are used, the analysis is called two-way analysis of variance. If three independent variables are used, it is three-way analysis of variance, and so on. The number of factors and their levels are identified in the "design label." For instance, a 3 × 3 (three-by-three) design, uses two factors which each have three levels. A 4 × 2 × 2 design has three factors; the first has four levels, the second and third each have two levels. Higher-order analyses of variance test for differences due to each factor alone ("main effects") and in combination ("interaction").

See: Blalock (1979:352–367), Minium (1978:404–413), and Hays (1973:457ff).

CHAPTER 12

INFERENCES OF ASSOCIATION

In this final chapter we are going to examine inferences of association. In the three preceding chapters we focused on inferences of differences: differences between one sample and one population (Chapter 9), differences between two samples and two populations (Chapter 10), and differences between multiple samples and multiple populations (Chapter 11). There is a close link between the processes of reaching inferences for differences and those for inferences of associations. The basic steps, the sampling distributions, and the logic of decision making are all the same. Further, we already know how to calculate most of the test statistics. They are often the same as the descriptive measures of association in Chapter 5 for nominal and ordinal data and in Chapter 6 for interval and ratio data. We will cover them in the same order as then.

Association Inferences at the Nominal Level

Chi-Square-Based Measures

In Chapter 5 we discussed two types of association measures for nominal variables. The first of these were the chi-square-based measures of phi, *C*, and *V*. There are no appropriate significance tests for these per se. They measure the extent of a nominal

TABLE 12.1 Ethnic Background and Interest in Sports

Interest in Sports	Ethnic Background: European	Asian	African	Other	
Yes	20 (18.0)	8 (11.2)	15 (11.2)	2 (4.5)	45
No	20 (22.0)	17 (13.8)	10 (13.8)	8 (5.5)	55
	40	25	25	10	100

$$\chi^2 = \frac{(20-18)^2}{18} + \frac{(8-11.2)^2}{11.2} + \frac{(15-11.2)^2}{11.2} + \frac{(2-4.5)^2}{4.5} + \frac{(20-22)^2}{22} + \frac{(17-13.8)^2}{13.8} + \frac{(10-13.8)^2}{13.8} + \frac{(8-5.5)^2}{5.5}$$

$$= \frac{(2)^2}{18} + \frac{(-3.2)^2}{11.2} + \frac{(3.8)^2}{11.2} + \frac{(-2.5)^2}{4.5} + \frac{(-2)^2}{22} + \frac{(3.2)^2}{13.8} + \frac{(-3.8)^2}{13.8} + \frac{(2.5)^2}{5.5}$$

$$= \frac{4}{18} + \frac{10.24}{11.2} + \frac{14.44}{11.2} + \frac{6.25}{4.5} + \frac{4}{22} + \frac{10.24}{13.8} + \frac{14.44}{13.8} + \frac{6.25}{5.5}$$

$$= .222 + .914 + 1.289 + 1.389 + .182 + .742 + 1.046 + 1.136$$

$$= 6.92$$

association in a descriptive sense and are not part of any sampling distribution. Instead, chi-square itself is the appropriate test of significance used to determine what inference can be drawn to a population. Thus a researcher uses the chi-square test of independence to test the null hypothesis of no association ($\chi^2 = 0$). You have already seen several illustrations of this in earlier chapters. If the decision is to reject that null hypothesis, the researcher concludes that there is an association in the population. Its strength is described in terms of the calculated phi, C, or V.

Consider the following example. A researcher wants to make an inference about the possible association between ethnic background and interest in sports. The researcher draws a random sample of 100 persons from a large metropolitan area and determines the ethnic background and interest in sports for each individual as shown in Table 12.1.

Step 1. The hypotheses.

H_0: $\chi^2 = 0$ — There is no association between ethnic background and interest in sports in the population. A random sample may show some association, but it will be within the expected limits of sampling error.

H_1: $\chi^2 \neq 0$ — There is an association between ethnic background and interest in sports in the population. A random sample will reflect this association beyond the expected limits of sampling error.

Notice that the only difference between these hypotheses for association and those we have used before for differences is one of conceptualization. Before we would have thought of the categories of ethnic background as representing separate samples (or subsamples) and we would have determined whether there was a significant difference between them in interest in sports. Now we think of the ethnic types as the measurement categories of a nominal variable called ethnic background and we want to know if there is a relationship between this variable and the interest-in-sports variable. The two conceptualizations are identical for the purposes of the statistical analysis. Different versions of chi-square tests are not required.

Step 2. The criteria for rejecting H_0.

The degrees of freedom are found as usual: $df = (r - 1)(c - 1) = (2 - 1)(4 - 1) = 3$. If the researcher uses the .05 level of significance, the tabular statistic found from Table D4 is 7.815. The test statistic must equal or exceed this value to allow the researcher to reject the null hypothesis.

Step 3. The test statistic.

The test statistic is calculated as before. The actual computations are shown below Table 12.1. There you see that the value of chi-square from the sample data is 6.92.

Step 4. The decision about H_0.

The test statistic (6.92) does not exceed the tabular statistic (7.815), so the null hypothesis cannot be rejected. The data show an association that falls within the expected limits of sampling error; $\chi^2 = 6.92$, $df = 3$, $p > .05$, n.s.

Step 5. The conclusion.

The inference to be drawn from this analysis is that there is no significant association between ethnic background and interest in sports. The population (the metropolitan area) from which the sample was drawn appears to be one in which interest in sports is not associated with ethnicity. Since the researcher draws the inference that an association does not exist in the population, there is no reason to calculate a measure of association strength. Although Cramér's V would be appropriate for a 2×4 table like the one in Table 12.1, there is no point in calculating it for inference purposes. Of course, a researcher could calculate V for descriptive reasons. He or she might be interested in the strength of the association that does exist in the sample. However, that is not a concern of inferential statistics, which we are now discussing.

Undaunted by this failure to reject the last null hypothesis, our researcher presses on to examine another possible association. Suppose the researcher also asked the sex of each respondent in the survey on interest in sports. Now an inference about the relationship between sex and interest in sports can be pursued. Table 12.2 shows the data.

TABLE 12.2 Sex and Interest in Sports

Interest in Sports	Sex: Male	Sex: Female	
Yes	35 (22.5)	10 (22.5)	45
No	15 (27.5)	40 (27.5)	55
	50	50	100

$$\chi^2 = \frac{(35 - 22.5)^2}{22.5} + \frac{(10 - 22.5)^2}{22.5} + \frac{(15 - 27.5)^2}{27.5} + \frac{(40 - 27.5)^2}{27.5}$$

$$= \frac{(12.5)^2}{22.5} + \frac{(-12.5)^2}{22.5} + \frac{(-12.5)^2}{27.5} + \frac{(12.5)^2}{27.5}$$

$$= \frac{156.25}{22.5} + \frac{156.25}{22.5} + \frac{156.25}{27.5} + \frac{156.25}{27.5}$$

$$= 6.944 + 6.944 + 5.682 + 5.682$$

$$= 25.252$$

Step 1. The hypotheses.

H_0: $\chi^2 = 0$ There is no association between sex and interest in sports in the population. A random sample may show some association, but it will be within the expected limits of sampling error.

H_1: $\chi^2 \neq 0$ There is an association between sex and interest in sports in the population. A random sample will reflect this association beyond the expected limits of sampling error.

Step 2. The criteria for rejecting H_0.

The *df* value is 1 [(2 − 1)(2 − 1)]. Suppose that the level of significance is again .05. The tabular value of chi-square from Table D4 is 3.841. This value must be equaled or exceeded by the test statistic in order for the null hypothesis to be rejected.

Step 3. The test statistic.

The calculation of the test value of chi-square is shown below Table 12.2. There you see it was found to be 25.252.

Step 4. The decision about H_0.

The test statistic (25.252) does exceed the tabular statistic (3.841). Therefore, the null hypothesis can be rejected; $\chi^2 = 25.252$, $df = 1$, $p < .05$.

Step 5. The conclusion.

The chi-square test allows the researcher to make the inference that there is an association between sex and interest in sports in the metropolitan area represented by this sample. Having decided that there is such an association, the researcher now turns to describing its strength. Since the data are in a 2 × 2 table, the appropriate chi-square-based measure of strength is phi. It is calculated to be

$$\phi = \sqrt{\frac{\chi^2}{N}} = \sqrt{\frac{25.252}{100}} = \sqrt{.25252} = .5025$$

The researcher now rounds out the inference by saying that there is a significant association between sex and interest in sports which the data suggest is of moderate strength. Additionally, phi from 2 × 2 tables can be given a PRE interpretation by squaring it. So the researcher can add to the generalization the idea that sex explains approximately 25% of the differences in interest in sports in this example ($.5025^2 = .2525$.)

The procedure is the same with all the chi-square-based measures of association. Making an inference first requires the use of the significance test for chi-square. If that results in the conclusion that there is an association in the population, the researcher calculates one of the measures of strength and uses it to estimate the extent of that association in the population. If the chi-square significance test results in the conclusion that there is no association in the population, the researcher's inference is finished.

Earlier, R.E. 5.1 illustrated the use of the chi-square and phi measures. In Chapter 5 we considered these only in their descriptive forms. Now we can add the relevant information for an inference test of association. The example concerned 567 unwed mothers and racial stereotypes. Table 1 of that example is reprinted here as R.E. 12.1. It shows the calculated chi-square to be 22.8 and phi to be .20 for a 2 × 2 distribution of place of father's residence with race. That test chi-square is noted as being significant beyond the .001 level. With $df = 1$, the tabular chi-square would be 10.827 at the .001 level.* Since the

*This tabular value is not shown in Table D4.

RESEARCH EXAMPLE 12.1 Inference at the Nominal Level Using Chi-Square

Comparison of the Place Where the Father and Unwed Mother Lived When She Met Him, by Race

Race	Place Where Father Lived: Same as Unwed Mother	Different from Unwed Mother	Total
White	212	101	313
Negro	216	38	254

$\chi^2 = 22.8$; $df = 1$; $p < .001$; $\phi = .20$

Source: Hallowell Pope, "Unwed Mothers and Their Sex Partners," *Journal of Marriage and the Family,* 29 (3), August 1967:555ff. Adapted from Table 1.

calculated chi-square exceeds this tabular value, a null hypothesis of no association would be rejected. The phi value of .20 suggests that this significant association is weak. Squaring phi, we find that $(.20)^2 = .04$. Only 4% of the differences in father's place of residence can be accounted for by racial differences. Thus, although the inference can be drawn that there was an association in the population, it was an extremely weak one in a predictive sense.

This is a common outcome for a chi-square test. It tends to lead to the conclusion that even very weak associations are significant with large samples. For example, suppose that we cut the observed frequencies in half for the data in R.E. 12.1. This will preserve the same proportional distributions for each variable but simulate their occurrence in a sample that is only half as large as the actual one ($N = 283.5$ versus 567). The test value of chi-square for such a table is 11.4, also exactly half the original. (You can verify the calculations yourself.) However, the tabular value of chi-square is based on the size of the table, which means that it is still 10.827 at the .001 level of significance. The conclusion would be the same (reject the null hypothesis of no association); however, it was a much closer call than before. If we would again cut the frequencies in half, the resulting test value of chi-square would also be halved again, now equal to 5.7. But the tabular chi-square would remain 10.827. In this last circumstance we would fail to reject a null hypothesis of no association even though the proportional distributions remained constant and we were still working with a reasonable sample size ($N = 141.75$).

This illustrates the relatively extreme sensitivity of chi-square to large sample sizes compared to other inference tests. It suggests how easy it is to misunderstand the appropriate inference that chi-square produces. You should not infer strength of an association on the basis of the margin by which a test chi-square exceeds a tabular chi-square. Chi-square tests only for the existence (weak or strong) of an association. The simplest solution to this problem is to always calculate a measure of association strength when the chi-square test has suggested that there is an association in the population. That strength measure can then be used to judge whether the association has any real theoretical or practical importance. This is exactly what the author of R.E. 12.1 did and it is the reason he concluded that there was no important evidence to support a racial stereotype concerning unwed mothers and the place of residence for their sex partners, even though the test chi-square was found to be significant.

Lambda and Tau-y

The second type of association measure for nominal data presented in Chapter 5 was the PRE statistic. The two examples of this that we discussed were lambda and tau-y.* Both provide a description of association strength between two nominal variables which is based directly on a proportional reduction in error technique. They are independent of chi-square. For these two reasons they are preferred over the chi-square-based measures. However, neither is adequately described by any known sampling distribution. Therefore, they cannot be used to draw inferences about association existence from sample data to a population. Instead, they represent normed measures of association strength for descriptive purposes.

Association Inferences at the Ordinal Level

Chapter 5 presented two descriptive measures of association for ordinal variables. These were gamma and Spearman's coefficient of correlation. They are often used to draw inferences about association since they have known sampling distributions. Let's take them in turn.

Gamma

Recall that gamma is a symmetrical PRE measure of association for ordinal variables. It is based on the numbers of agreement and disagreement pairs in a bivariate table. Beyond having ordinal level of measurement, the requirements for an inference test with gamma are that the sample data be independent and randomly selected and that the sample size not be too small. The exact minimum is debated, ranging from 10 to 200, and depends on the existence of tied pairs in the sample. Here we will use $N \geq 50$ as a minimum for assuming a normal (z) sampling distribution. A z-transformation formula is used to make the comparison between the value of gamma found from the sample data and a tabular value of z.

Suppose that a teacher wants to make an inference about ordinal association between education and optimism using gamma. The teacher selects 60 adults at random, determines whether or not they have had any college education, and measures their general level of optimism according to three ordered categories. The data are in Table 12.3.

Step 1. The hypotheses.†

H_0: $\gamma = 0$	The population of adults has no association between education and optimism. A random sample may show some association within the expected limits of sampling error.
H_1: $\gamma \neq 0$	The population of adults has an association between education and optimism. A random sample will reflect this association beyond the expected limits of sampling error.

Step 2. The criteria for rejecting H_0.

If we use the .05 level of significance for this nondirectional test, the tabular gamma is equivalent to a z score of ± 1.96. The test statistic must be beyond the tabular value (+ or −) to warrant rejection of the null hypothesis.

*Tau-y was presented in an optional section (p. 164).

†Note that the hypotheses are expressed using the population symbol for gamma, γ.

TABLE 12.3 Education by Optimism

	Education		
Optimism	Some College or Higher	No College	
High	20	6	26
Middle	10	16	26
Low	4	4	8
	34	26	60

$$P_a = 20(16 + 4) = 20(20) = 400$$
$$10(4) = \underline{40}$$
$$440$$

$$P_d = 6(10 + 4) = 6(14) = 84$$
$$16(4) = \underline{64}$$
$$148$$

$$G = \frac{P_a - P_d}{P_a + P_d} = \frac{440 - 148}{440 + 148} = \frac{292}{558} = .497$$

Step 3. The test statistic.

The calculation of the test value of gamma is shown below Table 12.3. There you see that the numbers of agreement and disagreement pairs in the table are found and entered in the gamma formula. Gamma is found to be .497. But this value cannot be compared directly to the test statistic. It must first be transformed to a z score. That is done by the following formula:

$$z = G\sqrt{\frac{P_a + P_d}{N(1 - G^2)}}$$ z Transformation Formula for Gamma

where P_a = number of pairs of agreement
P_d = number of pairs of disagreement
G = gamma calculated from sample data
N = number of cases

Using this formula, the teacher finds that

$$z = .497\sqrt{\frac{440 + 148}{60[1 - (.497)^2]}} = .497\sqrt{\frac{588}{60(1 - .2470)}}$$
$$= .497\sqrt{\frac{588}{60(.7530)}} = .497\sqrt{\frac{588}{45.18}} = .497\sqrt{13.0146}$$
$$= .497(3.6076) = 1.79$$

Step 4. The decision about H_0.

The test statistic (1.79) does not exceed the tabular statistic (1.96), so the null hypothesis cannot be rejected, $G = .497$, $p > .05$, n.s.

Step 5. The conclusion.

The teacher concludes that there is no significant association between education and optimism as they have been measured here. The association in the sample data could be merely the consequence of sampling error.

Consider another example using gamma. A researcher wants to draw an inference about the possible relationship between income level and work satisfaction. The researcher suspects that there is a positive association. A random sample of 200 workers is selected and their incomes and degrees of work satisfaction are determined. The sample data are shown in Table 12.4.

Step 1. The hypotheses.

H_0: $\gamma \leq 0$ The population of workers either has no association between income level and work satisfaction or the association is negative. A random sample may show a positive association, but it would be within the expected limits of sampling error.

H_1: $\gamma > 0$ The population of workers has a positive association between income level and work satisfaction. A random sample will reflect this positive association beyond the limits of expected sampling error.

Step 2. The criteria for rejecting H_0.

Notice that the researcher is using a directional test. A normal approximation of the sampling distribution is appropriate since there are 50 or more cases. If the level of significance is .05, the tabular value of z is $+1.645$. This value leaves 5% of the area of the theoretical distribution in one positive tail creating the rejection region.

TABLE 12.4 Income by Work Satisfaction

Work Satisfaction	**Income** High	Moderate	Low	
High	25	15	10	50
Moderate	40	60	20	120
Low	5	10	15	30
	70	85	45	200

$$
\begin{aligned}
P_a &= 25(60 + 20 + 10 + 15) = 25(105) = 2625 \\
&\quad 15(20 + 15) = 15(35) = 525 \\
&\quad 40(10 + 15) = 40(25) = 1000 \\
&\quad 60(15) = 900 \\
&\quad \overline{5050}
\end{aligned}
$$

$$
\begin{aligned}
P_d &= 10(60 + 40 + 10 + 5) = 10(115) = 1150 \\
&\quad 15(40 + 5) = 15(45) = 675 \\
&\quad 20(10 + 5) = 20(15) = 300 \\
&\quad 60(5) = 300 \\
&\quad \overline{2425}
\end{aligned}
$$

$$G = \frac{P_a - P_d}{P_a + P_d} = \frac{5050 - 2425}{5050 + 2425} = \frac{2625}{7475} = .351$$

Step 3. The test statistic.

The calculation of the test value of gamma is shown below the data in Table 12.4. There you see that gamma was .351. This is transformed to a z score as before:

$$z = .351\sqrt{\frac{5050 + 2425}{200[1 - (.351)^2]}} = .351\sqrt{\frac{7475}{200(1 - .1232)}}$$

$$= .351\sqrt{\frac{7475}{200(.8768)}} = .351\sqrt{\frac{7475}{175.36}} = .351\sqrt{42.6266}$$

$$= .351(6.5289) = 2.29$$

Step 4. The decision about H_0.

The test statistic (2.29) does exceed the tabular statistic (1.645). Therefore, the null hypothesis is rejected; $G = .351$, $p < .05$.

Step 5. The conclusion.

Since the researcher rejects the null hypothesis, the conclusion is drawn that there is an association between income level and work satisfaction in the population. The sample data suggest that this association is positive in direction. Those with higher income levels tend to have somewhat higher work satisfaction.

Spearman's r_s

The Spearman coefficient of correlation (r_s) is a symmetrical PRE measure of association for ordinal variables that have been ranked. It is based on a comparison of the pairs of ranks on two variables for each case. Other than the level of measurement assumption, a significance test for r_s requires a random sample of independent cases. There should be few tied ranks if r_s is calculated using the formula presented in Chapter 5. The sampling distributions for r_s can be evaluated by the t distributions for small samples and by the z distribution for large samples. However, exact tabular values of r_s have been calculated for use in the significance test with small samples ($N \leq 30$). (See Table D9 in Appendix D.) For larger samples, a z-transformation formula is used to convert the test value of r_s to a z score which allows a comparison to be made with a tabular value of z.

Suppose that a political scientist wants to test for an association between age and support for the concept of a woman becoming U.S. president using r_s. The scientist expects there to be a negative association. She draws a random sample of 12 adults, and determines their ages and levels of support for a female president. The age and support variables are converted to ranks, with the oldest age and highest support each given rank = 1. The resulting data are shown in Table 12.5.

Step 1. The hypotheses.*

H_0: $\rho_s \geq 0$ — The population of adults either has no association between ranks in age and ranks in support for a female U.S. president or the association is positive. A random sample may show a negative association but only within the expected limits of sampling error.

H_1: $\rho_s < 0$ — The population of adults has a negative association between ranks in age and ranks in support for a female U.S. president. A random sample will reflect this negative association beyond the expected limits of sampling error.

*Note that the hypotheses are expressed using the population symbol for the Spearman coefficient, ρ_s.

TABLE 12.5 Age and Support for Female President in Ranks

Subject	Rank in Age	Rank in Support for Female President	D	D^2
A	5	5	0	0
B	7	8	−1	1
C	3	9	−6	36
D	9	4	5	25
E	2	10	−8	64
F	8	6	2	4
G	1	12	−11	121
H	12	3	9	81
I	4	11	−7	49
J	6	7	−1	1
K	10	2	8	64
L	11	1	10	100
				546

$$r_s = 1 - \left[\frac{6D^2}{N(N^2 - 1)}\right] = 1 - \left[\frac{6(546)}{12[(12)^2 - 1]}\right]$$

$$= 1 - \left[\frac{3276}{12(144 - 1)}\right] = 1 - \left[\frac{3276}{12(143)}\right] = 1 - \left[\frac{3276}{1716}\right]$$

$$= 1 - 1.909 = -.909$$

Step 2. The criteria for rejecting H_0.

Since the sample size is small, Table D9 can be used to find the tabular statistic. If we use the .01 level of significance for this directional test, the tabular value of r_s is −.712. The sample must have a negative correlation of at least −.712 in order for the null hypothesis to be rejected. Notice that Table D9 is entered according to the sample size (N) rather than a degrees of freedom value.

Step 3. The test statistic.

The calculation of the test value of r_s is shown below Table 12.5. There you see that the sum of squared differences in ranks is 546. This leads to finding a test r_s of −.909.

Step 4. The decision about H_0.

The absolute value of the test statistic (.909) exceeds the tabular statistic (.712). Therefore, the null hypothesis is rejected; $r_s = -.909$, $p < .01$.

Step 5. The conclusion.

The political scientist concludes that there is a significant association between ranks in age and ranks in support for a female president. The inference can be made that this association is negative in the population represented by the sample. Those with higher ranks in age (older) tend to have lower ranks in support (less support).

Consider another example. Suppose that an office supervisor wants to test for an association between education and productivity. The supervisor draws a random sample of 62 office workers and determines their ages and levels of productivity. The two variables are converted to ranks and the Spearman coefficient is found to be .22. Does this mean that there is a significant association between the ranks for the two variables at the .05 level?

Step 1. The hypotheses.

H_0: $\rho_s = 0$	The population of office workers has no association between ranks in age and ranks in productivity. A random sample may show some association but it will be within the expected limits of sampling error.
H_1: $\rho_s \neq 0$	The population of office workers has an association between ranks in age and ranks in productivity. A random sample will reflect this association beyond the expected limits of sampling error.

Step 2. The criteria for rejecting H_0.

Notice that this is a nondirectional test. Since the sample size is greater than 30 cases, a value from the z distribution is used as the tabular statistic. The tabular value of z at the .05 level of significance for a nondirectional test is ± 1.960. The test statistic must equal or fall beyond either of these tabular values (+ or −) in order for H_0 to be rejected.

Step 3. The test statistic.

The test value of r_s has already been found to be .22. This is transformed into a z score by the following formula:

$$z = \frac{r_s}{1/\sqrt{N-1}} \quad \text{or} \quad r_s\sqrt{N-1}$$

z Transformation Formula for r_s

where r_s = value of r_s calculated from sample data
N = number of cases

Applied to this example, we find that

$$z = .22\sqrt{62-1} = .22\sqrt{61} = .22(7.8102)$$
$$= 1.718$$

Step 4. The decision about H_0.

The test statistic (1.718) does not exceed the tabular statistic (1.960). Therefore, the null hypothesis cannot be rejected. The test statistic is within the limits of sampling error at the .05 level; $r_s = .22$, $p > .05$, n.s.

Step 5. The conclusion.

Having failed to reject the null hypothesis, the supervisor concludes that there is no significant association between the ranks in age and ranks in productivity. Although there was a weak and positive association in the sample of 62 workers, it was not sufficient to warrant the conclusion that any real association exists in the population of workers from which the sample was drawn. This conclusion is reached with the acknowledged possibility that it represents a Type II error.

Association Inferences at the Interval Level

Chapter 6 presented several interrelated measures of association at the interval and ratio levels. Of these, only one is reintroduced here for a discussion of inferences. This is the Pearson correlation coefficient (r) for zero-order association. Inferences concerning some of the others are mentioned in the section *Related Statistics* at the end of this chapter. Mostly, they are too complex for general consideration in this text, although you may have reason to refer to them in the future.

Pearson's r

The Pearson r statistic measures association between two interval variables. An inference from a calculated r requires a parametric test. It makes the usual assumptions of a random sample containing independent observations. Beyond these it requires that both of the variables are jointly and normally distributed in the population being sampled. Such a population distribution is called *bivariate normal*. Further, the test assumes that the variances for all X–Y combinations are homogeneous across the range of values being considered. This is known as *homoscedasticity*. Figure 12.1 illustrates this and one example of its opposite, *heteroscedasticity*. Finally, the test considers only a linear pattern of relationship.

Like other statistics, sample values of r are estimates of their population parameter, ρ (Greek lowercase rho). Because of sampling error, observed values of r will be distributed about the value of ρ in a predictable pattern, with a mean equal to ρ and a standard error equal to $(1 - \rho^2)/\sqrt{(N - 1)}$. The shape of the r sampling distributions depend on the value of ρ. When $\rho = 0$, the sampling distribution of r is unimodal, symmetrical, and nearly normal. When $\rho \neq 0$, the sampling distributions are skewed. The more ρ differs from 0, the more skewed are the sampling distributions. Figure 12.2 illustrates this when $\rho = -.70$, 0, and $+.70$. The sampling distribution when $\rho = -.70$ shows how sample values are clustered around the assumed value of the parameter ($-.70$) but can take on a range of values from a low of -1.0 (the lower limit of any Pearson correlation coefficient) to a very few as high as $+1.0$ (the upper limit of r). It is thus severely and positively skewed. A similar pattern is shown when $\rho = +.70$, wherein the distribution is negatively skewed. However, when $\rho = 0$, the distribution is symmetrical. As a consequence of these characteristics of the sampling distributions, there are different procedures for testing the significance of r depending on the assumed value of ρ. The most common situation is a test of $\rho = 0$, which will be discussed here. The less common situation is when $\rho \neq 0$ and it uses a procedure described in the section *Related Statistics* at the end of the chapter.

Testing $\rho = 0$

Exact tabular values of r have been calculated for selected sample sizes of 1000 and less. (See Table D10 in Appendix D.) These are based on t sampling distributions at $df = N - 2$. For larger samples, a tabular value is approximated by the z distribution. A

FIGURE 12.1 Patterns in Bivariate Variances

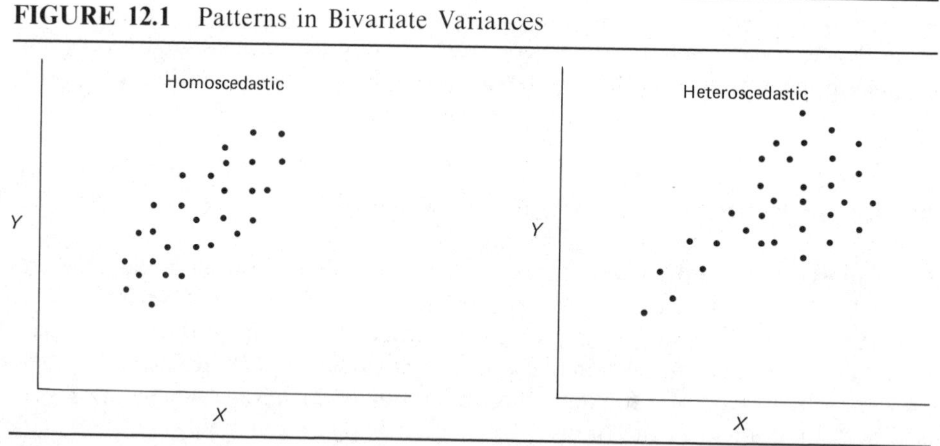

FIGURE 12.2 Sampling Distributions of r when $\rho = -.70, 0, +.70$

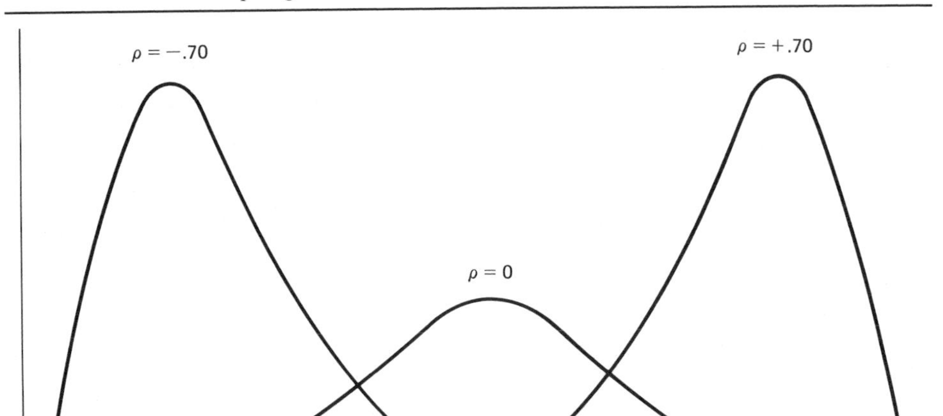

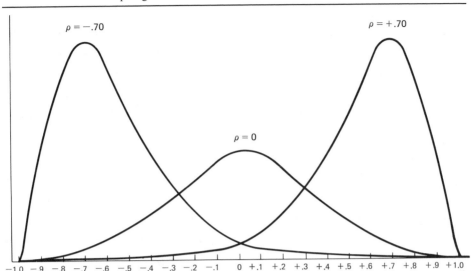

z-transformation formula is used to convert the test value of r to a z score. The transformed r is compared to the tabular z to reach a decision about the null hypothesis. The latter procedure is exactly like that shown above for Spearman's r. Indeed Spearman's r and its significance test are merely a special case of the r procedures whereby the scores are converted to ranks.

Imagine that a researcher wants to make an inference about the possible association between the amount of time children watch television and their creativity. The researcher draws a random sample of 20 children, determines the time they spend watching television during a month, and administers a standardized test of creativity. The value of Pearson's r calculated for these data turns out to be $-.50$. What inference can be drawn from this at the .05 level?

Step 1. The hypotheses.

H_0: $\rho = 0$ — The population of children has no association between time spent watching television and creativity. A random sample may have some association but it will be within the expected limits of sampling error.

H_1: $\rho \neq 0$ — The population of children has an association between time spent watching television and creativity. A random sample will reflect this association beyond the expected limits of sampling error.

Step 2. The criteria for rejecting H_0.

The sample is small enough that a tabular value of r can be found from Table D10. The degrees of freedom are $N - 2 = 18$. At the .05 level of significance and $df = 18$ for this nondirectional test, the tabular value of r is $\pm.444$. The null hypothesis will be rejected only if the test value of r is $\pm.444$ or beyond.

Step 3. The test statistic.

The test statistic has already been found to be $-.50$.

FIGURE 12.3 Testing H_0: $\rho = 0$

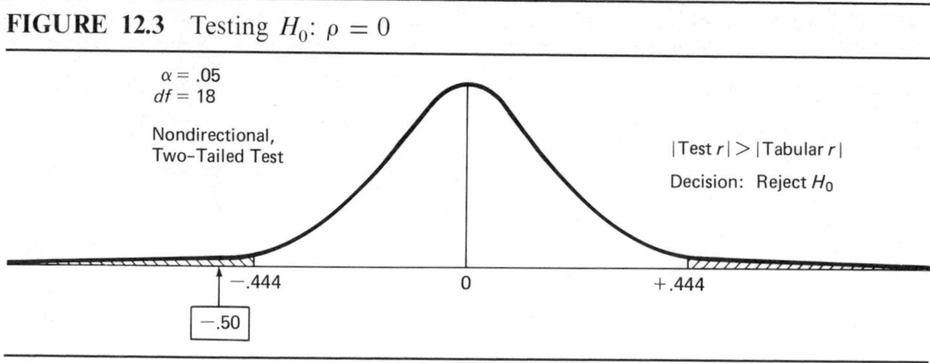

Step 4. The decision about H_0.

The absolute value of the test statistic (.50) does exceed the tabular statistic (.444). Thus the null hypothesis is rejected; $r = -.50$, $df = 18$, $p < .05$. This test is sketched in Figure 12.3.

Step 5. The conclusion.

The researcher can conclude that the observed association is significant. Excepting the occurrence of a Type I error, there is a relationship between time spent watching television and creativity in the population represented by this sample.

Consider another example. A social scientist wants to know whether she would be justified in drawing the inference that the degree of social criticism is positively related to the consumption of print journalism. She draws a random sample of 2000 adults and determines their scores on a scale measuring the degree to which they think critically about the nature of American society. She also measures their total consumption of print journalism in terms of the number of publications read and depth with which each is read. The calculated value of Pearson's r for these data is $+.43$. What inference concerning association can be reached at the .01 level?

Step 1. The hypotheses.

H_0: $\rho \leq 0$	The population of adults has either no association or a negative association between degree of social criticism and consumption of print journalism. A random sample may have some positive association but it will be within the expected limits of sampling error.
H_1: $\rho > 0$	The population of adults has a positive association between degree of social criticism and consumption of print journalism. A random sample will reflect this positive association beyond the expected limits of sampling error.

Step 2. The criteria for rejecting H_0.

The sample exceeds 1000 cases and therefore a tabular value from the z distribution can be determined. The tabular value of z at the .01 level of significance for this nondirectional test is $+2.33$. The test statistic must be at or beyond this value for H_0 to be rejected.

Step 3. The test statistic.

The value of Pearson's r for the sample data is .43. To compare this to the tabular statistic, it must first be transformed to a z score. The following formula is used when the population correlation (ρ) is assumed to be zero:

$$z = \frac{r}{1/\sqrt{N-1}} \quad \text{or} \quad r\sqrt{N-1}$$ z Transformation Formula for r

where r = value of r calculated from sample data
N = number of cases

Used here, we find that

$$z = .43\sqrt{2000 - 1} = .43\sqrt{1999} = .43(44.7102)$$
$$= 19.23$$

Step 4. The decision about H_0.

The test statistic (19.23) exceeds the tabular statistic (2.33). Therefore, the null hypothesis can be rejected; $r = +.43$, $p < .01$.

Step 5. The conclusion.

Having rejected the null hypothesis, the social scientist concludes that there is a significant and positive association between the degree of social criticism and the consumption of print journalism. Like all other significance tests, this does not demonstrate that either one of these variables causes the other. It merely suggests that they are associated in the appropriate population.

Earlier, R.E. 6.3 demonstrated the use of Pearson's r for describing association. We can now use this for an inference test. The example concerned the association between a state's maximum monthly AFDC payment and eight other structural variables describing the state. Table 2 of that example showed the intercorrelations among all nine of these variables. It is reprinted here as R.E. 12.2.

The correlations are based on 48 cases (states). Suppose that we use the .05 level of significance for nondirectional tests of a series of null hypotheses of no association between the pairs of variables. At $df = 46$ $(N - 2)$, the tabular r from Table D10 is $\pm.285$. The correlations that reach or exceed this value are marked in the table. Notice that 16 of the 36 intercorrelations are significant. Since the level of significance was set at .05, we must also note that in the long run 5% of the observed associations found to be statistically significant would actually represent a population in which there was no association. This means that approximately 1 of 16 $(.05 \times 16 = .8 \simeq 1)$ associations judged to be significant, such as those in R.E. 12.2, would represent a Type I error. Of course, none or more than one of these exact 16 associations may represent such an error. It is over the long run that 5% of the decisions made to reject $\rho = 0$ under these same circumstances would be incorrect.

Keys to Understanding

Many of the keys to understanding presented before are relevant to inferences of association. All of them will not be repeated, but it is recommended that you reread the keys given in Chapters 5 and 6 concerning descriptions of association and in Chapters 9 to 11 for inferences of difference. Here, let us emphasize three major points.

Key 12.1. Be sure to appreciate the importance of the model used to make an inference of association. Like those for inferences of difference, these models tell you the assumptions and techniques used to make decisions about null hypotheses of association and to reach defensible inferences. Violations of the assumptions and misuses of those

RESEARCH EXAMPLE 12.2 Inference at the Interval Level Using Pearson *r*

Zero-Order Correlation Matrix

	1	2	3	4	5	6	7	8	9
1	1.00	.57*	.44*	−.64*	.02	−.23	−.10	.10	.27
2		1.00	.22	−.29*	.18	−.54*	.16	.60*	.49*
3			1.00	−.20	.24	−.07	.22	.01	.19
4				1.00	.13	.03	.38*	.13	−.03
5					1.00	−.27	.51*	.49*	.56*
6						1.00	−.12	−.49*	−.39*
7							1.00	.41*	.47*
8								1.00	.56*
9									1.00

X_1 = states' maximum monthly AFDC payments.
X_2 = state per capita personal income.
X_3 = tax effort in the state.
X_4 = nonwhite percentage of state population.
X_5 = state unemployment rate.
X_6 = combined state percentage of children aged 1–17 and of the elderly aged 65 and over.
X_7 = AFDC child recipient rate in the state.
X_8 = urbanization in the state.
X_9 = density of population in the state.

*p. <.05.

Source: Martha N. Ozawa, "An Exploration into States' Commitment to AFDC," *Journal of Social Service Research,* 1 (3), Spring 1978:245ff. Table 2.

techniques can be disastrous. Many pieces of research are seriously flawed by their weak compliance with the requirements.

However, some points can be made on the other side of this issue. The models are ideal sets of conditions which can only be approximated in actual research. Some statistics are robust; they can tolerate surprisingly large shortcomings and still yield accurate results. This is not a license to violate the models deliberately. A researcher should always try to meet the conditions expressed in the model and to select an alternative technique if it fits the situation better. The models for the four main association techniques discussed in this chapter are presented in Inference Models 12.1 to 12.4. Use them as a guide to interpret and do this kind of research.

Inference Model 12.1 Nominal-Level Association—Test of Chi-Square

Purpose:

To determine whether two nominal variables are associated in the population that has been sampled.

Sampling Distribution:

Chi-square distributions.

Assumptions of the Test:

1. Independent and random observations.
2. One random sample.

3. Nominal level of measurement in two variables.
4. N must not be extremely small.

Typical Hypotheses:

Nondirectional Test	Directional Test
H_0: $\chi^2 = 0$	(Not ordinarily used)
H_1: $\chi^2 \neq 0$	

Tabular Statistic:

Chi-square at $df = (r - 1)(c - 1)$ and a predetermined level of significance from Table D4.

Test Statistic:

$$\chi^2 = \Sigma \frac{(f_o - f_e)^2}{f_e}$$

Inference Model 12.2 Ordinal-Level Association—Test of Gamma

Purpose:

To determine whether two ordinal variables are associated in the population that has been sampled.

Sampling Distribution:

Normal distribution (z).

Assumptions of the Test:

1. Independent and random observations.
2. One random sample.
3. Ordinal level of measurement in two variables.
4. $N \geq 50$.

Typical Hypotheses:

Nondirectional Test	Directional Test
H_0: $\gamma = 0$	H_0: $\gamma \geq$ or ≤ 0
H_1: $\gamma \neq 0$	H_1: $\gamma <$ or > 0

Tabular Statistic:

Tabular value of z from Table D2 at a predetermined level of significance.

Test Statistic:

$$z = G\sqrt{\frac{P_a + P_d}{N(1 - G^2)}} \qquad \text{where } G = \frac{P_a - P_d}{P_a + P_d}$$

Inference Model 12.3 Ranked Ordinal-Level of Association—Test of Spearman's r_s

Purpose:

To determine whether two ranked ordinal variables are associated in the population that has been sampled.

Sampling Distribution:

A. Student's t distributions for small samples ($N \leq 30$).
B. Normal distribution (z) for large samples ($N > 30$).

Assumptions of the Test:

1. Independent and random observations.

2. One random sample.
3. Rank-ordered measurement in two variables.
4. Few tied ranks.

Typical Hypotheses:

Nondirectional Test	Directional Test
H_0: $\rho_s = 0$	H_0: $\rho_s \geq$ or ≤ 0
H_1: $\rho_s \neq 0$	H_1: $\rho_s <$ or > 0

Tabular Statistic:

A. Tabular value of r_s from Table D9 at a predetermined level of significance and N for small samples based on t.
B. Tabular value of z from Table D2 at a predetermined level of significance for large samples.

Test Statistic:

A. (For small samples) $$r_s = 1 - \left[\frac{6D^2}{N(N^2 - 1)}\right]$$

B. (For large samples) $$z = r_s\sqrt{N - 1}$$

Inference Model 12.4 Interval-Level Association—Test of Pearson's r

Purpose:

To determine whether two interval variables are associated in the population that has been sampled.

Sampling Distribution:

A. Student's t distributions for small samples ($N \leq 1000$).
B. Normal distribution (z) for large samples ($N > 1000$).

Assumptions of the Test:

1. Independent and random observations.
2. One random sample.
3. Interval level of measurement in two variables.
4. Bivariate normal distribution in population.
5. Homoscedastic variances.
6. Linear relationship.

Typical Hypotheses:

Nondirectional Test	Directional Test
H_0: $\rho = 0$	H_0: $\rho \geq$ or ≤ 0
H_1: $\rho \neq 0$	H_1: $\rho <$ or > 0

Tabular Statistic:

A. Tabular value of r from Table D10 at a predetermined level of significance and $df = N - 2$ for small samples based on t.
B. Tabular value of z from Table D2 at a predetermined level of significance for large samples.

Test Statistic:

A. (For small samples) $$r = \frac{N(\Sigma XY) - (\Sigma X)(\Sigma Y)}{\sqrt{[N(\Sigma X^2) - (\Sigma X)^2][N(\Sigma Y^2) - (\Sigma Y)^2]}}$$

B. (For large samples) $$z = r\sqrt{N - 1}$$

Key 12.2. Know what the term "association" really means. There are three meanings that are often attributed incorrectly to it. First, as mentioned before, association is not causation. The inference that two variables are associated does not mean that they have a cause-and-effect relationship. It is so easy to equate the two that you must be on the alert to distinguish them. It is not uncommon to read a research report and find the authors slipping into a cause-and-effect discussion when they have only examined association.

Second, association in a sample does not mean that there is association in a population. That is the reason a significance test must be performed! This may seem obvious to you. But you can be misled if you look only to the descriptive aspects of measuring association. You can become so engrossed in the details by which a sample's association is examined for existence, strength, direction, and predictive capacity that you forget to attend to the bigger issue of whether the population has any association at all. Unless you are interested only in the exact sample being viewed, the more important question is the one of inference, not of description.

Third, statistically significant association must not be confused with meaningful association. Like all inferences, those for association are based on a determination of the probability that sampling error could cause the patterns noted in the sample. A statistically significant association is merely a pattern that is unlikely to be the result of sampling error at some prescribed level of significance. Hopefully, that significant association is also important and meaningful. But it does not have to be. Only the context of the research, the related theory, and the practical consequences give it true social meaning.

Key 12.3. Determine that an appropriate statistical test of association has been selected. There is no need for a formal guide covering the four techniques that we have discussed here, because they are so few. They can be distinguished on the basis of their measurement requirements. The chi-square test uses nominal measurement, gamma uses ordinal measurement, Spearman's r_s uses rank-ordered measurement, and Pearson's r uses interval measurement.

You may want to consult a comprehensive guide to selecting an appropriate statistic for any common social research situation and analysis purpose.* This can give you an overview that is very helpful.

A Concluding Comment

As you finish this final chapter of the book, you should reflect on the goals we set at the beginning and your experiences with statistics since then. If you are like most beginners, you found some things easy to understand and others difficult. You experienced some moments of discovery and some of utter befuddlement. You may be pleased with your accomplishments or pained by their absence. In fairness to yourself, try to recall where you were when you began and compare it to where you are now. Only the most precocious among you will be at the expert level already. Most of you will have made great progress in understanding social statistics and you have a right to take pride in this. Thinking back to the analogy with a first exposure to a course in the Russian language, you may not speak like a native of Moscow, but you sure know more than when you began. However, is it enough?

Whether you go on to improve your understanding of statistics is largely for you to

*One of these is produced by the Survey Research Center at the University of Michigan (Andrews et al., 1981). It covers both descriptive and inferential statistics.

determine. As we said at the outset, the decision you make about this is likely to have a profound influence on your life as a professional in the social sciences or as an ordinary individual. People who ignore social statistics and claim to be professionals in most areas of the social sciences are living in an imaginary world that does not exist. They may spend their entire careers making excuses for their lack of knowledge and, in consequence, be segregated to the more peripheral issues and mundane tasks of their fields. They are less likely to be contributors to an improved understanding of the social world but instead are more likely to perpetuate myths and inappropriate generalizations based on their own idiosyncratic experiences. That is not to say that one must conduct original research based on empirical data to which statistics are applied in order to be a contributor or demythodogizer. Rather, it means that an inability to interpret statistics closes the door on the capacity to evaluate most social research personally and to integrate one's own thoughts with it. There really is little good reason to deliberately place yourself in such a limiting circumstance.

Look at the message contained in the feature On the Side 12.1 in this chapter. It reports the results of one study exploring the extent of statistical literacy within a sample of social workers. That level was not very high. The point of the article is not that social workers, in particular, are illiterate about statistics. People in some other fields in the social sciences may be slightly higher and others slightly lower. The real issue raised by the study concerns the correlates (if not consequences) of such low levels of statistical knowledge. People who do not know much about statistics do not often *read* research as sources of information. As the author remarks, if they do not expose themselves to the research, by what criteria do they evaluate information that is based on research? The probable answers are not very comforting. Do they realize what information is based on research? Do they use a mystical criterion composed of the appearances of systematic research (statistical symbols, common phrases, and so forth) to evaluate it? Do they revert to their own experiences and then generalize from these to the remainder of the world? What a dismal state of affairs any of these makes! Further, as the author notes, these people deprive the researchers of the feedback that would improve both research and practice.

ON THE SIDE 12.1 Statistical Literacy

Statistical analysis has become a routine part of social science research. In social work, demands for more "objective" methods of accountability have combined with a growing array of statistical procedures and easily used "canned" computer programs (e.g., Lindsey, 1977) to make statistical data analysis increasingly popular. As a consequence of this shift toward the use of quantitative analysis, social workers are finding it necessary to interpret and understand the language of statistics in order to remain critical consumers of the research literature.

At the present time, little is known about the typical social worker's level of competency in the statistical area. Previous surveys of social workers' attitudes toward research have generally revealed an estrangement between practitioners and the research process (Casselman, 1972; Rosenblatt, 1968; Kirk, Osmalov, & Fischer, 1976). For example, Kirk, Osmalov, and Fischer (1976) found that 56.1% of their 470 respondents did not consult the research literature when confronted with a difficult problem. . . .

The present study sought to further investigate social workers' knowledge and attitudes toward research and statistics. Using a methodology similar to the Weed and Greenwald study, a sample of BA and

MSW workers in the state of Wisconsin were asked to identify eight common statistical symbols. Opinions of the importance of research and statistics to social work practice were also solicited. . . .

Social workers who participated in this survey had little formal training in research and statistics (see Table 1). At the undergraduate level, fewer than 25% had more than one course in research and less than 15% had more than one statistics course. Thirty-three percent had never taken a research course and 29% had never taken a statistics course. Among MSWs in the sample, less than half had taken a statistics course during their graduate studies. However, all but two of the MSWs reported taking at least one graduate course where presumably they may have been introduced to elementary statistics.

TABLE 1 Comparison of Research and Statistics Preparation at the Undergraduate and Graduate Level[a]

	Research		Statistics	
	Undergraduate	Graduate[b]	Undergraduate	Graduate
0 courses	34% (42)	4% (2)	29% (35)	52% (24)
1 course	44% (55)	45% (22)	57% (69)	35% (16)
2 courses	16% (20)	35% (17)	10% (12)	13% (6)
3 or more courses	6% (7)	16% (8)	4% (5)	0% (0)

[a]Numbers in parentheses equal frequencies.
[b]Based on 49 Masters level workers in the sample.

Once completing their formal education, only 6.3% of the respondents took additional courses in research or statistics. Reading of research articles was also infrequent. Mean number of articles read per month was 1.8 and 30% reported reading no research articles.

The respondents were also asked, "Do you believe that the use of statistics in research has value for social work practice?" Possible responses ranged from "no value" (with a score of 0) to "extremely valuable" (with a score of 4). The mean result for the sample was 1.99, indicating an overall assessment of "moderately valuable." . . .

The mean number of correct responses to the eight statistical symbols (excluding %) was 1.74, or almost 22% correct. Table 2 shows a breakdown of correct and incorrect responses to the eight symbols. Standard deviation (SD, *S*) was the most frequently recognized symbol, correctly identified by 51.6% of the respondents. Conversely, only 2.3% (n = 3) could correctly identify the

TABLE 2 Number of Correct and Incorrect Responses to Eight Statistical Symbols[a]

Symbol	Correct	Incorrect
SD, S	66 (51.6)	62 (48.4)
$\bar{X}$, M	54 (42.2)	74 (57.8)
$p < .05$	45 (35.2)	83 (64.8)
χ^2	20 (15.6)	108 (84.4)
r	13 (10.2)	115 (89.8)
t	12 (9.4)	116 (90.6)
df	10 (7.8)	118 (92.2)
F	3 (2.3)	125 (97.7)

[a]Numbers in parentheses equal percentages.

symbol *F* as denoting an "*F*-test." Thirty-six percent of the sample ($n = 47$) did not correctly identify any of the symbols, while 9% ($n = 12$) had five or more correct answers. Only two subjects correctly identified all eight symbols. . . .

The results of this study corroborate the earlier findings of the Weed and Greenwald study concerning social workers' inability to identify common statistical symbols. Moreover, this lack of statistical knowledge appeared related to a general avoidance of the research literature. If this latter assertion is true, it raises serious questions regarding the sources and methods of knowledge development in the profession. For example, aside from experiential data, by what criteria do social workers evaluate research-based information? Research findings are too important to the development of the social work profession to be left to the researcher alone. Critical evaluation and feedback from the practitioner is needed if such research is to have maximum relevancy and contribute to a growing empirical knowledge base for the profession. How may this situation be changed? Although workers with the most statistical training identified the most symbols, training alone was no guarantee of competency. Perhaps the more crucial factors are the perceived relevancy of statistics for practitioners and the ability to utilize this knowledge following the termination of formal statistical training. The consumerism approach to research advocated by a number of authors (e.g., Tripodi, 1974) seems like a step in the right direction, since most practitioners never conduct an actual research study. A limitation of this approach, however, is that a critical understanding of statistical analysis is often not included (as for example, when research and statistics are taught as separate courses). . . .

Adapted from Stanley L. Witkin, Jeffrey L. Edelson, and Duncan Lindsey, "Social Workers and Statistics: Preparation, Attitudes and Knowledge," *Journal of Social Service Research,* 3 (3), Spring 1980:313–322. Reprinted by permission.

What, then, should you have tried to accomplish from an introduction to social statistics? The primary goal should *not* have been the memorization of statistical symbols or the execution of formulas by rote. Above everything else you should have focused on an understanding of the basic logic and strategy behind statistical examinations of information. This is what we refer to as becoming a knowledgeable and critical consumer of social statistics in use. For this we must know the larger contexts in which statistics are applied. We must know the relationships between the statistical methods used and the questions researchers ask. We must know the limitations to those questions and methods in terms of the alternative conceptions that are not being utilized. That is the reason this book has included examples from actual research which usually showed more than a single table or sentence reporting the results of some analysis. These results cannot stand alone from the statistics used to produce them or from the entire research process that was employed. Their real meaning is contained within the conceptual and experiential environment in which research questions are asked and statistics are applied. If you have been concentrating on this type of issue as you read this book, you are on your way to a level of understanding that will serve you well regardless of your professional and personal goals.

KEY TERMS

Chi-Square Test of Association
Test for Ordinal Association Using Gamma
Test for Rank Ordinal Association Using Spearman's r_s
Test for Interval Association Using Pearson's r
Bivariate Normal Distribution
Homoscedasticity
Heteroscedasticity

Association in a Sample Versus Association in a Population
G Versus γ
r_s Versus ρ_s
r Versus ρ

EXERCISES

For Exercises 1 to 8: (a) state the hypotheses, (b) determine the tabular statistic, and (c) decide whether an inference that an association exists or does not exist is appropriate.

1. A researcher has examined a possible association between sex and voting behavior in a 2×2 table. Chi-square was calculated to be 2.50 and phi was .22. Let the level of significance be .05.
2. A researcher has examined a possible relationship between religious affiliation and political party affiliation in a 3×4 table. Chi-square was calculated to be 17.65 and Cramér's V was .57. Let the level of significance be .05.
3. A researcher has examined a possible relationship between population size and median family income for 12 cities. Pearson r was calculated to be .31. Use a nondirectional test and let the level of significance be .05.
4. A researcher has examined a possible relationship between years of education and feelings of alienation for 226 individuals. Pearson r was calculated to be $-.10$. Use a nondirectional test and let the level of significance be .05.
5. A researcher has examined a possible positive relationship between level of income and level of health for 65 persons. Gamma was calculated to be .67 with 1200 agreement pairs and 300 disagreement pairs. Use a directional test and let the level of significance be .01.
6. A researcher has examined a possible relationship between perceived self-esteem and number of close friends for 22 persons. Spearman's r_s was calculated to be .54. Use a nondirectional test and let the level of significance be .05.
7. A researcher has examined a possible negative relationship between socioeconomic status and feelings of success for 90 persons. Gamma was calculated to be $-.81$ with 1200 agreement pairs and 2100 disagreement pairs. Use a directional test and let the level of significance be .05.
8. A researcher has examined the possible positive relationship between age of house and age of resident for 59 persons. Spearman's r_s was calculated as $+.17$. Use a directional test and let the level of significance be .05.
9. Select two of the Exercises above (1–8) in which you drew the inference that an association existed. For each write a short narrative creating a realistic context to the problem and discuss the analysis procedure and conclusion as you think it might appear in published form.
10. Search the social science literature for an example of an intercorrelation matrix showing the calculated Pearson's r values between several variables. Find the sample size and level of significance used by the author. Determine the tabular statistic and indicate which of the correlation values are significant.
11. Search the social science literature for an example of any inference of association using the statistics discussed in this chapter. In your own words explain the researcher's goals and methods in drawing the inference reported. Critique the researcher's work by noting its strengths, weaknesses, and the alternatives that seem plausible to you.

RELATED STATISTICS

Test for a Nonzero Pearson r

A researcher may have occasion to hypothesize that the population value of Pearson r (ρ) is not zero. In this situation the sampling distribution will not be symmetrical and the significance test shown above is not appropriate. Instead, a new transformation known as Fisher's z' is used to convert the calculated r to a normally distributed variable. The observed r and assumed value of ρ are converted to the new z' scores by formula or a table, such as Table D11 in Appendix D. The appropriate standard error ($\sigma_{z'}$) is estimated by the formula $1/\sqrt{N-3}$. The test statistic is then found: $(z_{obs} - z'_{\rho})/\sigma_{z'}$. This value is compared to a tabular value of the usual z distribution (e.g., tabular $z = \pm 1.96$ when $\alpha = .05$.) If the test statistic falls into the rejection region(s), the null hypothesis is rejected, as always. See: Tai (1978:342–344) and Loether and McTavish (1980:597ff).

Test for a Difference Between Two Pearson *r* Values

A significance test exists to determine whether two independent Pearson r values differ. Like the test immediately above, it is based on the Fisher z transformation.
See: Champion (1981:338–339).

Test for Significance of Regression Coefficient *b*

In bivariate regression the significance of the coefficient b is equivalent to a test of the bivariate r. The b coefficient is transformed to its standardized form, in which it and r are equivalent. Therefore, the usual procedure for testing bivariate r substitutes as a significance test of b.
See: Loether and McTavish (1980:602) and Freund (1979:384–387).

Test for Significance of Partial and Multiple Correlations

Partial and multiple correlation coefficients can be tested for significance using the Fisher's z transformation or by a t test.
See: Loether and McTavish (1980:602, 604–605) and Mueller et al. (1977:451–453).

Confidence Intervals Around Values of Pearson *r*

A convenient method to estimate an unknown population correlation value, ρ, for existence, direction, and strength is with a confidence interval. Since the parameter is unknown, the Fisher's z' transformation is used to construct an interval estimate much like those shown earlier in Chapter 9 for means and proportions. The basic formula used is: $CI = z' \pm z(\sigma_{z'})$, where z' is the Fisher transformation of r, z is the standard normal score at the desired level of significance, and $\sigma_{z'}$ is the standard error as defined above in the section on testing a nonzero Pearson r.
See: Minium (1978:356–357).

Confidence Intervals Around *a* and *b* Regression Values

A sense of the accuracy in the a and b values from a bivariate regression for inference purposes can be gained by constructing confidence intervals around them. This is done like confidence intervals discussed before, although the computations are more involved.
See: Freund (1979:387–388).

APPENDIX

A REVIEW OF BASIC MATHEMATICS

This appendix reviews some of the basic skills of arithmetic and algebra necessary to follow the computations in this book. It should help to refresh your memory about those procedures you have not been using recently.

Symbols and Basic Arithmetic Operations

Symbols	Meaning
X, Y, X_1, X_2, X_3	Different ways to represent any one of a number of variables.
a, b, c, b_1, b_2, b_3	Different ways to represent any one of a number of constants.
$X = Y$	X is equal to Y.
$X \cong Y$ or $X \simeq Y$	X is approximately equal to Y.
$X \neq Y$	X is not equal to Y.
$X > Y$	X is greater than Y.
$X < Y$	X is less than Y.
$X \geq Y$ or $X \geqq Y$	X is equal to or greater than Y.
$X \leq Y$ or $X \leqq Y$	X is equal to or less than Y.
$X \pm a$	Two quantities, one equal to $X + a$ and the other equal to $X - a$.
$\|X\|$ (e.g., $\|8\| = 8$; $\|-8\| = 8$)	The absolute value of X; the value of X expressed as a positive value regardless of X's original sign.
$X + Y$	The addition of X and Y.
$X - Y$	The subtracted difference between X and Y.
$X \cdot Y, (X)(Y), X * Y, XY$	The multiplication of X times Y.
$X \div Y, \frac{X}{Y}, X/Y$	The division of X by Y.
$\frac{1}{X}$	The reciprocal of X: 1 divided by X.
$X = \frac{X}{1}$	Division by 1 is assumed even if not specifically expressed.

X^a (e.g., $X^3 = X \cdot X \cdot X$; if $X = 2$, $2^3 = 2 \cdot 2 \cdot 2 = 8$)	X multiplied by itself a times.
X^2	X times X, known as "X squared" or "X square."
$(XY)^2$	$(XY)(XY)$; the square of the product of X times Y.
X^2Y	$(X^2)(Y)$; the product of X square times Y.
XY^2	$(X)(Y^2)$; the product of X times Y square.
X^2Y^2	$(X^2)(Y^2)$; the product of X square times Y square.
$a = +a$	A number written without a sign is assumed to be a positive number.
$-a$	A negative number must be written with a negative sign.
$a!$ (e.g., $5! = 5 \cdot 4 \cdot 3 \cdot 2 \cdot 1 = 120$)	! is the factorial sign, indicating a series of multiplications of a times all successively smaller integers to 1.
$X(0) = 0$	The product of X times 0 equals 0.
$\frac{0}{X} = 0$	The division of 0 by X equals 0.
$\frac{X}{0} = \text{undefined}$	The division of X by 0 is impossible and yields an undefined result.
$X^0 = 1$	X raised to the 0 power equals 1.
$\sqrt{X}$ (e.g., $\sqrt{25} = 5$)	The square root of X; that number which squared would equal X; the root is always assumed to be positive in this book.

Combined Arithmetic Operations

Operation	Meaning
$5 - 10 = -5$	To subtract a larger number from a smaller one, find the difference and attach the sign of the larger number.
$5 + (-10) = 5 - 10 = -5$	Adding a negative number to a positive number is the same as subtracting the second from the first.
$-5 - 10 = -(5 + 10) = -15$	Subtracting a negative number from another negative number is the sum of the two numbers with the negative sign attached.
$(5)(-10) = -50$	The product of a positive and a negative number is negative in sign.
$(-5)(-10) = 50$	The product of two negative numbers is positive in sign.
$(-5)^2 = (-5)(-5) = 25$	The square of a negative number is positive in sign.
$\frac{-5}{10} = -.50$	The division of a negative number by a positive number is negative in sign.
$\frac{5}{-10} = -.50$	The division of a positive number by a negative number is negative in sign.
$\frac{-5}{-10} = .50$	The division of a negative number by a negative number is positive in sign.
$2 + (5 + 10) = 2 + 5 + 10 = 17$	Parentheses can be removed when their preceding sign is positive.
$2 + (5 - 10) = 2 + 5 - 10 = -3$	

$2 - (5 + 10) = 2 - 5 - 10$ $= -13$ $2 - (5 - 10) = 2 - 5 + 10$ $= 7$	The signs of terms within parentheses are reversed when removing parentheses preceded by a negative sign.
$2(5 + 10) = 2(5) + 2(10)$ $= 10 + 20$ $= 30$ or $= 2(15)$ $= 30$	Terms outside parentheses apply to those within separately or in combined form when the outside term is being multiplied by those within.
$\frac{5 + 10}{2} = \frac{1}{2}(5 + 10)$ $= \frac{1}{2}(15)$ $= 7.5$ or $= \frac{5}{2} + \frac{10}{2}$ $= 2.5 + 5$ $= 7.5$	Division of combined terms by another term is equal to their combination divided in one operation or their separate divisions and then combined.
$2(5(-10 + 2)) = 2(5(-8))$ $= 2(-40)$ $= -80$ $4(2 + 3(4 - 1)) = 4(2 + 3(3))$ $= 4(2 + 9)$ $= 4(11)$ $= 44$	With nested parentheses, work from the most interior set first toward to the most outside set last.
$\left(\frac{1}{2}\right)\left(\frac{1}{4}\right)\left(\frac{3}{5}\right) = \frac{3}{40}$ $= .075$	The product of several fractions equals the product of their numerators divided by the product of their denominators.

Summation Notation

Notation	Meaning
Σ	The process of summing together a set of numbers.
$\sum_{i=1}^{n} X_i = X_1 + X_2 + \cdots + X_n$ $\sum_{i=1}^{4} X_i = X_1 + X_2 + X_3 + X_4$ $\sum_{i=5}^{8} X_i = X_5 + X_6 + X_7 + X_8$	The set of numbers to be summed can be specified by use of the subscript i which is given a beginning point below the summation sign and an ending point above that sign.
$\sum_{i=1}^{n} X_i = \Sigma X$	When all numbers in a set are to be summed the use of the subscript i and its specification can be omitted.
$\Sigma X^2 = X_1^2 + X_2^2 + \cdots + X_n^2$	Squaring precedes summation without parentheses.

$(\Sigma X)^2 = (X_1 + X_2 + \cdots + X_n)^2$	Summing precedes squaring when parentheses are present as illustrated.
$\Sigma XY = X_1Y_1 + X_2Y_2 + \cdots + X_nY_n$	Multiplication of each pair of X and Y values precedes the summing of their products.
$(\Sigma X)(\Sigma Y) = (X_1 + X_2 + \cdots + X_n)(Y_1 + Y_2 + \cdots + Y_n)$	Summing of separate sets of numbers precedes the multiplication of their sums.
$\frac{\Sigma X}{N} = \frac{X_1 + X_2 + \cdots + X_n}{N}$	Summation precedes division of the sum by N.
$\frac{\Sigma X^2}{N} = \frac{X_1^2 + X_2^2 + \cdots + X_n^2}{N}$	Squaring the individual values of X precedes their summing and division by N.
$\left(\frac{\Sigma X}{N}\right)^2 = \left(\frac{X_1 + X_2 + \cdots + X_n}{N}\right)^2$	Summing individual values of X and dividing their sum by N precedes squaring the result.

Solving Equations

Equation	Meaning
$Y = a + bX$ if $a = 2$, $b = 3$, and let $X = 5$; $Y = 2 + 3(5)$ $= 2 + 15$ $= 17$	When the unknown is alone on one side of the equation, simply substitute and follow the usual rules of arithmetic and algebra.
$Y - 3 = 10$ $Y - 3 + 3 = 10 + 3$ $Y = 13$	To isolate the unknown when it is previously combined with a negative number, add that number to both sides of the equation.
$Y + 3 = 10$ $Y + 3 - 3 = 10 - 3$ $Y = 7$	To isolate the unknown when it is previously combined with a positive number, subtract that number from both sides of the equation.
$3Y = 10$ $\frac{3Y}{3} = \frac{10}{3}$ $Y = 3.3$	To isolate the unknown when it is previously being multiplied by a number, divide each side of the equation by that number.
$\frac{Y}{3} = 10$ $\frac{Y}{3}(3) = 10(3)$ $Y = 30$	To isolate the unknown when it is previously being divided by a number, multiply each side of the equation by that number.
$Y^2 = 100$ $\sqrt{Y^2} = \sqrt{100}$ $Y = 10$	To isolate the unknown when it is previously being squared, take the square root of each side of the equation.
$\sqrt{Y} = 10$ $(\sqrt{Y})^2 = (10)^2$ $Y = 100$	To isolate the unknown when its square root is previously being found, square each side of the equation.

Rounding

Calculations often require rounding. Since this involves the introduction of error into the work, it must be done carefully. For problems that involve several steps, maintain far more accuracy than you want the final answer to have so that you round only once to the desired degree of accuracy. When you round numbers that end in 5, look at the number now in the place to which you want to round. If it is odd, round the answer up to the next even number. If it is even, round the answer down to the last even number. This avoids any systematic bias that would otherwise tend to over- or underestimate the true answer.

Example	Explanation
Rounding to nearest whole number:	
4.5 → 4	Because 4 is an even number.
17.5 → 18	Because 17 is an odd number.
12.8 → 13	Because 12.8 is closer to 13 than to 12.
Rounding to nearest tenth:	
13.65 → 13.6	Because 6 is an even number.
111.75 → 111.8	Because 7 is an odd number.
3.22 → 3.2	Because 3.22 is closer to 3.2 than to 3.3.
Rounding to nearest hundredth:	
44.695 → 44.70	Because 9 is an odd number.
22.385 → 22.38	Because 8 is an even number.
.998 → 1.00	Because .998 is closer to 1.000 than to .99.

APPENDIX

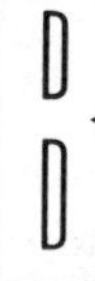

GLOSSARY OF TERMS

Alternative hypothesis (H_1) a statistical hypothesis which proposes a description of a population parameter that is opposite to that of the null hypothesis

Analysis systematic examination of data

Analysis of variance (ANOVA) a parametric investigation of differences between multiple sample means compared to expected population means

Area graph a figure that uses some geometric shape to illustrate a distribution

Association a description of the joint occurrence between two or more variables

Association, no a pattern in the joint distribution between variables in which there are no percentage differences between conditional distributions

Association, perfect a pattern in the joint distribution between variables in which there is a completely consistent pattern of association

Association direction ordered pattern of the joint distribution between variables; may be monotonic (consistent) or polytonic (containing reversals) and either positive or negative

Association existence a pattern in the joint distribution between variables in which there is some percentage difference between conditional distributions

Association strength extent to which an association approaches a perfect pattern

Asymmetrical measure a measure of association in which a distinction is made between an independent and dependent variable

Bar graph a figure that uses a series of separated rectangles (bars) to illustrate a distribution

Bar graph, horizontal a bar graph that uses a horizontal (X) axis to scale the illustrated distribution(s) while resting the bars on the vertical (Y) axis

Bar graph, vertical a bar graph that uses a vertical (Y) axis to scale the illustrated distribution(s) while resting the bars on the horizontal (X) axis

Bias methods of measurement or sets of data that are unreliable

Binomial probability of success probability that a particular outcome within a binomial distribution will occur

Binomial sampling distribution theoretical probability distribution for a dichotomized variable

Bivariate concurrent measurement or analysis of two variables

Bivariate normal distribution a population distribution in which two variables are jointly distributed normally

Cartesian coordinate graph a two-dimensional system of graphing data using a horizontal (X) and vertical (Y) axis

Case a research unit

Case number a numerical code assigned to each research unit to maintain identity and separation during data processing

Category midpoint (m) value exactly halfway between a category's true upper and lower limits

Category width (w) difference between a category's true upper and lower limits

Census a description of all research units in a population
Central Limit Theorem mathematical principle that a sampling distribution will approach a normal form with large sample sizes even though the population being sampled is not distributed normally
Central tendency description of a distribution's center or point(s) of cluster
Chi-square (χ^2) a measure of association existence for nominal data based on a comparison of observed and expected frequencies in a bivariate distribution
Chi-square goodness-of-fit test (χ^2-*I*) a nonparametric test of the concurrence between an observed set of frequencies and a theoretical expectation of that set of frequencies in a population
Chi-square *k*-sample test of independence (χ^2-*k*) a nonparametric test of the concurrence between multiple sets of sample frequencies and a theoretical expectation of those sets of frequencies in multiple populations
Chi-square two-sample test of independence (χ^2-*II*) a nonparametric test of the concurrence between two sets of sample frequencies and a theoretical expectation of those sets of frequencies in two populations
Closed category a measurement category which has defined limits
Code a description of the manner in which numbers are assigned to observations
Coefficient of variation (*CV*) a measure of variation for ratio data that are standardized for scale
Cohort statistical groups of persons who have similar characteristics but who do not necessarily interact
Computational formula a formula used to ease computation of a statistical measure but which does not define it
Concept abstraction represented by empirical observations
Conditional association association between two variables within a category of a third variable that is being controlled
Conditional distribution distribution of one variable under a fixed value of a second variable
Confidence, level of probability (in percentage format) that a confidence interval is one of a set of such intervals that contain the parameter being estimated
Confidence interval a range of values centering on a sample statistic used to estimate an unknown population parameter with a set level of confidence
Continuity of measurement a variable's use of a necessarily limited (discrete) or unlimited (continuous) set of numerical values in its representation
Correlation measurement of association at the interval/ratio level
Covariation the unstandardized description of association between interval variables
Cramér's *V* a measure of association strength for nominal data based on chi-square for tables of any size
Cumulative data form a data form that indicates the number (or percentage) of scores that are at and below (less-than) or at and above (more-than) each measurement category
Curvilinear association a pattern of association for interval variables in which there is a reversal of direction
Data coded observations collected from the research units
Data collection process of applying measurement to research units to gather observations
Definitional formula a formula that defines a statistical measure
Degrees of freedom (*df*) number of observations in sample data which are free to vary after some restrictions have been placed on the data in performing a statistical test
Dependent samples samples drawn from population(s) in such a manner that there is an a priori correspondence between the cases selected for each sample, (e.g., before–after or matched subjects methods)
Dependent variable a variable that is thought to be influenced or predicted by another variable
Descriptive statistics statistical methods that summarize data
Determination, coefficient of (r^2) a symmetrical PRE measure of association for interval data which indicates the proportion of total variation in one variable that is explained by a second variable
Deviation score (x) the difference between a raw score and the mean of a distribution
Dichotomy measurement having only two opposite categories

Directional test test of a null hypothesis in which the alternative hypothesis describes one directional option to that null hypothesis; often corresponds to a one-tailed test as graphically examined

Elaborating a relationship examination of a bivariate association under conditions representing the categories of a third variable

Empirical based on direct experience, not theoretical, speculative, or authority-based

***F* ratio** a ratio of two variances (between and within groups) used in analysis of variance

Form of a distribution the shape of a distribution described in terms of symmetry, kurtosis, and modality

Frequency distribution a distribution in which the data form is frequencies; may show frequencies either for simple (ungrouped) or grouped (combined) measurement categories

Friedman's two-way analysis of variance a nonparametric test of differences between multiple dependent samples' sets of ranked scores compared to a theoretical expectation of (no) such differences in multiple dependent populations

Goodman and Kruskal's gamma (*G*) a PRE measure of association for data grouped in ordered categories

Goodman and Kruskal's tau-*y* (T_y) a PRE measure of association for nominal data

Graph a visual representation of a distribution

Grouped data distribution a tabular display of a distribution which combines measurement categories that were previously separated; e.g., a grouped frequency distribution

Histogram an area graph positioned on a Cartesian grid that uses joined rectangles to illustrate the distribution of an interval or ratio variable with continuous measurement

Homogeneous variances an assumption that the variances in the populations being sampled are equal; used in parametric tests of differences between means; the opposite is called heterogeneous variances

Homoscedasticity an assumption that the variances in the dependent variable at each value of the independent variable are equal; used in parametric tests of association for interval data; the opposite is called heteroscedasticity

Hypothesis specific statement proposing to describe the state of or the relationship between certain aspects of a research problem

Hypothesis testing an inferential technique designed to test a specified value of a parameter that has been proposed a priori to the test

Independent samples samples drawn from population(s) without any a priori correspondence between the cases selected for each sample

Independent variable a variable that is thought to influence or predict another variable but for which no outside or previous influence is being investigated

Index of dispersion (*D*) a measure of variation for nominal data based on an analysis of pairs of scores

Inference an estimated description of a parameter based on a description of a statistic

Inferential standard deviation (*s*) inferential estimate of a population's standard deviation based on a sample of observations

Inferential statistics statistical methods that make generalizations from sample data to a population

Interquartile range (*Q*) a measure of variation for ordinal or interval data which identifies that part of an ordered distribution containing the middle 50% of the scores

Interval measurement measurement that classifies observations by kind, order, and numerical difference between categories

Kolmogorov–Smirnov single-sample test (*KS-I*) a nonparametric test used to determine the concurrence between a sample's set of ordered proportions and a theoretical expectation of that set of ordered proportions in a population

Kolmogorov–Smirnov two-sample test (*KS-II*) a nonparametric test used to determine the concurrence between two samples' sets of ordered proportions and a theoretical expectation of those sets in two populations

Kruskal–Wallis *H* Test a nonparametric test used to determine the concurrence between multiple samples' sets of ranked scores and those theoretically expected for multiple populations

Kurtosis that feature of a distribution's form which describes the extent of clustering or peakedness of the scores

Lambda (*L*) a PRE measure of association for nominal data

Law of large numbers the mathematical principle that an ever-larger number of trials (observations) will more closely approximate a theoretically expected distribution

Least squares, principle of mathematical principle that the sum of the squared deviations around a mean is the smallest possible sum of squared deviations in that distribution

Level of measurement the extent of mathematical meaning implied in the categories used to measure a variable

Level of significance (α) the probability value which is used to define that set of sample outcomes which are so unusual as to warrant rejection of a null hypothesis if any occur; also known as the alpha level

Line graph a graph illustrating a time-series distribution

Linear association a pattern of association between interval variables in which there is a constant direction

Longitudinal study a research study that collects observations at several points in time

Mean ($\bar{X}$) the sum of the scores in a distribution divided by the number of scores; used as a measure of central tendency; also known as the arithmetic average

Mean deviation (*MD*) a measure of variation for interval data that calculates the arithmetic average of the absolute value of the deviation scores

Mean of differences in means ($\mu_{\bar{X}_1-\bar{X}_2}$) mean of the sampling distribution of differences in means

Mean of differences in proportions ($\mu_{p_1-p_2}$) mean of the sampling distribution of differences in proportions

Mean of means ($\mu_{\bar{X}}$) mean of the sampling distribution of means

Mean of proportions (μ_p) mean of the sampling distribution of proportions

Mean square between (*MSB*) variance between groups in analysis of variance

Mean square within (*MSW*) variance within groups in analysis of variance

Measurement the process of operationalizing concepts as observable variables

Median (*Mdn*) middlemost value in an ordered distribution; used as a measure of central tendency

Median *k*-sample test (*Mdn-k*) a nonparametric test of differences between multiple samples' medians compared to a theoretical expectation of (no) differences in multiple populations

Median two-sample test (*Mdn-II*) a nonparametric test of difference between two samples' medians compared to a theoretical expectation of (no) difference in two populations

Modality that feature of a distribution's form which describes the number of points at which scores are clustered

Mode (*Mo*) the most frequently occurring value in a distribution; used as a measure of central tendency

Mode, major value in a distribution indicating the point of greatest concentration of scores

Mode, minor value in a distribution indicating a point of lesser concentration of scores than the major mode

Model of a statistical test the set of assumptions and methods of a statistical test used to draw inferences from a sample to a population

Multiple correlation simultaneous association between three or more interval variables

Multiple correlation, coefficient of (*R*) a measure of linear correlation between three or more interval variables

Multiple determination, coefficient of (R^2) a PRE measure of multiple association indicating the proportion of total variation in one variable that is explained by two or more other variables

Multiple regression the regression between one dependent variable and two or more independent variables

Multiplier a power of the number 10 that is applied to a ratio to move its decimal point

Multiresponse variable a variable measured such that more than one response is possible for each research unit

Multivariate concurrent measurement or analysis of three or more variables

Nominal measurement measurement that classifies observations by kind

Nondirectional test test of a null hypothesis in which the alternative hypothesis describes two directional options to that null hypothesis; often corresponds to a two-tailed test as graphically examined
Nonparametric test a test of a null hypothesis for which no assumptions about a population's distribution or parameters must be satisfied
Normal distribution a bell-shaped, symmetrical distribution
Normed measure a measure that has a fixed scale of values, e.g., 0 to 1
Null hypothesis (H_0) a statistical hypothesis proposing a description of a population that is tested for likely accuracy using sample data; it attributes to sampling error any unpredicted discrepency between sample outcomes and the assumed population parameter(s)
Objective measurement or data methods of measurement or sets of data that are reliable; the opposite of biased
Observation an empirical item of information for a research unit
Open categories measurement categories that have undefined limits
Ordered array a tabular display of coded data listed in order by magnitude of the scores
Ordinal measurement measurement that classifies observations by kind and order by magnitude of the scores
Parameter description of a characteristic of a population
Parameter estimation an inferential technique designed to estimate the value of a parameter that is unknown
Parametric test a test of a null hypothesis in which certain assumptions about a population's distribution or parameters must be satisfied
Partial association a description of the joint occurrence between two variables while controlling for one or more other variables
Partial correlation, coefficient of a measure of correlation for a partial association
Partial distribution a distribution containing only selected parts of an entire distribution
Pearson's contingency coefficient (C) a measure of association strength for nominal data based on chi-square; most useful for square tables larger than 2×2
Pearson's phi (ϕ) a measure of association strength for nominal data based on chi-square for $2 \times k$ tables
Pearson's product-moment coefficient of correlation (r) a symmetrical measure of linear association between two interval variables
Percentage (%) a frequency divided by the total number of cases and then multiplied by 100
Percentage change a ratio which compares the amount of change that has occurred between two points in time to the quantity at the first point as a percentage
Pictograph a figure or likeness used to illustrate a distribution
Point estimate a single estimated value of an unknown parameter based on a sample statistic
Polygon a graph showing a continuous distribution as a multisided closed figure
Population the entire set of observations or cases that is the subject of interest in research
Population standard deviation (σ) descriptive measure of a population's standard deviation
Power of a test the probability that a decision to reject a null hypothesis is a correct decision
PRE measure a proportional-reduction-in-error measure which provides a normed indicator of association strength based on a comparison of two methods of predicting scores on a variable
Precision of measurement the extent of accuracy maintained within a measurement or its subsequent analysis
Primary sampling units (PSU) the initial groupings of research units in cluster sampling
Probability the likelihood that certain outcomes will occur in a situation over the long run
Probability, cumulative the probability that one or more outcomes and all others within a defined range of possibilities will occur
Probability, point the probability that one or more exact outcomes will occur
Probability, simple probability for discrete distributions
Proportion (P) a frequency divided by the total number of cases
Range (R) a measure of variation for ordinal or interval data that calculates the difference between the highest and lowest scores in a distribution

Rate a ratio that compares the observed frequency of an event to the potential frequency of that event

Ratio a comparison between two quantities

Ratio measurement measurement that classifies observations by kind, order, and numerical differences on a scale having an absolute zero point

Raw data distribution a tabular presentation of observations as originally coded

Regression an asymmetrical measure of linear association between interval variables useful for making prediction-type explanations

Regression intercept (*a*) the regression-predicted value of the dependent variable when the independent variable is zero; value at which the regression line intersects the Y axis when graphed

Regression line graphic line representing the best linear fit between interval variables

Regression slope or coefficient (*b*) the incremental change in the dependent variable that is predicted by regression to occur for every unit change in the independent variable; the angle of the regression line when graphed

Rejection region an extreme area (or two such areas) within a graphic display of a sampling distribution which describes all those sample outcomes whose cumulative probability is equal to the level of significance chosen for a hypothesis test

Reliability the extent to which a measurement consistently represents an intended characteristic

Research unit the individual, social group, community, society, or other entity for which observations are collected in social research

Residual a measure of the deviation between an actual score and its value as predicted by regression

Robust the tendency for a statistical test to produce an accurate inference in spite of violations to its assumptions

Sample the set of observations or cases selected for study from a larger set (population)

Sample size (*N*) number of research units included in a sample

Sample standard deviation (*S*) descriptive measure of a sample's standard deviation

Sampling, cluster a probability sampling method which selects research units in several stages using a technique of grouping those units on the basis of geographic location

Sampling, convenience a nonprobability sampling method in which research units are selected from a population based on their ease of inclusion

Sampling, judgment a nonprobability sampling method in which research units are selected based on a researcher's opinion of their desirability for inclusion

Sampling, nonprobability a sampling method in which the probabilities that the research units will be selected from a population are unknown

Sampling, probability a sampling method in which the probabilities that the research units will be selected from a population are known

Sampling, quota a nonprobability sampling method in which research units are selected nonrandomly from strata within a population

Sampling, simple random a sampling method in which every research unit has an equal and known probability of being selected from a population

Sampling, stratified a probability sampling method in which research units are selected randomly from strata within a population

Sampling, systematic a probability sampling method in which research units are selected at a set interval from a population list

Sampling, volunteer a nonprobability sampling method in which research units are selected from a population based on their willingness to be included

Sampling distribution a theoretical probability distribution describing all possible sample outcomes for a statistic

Sampling distribution of differences in means the theoretical probability distribution of all possible differences between pairs of sample means based on samples of a set size drawn from a population

Sampling distribution of differences in proportions the theoretical probability distribution of all possible differences between pairs of sample proportions based on samples of a set size drawn from a population

Sampling distribution of means the theoretical probability distribution of all possible sample means based on samples of a set size drawn from a population
Sampling distribution of proportions the theoretical probability distribution of all possible sample proportions based on samples of a set size drawn from a population
Sampling error the error that naturally occurs when a sample is drawn from a population whereby some predicatable degree of misrepresentation of that population results
Scale the set of numerical values used in the measurement of a variable
Scatter diagram a Cartesian graph of the joint distribution of two interval variables displayed for individual cases
Skew lack of symmetry in a distribsution
Spearman's coefficient of correlation (r_s) a symmetrical measure of linear association for data in rank order
Spurious association an observed association which can be completely explained by a controlled variable and is therefore considered not to be genuine
Standard deviation a measure of variation for interval data that calculates the square root of the arithmetic average of the sum of squared deviation scores
Standard error standard deviation of a sampling distribution
Standard error of differences in means ($\sigma_{\bar{X}_1-\bar{X}_2}$) standard deviation of the sampling distribution of differences in means
Standard error of differences in proportions ($\sigma_{p_1-p_2}$) standard deviation of the sampling distribution of differences in proportions
Standard error of means ($\sigma_{\bar{X}}$) standard deviation of the sampling distribution of means
Standard error of proportions (σ_p) standard deviation of the sampling distribution of proportions
Standard normal distribution a bell-shaped, symmetrical distribution having a mean of 0 and a standard deviation of 1
Standard score a transformed score expressed as a deviation from an expected value that has a standard deviation as its unit of measurement (e.g., *z* or *t* score)
Standardized measure a measure whose original units of measurement have been transformed to make the measure comparable for some feature of measurement (e.g., case size or units of measurement) between distributions
Standardized regression coefficient (b^*) the standardized measure of a regression coefficient which is fully comparable between regressions; also known as beta or beta weight
Statistic description of a characteristic of a sample
Statistical methods procedures for interpreting data based on principles of mathematics and scientific inquiry
Student's *t* distributions a set of sampling distributions describing sample outcomes when the population's standard deviation is unknown but the population is normally distributed; often applied to inferences from small samples
Sum of squares between (*SSB*) variation between groups in analysis of variance
Sum of squares total (*SST*) total variation in a dependent variable for all groups combined in analysis of variance
Sum of squares within (*SSW*) variation within groups in analysis of variance
Symmetrical measure a measure of association in which no distinction is made between independent and dependent variables
Symmetry that feature of a distribution's form which describes the extent to which the scores can be divided into two mirror images graphically
***t* score** standard score belonging to one of the Student's *t* distributions
Table a tabulated description of a distribution
Tabular limits limits of measurement categories as they appear in a table displaying a distribution of scores
Tabular statistic value in a sampling distribution that defines the rejection region(s); also known as the critical value
Test statistic standardized score transforming an observed sample outcome to a value within a known sampling distribution

Theory an integrated set of general statements which together offer an explanation for some portion of concrete reality as it is experienced

Time-series distribution a distribution containing time as one of its variables

Traditional research model a model of social research which originates with theory and uses observations as the test of that theory

True limits exact numerical limits of measurement categories

True limits range (*TLR*) a measure of variation for ordinal or interval data that calculates the difference between the true upper limit of the highest score and the true lower limit of the lowest score in a distribution

Tukey *HSD* test a parametric test used to identify specific pairs of group means that differ significantly following an analysis of variance

Type I error the error that results from rejecting a null hypothesis that is true

Type II error the error that results from failing to reject a null hypothesis that is false

Ultimate sampling units the final groupings of research units in cluster sampling

Unit of measurement the smallest difference being quantified by a measurement; appropriate only at the interval and ratio levels of measurement

Univariate measurement or analysis of one variable

Univariate or simple frequency distribution a two-column tabular display of the distribution of a single variable showing the measurement categories and the frequency of scores in each category

Validity the extent to which a measurement truly represents an intended characteristic

Variable a measured characteristic that can take on different values for the research units

Variance a measure of variation for interval data that calculates the arithmetic average of the sum of squared deviation scores

Variation the extent and manner in which the scores in a distribution differ from each other

Variation, explained that part of the total variation in a variable that is explained (predicted) by regression; also known as regression variation

Variation, total sum of squared deviations around the mean of a distribution

Variation, unexplained that part of the total variation in a variable that is not explained (predicted) by regression; also known as error or residual variation

Weighted mean ($\bar{X}_w$) aggregated mean for more than one group of scores calculated to give proportional weight to each group's separate mean according to its sample size

***z* standard score** a standard score belonging to the standard normal distribution

APPENDIX 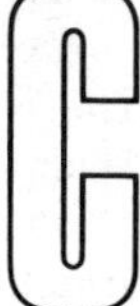

GLOSSARY OF FORMULAS

Page		
37	Category width	$w = l_u - l_l$
37	Category midpoint	$m = \frac{l_u - l_l}{2}$
39	Proportion	$P = \frac{f}{N}$
40	Percentage	$\% = \frac{f}{N}(100)$
42	Cumulative frequency	$cf = f_1 + f_2 + \cdots + f_k$
42	Cumulative percentage	$c\% = \frac{cf}{N}(100)$
44	Ratio	$\frac{f_a}{f_b}$
44	Sex ratio	$\frac{f_m}{f_f}(100)$
44	Crude birthrate	$\frac{\text{number of live births}}{\text{total population at midyear}}(1000)$
45	General fertility rate	$\frac{\text{number of live births}}{\text{population of women between 15 and 44 at midyear}}(1000)$
54	Percentage change	$\left[\frac{f_{t2} - f_{t1}}{f_{t1}}\right](100)$
90	Estimated median	$\widehat{Mdn} = l_m + \left[\frac{\frac{N}{2} - \text{cf}_{bm}}{f_m}\right] w_m$
82	Mean (for raw data)	$\bar{X} = \frac{\Sigma X}{N}$
87	(for simple frequencies)	$\bar{X} = \frac{\Sigma fX}{N}$

Page		
92	(for grouped frequencies)	$\hat{\bar{X}} = \frac{\Sigma fm}{N}$
83	Weighted mean (for two groups)	$\bar{X}_w = \frac{N_a(\bar{X}_a) + N_b(\bar{X}_b)}{N_a + N_b}$
84	Deviation score	$x = X - \bar{X}$
107	Index of Dispersion	$D = \frac{k(N^2 - \Sigma f^2)}{N^2(k-1)}$
110	Range	$R = X_h - X_l$
110	True limits range	$TLR = U_h - L_l$
113	Interquartile range	$Q = Q_3 - Q_1$
116	Mean deviation	$MD = \frac{\Sigma \lvert x \rvert}{N}$
118, 119, 125	Standard deviation (for raw data)	$\sigma = \sqrt{\frac{\Sigma x^2}{N}}$ or $\sqrt{\frac{\Sigma X^2 - \frac{(\Sigma X)^2}{N}}{N}}$
		$S = \sqrt{\frac{\Sigma x^2}{N}}$ or $\sqrt{\frac{\Sigma X^2 - \frac{(\Sigma X)^2}{N}}{N}}$
		$s = \sqrt{\frac{\Sigma x^2}{N-1}}$ or $\sqrt{\frac{\Sigma X^2 - \frac{(\Sigma X)^2}{N}}{N-1}}$
132	Coefficient of variation	$CV = \frac{S}{\bar{X}}$
149	Chi-square	$\chi^2 = \Sigma \frac{(f_o - f_e)^2}{f_e}$
150	Expected frequency	$f_e = \frac{(T_r)(T_c)}{N}$
152	Maximum chi-square	$\text{Max } \chi^2 = N(k-1)$
152	Pearson's phi	$\phi = \sqrt{\frac{\chi^2}{N}}$
153	Pearson's contingency coefficient	$C = \sqrt{\frac{\chi^2}{\chi^2 + N}}$
153	Maximum *C* (for square tables)	$\text{Max } C = \sqrt{\frac{k-1}{k}}$
154	Cramér's *V*	$V = \sqrt{\frac{\chi^2}{N(k-1)}}$
160	PRE (basic formula)	$\text{PRE value} = \frac{E_1 - E_2}{E_1}$
162	Lambda	$L = \frac{E_1 - E_2}{E_1}$

Page		
165	Goodman and Kruskal's tau-y	$T_y = \frac{E_1 - E_2}{E_1}$ where $E_1 = \Sigma \left[\frac{N - F_y}{N}(F_y)\right]$, $E_2 = \Sigma \left[\frac{F_x - f}{F_x}(f)\right]$
168	Total number of pairs	$P_t = \frac{N(N-1)}{2}$
170	Goodman and Kruskal's gamma	$G = \frac{P_a - P_d}{P_a + P_d}$
175	Spearman's coefficent of correlation	$r_s = 1 - \left[\frac{6 \Sigma D^2}{N(N^2 - 1)}\right]$
194	Straight-line equation	$Y = a + bX$
195	Linear regression equation	$\hat{Y} = a + bX$
196	Regression slope or coefficient	$b = \frac{\Sigma (X - \bar{X})(Y - \bar{Y})}{\Sigma (X - \bar{X})^2}$ or $\frac{\Sigma xy}{\Sigma x^2}$ or $\frac{N(\Sigma XY) - (\Sigma X)(\Sigma Y)}{N(\Sigma X^2) - (\Sigma X)^2}$
196	Covariation	$\Sigma (X - \bar{X})(Y - \bar{Y})$
196	Regression intercept	$a = \bar{Y} - b\bar{X}$
199	Standardized regression coefficient	$b^* = b\left(\frac{S_x}{S_y}\right)$
203	Residual	$(Y - \hat{Y})$
204	Unexplained variation	$\Sigma (Y - \hat{Y})^2$
204	Explained variation	$\Sigma (\hat{Y} - \bar{Y})^2$
206	Pearson's coefficient of correlation	$r = \frac{\Sigma (X - \bar{X})(Y - \bar{Y})}{\sqrt{[\Sigma (X - \bar{X})^2][\Sigma (Y - \bar{Y})^2]}}$ or $\frac{\Sigma xy}{\sqrt{(\Sigma x^2)(\Sigma y^2)}}$ or $\frac{N \Sigma XY - (\Sigma X)(\Sigma Y)}{\sqrt{[N(\Sigma X)^2 - (\Sigma X)^2][N(\Sigma Y)^2 - (\Sigma Y)^2]}}$
207	Coefficient of determination	$r^2 = (r)^2$
213	First-order partial correlation coefficient	$r_{y1.2} = \frac{r_{y1} - r_{y2}r_{12}}{\sqrt{(1 - r_{y2}^2)(1 - r_{12}^2)}}$
214	Multiple regression	$\hat{Y} = a + b_1X_1 + b_2X_2 + \cdots + b_kX_k$
232	Probability of an outcome (A)	$\Pr = \text{expected}\left(\frac{f}{N}\right)$
238	Binomial probability of success	$\Pr(r \text{ successes}) = (C_r^N)(p^r)(q^{N-r})$

Page		
243	z standard score (for raw data)	$z = \dfrac{X - \bar{X}}{S}$
294	Standard error of proportions	$\sigma_p = \sqrt{\dfrac{PQ}{N}}$
294	z transformation for single sample proportions	$z = \dfrac{p - P}{\sigma_p}$
303	Standard error of means	$\sigma_{\bar{X}} = \dfrac{\sigma}{\sqrt{N}}$
303	z transformation for single sample means	$z = \dfrac{\bar{X} - \mu}{\sigma_{\bar{X}}}$
306	Estimated standard error of means	$\hat{\sigma}_{\bar{X}} = \dfrac{s}{\sqrt{N}}$
310	t transformation of single sample means	$t = \dfrac{\bar{X} - \mu}{\hat{\sigma}_{\bar{X}}}$
314	General formula for a confidence interval	$CI = \text{observed statistic} \pm \begin{pmatrix}\text{tabular}\\ \text{standard}\\ \text{score}\end{pmatrix}\begin{pmatrix}\text{standard}\\ \text{error}\end{pmatrix}$
314	Level of confidence	$(1 - \alpha)100$
315	Confidence interval for a proportion	$CI = p \pm (z)(\sigma_p \text{ or } \hat{\sigma}_p)$
320	Confidence interval for a mean	$CI = \bar{X} \pm (z \text{ or } t)(\sigma_{\bar{X}} \text{ or } \hat{\sigma}_{\bar{X}})$
325	Chi-square-I test statistic	$\chi^2 = \Sigma \dfrac{(f_o - f_e)^2}{f_e}$
330	*KS-I* test statistic	$LD = \lvert <cp_o - <cp_e \rvert$
342	Estimate of P and Q from two samples	$\hat{P} = p = \dfrac{N_1 p_1 + N_2 p_2}{N_1 + N_2}$
		$\hat{Q} = q = 1 - p$
343	Standard error of differences in proportions	$\sigma_{p_1 - p_2} = \sqrt{PQ\left(\dfrac{1}{N_1} + \dfrac{1}{N_2}\right)}$
343	Estimated standard error or differences in proportions	$\hat{\sigma}_{p_1 - p_2} = \sqrt{pq\left(\dfrac{1}{N_1} + \dfrac{1}{N_2}\right)}$
343	z transformation of a difference in proportions	$z = \dfrac{(p_1 - p_2) - (P_1 - P_2)}{\sigma_{p_1 - p_2}}$
346	z transformation of a difference in means	$z = \dfrac{(\bar{X}_1 - \bar{X}_2) - (\mu_1 - \mu_2)}{\sigma_{\bar{X}_1 - \bar{X}_2}}$
347	Standard error of differences in means	$\sigma_{\bar{X}_1 - \bar{X}_2} = \sqrt{\dfrac{\sigma_1^2}{N_1} + \dfrac{\sigma_2^2}{N_2}}$
348	Pooled estimate of population variance	$s_p^2 = \dfrac{(N_1 - 1)s_1^2 + (N_2 - 1)s_2^2}{N_1 + N_2 - 2}$

Page		
349	Estimated standard error of differences in means	
	(when $\sigma_1^2 = \sigma_2^2$)	$\hat{\sigma}_{\bar{X}_1-\bar{X}_2} = \sqrt{s_p^2\left(\frac{1}{N_1} + \frac{1}{N_2}\right)}$
	(when $\sigma_1^2 \neq \sigma_2^2$)	$\hat{\sigma}_{\bar{X}_1-\bar{X}_2} = \sqrt{\frac{s_1^2}{N_1} + \frac{s_2^2}{N_2}}$
349	Estimated *df* for test of difference in means (when $\sigma_1^2 \neq \sigma_2^2$)	$df = \frac{N_1 df_1 + N_2 df_2}{N_1 + N_2}$
360	Standard error of difference in dependent means (definitional)	$\hat{\sigma}_{\bar{X}_1-\bar{X}_2} = \sqrt{\sigma_{\bar{X}_1}^2 + \sigma_{\bar{X}_2}^2 - 2\rho\sigma_{\bar{X}_1}\sigma_{\bar{X}_2}}$
361	(computational with population data)	$\hat{\sigma}_{\bar{X}_1-\bar{X}_2} = \sqrt{\left[\frac{\Sigma D^2}{N} - \left(\frac{\Sigma D}{N}\right)^2\right]\frac{1}{N}}$
361	(computational with sample data)	$\hat{\sigma}_{\bar{X}_1-\bar{X}_2} = \sqrt{\frac{\Sigma D^2 - \frac{(\Sigma D)^2}{N}}{N(N-1)}}$
366	Degrees of freedom for χ^2-*II* test	$df = (r-1)(c-1)$
366	Chi-square-II test statistic	$\chi^2 = \Sigma \frac{(f_o - f_e)^2}{f_e}$
374	*KS-II* tabular statistic	$D = (\text{constant})\sqrt{\frac{N_1 + N_2}{N_1 N_2}}$
394	Degrees of freedom between groups	$dfb = k - 1$
394	Degrees of freedom within groups	$dfw = N_T - k$
394	Degrees of freedom total	$dft = N_T - 1$
	Sum of squares total	
396	(definitional)	$SST = \Sigma (X - \bar{X}_T)^2$
398	(computational)	$SST = \Sigma X_T^2 - \frac{(\Sigma X_T)^2}{N_T}$
	Sum of squares between	
397	(definitional)	$SSB = \Sigma [N_i(\bar{X}_i - \bar{X}_T)^2]$
398	(computational)	$SSB = \left[\sum \frac{(\Sigma X_i)^2}{N_i}\right] - \frac{(\Sigma X_T)^2}{N_T}$
398	Sum of squares within	
	(definitional)	$SSW = \Sigma x_1^2 + \Sigma x_2^2 + \cdots + \Sigma x_k^2$
	(computational)	$SSW = \Sigma\left[\Sigma X_i^2 - \frac{(\Sigma X_i)^2}{N_i}\right]$
400	Mean square between	$MSB = \frac{SSB}{dfb}$
400	Mean square within	$MSW = \frac{SSW}{dfw}$

Page		
400	Test statistic of F	$F = \frac{MSB}{MSW}$
411	Tukey HSD statistic	$HSD = q_{\alpha}\sqrt{\frac{MSW}{N_i}}$
414	Chi-square-k sample test statistic	$\chi^2\text{-}k = \Sigma \frac{(f_o - f_e)^2}{f_e}$
418	Kruskal–Wallis H statistic	$H = \frac{12}{N(N+1)}\left[\sum \frac{(\Sigma R_i)^2}{N_i}\right] - [3(N+1)]$
420	Friedman's two-way analysis of variance statistic	$\chi_r^2 = \frac{12}{Nk(k+1)}[\Sigma (\Sigma R_i)^2] - [3N(k+1)]$
439	z transformation for gamma	$z = G\sqrt{\frac{P_a + P_d}{N(1 - G^2)}}$
443	z transformation for r_s	$z = r_s\sqrt{N-1}$
447	z transformation for r	$z = r\sqrt{N-1}$

APPENDIX D

TABLES

Appendix D Acknowledgments

The author is grateful to the authors and publishers listed below for permission to adapt from the following tables.

Table D1 RAND Corporation, *A Million Random Digits.* Glencoe, Ill: The Free Press of Glencoe, 1955.

Table D2 Adapted from Table A of R. P. Runyon and A. Haber, *Fundamentals of Behavioral Statistics,* 4th ed. Reading, Mass.: Addison-Wesley Publishing Co., Inc., 1980, with additional entries adapted from Table B of H. Loether and D. McTavish, *Descriptive and Inferential Statistics: An Introduction,* 2nd ed. Boston: Allyn and Bacon, Inc., 1980, by permission of the publishers.

Table D3 Adapted from Table 2.1 of D. B. Owen, *Handbook of Statistical Tables.* Reading, Mass.: Addison-Wesley Publishing Co., Inc., 1962, with additional values from E. T. Federighi, "Extended Tables of the Percentage Points of Student's t-distribution," *Journal of the American Statistical Association,* 54; 1959:683–688, by permission of the publishers.

Table D4 Adapted from Table 3.1 of D. B. Owen, *Handbook of Statistical Tables.* Reading, Mass.: Addison-Wesley Publishing Co., Inc., 1962, by permission of the publisher.

Table D5 Adapted from F. J. Massey, Jr., "The Kolmogorov–Smirnov Test for Goodness of Fit," *Journal of the American Statistical Association,* 46; 1951:70, with permission of the publisher. Reprinted from Table E of S. Siegel, *Nonparametric Statistics for the Behavioral Sciences,* New York: McGraw-Hill Book Company, 1956, by permission of the publisher.

Table D6 Adapted from N. Smirnov, "Tables for Estimating the Goodness of Fit of Empirical Distributions," *Annals of Mathematical Statistics,* 19; 1948:280–281, with permission of the Institute of Mathematical Statistics. Reprinted from Table M of S. Siegel, *Nonparametric Statistics for the Behavioral Sciences.* New York: McGraw-Hill Book Company, 1956, by permission of the publisher.

Table D7 Reprinted from G. W. Snedecor and W. G. Cochran, *Statistical Methods,* 6th ed. Ames, Iowa: Iowa State University Press, copyright 1967, with permission of the publisher.

Table D8 Adapted from Table 29 of E. S. Pearson and H. O. Hartley, *Biometrika Tables for Statisticians,* Vol. 1, 2nd ed. New York: Cambridge University Press, 1958, by permission of the *Biometrika Trustees.*

Table D9 Adapted from E. G. Olds, "The 5 Percent Significance Levels of Sums of Squares of Rank Differences and a Correction," *Annals of Mathematical Statistics,* 20; 1949:117–118, and E. G. Olds, "Distributions of Sums of Squares of Rank Differences for Small Numbers of Individuals," *Annals of Mathematical Statistics,* 9; 1938:133–148, by permission of the Institute of Mathematical Statistics. Reprinted from Table G of R. P. Runyon and A. Haber, *Fundamentals of Behavioral Statistics,* 4th ed. Reading, Mass.: Addison-Wesley Publishing Co., Inc., 1980, by permission of the publisher.

Table D10 Adapted from Table V.A. of R. A. Fisher, *Statistical Methods for Research Workers,* 14th ed. Edinburgh: Oliver & Boyd Ltd., 1970, by permission of the publisher. Supplementary values were calculated at San Jose State University by K. Fernandes. Reprinted from E. Minium, *Statistical Reasoning in Psychology and Education,* 2nd ed. New York: John Wiley & Sons, Inc., 1978, by permission of the publisher.

Table D11 Reprinted from Table VII of A. L. Edwards, *Statistical Methods,* 3rd ed. New York: Holt, Rinehard and Winston, 1973, by permission of the author and publisher.

Table D12 From F. Wilcoxon, S. Katti, and R. A. Wilcox, *Critical Values and Probability Levels for the Wilcoxon Rank Sum Test and the Wilcoxon Signed Rank Test.* New York: American Cyanamid Co., 1963, by permission of the publisher. Reprinted from Table J of R. P. Runyon and A. Haber, *Fundamentals of Behavioral Statistics,* 4th ed. Reading, Mass.: Addison-Wesley Publishing Co., Inc., 1980, by permission of the publisher.

Table D13 Adapted from F. S. Swed and C. Eisenhart, "Tables for Testing Randomness of Grouping in a Sequence of Alternatives," *Annals of Mathematical Statistics,* 14; 1943:83–86, by permission of the Institute of Mathematical Statistics. Reprinted in modified form from Table F_I of S. Siegel, *Nonparametric Statistics for the Behavioral Sciences.* New York: McGraw-Hill Book Company, 1956, by permission of the publisher.

Table D14 Adapted from H. B. Mann and D. R. Whitney, "On a Test of Whether

One of Two Random Variables Is Stochastically Larger than the Other," *Annals of Mathematical Statistics,* 18; 1947:52–54, with permission of the Institute of Mathematical Statistics. Reprinted from Table I of R. P. Runyon and A. Haber, *Fundamentals of Behavioral Statistics,* 4th ed. Reading, Mass.: Addison-Wesley Publishing Co., Inc., 1980, by permission of the publisher.

Table D15 From H. M. Walker and J. Lev, *Statistical Inference.* New York: Henry Holt and Company, 1953, copyright 1953 by Holt, Rinehart and Winston, by permission of the authors and publisher. Reprinted from Table A18 of D. Champion, *Basic Statistics for Social Research,* 2nd ed. New York: Macmillan Publishing Co., Inc., by permission of the publisher.

TABLE D1 Random Numbers

19612	78430	11661	94770	77603	65669	86868	12665	30012	75989
39141	77400	28000	64238	73258	71794	31340	26256	66453	37016
64756	80457	08747	12836	03469	50678	03274	43423	66677	82556
92901	51878	56441	22998	29718	38447	06453	25311	07565	53771
03551	90070	09483	94050	45938	18135	36908	43321	11073	51803
98884	66209	06830	53656	14663	56346	71430	04909	19818	05707
27369	86882	53473	07541	53633	70863	03748	12822	19360	49088
59066	75974	63335	20483	43514	37481	58278	26967	49325	43951
91647	93783	64169	49022	98588	09495	49829	59068	38831	04838
83605	92419	39542	07772	71568	75673	35185	89759	44901	74291
24895	88530	70774	35439	46758	70472	70207	92675	91623	61275
35720	26556	95596	20094	73750	85788	34264	01703	46833	65248
14141	53410	38649	06343	57256	61342	72709	75318	90379	37562
27416	75670	92176	72535	93119	56077	06886	18244	92344	31374
82071	07429	81007	47749	40744	56974	23336	88821	53841	10536
21445	82793	24831	93241	14199	76268	70883	68002	03829	17443
72513	76400	52225	92348	62308	98481	29744	33165	33141	61020
71479	45027	76160	57411	13780	13632	52308	77762	88874	33697
83210	51466	09088	50395	26743	05306	21706	70001	99439	80767
68749	95148	94897	78636	96750	09024	94538	91143	96693	61886
05184	75763	47075	88158	05313	53439	14908	08830	60096	21551
13651	62546	96892	25240	47511	58483	87342	78818	07855	39269
00566	21220	00292	24069	25072	29519	52548	54091	21282	21296
50958	17695	58072	68990	60329	95955	71586	63417	35947	67807
57621	64547	46850	37981	38527	09037	64756	03324	04986	83666
09282	25844	79139	78435	35428	43561	69799	63314	12991	93516
23394	94206	93432	37836	94919	26846	02555	74410	94915	48199
05280	37470	93622	04345	15092	19510	18094	16613	78234	50001
95491	97976	38306	32192	82639	54624	72434	92606	23191	74693
78521	00104	18248	75583	90326	50785	54034	66251	35774	14692
96345	44579	85932	44053	75704	20840	86583	83944	52456	73766
77963	31151	32364	91691	47357	40338	23435	24065	08458	95366
07520	11294	23238	01748	41690	67328	54814	37777	10057	42332
38423	02309	70703	85736	46148	14258	29236	12152	05088	65825
02463	65533	21199	60555	33928	01817	07396	89215	30722	22102

TABLE D1 Random Numbers (continued)

15880	92261	17292	88190	61781	48898	92525	21283	88581	60098
71926	00819	59144	00224	30570	90194	18329	06999	26857	19238
64425	28108	16554	16016	00042	83229	10333	36168	65617	94834
79782	23924	49440	30432	81077	31543	95216	64865	13658	51081
35337	74538	44553	64672	90960	41849	93865	44608	93176	34851
05249	29329	19715	94082	14738	86667	43708	66354	93692	25527
56463	99380	38793	85774	19056	13939	46062	27647	66146	63210
96296	33121	54196	34108	75814	85986	71171	15102	28992	63165
98380	36269	60014	07201	62448	46385	42175	88350	46182	49126
52567	64350	16315	53969	80395	81114	54358	64578	47269	15747
78498	90830	25955	99236	43286	91064	99969	95144	64424	77377
49553	24241	08150	89535	08703	91041	77323	81079	45127	93686
32151	07075	83155	10252	73100	88618	23891	87418	45417	20268
11314	50363	26860	27799	49416	83534	19187	08059	76677	02110
12364	71210	87052	50241	90785	97889	81399	58130	64439	05614

12210	78708	26642	30263	18817	83563	43508	30067	00814	52666
08207	84246	54377	04106	27748	65845	68959	24971	48296	06660
02389	16728	53183	18029	47943	18426	54760	01032	05582	45145
45957	89199	48812	49184	72609	12674	63480	17832	59507	47912
60472	69188	47417	62238	21912	70058	58128	67803	06484	90108
62879	77961	72626	08175	18237	63515	82297	26351	16342	58494
98479	47773	90234	38261	63231	46778	34636	20862	28247	62686
02221	54757	16467	42549	28168	92746	48416	00562	72246	58246
61912	61559	05124	98560	20379	78940	39742	29119	15136	76956
53879	10206	77322	68066	04668	27566	32941	62471	71653	46763
17706	78400	39590	91343	68952	86244	32101	41970	41949	83142
68412	01172	53330	63158	79023	67487	92224	44119	80203	47668
38380	85484	78758	48901	15416	54075	60035	92619	53423	55260
33691	53591	82856	22508	02838	18926	48821	80106	93755	96540
30028	36404	38583	56854	70045	94757	29048	97631	76683	95703
84597	44324	24456	20798	44230	64591	81741	11738	43053	49624
57808	80613	75014	40263	16842	20538	65586	27306	88686	37190
15026	95952	25448	17970	81020	53419	15706	03239	61798	58922
74615	50050	83459	18039	08342	23463	00062	78817	81351	67053
55740	91278	06817	81766	44781	19601	49551	76171	29122	21903
36807	62588	19791	66911	65385	84379	62182	01684	38690	52325
74616	99059	26153	88367	69058	48902	12245	39588	91052	89755
28780	69061	96114	21835	11459	65341	00511	88624	09746	14555
08178	22109	56063	88340	55965	08752	25993	09996	46470	29267
02893	70718	06836	67302	42540	78583	84581	11820	66843	90930
86840	68495	31789	12190	85453	55099	32460	81494	25440	93376
78290	85310	48915	74363	21216	40536	69311	31228	69523	49379
57658	02695	07529	81325	38551	09341	70933	31770	11103	50566
57241	21735	05496	49587	25762	56783	72103	44942	35027	55472
81346	95718	52962	95122	33641	98424	94558	49682	00395	90835

TABLE D1 Random Numbers (continued)

26203	95922	34009	69651	01356	64264	61445	26597	04005	38318
75437	58382	36108	44359	98592	33148	41686	63480	84801	64312
73126	65038	02219	80983	66429	26432	92263	24923	72437	04278
22732	98009	62337	35191	44256	97591	31050	02342	30141	75975
85893	00162	29357	03746	70544	45658	43214	09614	32926	52149
13774	55916	86636	60615	20977	97863	19044	94411	61029	15045
88188	67131	39604	16544	53367	12097	85278	97970	37989	73168
06781	68222	29334	46018	06793	64226	47956	40942	67801	94233
95685	15658	90989	46006	60094	24631	39331	66466	98200	57210
95483	11390	79500	95929	21018	80234	11967	02100	15249	75466
98447	86088	38236	52711	97289	05929	34306	41298	20159	16429
86732	24965	57529	65692	45102	94170	72991	34499	25469	88295
00017	14428	13950	27867	96012	90020	05657	36040	32013	73593
19577	23379	32314	88972	80943	00810	24438	02910	26327	11837
99949	02938	04327	68140	96504	90920	43025	42032	59946	39301
07888	07495	49523	62903	72418	67769	73333	50603	40895	79813
71122	74229	02785	51996	17100	75967	17346	28091	15781	12987
66431	12160	32367	29420	02365	41955	78345	64987	69707	19315
07508	74254	15399	78034	41472	74756	36962	91117	24407	87084
08235	04811	09261	23772	48732	68675	23262	05142	21310	52588

99576	03093	87527	35848	71876	21027	42002	67787	17981	67468
87429	81377	22996	43586	44303	98724	85239	86036	92063	25693
69704	95249	20426	55681	48290	56229	51269	93367	13304	62412
58287	91145	74044	73706	43089	43285	52871	20600	15172	63205
44543	06981	64923	94700	18579	34495	92207	46201	83687	45565
51345	82679	85959	58707	74177	73530	44421	72805	72445	77638
60805	54760	49291	35370	31358	53728	48837	22398	78577	34699
94709	77225	49119	59636	32374	29280	14821	72847	37038	97215
40692	55578	31932	31631	99065	59480	11434	94820	33320	48576
85772	87197	77476	86304	21054	27750	54141	47031	91362	46553
64109	17486	07580	90241	25288	67441	63958	26040	72928	58371
44130	68856	84927	08921	16351	34826	81753	13036	68664	55103
47570	64977	11839	81049	14859	45800	93519	56783	26678	95403
71505	97496	44242	62072	02427	67070	61558	77425	33061	28573
33130	39631	56393	41573	90914	28526	40997	43082	84646	00789
64943	12781	40642	46853	12205	35390	29999	58113	87978	44549
11636	25414	07362	35505	53651	23549	59229	02636	94269	67293
10140	13163	92380	98525	55559	43111	43137	64856	78413	73472
86815	46792	68828	32135	09493	95685	40409	47342	21559	81035
02601	84804	30947	70588	31510	36512	86046	92035	80971	78461
84167	05822	78593	71579	71522	17517	34760	26696	68553	63587
71741	48249	03605	81257	52236	59017	81071	02322	93910	51656
20901	53017	63203	17806	46713	33609	28888	39065	45852	41570
15574	07082	19542	60725	97056	66901	32613	53817	63856	85991
49319	56431	40655	18391	61770	46077	35487	25332	03194	16694

TABLE D1 Random Numbers (continued)

72195	68767	60846	08353	57475	15280	52887	51368	15713	20758
44052	49228	15868	07676	14994	39430	71803	79499	34989	51312
43646	55384	76359	07399	60317	42285	52859	30206	00584	51279
78899	85420	93118	33521	11608	50190	09894	77582	90196	22745
50201	89343	95604	32522	55549	33205	58427	86504	79224	85404
53664	24220	82877	11621	35505	32810	59665	42479	75120	42638
21884	09397	95109	51780	35828	34928	42561	23413	87256	96098
60576	79805	24509	92685	75800	57765	08333	81168	50606	36877
27642	32199	80919	80059	48973	19362	48226	64877	06207	08396
40917	46217	04820	20523	24956	43686	40304	01939	26727	90083
89658	80329	02180	54181	83692	53855	97220	64569	93510	54849
83426	65760	00098	29270	54332	33983	08605	52378	97373	23578
01837	82029	98140	60385	34759	38317	96270	14697	55137	91876
00997	61617	06272	00283	67396	91682	15334	73062	04327	58763
22529	54432	50704	00393	90650	11412	19933	86194	97585	31679
05078	55974	42408	34593	59319	84423	15322	92885	53548	71959
89729	54175	98683	72919	53001	99886	54586	86866	30767	60654
15565	90647	55917	05217	11039	86111	26083	49670	60226	09534
25500	60560	61431	52963	93685	74015	93275	16247	57468	87710
17339	53708	34099	75573	86717	02524	19945	79265	02488	79953
80302	89644	46540	16807	46562	61799	22816	88015	46304	13336
56452	13100	23947	31330	97352	12165	01178	98728	13430	64012
35392	18601	48424	38352	98377	11871	45511	07625	50878	92715
52728	68665	57473	88285	23203	22955	22663	49092	42908	33008
57726	68825	61025	59688	46649	45254	48217	66220	46873	11208

11767	41715	96921	39440	91713	01252	95477	90091	74570	47814
10209	39469	04395	96949	68306	59930	43058	74922	84796	25716
88035	86619	10325	12403	59223	80176	03721	42071	86811	58308
01488	22623	75963	52672	08033	94360	23673	80897	68904	98109
85453	72323	32385	04738	68076	97592	19285	41760	42244	51078
94870	60073	79608	75998	05122	18855	98943	26426	84879	74388
92793	48899	87556	46924	20939	72246	53768	14024	09495	40912
02048	90603	58746	02696	73115	25929	51870	61184	65085	78295
99402	91657	18017	77315	39059	24381	21328	36197	89852	69923
52804	36184	32274	65204	35397	44192	06055	60314	10940	34627
74967	53533	93594	69969	16216	97183	95154	37719	90074	06928
15360	86221	06240	68606	05993	28257	80451	90422	20624	31777
53092	10820	71341	56926	48072	70936	33884	63004	81011	90241
78293	43707	65260	46559	21593	61172	05802	48592	96801	88425
24698	22898	26339	77971	47433	41759	74193	54506	04385	71026
36341	31972	45946	23649	94999	27633	01609	19345	83854	65117
70032	05924	91583	26358	42395	85283	36548	48571	90295	65056
58505	19927	12491	29716	88554	84748	75150	20811	92332	52934
73594	99247	47006	04316	15214	68184	50502	00906	58323	81340
76445	56970	29913	49436	49418	89565	90371	02911	18781	49995

TABLE D1 Random Numbers (continued)

25906	22504	11638	73444	50364	29434	37229	81270	17091	08384
30906	29090	62259	64926	45247	63145	97190	79048	25472	63993
65174	27188	78410	88819	89691	71654	85623	76958	93888	34121
95707	46159	04585	08264	31148	55333	66354	20732	83733	71859
86800	89098	64605	45971	11658	89650	14679	29076	11295	26756
27050	19535	80967	86429	53087	24524	45176	58099	21062	93277
10321	30484	20157	44523	17221	52332	02171	69984	24729	45615
81688	67931	83947	92274	47572	05921	73696	13267	12007	36060
87754	63269	20414	97007	85234	02765	25037	08013	58209	32893
61003	95084	61292	17324	93461	48671	92398	73811	32577	03895
92634	41808	82881	62269	33586	15206	06382	92818	07525	18506
43023	46101	12756	91289	97832	03007	11112	83193	38078	10928
56571	76602	83541	73035	54212	82112	39610	98224	02571	03610
02277	75958	49212	59055	91260	88372	70664	91292	17350	78812
26333	27670	05556	64366	57386	17327	60852	92021	15754	55988
58163	45678	12771	96906	76195	16623	54681	38763	44930	12599
24866	01275	00882	31104	25970	06468	59772	11493	25244	57906
86997	96175	69773	38015	23916	94438	43706	85667	87188	72475
64237	20191	40654	96516	81157	18779	32641	89064	70199	22278
50601	68108	35854	45951	96090	18262	21094	13284	03783	47529
76183	41550	85252	38048	61986	96431	58408	90223	36116	13558
59038	73354	29214	64984	42285	99793	79106	35548	59039	40442
08430	83898	82979	99138	72201	37238	62822	29049	66756	80217
28493	94724	84455	17948	67633	89500	41013	08398	42274	57451
00322	75838	71501	37001	36824	74950	60632	11372	43392	66804
99942	82603	23184	26296	50994	21524	46967	28332	19674	32822
97872	36870	47178	69926	36075	93302	04530	10172	04809	96867
64219	06580	26192	82666	74607	31539	77593	73076	41422	56992
58098	55623	26057	28619	79776	69449	97532	09986	68865	92882
89624	83227	06730	16023	23771	51774	40547	13335	80053	88160

TABLE D2 Percentages of Area Under the Standard Normal Curve

Column 1 gives the z score for a standard normal distribution having a mean of 0, a standard deviation of 1.00, and a total area of 100%.

Column 2 gives the percentage of the area under the entire curve that is between the mean ($z = 0$) and the positive value of z. Areas for the negative values of z are the same as those for positive values of z, since the curve is symmetrical.

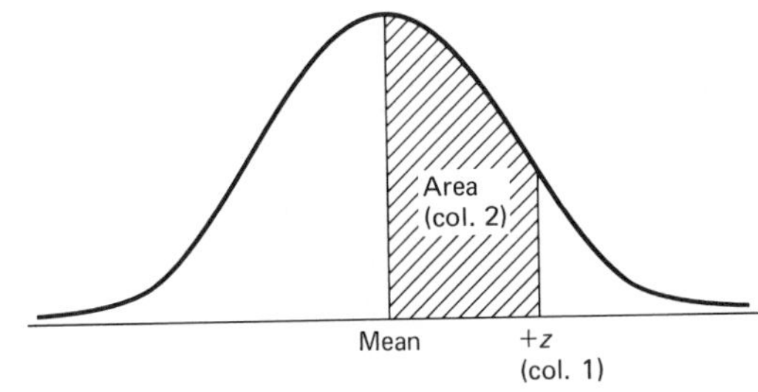

(1) z	(2) *Area between mean and* z	(1) z	(2) *Area between mean and* z	(1) z	(2) *Area between mean and* z
0.00	00.00	0.29	11.41	0.58	21.90
0.01	00.40	0.30	11.79	0.59	22.24
0.02	00.80	0.31	12.17	0.60	22.57
0.03	01.20	0.32	12.55	0.61	22.91
0.04	01.60	0.33	12.93	0.62	23.24
0.05	01.99	0.34	13.31	0.63	23.57
0.06	02.39	0.35	13.68	0.64	23.89
0.07	02.79	0.36	14.06	0.65	24.22
0.08	03.19	0.37	14.43	0.66	24.54
0.09	03.59	0.38	14.80	0.67	24.86
0.10	03.98	0.39	15.17	0.68	25.17
0.11	04.38	0.40	15.54	0.69	25.49
0.12	04.78	0.41	15.91	0.70	25.80
0.13	05.17	0.42	16.28	0.71	26.11
0.14	05.57	0.43	16.64	0.72	26.42
0.15	05.96	0.44	17.00	0.73	26.73
0.16	06.36	0.45	17.36	0.74	27.04
0.17	06.75	0.46	17.72	0.75	27.34
0.18	07.14	0.47	18.08	0.76	27.64
0.19	07.53	0.48	18.44	0.77	27.94
0.20	07.93	0.49	18.79	0.78	28.23
0.21	08.32	0.50	19.15	0.79	28.52
0.22	08.71	0.51	19.50	0.80	28.81
0.23	09.10	0.52	19.85	0.81	29.10
0.24	09.48	0.53	20.19	0.82	29.39
0.25	09.87	0.54	20.54	0.83	29.67
0.26	10.26	0.55	20.88	0.84	29.95
0.27	10.64	0.56	21.23	0.85	30.23
0.28	11.03	0.57	21.57	0.86	30.51

TABLE D2 Percentages of Area Under the Standard Normal Curve (continued)

(1) z	(2) *Area between mean and* z	(1) z	(2) *Area between mean and* z	(1) z	(2) *Area between mean and* z
0.87	30.78	1.35	41.15	1.83	46.64
0.88	31.06	1.36	41.31	1.84	46.71
0.89	31.33	1.37	41.47	1.85	46.78
0.90	31.59	1.38	41.62	1.86	46.86
0.91	31.86	1.39	41.77	1.87	48.68
0.92	32.12	1.40	41.92	1.88	48.71
0.93	32.38	1.41	42.07	1.89	48.75
0.94	32.64	1.42	42.22	1.90	48.78
0.95	32.89	1.43	42.36	1.91	48.81
0.96	33.15	1.44	42.51	1.92	48.84
0.97	33.40	1.45	42.65	1.93	48.87
0.98	33.65	1.46	42.79	1.94	48.90
0.99	33.89	1.47	42.92	1.95	48.93
1.00	34.13	1.48	43.06	1.96	48.96
1.01	34.38	1.49	43.19	1.97	48.98
1.02	34.61	1.50	43.32	1.98	49.01
1.03	34.85	1.51	43.45	1.99	49.04
1.04	35.08	1.52	43.57	2.00	49.06
1.05	35.31	1.53	43.70	2.01	49.09
1.06	35.54	1.54	43.82	2.02	49.11
1.07	35.77	1.55	43.94	2.03	49.13
1.08	35.99	1.56	44.06	2.04	49.16
1.09	36.21	1.57	44.18	2.05	49.18
1.10	36.43	1.58	44.29	2.06	49.20
1.11	36.65	1.59	44.41	2.07	49.22
1.12	36.86	1.60	44.52	2.08	49.25
1.13	37.08	1.61	44.63	2.09	49.74
1.14	37.29	1.62	44.74	2.10	49.74
1.15	37.49	1.63	44.84	2.11	49.75
1.16	37.70	1.64	44.95	2.12	49.76
1.17	37.90	1.65	45.05	2.13	49.77
1.18	38.10	1.66	45.15	2.14	49.77
1.19	38.30	1.67	45.25	2.15	49.78
1.20	38.49	1.68	45.35	2.16	49.79
1.21	38.69	1.69	45.45	2.17	49.79
1.22	38.88	1.70	45.54	2.18	49.80
1.23	39.07	1.71	45.64	2.19	49.81
1.24	39.25	1.72	45.73	2.20	49.81
1.25	39.44	1.73	45.82	2.21	49.82
1.26	39.62	1.74	45.91	2.22	49.82
1.27	39.80	1.75	45.99	2.23	49.83
1.28	39.97	1.76	46.08	2.24	49.84
1.29	40.15	1.77	46.16	2.25	49.84
1.30	40.32	1.78	46.25	2.26	49.85
1.31	40.49	1.79	46.33	2.27	49.85
1.32	40.66	1.80	46.41	2.28	49.86
1.33	40.82	1.81	46.49	2.29	49.86
1.34	40.99	1.82	46.56	2.30	49.87

TABLE D2 Percentages of Area Under the Standard Normal Curve (continued)

(1) z	(2) *Area between mean and* z	(1) z	(2) *Area between mean and* z	(1) z	(2) *Area between mean and* z
2.31	46.93	2.67	49.29	3.03	49.88
2.32	46.99	2.68	49.31	3.04	49.88
2.33	47.06	2.69	49.32	3.05	49.89
2.34	47.13	2.70	49.34	3.06	49.89
2.35	47.19	2.71	49.36	3.07	49.89
2.36	47.26	2.72	49.38	3.08	49.90
2.37	47.32	2.73	49.40	3.09	49.90
2.38	47.38	2.74	49.41	3.10	49.90
2.39	47.44	2.75	49.43	3.11	49.91
2.40	47.50	2.76	49.45	3.12	49.91
2.41	47.56	2.77	49.46	3.13	49.91
2.42	47.61	2.78	49.48	3.14	49.92
2.43	47.67	2.79	49.49	3.15	49.92
2.44	47.72	2.80	49.51	3.16	49.92
2.45	47.78	2.81	49.52	3.17	49.92
2.46	47.83	2.82	49.53	3.18	49.93
2.47	47.88	2.83	49.55	3.19	49.93
2.48	47.93	2.84	49.56	3.20	49.93
2.49	47.98	2.85	49.57	3.21	49.93
2.50	48.03	2.86	49.59	3.22	49.94
2.51	48.08	2.87	49.60	3.23	49.94
2.52	48.12	2.88	49.61	3.24	49.94
2.53	48.17	2.89	49.62	3.25	49.94
2.54	48.21	2.90	49.63	3.30	49.95
2.55	48.26	2.91	49.64	3.35	49.96
2.56	48.30	2.92	49.65	3.40	49.97
2.57	48.34	2.93	49.66	3.45	49.97
2.58	48.38	2.94	49.67	3.50	49.98
2.59	48.42	2.95	49.68	3.60	49.98
2.60	48.46	2.96	49.69	3.70	49.99
2.61	48.50	2.97	49.70	3.80	49.99
2.62	48.54	2.98	49.71	3.90	49.995
2.63	48.57	2.99	49.72	4.00	49.997
2.64	48.61	3.00	49.73	4.50	49.99966
2.65	48.64	3.01	49.87	5.00	49.99997
2.66	49.27	3.02	49.87	5.50	49.99999

TABLE D3 Tabular Values for Student's t Distributions

Column 1 identifies the specific t distribution by its degrees of freedom.

The table shows the values of t corresponding to various levels of significance for one- and two-tailed tests. The test value of t is significant if it is equal to or greater than the value shown in the table.

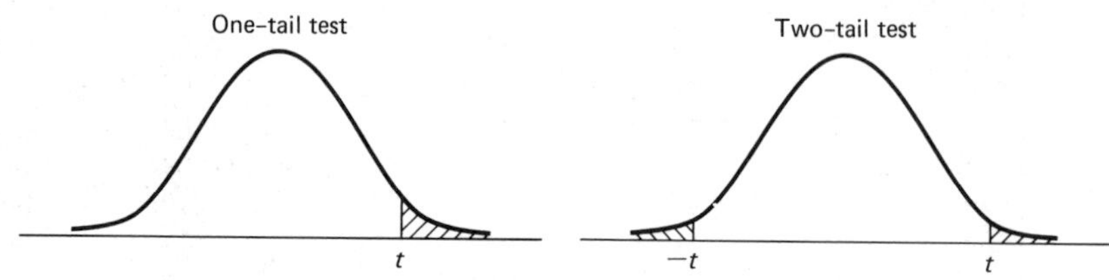

(1)	Level of Significance for a One-Tailed Test						
	.25	.10	.05	.025	.01	.005	.0005
	Level of Significance for a Two-Tailed Test						
df	.50	.20	.10	.05	.02	.01	.001
1	1.000	3.078	6.314	12.706	31.821	63.657	636.619
2	0.816	1.886	2.920	4.303	6.965	9.925	31.598
3	0.765	1.638	2.353	3.182	4.541	5.841	12.924
4	0.741	1.533	2.132	2.776	3.747	4.604	8.610
5	0.727	1.476	2.015	2.571	3.365	4.032	6.869
6	0.718	1.440	1.943	2.447	3.143	3.707	5.959
7	0.711	1.415	1.895	2.365	2.998	3.500	5.408
8	0.706	1.397	1.860	2.306	2.896	3.355	5.041
9	0.703	1.383	1.833	2.262	2.821	3.250	4.781
10	0.700	1.372	1.812	2.228	2.764	3.169	4.587
11	0.697	1.363	1.796	2.201	2.718	3.106	4.437
12	0.696	1.356	1.782	2.179	2.681	3.054	4.318
13	0.694	1.350	1.771	2.160	2.650	3.012	4.221
14	0.692	1.345	1.761	2.145	2.624	2.977	4.140
15	0.691	1.341	1.753	2.132	2.602	2.947	4.073
16	0.690	1.337	1.746	2.120	2.584	2.921	4.015
17	0.689	1.333	1.740	2.110	2.567	2.898	3.965
18	0.688	1.330	1.734	2.101	2.552	2.878	3.922
19	0.688	1.328	1.729	2.093	2.540	2.861	3.883
20	0.687	1.325	1.725	2.086	2.528	2.845	3.850
21	0.686	1.323	1.721	2.080	2.518	2.831	3.819
22	0.686	1.321	1.717	2.074	2.508	2.819	3.792
23	0.685	1.320	1.714	2.069	2.500	2.807	3.768
24	0.685	1.318	1.714	2.064	2.492	2.797	3.745
25	0.684	1.316	1.708	2.060	2.485	2.787	3.725

TABLE D3 Tabular Values for Student's *t* Distributions (continued)

	Level of Significance for a One-Tailed Test						
	.25	.10	.05	.025	.01	.005	.0005
(1)	Level of Significance for a Two-Tailed Test						
df	.50	.20	.10	.05	.02	.01	.001
26	0.684	1.315	1.706	2.056	2.479	2.779	3.707
27	0.684	1.314	1.703	2.052	2.473	2.771	3.690
28	0.683	1.312	1.701	2.048	2.467	2.763	3.674
29	0.683	1.311	1.700	2.045	2.462	2.756	3.659
30	0.683	1.310	1.697	2.042	2.457	2.750	3.646
31	0.682	1.310	1.696	2.040	2.453	2.744	
32	0.682	1.309	1.694	2.037	2.449	2.738	
33	0.682	1.308	1.692	2.034	2.445	2.733	
34	0.682	1.307	1.691	2.032	2.441	2.728	
35	0.682	1.306	1.690	2.030	2.438	2.724	3.591
36	0.681	1.306	1.688	2.028	2.434	2.720	
37	0.681	1.305	1.687	2.026	2.431	2.715	
38	0.681	1.304	1.686	2.024	2.429	2.712	
39	0.681	1.304	1.685	2.023	2.426	2.708	
40	0.681	1.303	1.684	2.021	2.423	2.704	3.551
41	0.680	1.302	1.683	2.020	2.421	2.701	
42	0.680	1.302	1.682	2.018	2.418	2.698	
43	0.680	1.302	1.681	2.017	2.416	2.695	
44	0.680	1.301	1.680	2.015	2.414	2.692	
45	0.680	1.301	1.679	2.014	2.412	2.690	3.520
46	0.680	1.300	1.679	2.013	2.410	2.687	
47	0.680	1.300	1.678	2.012	2.408	2.685	
48	0.680	1.299	1.677	2.011	2.407	2.682	
49	0.680	1.299	1.677	2.010	2.405	2.680	
50	0.679	1.299	1.676	2.009	2.403	2.678	3.496
51	0.679	1.298	1.675	2.008	2.402	2.676	
52	0.679	1.298	1.675	2.007	2.400	2.674	
53	0.679	1.298	1.674	2.006	2.399	2.672	
54	0.679	1.297	1.674	2.005	2.397	2.670	
55	0.679	1.297	1.673	2.004	2.396	2.668	3.476
56	0.679	1.297	1.672	2.003	2.395	2.666	
57	0.679	1.297	1.672	2.002	2.394	2.665	
58	0.679	1.296	1.672	2.002	2.392	2.663	
59	0.679	1.296	1.671	2.001	2.391	2.662	
60	0.679	1.296	1.671	2.000	2.390	2.660	3.460
61	0.678	1.296	1.670	2.000	2.389	2.659	
62	0.678	1.295	1.670	1.999	2.388	2.658	
63	0.678	1.295	1.669	1.998	2.387	2.656	
64	0.678	1.295	1.669	1.998	2.386	2.655	
65	0.678	1.295	1.669	1.997	2.385	2.654	

TABLE D3 Tabular Values for Student's *t* Distributions (continued)

	Level of Significance for a One-Tailed Test						
	.25	.10	.05	.025	.01	.005	.0005
(1)	Level of Significance for a Two-Tailed Test						
df	.50	.20	.10	.05	.02	.01	.001
66	0.678	1.294	1.668	1.997	2.384	2.652	
67	0.678	1.294	1.668	1.996	2.383	2.651	
68	0.678	1.294	1.668	1.996	2.382	2.650	
69	0.678	1.294	1.667	1.995	2.382	2.649	
70	0.678	1.294	1.667	1.994	2.381	2.648	3.435
71	0.678	1.294	1.667	1.994	2.380	2.647	
72	0.678	1.293	1.666	1.994	2.379	2.646	
73	0.678	1.293	1.666	1.993	2.378	2.645	
74	0.678	1.293	1.666	1.992	2.378	2.644	
75	0.678	1.293	1.665	1.992	2.377	2.643	
76	0.678	1.293	1.665	1.992	2.376	2.642	
77	0.678	1.293	1.665	1.991	2.376	2.641	
78	0.678	1.292	1.665	1.991	2.375	2.640	
79	0.678	1.292	1.664	1.990	2.374	2.640	
80	0.678	1.292	1.664	1.990	2.374	2.639	3.416
81	0.678	1.292	1.664	1.990	2.373	2.638	
82	0.678	1.292	1.664	1.989	2.373	2.637	
83	0.678	1.292	1.663	1.989	2.372	2.636	
84	0.677	1.292	1.663	1.989	2.372	2.636	
85	0.677	1.292	1.663	1.988	2.371	2.635	
86	0.677	1.292	1.663	1.988	2.370	2.634	
87	0.677	1.291	1.663	1.988	2.370	2.634	
88	0.677	1.291	1.662	1.987	2.370	2.633	
89	0.677	1.291	1.662	1.987	2.369	2.632	
90	0.677	1.291	1.662	1.987	2.368	2.632	3.402
91	0.677	1.291	1.662	1.986	2.368	2.631	
92	0.677	1.291	1.662	1.986	2.368	2.630	
93	0.677	1.291	1.661	1.986	2.367	2.630	
94	0.677	1.291	1.661	1.986	2.367	2.629	
95	0.677	1.290	1.661	1.985	2.366	2.629	
96	0.677	1.290	1.661	1.985	2.366	2.628	
97	0.677	1.290	1.661	1.985	2.365	2.628	
98	0.677	1.290	1.661	1.984	2.365	2.627	
99	0.677	1.290	1.660	1.984	2.365	2.626	
100	0.677	1.290	1.660	1.984	2.364	2.626	3.390
102	0.677	1.290	1.660	1.984	2.364	2.625	
104	0.677	1.290	1.660	1.983	2.363	2.624	
106	0.677	1.290	1.659	1.983	2.362	2.623	
108	0.677	1.289	1.659	1.982	2.361	2.622	
110	0.677	1.289	1.659	1.982	2.361	2.621	

TABLE D3 Tabular Values for Student's t Distributions (continued)

	Level of Significance for a One-Tailed Test						
	.25	.10	.05	.025	.01	.005	.0005
(1)	Level of Significance for a Two-Tailed Test						
df	.50	.20	.10	.05	.02	.01	.001
112	0.677	1.289	1.659	1.981	2.360	2.620	
114	0.677	1.289	1.658	1.981	2.360	2.620	
116	0.677	1.289	1.658	1.981	2.359	2.619	
118	0.677	1.289	1.658	1.980	2.358	2.618	
120	0.676	1.289	1.658	1.980	2.358	2.617	
122	0.676	1.288	1.657	1.980	2.357	2.617	
124	0.676	1.288	1.657	1.979	2.357	2.616	
126	0.676	1.288	1.657	1.979	2.356	2.615	
128	0.676	1.288	1.657	1.979	2.356	2.615	
130	0.676	1.288	1.657	1.978	2.355	2.614	
132	0.676	1.288	1.656	1.978	2.355	2.614	
134	0.676	1.288	1.656	1.978	2.354	2.613	
136	0.676	1.288	1.656	1.978	2.354	2.612	
138	0.676	1.288	1.656	1.977	2.354	2.612	
140	0.676	1.288	1.656	1.977	2.353	2.611	
142	0.676	1.288	1.656	1.977	2.353	2.611	
144	0.676	1.288	1.656	1.977	2.352	2.610	
146	0.676	1.287	1.655	1.976	2.352	2.610	
148	0.676	1.287	1.655	1.976	2.352	2.610	
150	0.676	1.287	1.655	1.976	2.352	2.609	
200	0.676	1.286	1.652	1.972	2.345	2.601	3.340
300	0.675	1.284	1.650	1.968	2.339	2.592	
400	0.675	1.284	1.649	1.966	2.336	2.588	
500	0.675	1.283	1.648	1.965	2.334	2.586	3.310
600	0.675	1.283	1.647	1.964	2.333	2.584	
700	0.675	1.283	1.647	1.963	2.332	2.583	
800	0.675	1.283	1.647	1.963	2.331	2.582	
900	0.675	1.282	1.646	1.963	2.330	2.581	
1000	0.675	1.282	1.646	1.962	2.330	2.581	3.300
∞	0.674	1.282	1.645	1.960	2.326	2.576	3.291

TABLE D4 Tabular Values for the Chi-Square Distributions

Column 1 identifies the specific chi-square distribution by its degrees of freedom.

The table shows the values of chi-square corresponding to various levels of significance for a test using the upper tail of the distributions (i.e., a nondirectional test). The test value of chi-square is significant if it is equal to or greater than the value shown in the table.

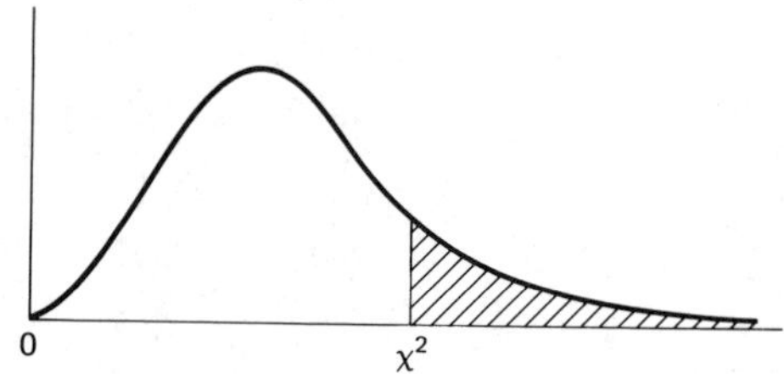

(1)	Level of Significance for Upper Tail					
df	.25	.10	.05	.025	.01	.005
1	1.323	2.706	3.841	5.024	6.635	7.879
2	2.773	4.605	5.991	7.378	9.210	10.597
3	4.108	6.251	7.815	9.348	11.345	12.838
4	5.385	7.779	9.488	11.143	13.277	14.860
5	6.626	9.236	11.071	12.833	15.086	16.750
6	7.841	10.645	12.592	14.449	16.812	18.548
7	9.037	12.017	14.067	16.013	18.475	20.278
8	10.219	13.362	15.507	17.535	20.090	21.955
9	11.389	14.684	16.919	19.023	21.666	23.589
10	12.549	15.987	18.307	20.483	23.209	25.188
11	13.701	17.275	19.675	21.920	24.725	26.757
12	14.845	18.549	21.026	23.337	26.217	28.299
13	15.984	19.812	22.362	24.736	27.688	29.819
14	17.117	21.064	23.685	26.119	29.141	31.319
15	18.245	22.307	24.996	27.488	30.578	32.801
16	19.369	23.542	26.296	28.845	32.000	34.267
17	20.489	24.769	27.587	30.191	33.409	35.718
18	21.605	25.989	28.869	31.526	34.805	37.156
19	22.718	27.204	30.144	32.852	36.191	38.582
20	23.828	28.412	31.410	34.170	37.566	39.997
21	24.935	29.615	32.671	35.479	38.932	41.401
22	26.039	30.813	33.924	36.781	40.289	42.796
23	27.141	32.007	35.172	38.076	41.638	44.181
24	28.241	33.196	36.415	39.364	42.980	45.559
25	29.339	34.382	37.652	40.646	44.314	46.928
26	30.435	35.563	38.885	41.923	45.642	48.290
27	31.528	36.741	40.113	43.194	46.963	49.645
28	32.620	37.916	41.337	44.461	48.278	50.993
29	33.711	39.087	42.557	45.722	49.588	52.336
30	34.800	40.256	43.773	46.979	50.892	53.672

TABLE D4 Tabular Values for the Chi-Square Distributions (continued)

(1) *df*	Level of Significance for Upper Tail .25	.10	.05	.025	.01	.005
40	45.616	51.805	55.758	59.342	63.691	66.766
50	56.334	63.167	67.505	71.420	76.154	79.490
60	66.981	74.397	79.082	83.298	88.379	91.952
70	77.577	85.527	90.531	95.023	100.425	104.215
80	88.130	96.578	101.879	106.629	112.329	116.321
90	98.650	107.565	113.145	118.136	124.116	128.299
100	109.141	118.498	124.342	129.561	135.807	140.169
120	130.055	140.233	146.567	152.211	158.950	163.648
150	161.291	172.581	179.581	185.800	193.208	198.360
500	520.950	540.930	553.127	563.852	576.493	585.207
1000	1029.790	1057.724	1074.679	1089.531	1106.969	1118.948

Note: When *df* > 30, the critical value of χ^2 for distributions not listed may be found by interpolation or by the following approximate formula: $\chi^2 = df[1 - (2/9df) + z\sqrt{2/9df}]^3$, where z is the normal deviate above which lies the same proportionate area in the normal curve.

TABLE D5 Tabular Values of D for the Kolmogorov–Smirnov One-Sample Test

Column 1 identifies the specific distribution by the sample size (N).

The table shows the critical values of D (differences in cumulative proportions) at various levels of significance for nondirectional tests. The levels must be halved for directional tests. The test difference (LD) in cumulative proportions is significant if it is equal to or greater than the value shown in the table.

(1) Sample size (N)	Level of Significance for a Nondirectional Test				
	.20	.15	.10	.05	.01
1	.900	.925	.950	.975	.995
2	.684	.726	.776	.842	.929
3	.565	.597	.642	.708	.828
4	.494	.525	.564	.624	.733
5	.446	.474	.510	.565	.669
6	.410	.436	.470	.521	.618
7	.381	.405	.438	.486	.577
8	.358	.381	.411	.457	.543
9	.339	.360	.388	.432	.514
10	.322	.342	.368	.410	.490
11	.307	.326	.352	.391	.468
12	.295	.313	.338	.375	.450
13	.284	.302	.325	.361	.433
14	.274	.292	.314	.349	.418
15	.266	.283	.304	.338	.404
16	.258	.274	.295	.328	.392
17	.250	.266	.286	.318	.381
18	.244	.259	.278	.309	.371
19	.237	.252	.272	.301	.363
20	.231	.246	.264	.294	.356
25	.21	.22	.24	.27	.32
30	.19	.20	.22	.24	.29
35	.18	.19	.21	.23	.27
Over 35	$\frac{1.07}{\sqrt{N}}$	$\frac{1.14}{\sqrt{N}}$	$\frac{1.22}{\sqrt{N}}$	$\frac{1.36}{\sqrt{N}}$	$\frac{1.63}{\sqrt{N}}$

TABLE D6 Tabular Values of D for the Kolmogorov-Smirnov Two-Sample Test (Large Samples)

Column 1 shows the levels of significance for various nondirectional tests. The levels must be halved for directional tests.

Column 2 shows the formula for D (differences in cumulative proportions) at the levels of significance. The test difference (LD) in cumulative proportions is significant if it is equal to or greater than the value determined from the table.

(1) Level of Significance for a Nondirectional Test	(2) Critical Value of D
.10	$1.22\sqrt{\frac{N_1 + N_2}{N_1 N_2}}$
.05	$1.36\sqrt{\frac{N_1 + N_2}{N_1 N_2}}$
.025	$1.48\sqrt{\frac{N_1 + N_2}{N_1 N_2}}$
.01	$1.63\sqrt{\frac{N_1 + N_2}{N_1 N_2}}$
.005	$1.73\sqrt{\frac{N_1 + N_2}{N_1 N_2}}$
.001	$1.95\sqrt{\frac{N_1 + N_2}{N_1 N_2}}$

TABLE D7 Tabular Values for the F Distributions

The column and row headings identify the specific F distributions by their degrees of freedom for the numerator (between groups), and denominator (within groups), respectively. The table shows the values of F corresponding to the .05 level of significance (regular type) and the .01 level of significance (**boldface** type) for a test using the upper tail of the distributions (i.e., a nondirectional test). The test value of F is significant if it is equal to or greater than the value shown in the table.

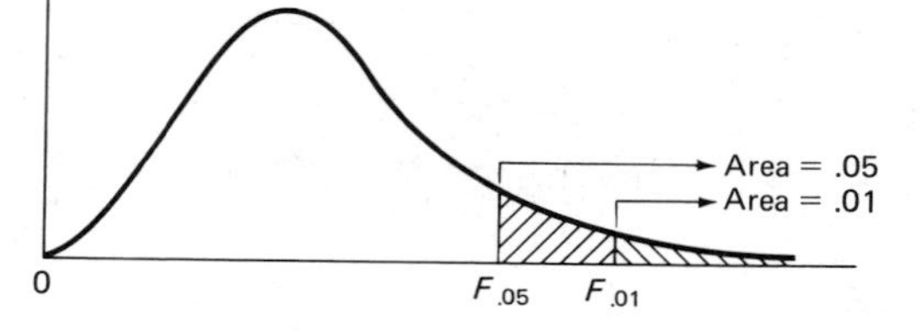

Degrees of Freedom: Denominator	α	Degrees of Freedom: Numerator 1	2	3	4	5	6	7	8	9	10	11	12	14	16	20	24	30	40	50	75	100	200	500	∞
1	.05	161	200	216	225	230	234	237	239	241	242	243	244	245	246	248	249	250	251	252	253	253	254	254	254
	.01	**4,052**	**4,999**	**5,403**	**5,625**	**5,764**	**5,859**	**5,928**	**5,981**	**6,022**	**6,056**	**6,082**	**6,106**	**6,142**	**6,169**	**6,208**	**6,234**	**6,258**	**6,286**	**6,302**	**6,323**	**6,334**	**6,352**	**6,361**	**6,366**
2	.05	18.51	19.00	19.16	19.25	19.30	19.33	19.36	19.37	19.38	19.39	19.40	19.41	19.42	19.43	19.44	19.45	19.46	19.47	19.47	19.48	19.49	19.49	19.50	19.50
	.01	**98.49**	**99.00**	**99.17**	**99.25**	**99.30**	**99.33**	**99.34**	**99.36**	**99.38**	**99.40**	**99.41**	**99.42**	**99.43**	**99.44**	**99.45**	**99.46**	**99.47**	**99.48**	**99.48**	**99.49**	**99.49**	**99.49**	**99.50**	**99.50**
3	.05	10.13	9.55	9.28	9.12	9.01	8.94	8.88	8.84	8.81	8.78	8.76	8.74	8.71	8.69	8.66	8.64	8.62	8.60	8.58	8.57	8.56	8.54	8.54	8.53
	.01	**34.12**	**30.82**	**29.46**	**28.71**	**28.24**	**27.91**	**27.67**	**27.49**	**27.34**	**27.23**	**27.13**	**27.05**	**26.92**	**26.83**	**26.69**	**26.60**	**26.50**	**26.41**	**26.35**	**26.27**	**26.23**	**26.18**	**26.14**	**26.12**
4	.05	7.71	6.94	6.59	6.39	6.26	6.16	6.09	6.04	6.00	5.96	5.93	5.91	5.87	5.84	5.80	5.77	5.74	5.71	5.70	5.68	5.66	5.65	5.64	5.63
	.01	**21.20**	**18.00**	**16.69**	**15.98**	**15.52**	**15.21**	**14.98**	**14.80**	**14.66**	**14.54**	**14.45**	**14.37**	**14.24**	**14.15**	**14.02**	**13.93**	**13.83**	**13.74**	**13.69**	**13.61**	**13.57**	**13.52**	**13.48**	**13.46**
5	.05	6.61	5.79	5.41	5.19	5.05	4.95	4.88	4.82	4.78	4.74	4.70	4.68	4.64	4.60	4.56	4.53	4.50	4.46	4.44	4.42	4.40	4.38	4.37	4.36
	.01	**16.26**	**13.27**	**12.06**	**11.39**	**10.97**	**10.67**	**10.45**	**10.27**	**10.15**	**10.05**	**9.96**	**9.89**	**9.77**	**9.68**	**9.55**	**9.47**	**9.38**	**9.29**	**9.24**	**9.17**	**9.13**	**9.07**	**9.04**	**9.02**
6	.05	5.99	5.14	4.76	4.53	4.39	4.28	4.21	4.15	4.10	4.06	4.03	4.00	3.96	3.92	3.87	3.84	3.81	3.77	3.75	3.72	3.71	3.69	3.68	3.67
	.01	**13.74**	**10.92**	**9.78**	**9.15**	**8.75**	**8.47**	**8.26**	**8.10**	**7.98**	**7.87**	**7.79**	**7.72**	**7.60**	**7.52**	**7.39**	**7.31**	**7.23**	**7.14**	**7.09**	**7.02**	**6.99**	**6.94**	**6.90**	**6.88**
7	.05	5.59	4.47	4.35	4.12	3.97	3.87	3.79	3.73	3.68	3.63	3.60	3.57	3.52	3.49	3.44	3.41	3.38	3.34	3.32	3.29	3.28	3.25	3.24	3.23
	.01	**12.25**	**9.55**	**8.45**	**7.85**	**7.46**	**7.19**	**7.00**	**6.84**	**6.71**	**6.62**	**6.54**	**6.47**	**6.35**	**6.27**	**6.15**	**6.07**	**5.98**	**5.90**	**5.85**	**5.78**	**5.75**	**5.70**	**5.67**	**5.65**

Degrees of Freedom: Denominator	α	Degrees of Freedom: Numerator 1	2	3	4	5	6	7	8	9	10	11	12	14	16	20	24	30	40	50	75	100	200	500	∞
8	.05	5.32	4.46	4.07	3.84	3.69	3.58	3.50	3.44	3.39	3.34	3.31	3.28	3.23	3.20	3.15	3.12	3.08	3.05	3.03	3.00	2.98	2.96	2.94	2.93
	.01	11.26	8.65	7.59	7.01	6.63	6.37	6.19	6.03	5.91	5.82	5.74	5.67	5.56	5.48	5.36	5.28	5.20	5.11	5.06	5.00	4.96	4.91	4.88	4.86
9	.05	5.12	4.26	3.86	3.63	3.48	3.37	3.29	3.23	3.18	3.13	3.10	3.07	3.02	2.98	2.93	2.90	2.86	2.82	2.80	2.77	2.76	2.73	2.72	2.71
	.01	10.56	8.02	6.99	6.42	6.06	5.80	5.62	5.47	5.35	5.26	5.18	5.11	5.00	4.92	4.80	4.73	4.64	4.56	4.51	4.45	4.41	4.36	4.33	4.31
10	.05	4.96	4.10	3.71	3.48	3.33	3.22	3.14	3.07	3.02	2.97	2.94	2.91	2.86	2.82	2.77	2.74	2.70	2.67	2.64	2.61	2.59	2.56	2.55	2.54
	.01	10.04	7.56	6.55	5.99	5.64	5.39	5.21	5.06	4.95	4.85	4.78	4.71	4.60	4.52	4.41	4.33	4.25	4.17	4.12	4.05	4.01	3.96	3.93	3.91
11	.05	4.84	3.98	3.59	3.36	3.20	3.09	3.01	2.95	2.90	2.86	2.82	2.79	2.74	2.70	2.65	2.61	2.57	2.53	2.50	2.47	2.45	2.42	2.41	2.40
	.01	9.65	7.20	6.22	5.67	5.32	5.07	4.88	4.74	4.63	4.54	4.46	4.40	4.29	4.21	4.10	4.02	3.94	3.86	3.80	3.74	3.70	3.66	3.62	3.60
12	.05	4.75	3.88	3.49	3.26	3.11	3.00	2.92	2.85	2.80	2.76	2.72	2.69	2.64	2.60	2.54	2.50	2.46	2.42	2.40	2.36	2.35	2.32	2.31	2.30
	.01	9.33	6.93	5.95	5.41	5.06	4.82	4.65	4.50	4.39	4.30	4.22	4.16	4.05	3.98	3.86	3.78	3.70	3.61	3.56	3.49	3.46	3.41	3.38	3.36
13	.05	4.67	3.80	3.41	3.18	3.02	2.92	2.84	2.77	2.72	2.67	2.63	2.60	2.55	2.51	2.46	2.42	2.38	2.34	2.32	2.28	2.26	2.24	2.22	2.21
	.01	9.07	6.70	5.74	5.20	4.86	4.62	4.44	4.30	4.19	4.10	4.02	3.96	3.85	3.78	3.67	3.59	3.51	3.42	3.37	3.30	3.27	3.21	3.18	3.16
14	.05	4.60	3.74	3.34	3.11	2.96	2.85	2.77	2.70	2.65	2.60	2.56	2.53	2.48	2.44	2.39	2.35	2.31	2.27	2.24	2.21	2.19	2.16	2.14	2.13
	.01	8.86	6.51	5.56	5.03	4.69	4.46	4.28	4.14	4.03	3.94	3.86	3.80	3.70	3.62	3.51	3.43	3.34	3.26	3.21	3.14	3.11	3.06	3.02	3.00
15	.05	4.54	3.68	3.29	3.06	2.90	2.79	2.70	2.64	2.59	2.55	2.51	2.48	2.43	2.39	2.33	2.29	2.25	2.21	2.18	2.15	2.12	2.10	2.08	2.07
	.01	8.68	6.36	5.42	4.89	4.56	4.32	4.14	4.00	3.89	3.80	3.73	3.67	3.56	3.48	3.36	3.29	3.20	3.12	3.07	3.00	2.97	2.92	2.89	2.87
16	.05	4.49	3.63	3.24	3.01	2.85	2.74	2.66	2.59	2.54	2.49	2.45	2.42	2.37	2.33	2.28	2.24	2.20	2.16	2.13	2.09	2.07	2.04	2.02	2.01
	.01	8.53	6.23	5.29	4.77	4.44	4.20	4.03	3.89	3.78	3.69	3.61	3.55	3.45	3.37	3.25	3.18	3.10	3.01	2.96	2.89	2.86	2.80	2.77	2.75
17	.05	4.45	3.59	3.20	2.96	2.81	2.70	2.62	2.55	2.50	2.45	2.41	2.38	2.33	2.29	2.23	2.19	2.15	2.11	2.08	2.04	2.02	1.99	1.97	1.96
	.01	8.40	6.11	5.18	4.67	4.34	4.10	3.93	3.79	3.68	3.59	3.52	3.45	3.35	3.27	3.16	3.08	3.00	2.92	2.86	2.79	2.76	2.70	2.67	2.65
18	.05	4.41	3.55	3.16	2.93	2.77	2.66	2.58	2.51	2.46	2.41	2.37	2.34	2.29	2.25	2.19	2.15	2.11	2.07	2.04	2.00	1.98	1.95	1.93	1.92
	.01	8.28	6.01	5.09	4.58	4.25	4.01	3.85	3.71	3.60	3.51	3.44	3.37	3.27	3.19	3.07	3.00	2.91	2.83	2.78	2.71	2.68	2.62	2.59	2.57
19	.05	4.38	3.52	3.13	2.90	2.74	2.63	2.55	2.48	2.43	2.38	2.34	2.31	2.26	2.21	2.15	2.11	2.07	2.02	2.00	1.96	1.94	1.91	1.90	1.88
	.01	8.18	5.93	5.01	4.50	4.17	3.94	3.77	3.63	3.52	3.43	3.36	3.30	3.19	3.12	3.00	2.92	2.84	2.76	2.70	2.63	2.60	2.54	2.51	2.49
20	.05	4.35	3.49	3.10	2.87	2.71	2.60	2.52	2.45	2.40	2.35	2.31	2.28	2.23	2.18	2.12	2.08	2.04	1.99	1.96	1.92	1.90	1.87	1.85	1.84
	.01	8.10	5.85	4.94	4.43	4.10	3.87	3.71	3.56	3.45	3.37	3.30	3.23	3.13	3.05	2.94	2.86	2.77	2.69	2.63	2.56	2.53	2.47	2.44	2.42
21	.05	4.32	3.47	3.07	2.84	2.68	2.57	2.49	2.42	2.37	2.32	2.28	2.25	2.20	2.15	2.09	2.05	2.00	1.96	1.93	1.89	1.87	1.84	1.82	1.81
	.01	8.02	5.78	4.87	4.37	4.04	3.81	3.65	3.51	3.40	3.31	3.24	3.17	3.07	2.99	2.88	2.80	2.72	2.63	2.58	2.51	2.47	2.42	2.38	2.36
22	.05	4.30	3.44	3.05	2.82	2.66	2.55	2.47	2.40	2.35	2.30	2.26	2.23	2.18	2.13	2.07	2.03	1.98	1.93	1.91	1.87	1.84	1.81	1.80	1.78
	.01	7.94	5.72	4.82	4.31	3.99	3.76	3.59	3.45	3.35	3.26	3.18	3.12	3.02	2.94	2.83	2.75	2.67	2.58	2.53	2.46	2.42	2.37	2.33	2.31
23	.05	4.28	3.42	3.03	2.80	2.64	2.53	2.45	2.38	2.32	2.28	2.24	2.20	2.14	2.10	2.04	2.00	1.96	1.91	1.88	1.84	1.82	1.79	1.77	1.76
	.01	7.88	5.66	4.76	4.26	3.94	3.71	3.54	3.41	3.30	3.21	3.14	3.07	2.97	2.89	2.78	2.70	2.62	2.53	2.48	2.41	2.37	2.32	2.28	2.26
24	.05	4.26	3.40	3.01	2.78	2.62	2.51	2.43	2.36	2.30	2.26	2.22	2.18	2.13	2.09	2.02	1.98	1.94	1.89	1.86	1.82	1.80	1.76	1.74	1.73
	.01	7.82	5.61	4.72	4.22	3.90	3.67	3.50	3.36	3.25	3.17	3.09	3.03	2.93	2.85	2.74	2.66	2.58	2.49	2.44	2.36	2.33	2.27	2.23	2.21

TABLE D7 Tabular Values for the F Distributions (continued)

Degrees of Freedom: Denominator	α	Degrees of Freedom: Numerator																							
		1	2	3	4	5	6	7	8	9	10	11	12	14	16	20	24	30	40	50	75	100	200	500	∞
25	.05	4.24	3.38	2.99	2.76	2.60	2.49	2.41	2.34	2.28	2.24	2.20	2.16	2.11	2.06	2.00	1.96	1.92	1.87	1.84	1.80	1.77	1.74	1.72	1.71
	.01	**7.77**	**5.57**	**4.68**	**4.18**	**3.86**	**3.63**	**3.46**	**3.32**	**3.21**	**3.13**	**3.05**	**2.99**	**2.89**	**2.81**	**2.70**	**2.62**	**2.54**	**2.45**	**2.40**	**2.32**	**2.29**	**2.23**	**2.19**	**2.17**
26	.05	4.22	3.37	2.98	2.74	2.59	2.47	2.39	2.32	2.27	2.22	2.18	2.15	2.10	2.05	1.99	1.95	1.90	1.85	1.82	1.78	1.76	1.72	1.70	1.69
	.01	**7.72**	**5.53**	**4.64**	**4.14**	**3.82**	**3.59**	**3.42**	**3.29**	**3.17**	**3.09**	**3.02**	**2.96**	**2.86**	**2.77**	**2.66**	**2.58**	**2.50**	**2.41**	**2.36**	**2.28**	**2.25**	**2.19**	**2.15**	**2.13**
27	.05	4.21	3.35	2.96	2.73	2.57	2.46	2.37	2.30	2.25	2.20	2.16	2.13	2.08	2.03	1.97	1.93	1.88	1.84	1.80	1.76	1.74	1.71	1.68	1.67
	.01	**7.68**	**5.49**	**4.60**	**4.11**	**3.79**	**3.56**	**3.39**	**3.26**	**3.14**	**3.06**	**2.98**	**2.93**	**2.83**	**2.74**	**2.63**	**2.55**	**2.47**	**2.38**	**2.33**	**2.25**	**2.21**	**2.16**	**2.12**	**2.10**
28	.05	4.20	3.34	2.95	2.71	2.56	2.44	2.36	2.29	2.24	2.19	2.15	2.12	2.06	2.02	1.96	1.91	1.87	1.81	1.78	1.75	1.72	1.69	1.67	1.65
	.01	**7.64**	**5.45**	**4.57**	**4.07**	**3.76**	**3.53**	**3.36**	**3.23**	**3.11**	**3.03**	**2.95**	**2.90**	**2.80**	**2.71**	**2.60**	**2.52**	**2.44**	**2.35**	**2.30**	**2.22**	**2.18**	**2.13**	**2.09**	**2.06**
29	.05	4.18	3.33	2.93	2.70	2.54	2.43	2.35	2.28	2.22	2.18	2.14	2.10	2.05	2.00	1.94	1.90	1.85	1.80	1.77	1.73	1.71	1.68	1.65	1.64
	.01	**7.60**	**5.42**	**4.54**	**4.04**	**3.73**	**3.50**	**3.33**	**3.20**	**3.08**	**3.00**	**2.92**	**2.87**	**2.77**	**2.68**	**2.57**	**2.49**	**2.41**	**2.32**	**2.27**	**2.19**	**2.15**	**2.10**	**2.06**	**2.03**
30	.05	4.17	3.32	2.92	2.69	2.53	2.42	2.34	2.27	2.21	2.16	2.12	2.09	2.04	1.99	1.93	1.89	1.84	1.79	1.76	1.72	1.69	1.66	1.64	1.62
	.01	**7.56**	**5.39**	**4.51**	**4.02**	**3.70**	**3.47**	**3.30**	**3.17**	**3.06**	**2.98**	**2.90**	**2.84**	**2.74**	**2.66**	**2.55**	**2.47**	**2.38**	**2.29**	**2.24**	**2.16**	**2.13**	**2.07**	**2.03**	**2.01**
32	.05	4.15	3.30	2.90	2.67	2.51	2.40	2.32	2.25	2.19	2.14	2.10	2.07	2.02	1.97	1.91	1.86	1.82	1.76	1.74	1.69	1.67	1.64	1.61	1.59
	.01	**7.50**	**5.34**	**4.46**	**3.97**	**3.66**	**3.42**	**3.25**	**3.12**	**3.01**	**2.94**	**2.86**	**2.80**	**2.70**	**2.62**	**2.51**	**2.42**	**2.34**	**2.25**	**2.20**	**2.12**	**2.08**	**2.02**	**1.98**	**1.96**
34	.05	4.13	3.28	2.88	2.65	2.49	2.38	2.30	2.23	2.17	2.12	2.08	2.05	2.00	1.95	1.89	1.84	1.80	1.74	1.71	1.67	1.64	1.61	1.59	1.57
	.01	**7.44**	**5.29**	**4.42**	**3.93**	**3.61**	**3.38**	**3.21**	**3.08**	**2.97**	**2.89**	**2.82**	**2.76**	**2.66**	**2.58**	**2.47**	**2.38**	**2.30**	**2.21**	**2.15**	**2.08**	**2.04**	**1.98**	**1.94**	**1.91**
36	.05	4.11	3.26	2.86	2.63	2.48	2.36	2.28	2.21	2.15	2.10	2.06	2.03	1.98	1.93	1.87	1.82	1.78	1.72	1.69	1.65	1.62	1.59	1.56	1.55
	.01	**7.39**	**5.25**	**4.38**	**3.89**	**3.58**	**3.35**	**3.18**	**3.04**	**2.94**	**2.86**	**2.78**	**2.72**	**2.62**	**2.54**	**2.43**	**2.35**	**2.26**	**2.17**	**2.12**	**2.04**	**2.00**	**1.94**	**1.90**	**1.87**
38	.05	4.10	3.25	2.85	2.62	2.46	2.35	2.26	2.19	2.14	2.09	2.05	2.02	1.96	1.92	1.85	1.80	1.76	1.71	1.67	1.63	1.60	1.57	1.54	1.53
	.01	**7.35**	**5.21**	**4.34**	**3.86**	**3.54**	**3.32**	**3.15**	**3.02**	**2.91**	**2.82**	**2.75**	**2.69**	**2.59**	**2.51**	**2.40**	**2.32**	**2.22**	**2.14**	**2.08**	**2.00**	**1.97**	**1.90**	**1.86**	**1.84**
40	.05	4.08	3.23	2.84	2.61	2.45	2.34	2.25	2.18	2.12	2.07	2.04	2.00	1.95	1.90	1.84	1.79	1.74	1.69	1.66	1.61	1.59	1.55	1.53	1.51
	.01	**7.31**	**5.18**	**4.31**	**3.83**	**3.51**	**3.29**	**3.12**	**2.99**	**2.88**	**2.80**	**2.73**	**2.66**	**2.56**	**2.49**	**2.37**	**2.29**	**2.20**	**2.11**	**2.05**	**1.97**	**1.94**	**1.88**	**1.84**	**1.81**
42	.05	4.07	3.22	2.83	2.59	2.44	2.32	2.24	2.17	2.11	2.06	2.02	1.99	1.94	1.89	1.82	1.78	1.73	1.68	1.64	1.60	1.57	1.54	1.51	1.49
	.01	**7.27**	**5.15**	**4.29**	**3.80**	**3.49**	**3.26**	**3.10**	**2.96**	**2.86**	**2.77**	**2.70**	**2.64**	**2.54**	**2.46**	**2.35**	**2.26**	**2.17**	**2.08**	**2.02**	**1.94**	**1.91**	**1.85**	**1.80**	**1.78**
44	.05	4.06	3.21	2.82	2.58	2.43	2.31	2.23	2.16	2.10	2.05	2.01	1.98	1.92	1.88	1.81	1.76	1.72	1.66	1.63	1.58	1.56	1.52	1.50	1.48
	.01	**7.24**	**5.12**	**4.26**	**3.78**	**3.46**	**3.24**	**3.07**	**2.94**	**2.84**	**2.75**	**2.68**	**2.62**	**2.52**	**2.44**	**2.32**	**2.24**	**2.15**	**2.06**	**2.00**	**1.92**	**1.88**	**1.82**	**1.78**	**1.75**
46	.05	4.05	3.20	2.81	2.57	2.42	2.30	2.22	2.14	2.09	2.04	2.00	1.97	1.91	1.87	1.80	1.75	1.71	1.65	1.62	1.57	1.54	1.51	1.48	1.46
	.01	**7.21**	**5.10**	**4.24**	**3.76**	**3.44**	**3.22**	**3.05**	**2.92**	**2.82**	**2.73**	**2.66**	**2.60**	**2.50**	**2.42**	**2.30**	**2.22**	**2.13**	**2.04**	**1.98**	**1.90**	**1.86**	**1.80**	**1.76**	**1.72**

Degrees of Freedom: Denominator	α	Degrees of Freedom: Numerator 1	2	3	4	5	6	7	8	9	10	11	12	14	16	20	24	30	40	50	75	100	200	500	∞
48	.05	4.04	3.19	2.80	2.56	2.41	2.30	2.21	2.14	2.08	2.03	1.99	1.96	1.90	1.86	1.79	1.74	1.70	1.64	1.61	1.56	1.53	1.50	1.47	1.45
	.01	**7.19**	**5.08**	**4.22**	**3.74**	**3.42**	**3.20**	**3.04**	**2.90**	**2.80**	**2.71**	**2.64**	**2.58**	**2.48**	**2.40**	**2.28**	**2.20**	**2.11**	**2.02**	**1.96**	**1.88**	**1.84**	**1.78**	**1.73**	**1.70**
50	.05	4.03	3.18	2.79	2.56	2.40	2.29	2.20	2.13	2.07	2.02	1.98	1.95	1.90	1.85	1.78	1.74	1.69	1.63	1.60	1.55	1.52	1.48	1.46	1.44
	.01	**7.17**	**5.06**	**4.20**	**3.72**	**3.41**	**3.18**	**3.02**	**2.88**	**2.78**	**2.70**	**2.62**	**2.56**	**2.46**	**2.39**	**2.26**	**2.18**	**2.10**	**2.00**	**1.94**	**1.86**	**1.82**	**1.76**	**1.71**	**1.68**
55	.05	4.02	3.17	2.78	2.54	2.38	2.27	2.18	2.11	2.05	2.00	1.97	1.93	1.88	1.83	1.76	1.72	1.67	1.61	1.58	1.52	1.50	1.46	1.43	1.41
	.01	**7.12**	**5.01**	**4.16**	**3.68**	**3.37**	**3.15**	**2.98**	**2.85**	**2.75**	**2.66**	**2.59**	**2.53**	**2.43**	**2.35**	**2.23**	**2.15**	**2.06**	**1.96**	**1.90**	**1.82**	**1.78**	**1.71**	**1.66**	**1.64**
60	.05	4.00	3.15	2.76	2.52	2.37	2.25	2.17	2.10	2.04	1.99	1.95	1.92	1.86	1.81	1.75	1.70	1.65	1.59	1.56	1.50	1.48	1.44	1.41	1.39
	.01	**7.08**	**4.98**	**4.13**	**3.65**	**3.34**	**3.12**	**2.95**	**2.82**	**2.72**	**2.63**	**2.56**	**2.50**	**2.40**	**2.32**	**2.20**	**2.12**	**2.03**	**1.93**	**1.87**	**1.79**	**1.74**	**1.68**	**1.63**	**1.60**
65	.05	3.99	3.14	2.75	2.51	2.36	2.24	2.15	2.08	2.02	1.98	1.94	1.90	1.85	1.80	1.73	1.68	1.63	1.57	1.54	1.49	1.46	1.42	1.39	1.37
	.01	**7.04**	**4.95**	**4.10**	**3.62**	**3.31**	**3.09**	**2.93**	**2.79**	**2.70**	**2.61**	**2.54**	**2.47**	**2.37**	**2.30**	**2.18**	**2.09**	**2.00**	**1.90**	**1.84**	**1.76**	**1.71**	**1.64**	**1.60**	**1.56**
70	.05	3.98	3.13	2.74	2.50	2.35	2.23	2.14	2.07	2.01	1.97	1.93	1.89	1.84	1.79	1.72	1.67	1.62	1.56	1.53	1.47	1.45	1.40	1.37	1.35
	.01	**7.01**	**4.92**	**4.08**	**3.60**	**3.29**	**3.07**	**2.91**	**2.77**	**2.67**	**2.59**	**2.51**	**2.45**	**2.35**	**2.28**	**2.15**	**2.07**	**1.98**	**1.88**	**1.82**	**1.74**	**1.69**	**1.62**	**1.56**	**1.53**
80	.05	3.96	3.11	2.72	2.48	2.33	2.21	2.12	2.05	1.99	1.95	1.91	1.88	1.82	1.77	1.70	1.65	1.60	1.54	1.51	1.45	1.42	1.38	1.35	1.32
	.01	**6.96**	**4.88**	**4.04**	**3.56**	**3.25**	**3.04**	**2.87**	**2.74**	**2.64**	**2.55**	**2.48**	**2.41**	**2.32**	**2.24**	**2.11**	**2.03**	**1.94**	**1.84**	**1.78**	**1.70**	**1.65**	**1.57**	**1.52**	**1.49**
100	.05	3.94	3.09	2.70	2.46	2.30	2.19	2.10	2.03	1.97	1.92	1.88	1.85	1.79	1.75	1.68	1.63	1.57	1.51	1.48	1.42	1.39	1.34	1.30	1.28
	.01	**6.90**	**4.82**	**3.98**	**3.51**	**3.20**	**2.99**	**2.82**	**2.69**	**2.59**	**2.51**	**2.43**	**2.36**	**2.26**	**2.19**	**2.06**	**1.98**	**1.89**	**1.79**	**1.73**	**1.64**	**1.59**	**1.51**	**1.46**	**1.43**
125	.05	3.92	3.07	2.68	2.44	2.29	2.17	2.08	2.01	1.95	1.90	1.86	1.83	1.77	1.72	1.65	1.60	1.55	1.49	1.45	1.39	1.36	1.31	1.27	1.25
	.01	**6.84**	**4.78**	**3.94**	**3.47**	**3.17**	**2.95**	**2.79**	**2.65**	**2.56**	**2.47**	**2.40**	**2.33**	**2.23**	**2.15**	**2.03**	**1.94**	**1.85**	**1.75**	**1.68**	**1.59**	**1.54**	**1.46**	**1.40**	**1.37**
150	.05	3.91	3.06	2.67	2.43	2.27	2.16	2.07	2.00	1.94	1.89	1.85	1.82	1.76	1.71	1.64	1.59	1.54	1.47	1.44	1.37	1.34	1.29	1.25	1.22
	.01	**6.81**	**4.75**	**3.91**	**3.44**	**3.14**	**2.92**	**2.76**	**2.62**	**2.53**	**2.44**	**2.37**	**2.30**	**2.20**	**2.12**	**2.00**	**1.91**	**1.83**	**1.72**	**1.66**	**1.56**	**1.51**	**1.43**	**1.37**	**1.33**
200	.05	3.89	3.04	2.65	2.41	2.26	2.14	2.05	1.98	1.92	1.87	1.83	1.80	1.74	1.69	1.62	1.57	1.52	1.45	1.42	1.35	1.32	1.26	1.22	1.19
	.01	**6.76**	**4.71**	**3.88**	**3.41**	**3.11**	**2.90**	**2.73**	**2.60**	**2.50**	**2.41**	**2.34**	**2.28**	**2.17**	**2.09**	**1.97**	**1.88**	**1.79**	**1.69**	**1.62**	**1.53**	**1.48**	**1.39**	**1.33**	**1.28**
400	.05	3.86	3.02	2.62	2.39	2.23	2.12	2.03	1.96	1.90	1.85	1.81	1.78	1.72	1.67	1.60	1.54	1.49	1.42	1.38	1.32	1.28	1.22	1.16	1.13
	.01	**6.70**	**4.66**	**3.83**	**3.36**	**3.06**	**2.85**	**2.69**	**2.55**	**2.46**	**2.37**	**2.29**	**2.23**	**2.12**	**2.04**	**1.92**	**1.84**	**1.74**	**1.64**	**1.57**	**1.47**	**1.42**	**1.32**	**1.24**	**1.19**
1000	.05	3.85	3.00	2.61	2.38	2.22	2.10	2.02	1.95	1.89	1.84	1.80	1.76	1.70	1.65	1.58	1.53	1.47	1.41	1.36	1.30	1.26	1.19	1.13	1.08
	.01	**6.66**	**4.62**	**3.80**	**3.34**	**3.04**	**2.82**	**2.66**	**2.53**	**2.43**	**2.34**	**2.26**	**2.20**	**2.09**	**2.01**	**1.89**	**1.81**	**1.71**	**1.61**	**1.54**	**1.44**	**1.38**	**1.28**	**1.19**	**1.11**
∞	.05	3.84	2.99	2.60	2.37	2.21	2.09	2.01	1.94	1.88	1.83	1.79	1.75	1.69	1.64	1.57	1.52	1.46	1.40	1.35	1.28	1.24	1.17	1.11	1.00
	.01	**6.64**	**4.60**	**3.78**	**3.32**	**3.02**	**2.80**	**2.64**	**2.51**	**2.41**	**2.32**	**2.24**	**2.18**	**2.07**	**1.99**	**1.87**	**1.79**	**1.69**	**1.59**	**1.52**	**1.41**	**1.36**	**1.25**	**1.15**	**1.00**

TABLE D8 Tabular Values of q_α for the Tukey Test

The row and column headings identify the appropriate values of q_α according to the degrees of freedom within (*dfw*) and the number of means (k), respectively. These values are formally known as the percentage points of the Studentized range. The table shows the values of q_α corresponding to the .05 level of significance (regular type) and the .01 level of significance (**boldface** type). These values of q_α are entered in the equation for the *HSD* statistic of Tukey to determine the minimum difference between pairs of means that would be significant. The difference between a pair of means is significant if it is equal to or greater than the *HSD* value determined by the equation.

		k = number of means																		
dfw	α	2	3	4	5	6	7	8	9	10	11	12	13	14	15	16	17	18	19	20
1	.05	18.0	27.0	32.8	37.1	40.4	43.1	45.4	47.4	49.1	50.6	52.0	53.2	54.3	55.4	56.3	57.2	58.0	58.8	59.6
	.01	**90.0**	**135**	**164**	**186**	**202**	**216**	**227**	**237**	**246**	**253**	**260**	**266**	**272**	**277**	**282**	**286**	**290**	**294**	**298**
2	.05	6.09	8.3	9.8	10.9	11.7	12.4	13.0	13.5	14.0	14.4	14.7	15.1	15.4	15.7	15.9	16.1	16.4	16.6	16.8
	.01	**14.0**	**19.0**	**22.3**	**24.7**	**26.6**	**28.2**	**29.5**	**30.7**	**31.7**	**32.6**	**33.4**	**34.1**	**34.8**	**35.4**	**36.0**	**36.5**	**37.0**	**37.5**	**37.9**
3	.05	4.50	5.91	6.82	7.50	8.04	8.48	8.85	9.18	9.46	9.72	9.95	10.15	10.35	10.52	10.69	10.84	10.98	11.11	11.24
	.01	**8.26**	**10.6**	**12.2**	**13.3**	**14.2**	**15.0**	**15.6**	**16.2**	**16.7**	**17.1**	**17.5**	**17.9**	**18.2**	**18.5**	**18.8**	**19.1**	**19.3**	**19.5**	**19.8**
4	.05	3.93	5.04	5.76	6.29	6.71	7.05	7.35	7.60	7.83	8.03	8.21	8.37	8.52	8.66	8.79	8.91	9.03	9.13	9.23
	.01	**6.51**	**8.12**	**9.17**	**9.96**	**10.6**	**11.1**	**11.5**	**11.9**	**12.3**	**12.6**	**12.8**	**13.1**	**13.3**	**13.5**	**13.7**	**13.9**	**14.1**	**14.2**	**14.4**
5	.05	3.64	4.60	5.22	5.67	6.03	6.33	6.58	6.80	6.99	7.17	7.32	7.47	7.60	7.72	7.83	7.93	8.03	8.12	8.21
	.01	**5.70**	**6.97**	**7.80**	**8.42**	**8.91**	**9.32**	**9.67**	**9.97**	**10.24**	**10.48**	**10.70**	**10.89**	**11.08**	**11.24**	**11.40**	**11.55**	**11.68**	**11.81**	**11.93**
6	.05	3.46	4.34	4.90	5.31	5.63	5.89	6.12	6.32	6.49	6.65	6.79	6.92	7.03	7.14	7.24	7.34	7.43	7.51	7.59
	.01	**5.24**	**6.33**	**7.03**	**7.56**	**7.97**	**8.32**	**8.61**	**8.87**	**9.10**	**9.30**	**9.49**	**9.65**	**9.81**	**9.95**	**10.08**	**10.21**	**10.32**	**10.43**	**10.54**
7	.05	3.34	4.16	4.68	5.06	5.36	5.61	5.82	6.00	6.16	6.30	6.43	6.55	6.66	6.76	6.85	6.94	7.02	7.09	7.17
	.01	**4.95**	**5.92**	**6.54**	**7.01**	**7.37**	**7.68**	**7.94**	**8.17**	**8.37**	**8.55**	**8.71**	**8.86**	**9.00**	**9.12**	**9.24**	**9.35**	**9.46**	**9.55**	**9.65**
8	.05	3.26	4.04	4.53	4.89	5.17	5.40	5.60	5.77	5.92	6.05	6.18	6.29	6.39	6.48	6.57	6.65	6.73	6.80	6.87
	.01	**4.74**	**5.63**	**6.20**	**6.63**	**6.96**	**7.24**	**7.47**	**7.68**	**7.87**	**8.03**	**8.18**	**8.31**	**8.44**	**8.55**	**8.66**	**8.76**	**8.85**	**8.94**	**9.03**
9	.05	3.20	3.95	4.42	4.76	5.02	5.24	5.43	5.60	5.74	5.87	5.98	6.09	6.19	6.28	6.36	6.44	6.51	6.58	6.64
	.01	**4.60**	**5.43**	**5.96**	**6.35**	**6.66**	**6.91**	**7.13**	**7.32**	**7.49**	**7.65**	**7.78**	**7.91**	**8.03**	**8.13**	**8.23**	**8.32**	**8.41**	**8.49**	**8.57**
10	.05	3.15	3.88	4.33	4.65	4.91	5.12	5.30	5.46	5.60	5.72	5.83	5.93	6.03	6.11	6.20	6.27	6.34	6.40	6.47
	.01	**4.48**	**5.27**	**5.77**	**6.14**	**6.43**	**6.67**	**6.87**	**7.05**	**7.21**	**7.36**	**7.48**	**7.60**	**7.71**	**7.81**	**7.91**	**7.99**	**8.07**	**8.15**	**8.22**
11	.05	3.11	3.82	4.26	4.57	4.82	5.03	5.20	5.35	5.49	5.61	5.71	5.81	5.90	5.99	6.06	6.14	6.20	6.26	6.33
	.01	**4.39**	**5.14**	**5.62**	**5.97**	**6.25**	**6.48**	**6.67**	**6.84**	**6.99**	**7.13**	**7.25**	**7.36**	**7.46**	**7.56**	**7.65**	**7.73**	**7.81**	**7.88**	**7.95**

dfw	α	2	3	4	5	6	7	8	9	10	11	12	13	14	15	16	17	18	19	20
		k = number of means																		
12	.05	3.08	3.77	4.20	4.51	4.75	4.95	5.12	5.27	5.40	5.51	5.62	5.71	5.80	5.88	5.95	6.03	6.09	6.15	6.21
	.01	**4.32**	**5.04**	**5.50**	**5.84**	**6.10**	**6.32**	**6.51**	**6.67**	**6.81**	**6.94**	**7.06**	**7.17**	**7.26**	**7.36**	**7.44**	**7.52**	**7.59**	**7.66**	**7.73**
13	.05	3.06	3.73	4.15	4.45	4.69	4.88	5.05	5.19	5.32	5.43	5.53	5.63	5.71	5.79	5.86	5.93	6.00	6.05	6.11
	.01	**4.26**	**4.96**	**5.40**	**5.73**	**5.98**	**6.19**	**6.37**	**6.53**	**6.67**	**6.79**	**6.90**	**7.01**	**7.10**	**7.19**	**7.27**	**7.34**	**7.42**	**7.48**	**7.55**
14	.05	3.03	3.70	4.11	4.41	4.64	4.83	4.99	5.13	5.25	5.36	5.46	5.55	5.64	5.72	5.79	5.85	5.92	5.97	6.03
	.01	**4.21**	**4.89**	**5.32**	**5.63**	**5.88**	**6.08**	**6.26**	**6.41**	**6.54**	**6.66**	**6.77**	**6.87**	**6.96**	**7.05**	**7.12**	**7.20**	**7.27**	**7.33**	**7.39**
15	.05	3.01	3.67	4.08	4.37	4.60	4.78	4.94	5.08	5.20	5.31	5.40	5.49	5.58	5.65	5.72	5.79	5.85	5.90	5.96
	.01	**4.17**	**4.83**	**5.25**	**5.56**	**5.80**	**5.99**	**6.16**	**6.31**	**6.44**	**6.55**	**6.66**	**6.76**	**6.84**	**6.93**	**7.00**	**7.07**	**7.14**	**7.20**	**7.26**
16	.05	3.00	3.65	4.05	4.33	4.56	4.74	4.90	5.03	5.15	5.26	5.35	5.44	5.52	5.59	5.66	5.72	5.79	5.84	5.90
	.01	**4.13**	**4.78**	**5.19**	**5.49**	**5.72**	**5.92**	**6.08**	**6.22**	**6.35**	**6.46**	**6.56**	**6.66**	**6.74**	**6.82**	**6.90**	**6.97**	**7.03**	**7.09**	**7.15**
17	.05	2.98	3.63	4.02	4.30	4.52	4.71	4.86	4.99	5.11	5.21	5.31	5.39	5.47	5.55	5.61	5.68	5.74	5.79	5.84
	.01	**4.10**	**4.74**	**5.14**	**5.43**	**5.66**	**5.85**	**6.01**	**6.15**	**6.27**	**6.38**	**6.48**	**6.57**	**6.66**	**6.73**	**6.80**	**6.87**	**6.94**	**7.00**	**7.05**
18	.05	2.97	3.61	4.00	4.28	4.49	4.67	4.82	4.96	5.07	5.17	5.27	5.35	5.43	5.50	5.57	5.63	5.69	5.74	5.79
	.01	**4.07**	**4.70**	**5.09**	**5.38**	**5.60**	**5.79**	**5.94**	**6.08**	**6.20**	**6.31**	**6.41**	**6.50**	**6.58**	**6.65**	**6.72**	**6.79**	**6.85**	**6.91**	**6.96**
19	.05	2.96	3.59	3.98	4.25	4.47	4.65	4.79	4.92	5.04	5.14	5.23	5.32	5.39	5.46	5.53	5.59	5.65	5.70	5.75
	.01	**4.05**	**4.67**	**5.05**	**5.33**	**5.55**	**5.73**	**5.89**	**6.02**	**6.14**	**6.25**	**6.34**	**6.43**	**6.51**	**6.58**	**6.65**	**6.72**	**6.78**	**6.84**	**6.89**
20	.05	2.95	3.58	3.96	4.23	4.45	4.62	4.77	4.90	5.01	5.11	5.20	5.28	5.36	5.43	5.49	5.55	5.61	5.66	5.71
	.01	**4.02**	**4.64**	**5.02**	**5.29**	**5.51**	**5.69**	**5.84**	**5.97**	**6.09**	**6.19**	**6.29**	**6.37**	**6.45**	**6.52**	**6.59**	**6.65**	**6.71**	**6.76**	**6.82**
24	.05	2.92	3.53	3.90	4.17	4.37	4.54	4.68	4.81	4.92	5.01	5.10	5.18	5.25	5.32	5.38	5.44	5.50	5.54	5.59
	.01	**3.96**	**4.54**	**4.91**	**5.17**	**5.37**	**5.54**	**5.69**	**5.81**	**5.92**	**6.02**	**6.11**	**6.19**	**6.26**	**6.33**	**6.39**	**6.45**	**6.51**	**6.56**	**6.61**
30	.05	2.89	3.49	3.84	4.10	4.30	4.46	4.60	4.72	4.83	4.92	5.00	5.08	5.15	5.21	5.27	5.33	5.38	5.43	5.48
	.01	**3.89**	**4.45**	**4.80**	**5.05**	**5.24**	**5.40**	**5.54**	**5.65**	**5.76**	**5.85**	**5.93**	**6.01**	**6.08**	**6.14**	**6.20**	**6.26**	**6.31**	**6.36**	**6.41**
40	.05	2.86	3.44	3.79	4.04	4.23	4.39	4.52	4.63	4.74	4.82	4.91	4.98	5.05	5.11	5.16	5.22	5.27	5.31	5.36
	.01	**3.82**	**4.37**	**4.70**	**4.93**	**5.11**	**5.27**	**5.39**	**5.50**	**5.60**	**5.69**	**5.77**	**5.84**	**5.90**	**5.96**	**6.02**	**6.07**	**6.12**	**6.17**	**6.21**
60	.05	2.83	3.40	3.74	3.98	4.16	4.31	4.44	4.55	4.65	4.73	4.81	4.88	4.94	5.00	5.06	5.11	5.16	5.20	5.24
	.01	**3.76**	**4.28**	**4.60**	**4.82**	**4.99**	**5.13**	**5.25**	**5.36**	**5.45**	**5.53**	**5.60**	**5.67**	**5.73**	**5.79**	**5.84**	**5.89**	**5.93**	**5.98**	**6.02**
120	.05	2.80	3.36	3.69	3.92	4.10	4.24	4.36	4.48	4.56	4.64	4.72	4.78	4.84	4.90	4.95	5.00	5.05	5.09	5.13
	.01	**3.70**	**4.20**	**4.50**	**4.71**	**4.87**	**5.01**	**5.12**	**5.21**	**5.30**	**5.38**	**5.44**	**5.51**	**5.56**	**5.61**	**5.66**	**5.71**	**5.75**	**5.79**	**5.83**
∞	.05	2.77	3.31	3.63	3.86	4.03	4.17	4.29	4.39	4.47	4.55	4.62	4.68	4.74	4.80	4.85	4.89	4.93	4.97	5.01
	.01	**3.64**	**4.12**	**4.40**	**4.60**	**4.76**	**4.88**	**4.99**	**5.08**	**5.16**	**5.23**	**5.29**	**5.35**	**5.40**	**5.45**	**5.49**	**5.54**	**5.57**	**5.61**	**5.65**

TABLE D9 Tabular Values of Spearman's r_s

Column 1 identifies the sample size (N) representing the number of pairs of scores.

The table shows the values of r_s corresponding to various levels of significance for one- and two-tailed tests. The test value of r_s is significant if it is equal to or greater than the value shown in the table. To interpolate, sum the critical values above and below the N of interest and divide by 2. Thus the critical value at $\alpha = 0.05$, two-tailed test, when $N = 21$, is $(.450 + .428)/2 = .439$.

	Level of Significance for One-Tailed Test			
	.05	**.025**	**.01**	**.005**
(1)	**Level of Significance for Two-Tailed Test**			
N*	**.10**	**.05**	**.02**	**.01**
5	.900	1.000	1.000	—
6	.829	.886	.943	1.000
7	.714	.786	.893	.929
8	.643	.738	.833	.881
9	.600	.683	.783	.833
10	.564	.648	.746	.794
12	.506	.591	.712	.777
14	.456	.544	.645	.715
16	.425	.506	.601	.665
18	.399	.475	.564	.625
20	.377	.450	.534	.591
22	.359	.428	.508	.562
24	.343	.409	.485	.537
26	.329	.392	.465	.515
28	.317	.377	.448	.496
30	.306	.364	.432	.478

*N = number of pairs

TABLE D10 Tabular Values of Pearson's r

Column 1 identifies the specific distribution of r by its degrees of freedom, $df = N - 2$.

The table shows the values of r corresponding to various levels of significance for one- and two-tailed tests when $H_0: \rho = 0$. The test value of r is significant if it is equal to or greater than the value shown in the table.

TABLE D10 Tabular Values of Pearson's *r* (continued)

	Levels of significance for a one-tailed test			
	.05	.025	.01	.005
	Levels of significance for a two-tailed test			
(1) *df*	.10	.05	.02	.01
1	.988	.997	.9995	.9999
2	.900	.950	.980	.990
3	.805	.878	.934	.959
4	.729	.811	.882	.917
5	.669	.754	.833	.874
6	.622	.707	.789	.834
7	.582	.666	.750	.798
8	.549	.632	.716	.765
9	.521	.602	.685	.735
10	.497	.576	.658	.708
11	.476	.553	.634	.684
12	.458	.532	.612	.661
13	.441	.514	.592	.641
14	.426	.497	.574	.623
15	.412	.482	.558	.606
16	.400	.468	.542	.590
17	.389	.456	.528	.575
18	.378	.444	.516	.561
19	.369	.433	.503	.549
20	.360	.423	.492	.537
21	.352	.413	.482	.526
22	.344	.404	.472	.515
23	.337	.396	.462	.505
24	.330	.388	.453	.496
25	.323	.381	.445	.487
26	.317	.374	.437	.479
27	.311	.367	.430	.471
28	.306	.361	.423	.463
29	.301	.355	.416	.456
30	.296	.349	.409	.449
32	.287	.339	.397	.436
34	.279	.329	.386	.424
36	.271	.320	.376	.413
38	.264	.312	.367	.403
40	.257	.304	.358	.393
42	.251	.297	.350	.384
44	.246	.291	.342	.376
46	.240	.285	.335	.368
48	.235	.279	.328	.361
50	.231	.273	.322	.354
55	.220	.261	.307	.339
60	.211	.250	.295	.325
65	.203	.240	.284	.313
70	.195	.232	.274	.302
75	.189	.224	.265	.292
80	.183	.217	.256	.283
85	.178	.211	.249	.275
90	.173	.205	.242	.267

TABLE D10 Tabular Values of Pearson's r (continued)

(1)	Levels of significance for a one-tailed test			
	.05	.025	.01	.005
	Levels of significance for a two-tailed test			
df	.10	.05	.02	.01
95	.168	.200	.236	.260
100	.164	.195	.230	.254
120	.150	.178	.210	.232
150	.134	.159	.189	.208
200	.116	.138	.164	.181
300	.095	.113	.134	.148
400	.082	.098	.116	.128
500	.073	.088	.104	.115
1000	.052	.062	.073	.081

TABLE D11 Transformation of Pearson's r to Fisher's z'

Column 1 identifies the value of Pearson's r and column 2 identifies the corresponding value of Fisher's z'. The transformation is used whenever ρ is assumed not to equal 0.

(1) r	(2) z'	(1) r	(2) z'	(1) r	(2) z'	(1) r	(2) z'	(1) r	(2) z'
.000	.000	.125	.126	.250	.255	.375	.394	.500	.549
.005	.005	.130	.131	.255	.261	.380	.400	.505	.556
.010	.010	.135	.136	.260	.266	.385	.406	.510	.563
.015	.015	.140	.141	.265	.271	.390	.412	.515	.570
.020	.020	.145	.146	.270	.277	.395	.418	.520	.576
.025	.025	.150	.151	.275	.282	.400	.424	.525	.583
.030	.030	.155	.156	.280	.288	.405	.430	.530	.590
.035	.035	.160	.161	.285	.293	.410	.436	.535	.597
.040	.040	.165	.167	.290	.299	.415	.442	.540	.604
.045	.045	.170	.172	.295	.304	.420	.448	.545	.611
.050	.050	.175	.177	.300	.310	.425	.454	.550	.618
.055	.055	.180	.182	.305	.315	.430	.460	.555	.626
.060	.060	.185	.187	.310	.321	.435	.466	.560	.633
.065	.065	.190	.192	.315	.326	.440	.472	.565	.640
.070	.070	.195	.198	.320	.332	.445	.478	.570	.648
.075	.075	.200	.203	.325	.337	.450	.485	.575	.655
.080	.080	.205	.208	.330	.343	.455	.491	.580	.662
.085	.085	.210	.213	.335	.348	.460	.497	.585	.670
.090	.090	.215	.218	.340	.354	.465	.504	.590	.678
.095	.095	.220	.224	.345	.360	.470	.510	.595	.685
.100	.100	.225	.229	.350	.365	.475	.517	.600	.693
.105	.105	.230	.234	.355	.371	.480	.523	.605	.701
.110	.110	.235	.239	.360	.377	.485	.530	.610	.709
.115	.116	.240	.245	.365	.383	.490	.536	.615	.717
.120	.121	.245	.250	.370	.388	.495	.543	.620	.725

TABLE D11 Transformation of Pearson's r to Fisher's z' (continued)

(1) r	(2) z'	(1) r	(2) z'	(1) r	(2) z'	(1) r	(2) z'	(1) r	(2) z'
.625	.733	.700	.867	.775	1.033	.850	1.256	.925	1.623
.630	.741	.705	.877	.780	1.045	.855	1.274	.930	1.658
.635	.750	.710	.887	.785	1.058	.860	1.293	.935	1.697
.640	.758	.715	.897	.790	1.071	.865	1.313	.940	1.738
.645	.767	.720	.908	.795	1.085	.870	1.333	.945	1.783
.650	.775	.725	.918	.800	1.099	.875	1.354	.950	1.832
.655	.784	.730	.929	.805	1.113	.880	1.376	.955	1.886
.660	.793	.735	.940	.810	1.127	.885	1.398	.960	1.946
.665	.802	.740	.950	.815	1.142	.890	1.422	.965	2.014
.670	.811	.745	.962	.820	1.157	.895	1.447	.970	2.092
.675	.820	.750	.973	.825	1.172	.900	1.472	.975	2.185
.680	.829	.755	.984	.830	1.188	.905	1.499	.980	2.298
.685	.838	.760	.996	.835	1.204	.910	1.528	.985	2.443
.690	.848	.767	1.008	.840	1.221	.915	1.557	.990	2.647
.695	.858	.770	1.020	.845	1.238	.920	1.589	.995	2.994

TABLE D12 Tabular Values of T for the Wilcoxon Matched-Pairs Test

Column 1 identifies the sample size (N) representing the number of ranked differences.

The table shows the values of T corresponding to various levels of significance for one- and two-tailed tests. The values of T denote the smaller sum of ranks associated with differences that are all of the same sign. The test value of T is significant if it is equal to or *less than* the value shown in the table. All entries are for the *absolute* value of T.

	Level of significance for one-tailed test			
	.05	.025	.01	.005
	Level of significance for two-tailed test			
(1) N	.10	.05	.02	.01
5	0	—	—	—
6	2	0	—	—
7	3	2	0	—
8	5	3	1	0
9	8	5	3	1
10	10	8	5	3
11	13	10	7	5
12	17	13	9	7
13	21	17	12	9
14	25	21	15	12
15	30	25	19	15
16	35	29	23	19
17	41	34	27	23
18	47	40	32	27
19	53	46	37	32
20	60	52	43	37
21	67	58	49	42
22	75	65	55	48
23	83	73	62	54
24	91	81	69	61
25	100	89	76	68
26	110	98	84	75
27	119	107	92	83
28	130	116	101	91
29	140	126	110	100
30	151	137	120	109
31	163	147	130	118
32	175	159	140	128
33	187	170	151	138
34	200	182	162	148
35	213	195	173	159
36	227	208	185	171
37	241	221	198	182
38	256	235	211	194
39	271	249	224	207
40	286	264	238	220
41	302	279	252	233
42	319	294	266	247
43	336	310	281	261
44	353	327	296	276
45	371	343	312	291
46	389	361	328	307
47	407	378	345	322
48	426	396	362	339
49	446	415	379	355
50	466	434	397	373

[Slight discrepancies will be found between the critical values appearing in the table above and in Table 2 of the 1964 revision of F. Wilcoxon and R. A. Wilcox. *Some Rapid Approximate Statistical Procedures.* (New York: Lederle Laboratories). The disparity reflects the latter's policy of selecting the critical value nearest a given significance level, occasionally overstepping that level. For example, for $N = 8$,

the probability of a T of 3 = 0.0390 (two-tail),

and

the probability of a T of 4 = 0.0546 (two-tail).

Wilcoxon and Wilcox select a T of 4 as the critical value at the 0.05 level of significance (two-tail), whereas Table D12 reflects a more conservative policy by setting a T of 3 as the critical value at this level.]

TABLE D13 Tabular Values of *R* for the Wald–Wolfowitz Runs Test

The row and column headings identify appropriate sample sizes, N_1 and N_2, respectively. The table shows the values of R for the .05 level of significance. For the Wald–Wolfowitz two-sample runs test, the test value of R is significant if it is equal to or *less than* the value shown in the table.

N_1 \ N_2	2	3	4	5	6	7	8	9	10	11	12	13	14	15	16	17	18	19	20
2											2	2	2	2	2	2	2	2	2
3					2	2	2	2	2	2	2	2	2	3	3	3	3	3	3
4				2	2	2	3	3	3	3	3	3	3	3	4	4	4	4	4
5			2	2	3	3	3	3	3	4	4	4	4	4	4	4	5	5	5
6		2	2	3	3	3	3	4	4	4	4	5	5	5	5	5	5	6	6
7		2	2	3	3	3	4	4	5	5	5	5	5	6	6	6	6	6	6
8		2	3	3	3	4	4	5	5	5	6	6	6	6	6	7	7	7	7
9		2	3	3	4	4	5	5	5	6	6	6	7	7	7	7	8	8	8
10		2	3	3	4	5	5	5	6	6	7	7	7	7	8	8	8	8	9
11		2	3	4	4	5	5	6	6	7	7	7	8	8	8	9	9	9	9
12	2	2	3	4	4	5	6	6	7	7	7	8	8	8	9	9	9	10	10
13	2	2	3	4	5	5	6	6	7	7	8	8	9	9	9	10	10	10	10
14	2	2	3	4	5	5	6	7	7	8	8	9	9	9	10	10	10	11	11
15	2	3	3	4	5	6	6	7	7	8	8	9	9	10	10	11	11	11	12
16	2	3	4	4	5	6	6	7	8	8	9	9	10	10	11	11	11	12	12
17	2	3	4	4	5	6	7	7	8	9	9	10	10	11	11	11	12	12	13
18	2	3	4	5	5	6	7	8	8	9	9	10	10	11	11	12	12	13	13
19	2	3	4	5	6	6	7	8	8	9	10	10	11	11	12	12	13	13	13
20	2	3	4	5	6	6	7	8	9	9	10	10	11	12	12	13	13	13	14

TABLE D14 Tabular Values of U for the Mann–Whitney Test

Tables 14.1 through 14.4 show values of U (and U') for various levels of significance with one- and two-tailed tests. To be significant for any given sample sizes, N_1 and N_2, the test value of U (or U') must be outside the lower and upper limits defined by the two values shown in the table.

TABLE 14.1 Critical Values for a One-Tailed Test at $\alpha = .005$ or a Two-Tailed Test at $\alpha = .01$

N_2 \ N_1	1	2	3	4	5	6	7	8	9	10	11	12	13	14	15	16	17	18	19	20
1	—	—	—	—	—	—	—	—	—	—	—	—	—	—	—	—	—	—	—	—
2	—	—	—	—	—	—	—	—	—	—	—	—	—	—	—	—	—	—	0 38	0 40
3	—	—	—	—	—	—	—	—	0 27	0 30	0 33	1 35	1 38	1 41	2 43	2 46	2 49	2 52	3 54	3 57
4	—	—	—	—	—	0 24	0 28	1 31	1 35	2 38	2 42	3 45	3 49	4 52	5 55	5 59	6 62	6 66	7 69	8 72
5	—	—	—	—	0 25	1 29	1 34	2 38	3 42	4 46	5 50	6 54	7 58	7 63	8 67	9 71	10 75	11 79	12 83	13 87
6	—	—	—	0 24	1 29	2 34	3 39	4 44	5 49	6 54	7 59	9 63	10 68	11 73	12 78	13 83	15 87	16 92	17 97	18 102
7	—	—	—	0 28	1 34	3 39	4 45	6 50	7 56	9 61	10 67	12 72	13 78	15 83	16 89	18 94	19 100	21 105	22 111	24 116
8	—	—	—	1 31	2 38	4 44	6 50	7 57	9 63	11 69	13 75	15 81	17 87	18 94	20 100	22 106	24 112	26 118	28 124	30 130
9	—	—	0 27	1 35	3 42	5 49	7 56	9 63	11 70	13 77	16 83	18 90	20 97	22 104	24 111	27 117	29 124	31 131	33 138	36 144
10	—	—	0 30	2 38	4 46	6 54	9 61	11 69	13 77	16 84	18 92	21 99	24 106	26 114	29 121	31 129	34 136	37 143	39 151	42 158
11	—	—	0 33	2 42	5 50	7 59	10 67	13 75	16 83	18 92	21 100	24 108	27 116	30 124	33 132	36 140	39 148	42 156	45 164	48 172
12	—	—	1 35	3 45	6 54	9 63	12 72	15 81	18 90	21 99	24 108	27 117	31 125	34 134	37 143	41 151	44 160	47 169	51 177	54 186
13	—	—	1 38	3 49	7 58	10 68	13 78	17 87	20 97	24 106	27 116	31 125	34 125	38 144	42 153	45 163	49 172	53 181	56 191	60 200
14	—	—	1 41	4 52	7 63	11 73	15 83	18 94	22 104	26 114	30 124	34 134	38 144	42 154	46 164	50 174	54 184	58 194	63 203	67 213
15	—	—	2 43	5 55	8 67	12 78	16 89	20 100	24 111	29 121	33 132	37 143	42 153	46 164	51 174	55 185	60 195	64 206	69 216	73 227
16	—	—	2 46	5 59	9 71	13 83	18 94	22 106	27 117	31 129	36 140	41 151	45 163	50 174	55 185	60 196	65 207	70 218	74 230	79 241
17	—	—	2 49	6 62	10 75	15 87	19 100	24 112	29 124	34 148	39 148	44 160	49 172	54 184	60 195	65 207	70 219	75 231	81 242	86 254

TABLE D14 Tabular Values of U for the Mann–Whitney Test (continued)

N_2 \ N_1	1	2	3	4	5	6	7	8	9	10	11	12	13	14	15	16	17	18	19	20
18	—	—	2 52	6 66	11 79	16 92	21 105	26 118	31 131	37 143	42 156	47 169	53 181	58 194	64 206	70 218	75 231	81 243	87 255	92 268
19	—	0 38	3 54	7 69	12 83	17 97	22 111	28 124	33 138	39 151	45 164	51 177	56 191	63 203	69 216	74 230	81 242	87 255	93 268	99 281
20	—	0 40	3 57	8 72	13 87	18 102	24 116	30 130	36 144	42 158	48 172	54 186	60 200	67 213	73 227	79 241	86 254	92 268	99 281	105 295

(Dashes in the body of the table indicate that no decision is possible at the stated level of significance.)

TABLE D14.2 Critical Values for a One-Tailed Test at $\alpha = .01$ or a Two-Tailed Test at $\alpha = .02$

N_2 \ N_1	1	2	3	4	5	6	7	8	9	10	11	12	13	14	15	16	17	18	19	20
1	—	—	—	—	—	—	—	—	—	—	—	—	—	—	—	—	—	—	—	—
2	—	—	—	—	—	—	—	—	—	—	—	—	0 26	0 28	0 30	0 32	0 34	0 36	1 37	1 39
3	—	—	—	—	—	—	0 21	0 24	1 26	1 29	1 32	2 34	2 37	2 40	3 42	3 45	4 47	4 50	4 52	5 55
4	—	—	—	—	0 20	1 23	1 27	2 30	3 33	3 37	4 40	5 43	5 47	6 50	7 53	7 57	8 60	9 63	9 67	10 70
5	—	—	—	0 20	1 24	2 28	3 32	4 36	5 40	6 44	7 48	8 52	9 56	10 60	11 64	12 68	13 72	14 76	15 80	16 84
6	—	—	—	1 23	2 28	3 33	4 38	6 42	7 47	8 52	9 57	11 61	12 66	13 71	15 75	16 80	18 84	19 89	20 94	22 98
7	—	—	0 21	1 27	3 32	4 38	6 43	7 49	9 54	11 59	12 65	14 70	16 75	17 81	19 86	21 91	23 96	24 102	26 107	28 112
8	—	—	0 24	2 30	4 36	6 42	7 49	9 55	11 61	13 67	15 73	17 79	20 84	22 90	24 96	26 102	28 108	30 114	32 120	34 126
9	—	—	1 26	3 33	5 40	7 47	9 54	11 61	14 67	16 74	18 81	21 87	23 94	26 100	28 107	31 113	33 120	36 126	38 133	40 140
10	—	—	1 29	3 37	6 44	8 52	11 59	13 67	16 74	19 81	22 88	24 96	27 103	30 110	33 117	36 124	38 132	41 139	44 146	47 153
11	—	—	1 32	4 40	7 48	9 57	12 65	15 73	18 81	22 88	25 96	28 104	31 112	34 120	37 128	41 135	44 143	47 151	50 159	53 167
12	—	—	2 34	5 43	8 52	11 61	14 70	17 79	21 87	24 96	28 104	31 113	35 121	38 130	42 138	46 146	49 155	53 163	56 172	60 180
13	—	0 26	2 37	5 47	9 56	12 66	16 75	20 84	23 94	27 103	31 112	35 121	39 130	43 139	47 148	51 157	55 166	59 175	63 184	67 193

TABLE D14 Tabular Values of *U* for the Mann–Whitney Test (continued)

N_2 \ N_1	1	2	3	4	5	6	7	8	9	10	11	12	13	14	15	16	17	18	19	20
14	—	0	2	6	10	13	17	22	26	30	34	38	43	47	51	56	60	65	69	73
		28	40	50	60	71	81	90	100	110	120	130	139	149	159	168	178	187	197	207
15	—	0	3	7	11	15	19	24	28	33	37	42	47	51	56	61	66	70	75	80
		30	42	53	64	75	86	96	107	117	128	138	148	159	169	179	189	200	210	220
16	—	0	3	7	12	16	21	26	31	36	41	46	51	56	61	66	71	76	82	87
		32	45	57	68	80	91	102	113	124	135	146	157	168	179	190	201	212	222	233
17	—	0	4	8	13	18	23	28	33	38	44	49	55	60	66	71	77	82	88	93
		34	47	60	72	84	96	108	120	132	143	155	166	178	189	201	212	224	234	247
18	—	0	4	9	14	19	24	30	36	41	47	53	59	65	70	76	82	88	94	100
		36	50	63	76	89	102	114	126	139	151	163	175	187	200	212	224	236	248	260
19	—	1	4	9	15	20	26	32	38	44	50	56	63	69	75	82	88	94	101	107
		37	53	67	80	94	107	120	133	146	159	172	184	197	210	222	235	248	260	273
20	—	1	5	10	16	22	28	34	40	47	53	60	67	73	80	87	93	100	107	114
		39	55	70	84	98	112	126	140	153	167	180	193	207	220	233	247	260	273	286

(Dashes in the body of the table indicate that no decision is possible at the stated level of significance.)

TABLE D14.3 Critical Values for a One-Tailed Test at $\alpha = .025$ or a Two Tailed Test at $\alpha = .05$

N_2 \ N_1	1	2	3	4	5	6	7	8	9	10	11	12	13	14	15	16	17	18	19	20
1	—	—	—	—	—	—	—	—	—	—	—	—	—	—	—	—	—	—	—	—
2	—	—	—	—	—	—	—	0	0	0	0	1	1	1	1	1	2	2	2	2
								16	18	20	22	23	25	27	29	31	32	34	36	38
3	—	—	—	—	0	1	1	2	2	3	3	4	4	5	5	6	6	7	7	8
					15	17	20	22	25	27	30	32	35	37	40	42	45	47	50	52
4	—	—	—	0	1	2	3	4	4	5	6	7	8	9	10	11	11	12	13	13
				16	19	22	25	28	32	35	38	41	44	47	50	53	57	60	63	67
5	—	—	0	1	2	3	5	6	7	8	9	11	12	13	14	15	17	18	19	20
			15	19	23	27	30	34	38	42	46	49	53	57	61	65	68	72	76	80
6	—	—	1	2	3	5	6	8	10	11	13	14	16	17	19	21	22	24	25	27
			17	22	27	31	36	40	44	49	53	58	62	67	71	75	80	84	89	93
7	—	—	1	3	5	6	8	10	12	14	16	18	20	22	24	26	28	30	32	34
			20	25	30	36	41	46	51	56	61	66	71	76	81	86	91	96	101	106
8	—	0	2	4	6	8	10	13	15	17	19	22	24	26	29	31	34	36	38	41
		16	22	28	34	40	46	51	57	63	69	74	80	86	91	97	102	108	111	119
9	—	0	2	4	7	10	12	15	17	20	23	26	28	31	34	37	39	42	45	48
		18	25	32	38	44	51	57	64	70	76	82	89	95	101	107	114	120	126	132
10	—	0	3	5	8	11	14	17	20	23	26	29	33	36	39	42	45	48	52	55
		20	27	35	42	49	56	63	70	77	84	91	97	104	111	118	125	132	138	145

TABLE D14 Tabular Values of *U* for the Mann–Whitney Test (continued)

N_2 \ N_1	1	2	3	4	5	6	7	8	9	10	11	12	13	14	15	16	17	18	19	20
11	—	0	3	6	9	13	16	19	23	26	30	33	37	40	44	47	51	55	58	62
		22	30	38	46	53	61	69	76	84	91	99	106	114	121	129	136	143	151	158
12	—	1	4	7	11	14	18	22	26	29	33	37	41	45	49	53	57	61	65	69
		23	32	41	49	58	66	74	82	91	99	107	115	123	131	139	147	155	163	171
13	—	1	4	8	12	16	20	24	28	33	37	41	45	50	54	59	63	67	72	76
		25	35	44	53	62	71	80	89	97	106	115	124	132	141	149	158	167	175	184
14	—	1	5	9	13	17	22	26	31	36	40	45	50	55	59	64	67	74	78	83
		27	37	47	51	67	76	86	95	104	114	123	132	141	151	160	171	178	188	197
15	—	1	5	10	14	19	24	29	34	39	44	49	54	59	64	70	75	80	85	90
		29	40	50	61	71	81	91	101	111	121	131	141	151	161	170	180	190	200	210
16	—	1	6	11	15	21	26	31	37	42	47	53	59	64	70	75	81	86	92	98
		31	42	53	65	75	86	97	107	118	129	139	149	160	170	181	191	202	212	222
17	—	2	6	11	17	22	28	34	39	45	51	57	63	67	75	81	87	93	99	105
		32	45	57	68	80	91	102	114	125	136	147	158	171	180	191	202	213	224	235
18	—	2	7	12	18	24	30	36	42	48	55	61	67	74	80	86	93	99	106	112
		34	47	60	72	84	96	108	120	132	143	155	167	178	190	202	213	225	236	248
19	—	2	7	13	19	25	32	38	45	52	58	65	72	78	85	92	99	106	113	119
		36	50	63	76	89	101	114	126	138	151	163	175	188	200	212	224	236	248	261
20	—	2	8	13	20	27	34	41	48	55	62	69	76	83	90	98	105	112	119	127
		38	52	67	80	93	106	119	132	145	158	171	184	197	210	222	235	248	261	273

(Dashes in the body of the table indicate that no decision is possible at the stated level of significance.)

TABLE D14.4 Critical Values for a One-Tailed Test at $\alpha = .05$ or a Two-Tailed Test at $\alpha = .10$

N_2 \ N_1	1	2	3	4	5	6	7	8	9	10	11	12	13	14	15	16	17	18	19	20
1	—	—	—	—	—	—	—	—	—	—	—	—	—	—	—	—	—	—	0	0
																			19	20
2	—	—	—	—	0	0	0	1	1	1	1	2	2	2	3	3	3	4	4	4
					10	12	14	15	17	19	21	22	24	26	27	29	31	32	34	36
3	—	—	0	0	1	2	2	3	3	4	5	5	6	7	7	8	9	9	10	11
			9	12	14	16	19	21	24	26	28	31	33	35	38	40	42	45	47	49
4	—	—	0	1	2	3	4	5	6	7	8	9	10	11	12	14	15	16	17	18
			12	15	18	21	24	27	30	33	36	39	42	45	48	50	53	56	59	62
5	—	0	1	2	4	5	6	8	9	11	12	13	15	16	18	19	20	22	23	25
		10	14	18	21	25	29	32	36	39	43	47	50	54	57	61	65	68	72	75
6	—	0	2	3	5	7	8	10	12	14	16	17	19	21	23	25	26	28	30	32
		12	16	21	25	29	34	38	42	46	50	55	59	63	67	71	76	80	84	88

TABLE D14 Tabular Values of *U* for the Mann–Whitney Test (continued)

N_2 \ N_1	1	2	3	4	5	6	7	8	9	10	11	12	13	14	15	16	17	18	19	20
7	—	0 14	2 19	4 24	6 29	8 34	11 38	13 43	15 48	17 53	19 58	21 63	24 67	26 72	28 77	30 82	33 86	35 91	37 96	39 101
8	—	1 15	3 21	5 27	8 32	10 38	13 43	15 49	18 54	20 60	23 65	26 70	28 76	31 81	33 87	36 92	39 97	41 103	44 108	47 113
9	—	1 17	3 24	6 30	9 36	12 42	15 48	18 54	21 60	24 66	27 72	30 78	33 84	36 90	39 96	42 102	45 108	48 114	51 120	54 126
10	—	1 19	4 26	7 33	11 39	14 46	17 53	20 60	24 66	27 73	31 79	34 86	37 93	41 99	44 106	48 112	51 119	55 125	58 132	62 138
11	—	1 21	5 28	8 36	12 43	16 50	19 58	23 65	27 72	31 79	34 87	38 94	42 101	46 108	50 115	54 122	57 130	61 137	65 144	69 151
12	—	2 22	5 31	9 39	13 47	17 55	21 63	26 70	30 78	34 86	38 94	42 102	47 109	51 117	55 125	60 132	64 140	68 148	72 156	77 163
13	—	2 24	6 33	10 42	15 50	19 59	24 67	28 76	33 84	37 93	42 101	47 109	51 118	56 126	61 134	65 143	70 151	75 159	80 167	84 176
14	—	2 26	7 35	11 45	16 54	21 63	26 72	31 81	36 90	41 99	46 108	51 117	56 126	61 135	66 144	71 153	77 161	82 170	87 179	92 188
15	—	3 27	7 38	12 48	18 57	23 67	28 77	33 87	39 96	44 106	50 115	55 125	61 134	66 144	72 153	77 163	83 172	88 182	94 191	100 200
16	—	3 29	8 40	14 50	19 61	25 71	30 82	36 92	42 102	48 112	54 122	60 132	65 143	71 153	77 163	83 173	89 183	95 193	101 203	107 213
17	—	3 31	9 42	15 53	20 65	26 76	33 86	39 97	45 108	51 119	57 130	64 140	70 151	77 161	83 172	89 183	96 193	102 204	109 214	115 225
18	—	4 32	9 45	16 56	22 68	28 80	35 91	41 103	48 114	55 123	61 137	68 148	75 159	82 170	88 182	95 193	102 204	109 215	116 226	123 237
19	0 19	4 34	10 47	17 59	23 72	30 84	37 96	44 108	51 120	58 132	65 144	72 156	80 167	87 179	94 191	101 203	109 214	116 226	123 238	130 250
20	0 20	4 36	11 49	18 62	25 75	32 88	39 101	47 113	54 126	62 138	69 151	77 163	84 176	92 188	100 200	107 213	115 225	123 237	130 250	138 262

(Dashes in the body of the table indicate that no decision is possible at the stated level of significance.)

TABLE D15 Cumulative Binomial Probabilities (P = .5) for the Sign Test

The table shows cumulative binomial probabilities for one-tailed tests. These must be doubled for two-tailed tests. The symbol N identifies the total number of pluses and minuses while the symbol m identifies the number of signs that occurs less frequently. The probabilities as shown are assumed to be preceded by a decimal point. For example, 031 is .031, and so on.

N \ m	0	1	2	3	4	5	6	7	8	9	10	11	12	13	14	15
5	031	188	500	812	969	*										
6	016	109	344	656	891	984	*									
7	008	062	227	500	773	938	992	*								
8	004	035	145	363	637	855	965	996	*							
9	002	020	090	254	500	746	910	980	998	*						
10	001	011	055	172	377	623	828	945	989	999	*					
11		006	033	113	274	500	726	887	967	994	*	*				
12		003	019	073	194	387	613	806	927	981	997	*	*			
13		002	011	046	133	291	500	709	867	954	989	998	*	*		
14		001	006	029	090	212	395	605	788	910	971	994	999	*	*	
15			004	018	059	151	304	500	696	849	941	982	996	*	*	*
16			002	011	038	105	227	402	598	773	895	962	989	998	*	*
17			001	006	025	072	166	315	500	685	834	928	975	994	999	*
18			001	004	015	048	119	240	407	593	760	881	952	985	996	999
19				002	010	032	084	180	324	500	676	820	916	968	990	998
20				001	006	021	058	132	252	412	588	748	868	942	979	994
21				001	004	013	039	095	192	332	500	668	808	905	961	987
22					002	008	026	067	143	262	416	584	738	857	933	974
23					001	005	017	047	105	202	339	500	661	798	895	953
24					001	003	011	032	076	154	271	419	581	729	846	924
25						002	007	022	054	115	212	345	500	655	788	885

* 1.0 or approximately 1.0

APPENDIX E

ANSWERS TO SELECTED EXERCISES

Answers are given to selected exercises from those at the ends of the chapters. You may not always find the same exact quantitative answer because of differences in rounding. Some answers have been abridged to conserve space.

Chapter 1

1. (a) Ordinal; (b) nominal; (c) ratio; (d) ordinal to most measurement theorists, but interval to many researchers
3. (a) Nominal; (b) ordinal
4. (a) Ratio, continuous; (c) ratio, discrete
5. (a) Sexual offense, nominal; (c) cause of death, nominal
6. (a) Inferential; (c) inferential

Chapter 2

1. Occupations of Mental Health Board Members

Occupation	%
Professional	43.8
Managerial/admin.	14.6
Sales	3.4
Clerical	5.6
Craftsman	4.5
Operative	2.2
Homemaker	10.1
Student	2.2
Unemployed	13.5
Total	99.9
	(89)

2. (a) Bar graph, because the variable is nominal and discontinuous
3. (a) 7.8; (c) 0.17
5. (a) 101.44; (c) 16.67

6. **Scores on Social Work License Exam**

Score	f
60–69	7
50–59	9
40–49	13
30–39	10
20–29	11
Total	50

8. (b) **Polygon of Recall Interval Between Pregnancy and Interview**

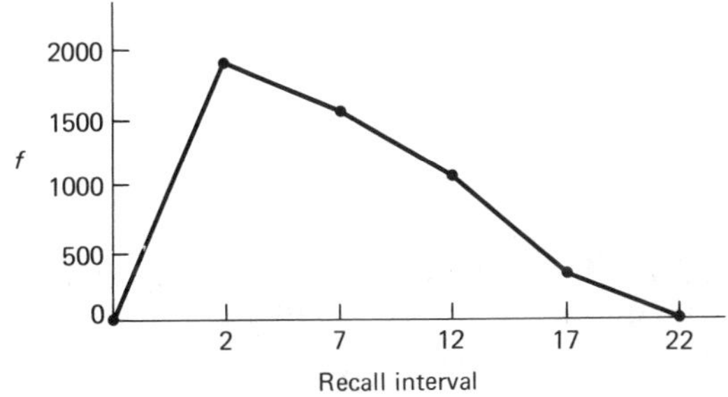

11. (a) Percentage on Row Totals

	Male	Female	Total
Yes	75.8	24.2	100.0 (281)
No	51.0	49.0	100.0 (567)
Total	59.2	40.8	100.0 (848)

(d) Part b, percentages by column totals

13. (a) 212.4; (c) −26.4 (1966–1967 to West Germany)

Chapter 3

1. (a) 19; (b) 19; (c) 21.4 3. (a) 5; (b) 5
4. (a) 2, 5.5 It is estimated that more businesses in community A had 2 employees than any one other number. It is estimated that more businesses in community B had 5.5 employees than any one other number.
   (d) 5.94 (e) 18.72
7. (a) False; the median does not necessarily fall in the category with the highest frequency.
   (b) True
   (c) False; there may be one or more scores equal to the median and, thus, less than exactly 50% of all the scores are below (or above) the median.
   (d) False; the median does not necessarily coincide with the mode.
9. The median is the most appropriate measure since it reacts most representatively to such skew, is appropriate to ordinal data which might remove some of the error in reported exact incomes, and matches well the usual research purpose of summarizing the central tendency of family income.

Chapter 4

1. (a) .884; (b) .723; (c) students
3. (a) 24, 36; (b) 10, 10
5. (a) 5.35, 7.5; (b) 6.40, 9.36; (c) yes (same means and similar scales); males are more variable by both measures.
11. (a) True ($6\sigma \simeq R$)
    (b) False; the standard deviation does not equal the mean deviation.
    (c) True ($3\sigma = 12$)
    (d) False; "much" is subjective and insufficient context is provided to defend such a characterization or to allow a valid comparison ("much," compared to what?).
13. 8.98

Chapter 5

1. (a) 8.129; (b) 0.285; an association does exist and is of weak to moderate strength; (c) 0.081; only 8% of the errors in predicting stigma are reduced from knowledge of public aid source, and the reverse.
3. .20 (same as published)
5. .029 (same as published)
7. (a) 23.528; (b) .396, an existing association of moderate strength
8. (a) .093; (b) an existing association of very weak strength (only 9% reduction in predictive errors)
11. (a) .489; (b) an existing association of moderate strength and positive direction; (c) 48.9% of the ordered differences in length of sentence can be explained by ordered differences in severity of crime, or the reverse.
13. (a) .350 (same as published)
15. (a) .215; (b) an existing association of weak strength and positive direction

17. (b) −.683; (c) an existing association of negative direction; (d) $(r_s)^2 = .466$, so 46.6% of the differences in population density ranks can be explained by ranks in percentage of revenue from local sources, and the reverse (moderate strength).

Chapter 6

1. (a)

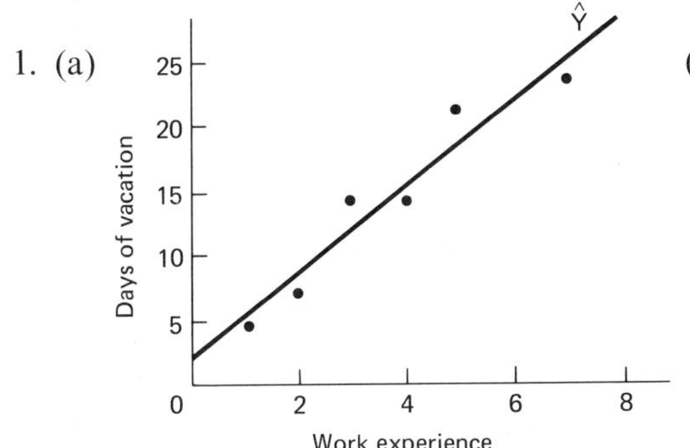

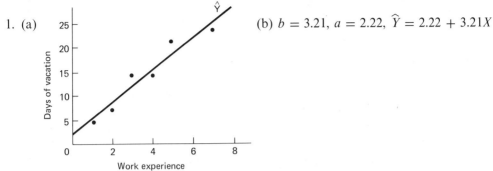

(b) $b = 3.21$, $a = 2.22$, $\widehat{Y} = 2.22 + 3.21X$

3. (a) 12; (b) 40; (c) 90
5. (a) .1 year; (b) +1.1 years
7. (a) For every additional mean mile run per week subject's lung capacity is predicted to increase by .55 cubic centimeter. For every additional cubic centimeter of mother's lung capacity, subject's lung capacity is predicted to increase .38 cubic centimeter.
   (b) .165, .304
   (c) Mother's lung capacity makes a greater contribution to increased subject's lung capacity, so inheritance is more important.
   (d) No, although mother's lung capacity is more important, jogging still makes a contribution to subject's lung capacity, at least as viewed in these two zero-order bivariate analyses.
9. (a) −.09; (b) The association is very weak and negative.
11. (a) 170; (b) .15; (c) 15% of the total variation in number of driving violations can be explained by differences in prejudice scores; (d) .387
13. (a) 90%; (b) yes, since 90% of the variation in either measure is shared by the other measure
15. (a) $110,500; (b) $238,250; (c) $262,750
17. (a) Yes; (b) no; (c) multiple or partial
19. Health is the greater contributor because it has the larger standardized partial.

Chapter 7

1. (a) .1515; (b) .2909; (c) .4424; (d) .0154; (e) .1137; (f) .7818
3. (a) Empirical
5. (a) .3333; (b) .1667; (c) .5000; (d) .6667; (e) .0556; (f) .0370; (g) .2778; (h) .0001
7. (a) .0192; (b) .0769; (c) .2500; (d) .1538; (e) .0000001; (f) .0036
8. (a) .60; (b) .40; (c) 1.00; (d) .36; (e) .064

10. (a) .03125; (b) .3125; (c) .15625; (d) .1875
11. (a) 1.00; (b) 34.13%; (c) .3413; (d) 15.87%; (e) .1587; (f) −2.00; (g) 47.72%; (h) .4772; (i) 2.28%; (j) .0228; (k) .2277
13. (a) −0.45; (b) 17.36%; (c) .1736; (d) 2.73; (e) 0.32%; (f) −1.82; (g) 3.44% (h) .0376
14. Yes, the z-score equivalents of 3.2 and 6.9 (1.27 and −1.27, respectively) would have approximately 80% of the scores between them (79.60%).
16. (a) Nonrandom; (c) nonrandom
17. (a) Systematic; (c) judgment

Chapter 8

1. H_0: $P = .60$ The proportion of the population supporting the freeway project is .60. A random sample may differ from this because of sampling error.
   H_1: $P \neq .60$ The proportion of the population supporting the freeway project is not .60. Sampling error will not be the cause of a random sample differing from this.
3. H_0: $\mu_1 = \mu_2$ The two populations of junior high students have the same mean achievement levels. Two random samples may differ from this because of sampling error.
   H_1: $\mu_1 \neq \mu_2$ The two populations of junior high school students do not have the same mean achievement levels. Sampling error will not be the cause of random samples differing from this.
5. H_0: $\rho = 0$ The population of communities is one in which there is no correlation between altitude and life expectancy. A random sample may differ from this because of sampling error.
   H_1: $\rho \neq 0$ The population of communities is one in which there is a correlation between altitude and life expectancy. Sampling error will not be the cause of a random sample differing from this.
7. Fail to reject
9. Fail to reject
11. Reject
13. Type I—could decide $\mu \neq 100$, but it is equal to 100. Type II—could decide $\mu = 100$, but it is not equal to 100.
15. Type I—could decide $P \neq .50$ but it is equal to .50. Type II—could decide $P = .50$, but it is not equal to .50.
19. Directional: H_0: $P \geq .75$; H_1: $P < .75$
22. Nondirectional: H_0: $\mu_1 = \mu_2$; H_1: $\mu_1 \neq \mu_2$

Chapter 9

1. (a) H_0: $P = .50$ (no difference), H_1: $P \neq .50$ (difference exists)
   (b) $NP = 45$, $NQ = 45$, so a normal (z) sampling distribution can be used. Tabular $z = \pm 1.96$, rejection regions are $\geq +1.96$ and ≤ -1.96.
   (c) $\sigma_p = .053$, $z = -2.83$
   (d) |Test z| $\geq$ |tabular z|, so reject H_0
   (e) The sample's proportion differs significantly from the hypothesized population's proportion at the .05 level. The proportion of the agency's work force that is female is not .50.

3. (a) $z = \pm 1.96$; (b) yes; (c) $z = \pm 2.575$; (d) no; (e) yes; (f) no (a directional test was not conducted); (g) no; (h) no
5. (a) $t = +1.833$; (b) yes; (c) $t = +2.821$; (d) yes; (e) yes; (f) yes (a directional test was conducted); (g) no; (h) no
7. (a) 14; (b) t distribution at 14 *df*; (c) $t = \pm 2.145$; (d) $t = \pm 2.624$
9. (a) 499; (b) normal (z) distribution; (c) $z = \pm 1.645$; (d) $z = 2.33$
11. (a) Yes, Np and Nq are over 5; (b) .032; (c) .637 to .763; (d) .647 to .753
13. d, e
15. (a) 3.87; (b) 1.96; (c) 162.41 to 177.58; (d) $t(59\ df) = 2.001$; (e) 162.26 to 177.74
17. (a) 2; (b) 5.991; (c) no; (d) 4.605; (e) no; (f) yes; (g) yes; (h) yes
19. (a) .155; (b) .194; (c) reject; (d) .139; (e) reject
20. (a) $df = 3$, tabular $\chi^2 = 7.815$; (b) 7.5 in each category; (c) 32.4; (d) reject
21. (a) .24; (b) .433; (c) reject

Chapter 10

1. (a) Means; (b) independent; (c) z
3. (a) Means; (b) dependent; (c) t
5. (a) proportions; (b) independent; (c) z
7. (a) H_0: $P_1 = P_2$ (no difference), H_1: $P_1 \neq P_2$ (difference exists)
   (b) $p = .46$, $q = .54$; $Np = 46$, $Nq = 54$
   (c) Normal (z) sampling distribution, tabular $z = \pm 1.96$, rejection regions are everything at and above 1.96 and at and below -1.96.
   (d) $\hat{\sigma}_{p_1-p_2} = .07$, $z = -1.71$
   (e) Fail to reject H_0
   (f) There is no significant difference between the sample proportions and the hypothesized population proportions at the .05 level. The proportion of adults in the Western state who wish pollution standards were better enforced is not different from the proportion of adults in the Eastern state who so wish.
9. (a) Normal sampling distribution, $z = \pm 2.575$, rejection regions are everything at and above 2.575 and at and below -2.575; fail to reject H_0; no difference in proportions between the two states.
   (b) Normal sampling distribution, $z = -2.33$, rejection region is everything at and below -2.33; fail to reject H_0; no difference in proportions between the two states.
11. (a) H_0: $P_1 = P_2$ (no difference), H_1: $P_1 \neq P_2$ (difference exists)
   (b) $p = .25$, $q = .75$; $Np = 250$, $Nq = 750$
   (c) Normal sampling distribution, $z = \pm 3.30$, rejection regions are everything at and above 3.30 and at and below -3.30.
   (d) $\hat{\sigma}_{p_1-p_2} = .019$, $z = 5.26$
   (e) Reject H_0
   (f) There is a significant difference between the sample proportions and expected population proportions at the .001 level. The proportion in the first population who believe nuclear energy production should be encouraged is different from the proportion in the second population who so believe.
13. (a) $z = \pm 1.96$; (b) yes; (c) $z = \pm 2.575$; (d) no; (e) yes; (f) no; (g) no; (h) no
15. (a) H_0: $\mu_1 = \mu_2$ (no difference), H_1: $\mu_1 \neq \mu_2$ (difference exists)

(b) $df = 150$, so normal sampling distribution can be used, tabular $z = \pm 1.96$, rejection regions are everything at and above 1.96 and at and below -1.96.
(c) $s_p^2 = 9$, $\hat{\sigma}_{\bar{X}_1 - \bar{X}_2} = .515$, $z = 2.14$
(d) Reject H_0
(e) There is a significant difference between the means for the two samples and those expected for the two populations at the .05 level. The mean achievement for the population of children attending integrated public schools is not equal to the mean for the population of children attending segregated schools.

20. (a) .426; (b) $t = -4.69$; (c) tabular $t = -1.812$; (d) yes; (e) tabular $t = -2.764$; (f) yes
22. (a) 3.841; (b) 6.635; (c) 4.16; (d) yes; (e) no; (f) there is a significant difference in union support between those with 0–4 and those with 5+ years of job tenure.
25. (a) 2.5; (b) .556; (c) no (tabular $\chi^2 = 3.841$); (d) no (tabular $\chi^2 = 2.706$); (e) there is a significant difference in number of children in Catholic and Protestant families.
27. (a) .279; (b) .334; (c) .138; (d) no; (e) no; (f) there is a significant difference in support for the candidate between northerners and southerners.

Chapter 11

1. (a) H_0: $\mu_1 = \mu_2 = \mu_3 = \mu_4$ (no differences), H_1: not $[\mu_1 = \mu_2 = \mu_3 = \mu_4]$ (at least one difference exists); (b) $dfb = 3$, $dfw = 20$, tabular $F = 3.10$; (c) $F = 11.667$; (d) reject H_0; (e) there is a difference in mean number of psychiatric admissions between at least two of the seasons.
2. (a)

	$\bar{X}_1$	$\bar{X}_2$	$\bar{X}_3$	$\bar{X}_4$
$\bar{X}_1 = 2$	—	1	1	3
$\bar{X}_2 = 1$		—	2	4
$\bar{X}_3 = 3$			—	2
$\bar{X}_4 = 5$				—

(b) 1.98
(c) $\bar{X}_1$ versus $\bar{X}_4$, $\bar{X}_2$ versus $\bar{X}_3$, $\bar{X}_2$ versus $\bar{X}_4$, $\bar{X}_3$ versus $\bar{X}_4$

3. (a) H_0: $Mdn_1 = Mdn_2 = Mdn_3 = Mdn_4$ (no differences), H_1: not $[Mdn_1 = Mdn_2 = Mdn_3 = Mdn_4]$ (at least one difference exists); (b) $df = 3$, tabular $\chi^2 = 7.815$; (c) 13.333; (d) reject H_0; (e) there is a difference in median admissions between at least two of the seasons.
4. (a) H_0: $H = 0$ (no differences), H_1: $H \neq 0$ (at least one difference exists); (b) $df = 3$, tabular $\chi^2 = 7.815$; (c) 14.535; (d) reject H_0; (e) there is a difference in mean ranked admissions between at least two of the seasons.
5. All concluded there were differences in admissions by seasons. Only *HSD* indicated which specific pairs of seasons differed.
11.

Source	***SS***	***df***	***MS***	***F***
Between	24	2	12	6.63
Within	76	42	1.81	
Total	100			

13. (a) 4; (b) 104; (c) 500; (d) yes, tabular $F = 2.70$; (e) yes, tabular $F = 3.98$
15. (a) H_0: $\chi^2 = 0$ (no differences), H_1: $\chi^2 \neq 0$ (difference exists); (b) $df = 4$, tabular $\chi^2 = 9.488$; (c) 20.81; (d) reject H_0; (e) there is a difference in job satisfaction by basis of authority.
17. (a) H_0: $\chi_r^2 = 0$ (no differences), H_1: $\chi_r^2 \neq 0$ (differences exist); (b) $df = 3$, tabular $\chi^2 = 7.815$; (c) 21.914; (d) reject H_0; (e) there is a difference in response by condition.
19. (a) Yes; (b) no; (c) no

Chapter 12

1. (a) H_0: $\chi^2 = 0$ (no association), H_1: $\chi^2 \neq 0$ (association exists); (b) $df = 1$, tabular $\chi^2 = 3.841$; (c) no association
3. (a) H_0: $\rho = 0$ (no association), H_1: $\rho \neq 0$ (association exists); (b) $df = 10$, tabular $r = \pm.576$; (c) no association
5. (a) H_0: $\gamma \leq 0$ (no positive association exists), H_1: $\gamma > 0$ (positive association exists); (b) tabular $z = +2.33$; (c) $z = 4.34$, positive association exists
7. (a) H_0: $\gamma \geq 0$ (no negative association exists), H_1: $\gamma < 0$ (negative association exists); (b) tabular $z = -1.645$; (c) $z = -8.36$, negative association exists

REFERENCES

Andrews, Frank M., et al. (1981), *A Guide for Selecting Statistical Techniques for Analyzing Social Science Data,* Second Edition. Ann Arbor, MI: Institute for Social Research, University of Michigan.

Atkins, Liz and David Jarrett (1979), "The Significance of 'Significance Tests,'" in John Irvine et al., ed., *Demystifying Social Statistics.* London: Pluto Press, 87–109.

Bailey, Kenneth D. (1982), *Methods of Social Research,* Second Edition. New York: The Free Press.

Blalock, Hubert M., Jr. (1979), *Social Statistics,* Revised Second Edition. New York: McGraw-Hill.

Bouden, Raymond (1965), "A Method of Linear Causal Analysis: Dependence Analysis," *American Sociological Review,* 30 (3), June: 365–374.

Bradley, Drake R., T. D. Bradley, Steven G. McGrath, and Stephen D. Cutcomb (1979), "Type I Error Rate of the Chi-Square Test of Independence in R × C Tables that Have Small Expected Frequencies," *Psychological Bulletin,* 86 (6) November: 1290–1297.

Camilli, Gregory and Kenneth D. Hopkins (1978), "Applications of Chi-Square to 2 × 2 Contingency Tables with Small Expected Frequencies," *Psychological Bulletin,* 85 (1) January: 163–167.

Camilli, Gregory and Kenneth D. Hopkins (1979), "Testing for Association in 2 × 2 Contingency Tables with Very Small Sample Sizes," *Psychological Bulletin,* 86 (5) September: 1011–1014.

Champion, Dean J. (1981), *Basic Statistics for Social Research,* Second Edition. New York: Macmillan.

Duncan, Otus Dudley (1966), "Path Analysis: Sociological Examples," *American Journal of Sociology,* 72 (1), July: 1–16.

Freund, John E. (1979), *Modern Elementary Statistics,* Fifth Edition. Englewood Cliffs, NJ: Prentice-Hall.

Hadden, Kenneth and Billie DeWalt (1974), "Path Analysis: Some Anthropological Examples," *Ethnology,* 13 (1), January: 105–128.

Hayes, William L. (1973), *Statistics for the Social Sciences,* Second Edition. New York: Holt, Rinehart and Winston.

Hempel, Carl G. (1966), *Philosophy of Natural Science.* Englewood Cliffs, NJ: Prentice-Hall.

Johnson, Robert R. (1980), *Elementary Statistics,* Third Edition. Belmont, CA: Wadsworth.

Kerlinger, Fred N. (1964), *Foundations of Behavioral Research.* New York: Holt, Rinehart, and Winston.

Kerlinger, Fred N. and Elazar J. Pedhazur (1973), *Multiple Regression in Behavioral Research.* New York: Holt, Rinehart and Winston.

Kim, Jae-On and Charles W. Mueller (1978), *Introduction to Factor Analysis.* Beverly Hills, CA: Sage Publications.

Kish, Leslie (1965), *Survey Sampling.* New York: Wiley.

Labovitz, Sanford (1970), "The Assignment of Numbers to Rank Order Categories," *American Sociological Review,* 35 (3) June: 515–524.

Loether, Herman J. and Donald G. McTavish (1980), *Descriptive and Inferential Statistics,* Second Edition. Boston: Allyn and Bacon.

Mattson, Dale E. (1981), *Statistics: Difficult Concepts, Understandable Explanations.* St. Louis: C. V. Mosby.

Minium, Edward W. (1978), *Statistical Reasoning in Psychology and Education,* Second Edition. New York: Wiley.

Motwani, Kewal (1967), *A Critique of Empiricism in Sociology.* New York: Allied Publishers Private Limited.

Mueller, John M., Karl F. Schuessler, and Herbert L. Costner (1977), *Statistical Reasoning in Sociology,* Third Edition. Boston: Houghton Mifflin.

Nagel, Ernest (1961), *The Structure of Science.* New York: Harcourt, Brace and World.

Neter, John, William Wasserman, and G. A. Whitmore (1978), *Applied Statistics.* Boston: Allyn and Bacon.

Ott, Lyman, William Mendenhall, and Richard F. Larson (1978), *Statistics: A Tool for the Social Sciences,* Second Edition. North Scituate, MA: Duxbury Press.

Reichenbach, Hans (1959), *The Rise of Scientific Philosophy.* Berkeley: University of California Press.

Runyon, Richard P. and Audrey Haber (1980), *Fundamentals of Behavioral Statistics,* Fourth Edition. Reading, MA: Addison-Wesley.

Schwartz, Howard and Jerry Jacobs (1979), *Qualitative Sociology.* New York: The Free Press.

Selvin, Hanan (1957, "A Critique of Tests of Significance in Survey Research," *American Sociological Review,* 22 (5), October: 519–527.

Skipper, James K., Anthony L. Guenther, and Gilbert Nass (1967), "The Sacredness of .05: A Note Concerning the Uses of Statistical Levels of Significance in Social Science," *American Sociologist,* 2 (1), February: 16–18.

Simon, Julian L. (1969), *Basic Research Methods in Social Research.* New York: Random House.

Spatz, Chris and James O. Johnston (1981), *Basic Statistics: Tales of Distributions.* Monterey, CA: Brooks/Cole Publishing Co.

Stevens, S. S. (1951), *Handbook of Experimental Psychology,* New York: Wiley.

Tai, Simon W. (1978), *Social Science Statistics: Its Elements and Applications.* Santa Monica, CA: Goodyear.

INDEX